TRUTH, PROOF AND INFINITY

SYNTHESE LIBRARY

STUDIES IN EPISTEMOLOGY,

LOGIC, METHODOLOGY, AND PHILOSOPHY OF SCIENCE

VOLUME 276

PETER FLETCHER

Department of Mathematics,
Keele University,
United Kingdom

TRUTH, PROOF AND INFINITY

A Theory of Constructions and Constructive Reasoning

KLUWER ACADEMIC PUBLISHERS
DORDRECHT / BOSTON / LONDON

A C.I.P. Catalogue record for this book is available from the Library of Congress.

ISBN 0-7923-5262-9

Published by Kluwer Academic Publishers,
P.O. Box 17, 3300 AA Dordrecht, The Netherlands.

Sold and distributed in North, Central and South America
by Kluwer Academic Publishers,
101 Philip Drive, Norwell, MA 02061, U.S.A.

In all other countries, sold and distributed
by Kluwer Academic Publishers,
P.O. Box 322, 3300 AH Dordrecht, The Netherlands.

Printed on acid-free paper

Printed in the Netherlands.

TABLE OF CONTENTS

PREFACE

Mathematics is largely concerned with infinities; yet throughout most of its history mathematicians have been uneasy about accepting infinity literally. 'Actual' infinities (such things as lines of infinite length, infinitesimal numbers, sums of infinitely many terms, points at infinity, infinite sets, and infinite quantifiers) have been viewed with suspicion, and as standards of rigour have advanced the tendency has been to rephrase all references to actual infinity in terms of 'potential' infinity (for example, defining an infinite sum as the limit of finite partial sums).

By the end of the nineteenth century, this process had reached the point where all varieties of actual infinity could be reduced to two basic types: infinite sets and infinite quantifiers. The task for the twentieth century has been to interpret infinite sets and quantifiers in a way that a finite human mind can understand.

Three rival philosophies of mathematics have emerged to do this: Logicism, Formalism and Intuitionism. I subscribe to all three. In this book, for expository convenience, I shall approach the subject from an intuitionist direction, but see Chapters 1, 9 and 35 for discussions of logicism and formalism.

My task is to present a philosophical interpretation of arithmetic and analysis, together with its mathematical implementation. Underlying arithmetic and analysis, I shall argue, is a mathematical Theory of Constructions, which serves as the foundation (or rather the ground floor) of mathematics. The Theory of Constructions is intended to shed light on the following questions.

- What is a construction?
- What is the source of our ability to grasp constructions and reason with them?
- What content does a constructive statement have?
- What is the relation between mathematical constructions and the physical world?
- What is a proof?
- How can infinite quantifiers be understood in constructive terms?

These questions are central to the intuitionistic account of mathematics; yet intuitionists have traditionally been very evasive or obscure about them. As George (1988) remarks, 'many of his [Brouwer's] sympathizers have contented themselves with a hurried mumble about the fundamental nature of it all and moved on to the reals, thus admirably heeding R.W. Emerson's advice that when skating over thin ice, best safety lies in speed.' A serious examination of these questions quickly leads into severe difficulties (Dummett, 1977, §7.2; Weinstein, 1983). I hope, however, to demonstrate that the ice is thicker than George thinks.

Answering these questions will take up Part I and require consideration of the views of most of the major figures in the philosophy of mathematics, notably Quine, Frege, Cantor, Wittgenstein, Brouwer, Dummett and Hilbert, from an intuitionistic perspective. The result of Part I is a programme for founding arithmetic and analysis, which is carried out in Parts II, III and IV. Part II describes the Theory of Constructions itself; from a purely mathematical point of view it is simply a functional programming language together with a program correctness calculus, in the style of Goodman (1972). Part III puts the Theory of Constructions to work, using it to interpret a sequence of logical theories, culminating in Peano Arithmetic; this follows in the tradition of Kreisel (1962, 1965) and Goodman (1970). Part IV extends the treatment of Part III to include second-order quantifiers, thus giving an interpretation of Second-Order Peano Arithmetic and hence analysis.

READERSHIP

This book is intended for three audiences: philosophers of mathematics, logicians and computer scientists.

1. For philosophers of mathematics, I am attempting to provide the *intended* interpretation of constructive arithmetic and analysis, in contrast with, say, Gödel's (1958) Dialectica interpretation, which is a philosophically minimal nonstandard interpretation. Thus my work is intended to clarify what intuitionists mean by the logical constants, proof and truth. See Chapter 7 in particular for further discussion of the issues that I hope it will illuminate. In addition my aim is to reconcile the essential insights of intuitionism, formalism and logicism.

2. For logicians, I am attempting to make the concept of intuitionistic proof more accessible to mathematical analysis and reveal the complex substructure underlying the constructive logical constants. Intuitionists have always stressed that their concept of proof transcends all formal systems, and this has made it difficult for logicians to study it. The central mechanisms of intuitionistic logic seem to be understood only by intuitionists themselves; Gödel (1933b, 1938, 1941) and others have complained that intuitionistic logic is not really constructive at all because of its reliance on a vague and unsurveyable totality of proofs. My theory of proof attempts to overcome this problem by construing proofs as well-founded trees. Well-foundedness gives proof its essential 'open-ended' character and yet is sufficiently precise and objective to be understandable to logicians of all philosophical persuasions. Accordingly, Chapter 26 gives a completely explicit account of the nature

of an intuitionistic proof, which is used in Chapters 25 and 29 to introduce some new intuitionistic logical constants. Chapter 28 presents a Calculus of Proof Functions, a formal system describing intuitionistic proof at a level of abstraction intermediate between direct constructive manipulation and predicate calculus. Chapters 37–43 give a new account of the substructure of second-order quantifiers. These developments show that, even at the level of arithmetic, intuitionism is not simply a subsystem of platonistic mathematics but involves substantial new mathematical theories that have no platonistic counterpart. These theories are described in sufficiently explicit a way that non-intuitionists can understand and use them.

3. For computer scientists, it is possible to regard my Expanded Term Language (Chapter 16) as a functional programming language with the Term Language (Chapter 14) as its machine code, Protologic (Chapters 17 & 20) as its correctness calculus, Logic (Chapter 26) as its specification language, and the Calculus of Proof Functions (Chapter 28) as a program transformation calculus. Hence the Theory of Constructions provides a complete integrated theory of programming. The distinctive feature of this approach is that the usual capabilities of functional programming (such as pattern matching, complex type systems, and interactive stream programming) can be supported in the most powerful and rigorous way by mapping them into a language with an extremely simple syntax (just constants, variables and function applications) and a firmly grounded informal semantics (see Chapter 13). These ideas are elaborated in Chapter 23.

ACKNOWLEDGEMENTS

I am grateful to John Mayberry for philosophical discussion and supervision and to the Science and Engineering Research Council for financial support, during the formative stages of this research (Fletcher, 1988), and also to Margaret MacDougall, Charles Day and the publisher's referee for reading and commenting on drafts of parts of this book.

Excerpts are reprinted from *Brouwer's Intuitionism*, by W.P. van Stigt, 1990, pp. 90, 197, 226, 418–422, 432, 434, 481, with kind permission from Elsevier Science.

Excerpts are reprinted from *Remarks on the Foundations of Mathematics*, by L. Wittgenstein, edited by G.H. von Wright, R. Rhees & G.E.M. Anscombe, translated by G.E.M. Anscombe, 1978, with kind permission from Blackwell Publishers.

PART I: PHILOSOPHICAL FOUNDATIONS

CHAPTER 1

INTRODUCTION AND STATEMENT OF THE PROBLEM

Geology is the study of rocks; astronomy is the study of stars; what is mathematics the study of? Mathematics has been variously construed as the science of numbers or abstract structures, as deductive reasoning in general, as a game with symbols, as a process of free mental 'construction', and as a very abstract branch of theoretical physics. Since we cannot agree on its subject matter it is not surprising that we differ on what philosophical questions to ask about it, what is required to justify it, or indeed whether it needs any sort of extra-mathematical foundation. In this chapter I shall try to formulate the correct question and to distinguish it from irrelevancies.

A THOUGHT EXPERIMENT

To focus our minds on the essentials, consider the following thought experiment, a myth in the history of mathematics.

A shepherd lives alone on a mountain with a large flock of sheep. Every night he collects his sheep into a pen to protect them from lions. A problem he faces each evening is knowing whether he has gathered all the sheep or whether some are missing. There are too many for him to keep track of them individually, so he counts them off against a tally of lines scratched into a rock at the entrance to the pen. He has no notion of arithmetic; he simply matches each line against a sheep as it enters the pen. If he reaches the last line as the last sheep enters the pen he concludes that all the sheep are present; otherwise he concludes that he needs to go and search for some lost sheep.

Now, the question is, what is the justification for this procedure? What have scratches on rock to do with sheep? Why should he bother with scratches when his real concern is with lost sheep? How does his procedure differ from other divination techniques involving line marks, such as *I Ching*?

The question is complicated by the fact that the procedure is not wholly reliable. There are many reasons why it may misdirect him as to whether to go looking for lost sheep: he may make an 'error' in matching the sheep against lines, a lamb may have been born or a sheep may have died during the day, or some supernatural agency may have made extra scratches on the rock. However, the procedure is still usable; his reasoning is as follows.

(1) I have made no error today, no sheep have appeared or disappeared, and no changes have been made to the rock.
(2) (1) necessarily implies that the procedure tells me correctly whether I need to look for lost sheep.
(3) Therefore the procedure tells me correctly whether I need to look for lost sheep.

Step (1) is justified (to a high degree of confidence) by the shepherd's general knowledge and competence in his job. Step (3) certainly follows from steps (1) and (2). So the problem is to justify step (2), and in particular to account for the strange *necessity* that the shepherd attributes to it.

Clearly it is irrelevant to enquire into how the shepherd acquired this procedure. He may have been taught it by a passing sage, he may have read it in a mathematical journal, he may have discovered it after laborious trial and error, he may have worked it out in his head, or it may have come to him in a dream. One might know everything there is to know about how he learned of the procedure, what makes him regard it as plausible, what role it plays in his life, and how it is represented in his brain; but one would still be none the wiser about whether it is *justified*, that is, whether he should go looking for lost sheep when it tells him to. Assume he acquired this procedure so long ago that he has forgotten its origin. However, he does have a strong conviction that it (or, at least, step (2)) is necessarily correct, based on an informal grasp of (what we would call) counting and one-to-one correspondences.

His belief is not based on experiment: he could have the same informal grasp of the situation even before trying it for the first time. Moreover, he believes that it would continue to work if he traded in his sheep and kept yaks instead, even though he has never seen a yak before and knows nothing about them.

Our task is to account for this belief, either by dismissing it as delusory or by justifying it. Thus our account will deal with either the psychology of superstition or the epistemology of mathematics. I shall assume the latter.

The required account will not be phrased in terms of whether step (2) is a consequence of the Zermelo-Fraenkel axioms, nor whether the shepherd could write up an account according to contemporary standards of rigour that would be accepted for publication in a mathematical journal, nor whether he would be successful in applying for a research grant to continue his studies. The shepherd, after all, does not care about the Zermelo-Fraenkel axioms, nor does he aspire to an academic career. All he cares about is whether he has lost any sheep.

FIRST CONCLUSIONS FROM THE THOUGHT EXPERIMENT

The point of the thought experiment is not to suggest that counting arose in this way, nor that any individual genius could have invented counting in isolation. The point is to pose clearly what I take to be the central problem in the philosophy of mathematics. The thought experiment enables us to dismiss a number of theories on the nature of mathematics as inadequate or beside the point.

First, consider *crude formalism*, the view that mathematics consists of proposing a list of axioms and seeing what can be derived from them by applying syntactic rules of inference. (I use the word 'crude' to distinguish this view from Hilbert's formalism, which is discussed in Chapter 9.) On this account, each axiom system is considered to define a branch of mathematics; axiom systems are valued for their aesthetic appeal (the pleasing shapes made by formulae on the page or the imagery they inspire in the mind) or abandoned as pointless if a contradiction is derived (just as chess would be rendered pointless by the discovery of a winning strategy). Most formalists also say that axiom systems should be valued for their applications to science: formal derivations can be used to predict the results of experiments, rather like reading tea-leaves. Curry (1954) includes empirical usefulness (in the tea-leaf reading sense) and consistency as two criteria for acceptability of a formal theory. But his position is a hybrid one: he also counts 'intuitive evidence' as a criterion, thus allowing that formulae may be interpreted as meaningful propositions.

There are two objections to the crude formalist view. The first is that it ignores the problem that I have introduced with the sheep-counting thought experiment: it asserts that mathematics is applicable but offers no explanation. Secondly, the proposition that a formula is derivable from a certain list of axioms and rules by a certain derivation is itself a necessary mathematical truth (or falsehood). If we are to accept formulae and derivations as 'given', in the straightforward way that the formalist does, then we might just as well accept numbers and the elementary operations of arithmetic as given.

A second philosophical position that appears questionable, although it cannot be dismissed at this point, is *platonism*, the view that mathematics is the science of 'mathematical objects' existing in a remote mathematical world. (I spell 'platonism' with a lower-case 'p' to indicate that it refers to a contemporary philosophical tendency rather than the views of Plato.) On this view, mathematicians study numbers and sets, just as astronomers study stars and planets; mathematicians use set-theoretic intuition just as astronomers use visual perception. Intuition justifies us in believing in mathematical objects in much the same way that perception justifies us in believing in physical objects (Gödel, 1964).

The reason this appears inadequate is that it leaves one wondering what the mathematical world has to do with the physical world. What has 50 to do with a flock of fifty sheep? How do one-to-one correspondences (in the mathematical world) justify the shepherd's counting procedure? To answer, the platonist has to posit a new supply of one-to-one correspondences relating mathematical objects to physical objects: these are not themselves mathematical objects as they have one foot in each world. If an astronomer made an analogous claim about links between stellar and terrestrial phenomena we would dismiss it as astrology. The platonist, I think, must appeal to general principles of reasoning that apply equally to mathematical, physical, or any other sort of object; in so doing, she turns herself into a logicist, and may as well jettison the mathematical world and carry out the whole argument in the physical world.

A third philosophical position, which is certainly inadequate, is the 'New Directions' tendency (Davis & Hersh 1981; Goodman, 1990, 1991; Hersh, 1991; Tymoczko, 1991), which seems to overlap with modern platonism (Maddy, 1989). This school of thought dismisses all the traditional philosophies of mathematics as hopelessly out of touch with current mathematical realities and shifts the emphasis to the psychological and social aspects of mathematics. It takes mathematics to be a branch of natural science, just like physics and chemistry, and denies its truths any special *a priori* status: that is, it denies that mathematics is necessarily true regardless of experience.

This school criticises philosophers for ignoring aspects of mathematical practice such as: the social nature of mathematics; the fallibility of mathematicians; computer-generated proofs that are too long for a human to read; the informality of proofs as produced and published by mathematicians; and historical changes in standards of proof.

Now, there is some justification in these complaints. The notion of *error*, for example, is central to an understanding of mathematical objectivity but was ignored by philosophers prior to Wittgenstein; I shall deal with it in Chapters 3 & 4. The relation between informal and formal proofs will be discussed in Chapter 9. However, the historical, psychological and social issues are wholly irrelevant, as I hope my thought experiment makes clear. The shepherd needs a guarantee that if his counting procedure says that all the sheep are in the pen then that is so, and it is our task to provide it. If the procedure misleads him into losing a sheep he will not be consoled by being told that it would be accepted as sound by the mathematical community.

Similarly the psychological processes by which theorems are discovered, generalised, corrected following counter-examples, clarified, checked, and finally accepted by the mathematical community are irrelevant to the question of how we know, at the end of the discovery process, that the theorem is necessarily true. This is the function of *proofs*; a proof is not intended to document the reasoning that led to the theorem's discovery but simply to convince the reader that the theorem is true. It is futile to complain that a philosophy of mathematics such as formalism 'is not compatible with the mode of thought of working mathematicians' (Davis & Hersh, 1981, p. 343): it is not intended as descriptive psychology (see Frege, 1884, Introduction). The social and psychological aspects of mathematics are inessential to its epistemology. It is, of course, immensely helpful to have colleagues with whom to swap ideas and check proofs. Still, our herd instincts are as much a hindrance as a help in the pursuit of truth. Mathematical rigour is ultimately a matter for the *individual conscience*, as George Orwell (1949, I, §7) saw clearly: 'Freedom is the freedom to say that two plus two make four. If that is granted, all else follows.'

To take the most glamorous weapon in the New Directions armoury, consider computer-generated proofs. Imagine a proof in which a theorem is shown to hold provided a decidable property holds in a large number of special cases, and suppose that a computer is used to check the special cases, thus completing the proof. According to the New Directions school (Tymoczko, 1979), this constitutes an *a posteriori* proof of the theorem since it relies on the computer's working correctly, and this shows the inadequacy of the traditional notion of proof. What is really going on is as follows. Let P_1 be the non-computerised part of the proof. Suppose that the computer program is written so as to output its deliberations as a formal proof, P_2, which a human could read, although in practice it is far too long to read. (It is straightforward to modify any program to produce such output.) Thus P_1 concatenated with P_2 is a complete proof of the theorem. Suppose also that we write a correctness proof, P_3, for the program, as we should always do when programming. Then we may reason as follows.

(1) The computer is in working order and is being operated correctly.
(2) (1) necessarily implies that P_1 concatenated with P_2 forms an *a priori* proof of the theorem.
(3) Therefore, we have produced an *a priori* proof of the theorem.

Here, step (1) is justified (to a high degree of confidence) by the empirical reliability of electronic hardware and experienced operators. Step (2) is justified by P_1 and P_3; both these proof are written and readable by a human mathematician. Step (3) clearly follows from steps (1) and (2). If all is well then we have an *a priori* proof (P_1 concatenated with P_2); if the computer is malfunctioning then we have no proof. In *no* case do we have an *a posteriori* proof. The proof, if it is correct at all, is *necessarily* correct.

This leads to two conclusions: computer-generated proofs do nothing to undermine the distinct notion of *a priori* proof; and the reasoning (steps (1), (2) and (3)) is just like the shepherd's in the thought experiment. In both cases, the argument has an empirical component, step (1), and an *a priori* mathematical component, step (2). Computer-generated proofs, therefore, pose no new problems for the philosophy of mathematics.

I hope the purpose of the sheep-counting thought experiment is now clear. It poses the question of how mathematics can be *meaningful reasoning*. This is not merely a problem for the philosophy of applied mathematics: it goes to the heart of the meaning of pure mathematics.

LOGICISM

In posing the problem in this way I have adopted a logicist position. *Logicism* is the doctrine that mathematics is logic. To explain its connection with the present discussion I must begin by clearing away the confusions surrounding the terms 'logic' and 'logicism'.

Logicism is often equated with a technical thesis: that a formal system of mathematics, such as Zermelo-Fraenkel set theory or some rival system, is derivable from first-order predicate calculus purely by definitions. This thesis is generally taken to be false, on the grounds that when Frege attempted to accomplish this his system turned out to be inconsistent, and when Whitehead and Russell repeated the attempt in *Principia Mathematica* they were forced to assume Axioms of Infinity, Choice and Reducibility (see, for example, Church, 1962, and Dummett, 1977, p. 2). Most of Whitehead and Russell's theorems, then, were not logical truths but merely logical consequences of these three axioms, so their position degenerated into what is known as 'if-thenism'. If one uses set theory instead of the *Principia Mathematica* system, one finds that the 'non-logical' notion of 'set' appears to be indefinable in purely 'logical' terms. Consequently, logicism is widely regarded as discredited.

However, these assumptions about what is mathematical and what is logical are highly tendentious. Mathematics cannot be equated, for philosophical purposes, with any formal system, since we cannot possibly predict what future branches of mathematics may arise, any more than Archimedes could predict category theory (which, by the way, does not fit inside Zermelo-Fraenkel set theory). Nor may logic be equated with first-order predicate calculus: we have no right to exclude second-order predicate calculus, modal logic, temporal logic, deontic logic, and so on, from 'logic'. The privileged status commonly accorded to first-order predicate calculus is a historical accident: it was simply the first formal logical system to be established. The exclusion of the concept of set from logic is equally arbitrary. It suggests to me a general bad conscience about sets: if we really believed that set theory was a general theory of pluralities or multiplicities then we would include it in logic without hesitation, as would any logician prior to the twentieth century. Shapiro (1992), for example, excludes set theory from logic (and hence regards logicism as unviable) on the grounds that set theory is not 'self-evident'. A more natural reaction to the lack of self-evidence of set theory is to regard it not as *external to logic* but as *bad logic*, a botched attempt at a general logic of plurality (see Chapter 2).

Musgrave (1977) suggests that second-order predicate calculus and set theory are regarded as non-logical because, unlike first-order predicate calculus, they are incomplete. However, as Musgrave points out, this depends on the unwarranted assumption that logic must be formalisable in a sound and complete axiom system.

A second attempt to characterise logicism is that it holds that logic is the theory of sound deductive reasoning in general, applicable to objects, events, processes, and anything else, and that so is mathematics. An opponent of logicism would then claim that mathematics has a special subject-matter, while logic is 'topic-neutral' (Detlefsen, 1990a). This distinction seems unpersuasive. Compare the following two arguments.

(1) *Modus ponens* is a rule of inference, part of the special subject-matter of logic (although admittedly it can be applied to propositions of any sort).
(2) Four is a number, part of the special subject-matter of mathematics (although admittedly it can be used to count objects of any sort).

Both arguments seem equally successful in showing that logic or mathematics, respectively, has a special subject-matter. It is also often claimed (Dummett, 1967; Parsons, 1979–80) that mathematics is committed to the existence of specific objects (numbers or sets) while logic has no ontological commitments. This is so only in a platonist semantics. In the intuitionistic semantics that I shall propose in Chapter 6, an existential formula, say 'there exists an even prime number', is neither true nor false but is a statement of a problem, solved by 2. The true statement that 2 solves the problem (that is, 2 is an even prime) means that if anyone carries out a certain computation process without error they will obtain the answer *true*. This involves no commitment to the existence of numbers, persons, computation tokens, or anything else.

I conclude that logic and mathematics cannot be distinguished on grounds of subject-matter or ontological commitment.

Some people distinguish logic from mathematics not by subject-matter but by certain allegedly non-logical axioms that mathematics relies on: for example the Axioms of Infinity, Choice and Reducibility in *Principia Mathematica* or (according to Poincaré) the principle of induction. But it is easy to think of branches of mathematics (group theory, for example) that do not use the axioms in question.

The boundary between logic and mathematics may appear somewhat arbitrary. It is nevertheless worth asserting that logic is mathematics: it saves us from wasting time looking for a specifically mathematical subject matter or a notion of mathematical necessity distinct from logical necessity. The real content of logicism, however, is not about the scope of the words 'mathematics' and 'logic': it is a thesis about the *application* of mathematics: 'it is applicability alone which elevates arithmetic from a game to the rank of a science' (Frege, 1903, §91; see also Dummett, 1991, §20). Mathematical arguments apply directly to the physical world: for example, '2+2=4' could be taken as simply meaning that a pair of anything together with another pair constitute a quadruple. This is in contrast to the commonly held view that the applicability of mathematics to science is a miracle. There are, indeed, some puzzling aspects in the success of applied mathematics, but the bare fact that mathematics applies to the world is not a puzzle; mathematics is about things in general and therefore applies to things in particular – such as flocks of sheep.

In this book I shall sometimes use the word 'logic' in the full sense of general topic-neutral deductive reasoning and sometimes in the sense of formal predicate calculus; but I shall always make clear which sense I mean. This terminology is far from ideal, but here as elsewhere I must conform to common usage.

THE NECESSITY OF NECESSITY

I said above that we have a choice between dismissing mathematical necessity as delusory or justifying it, and that I took the latter road. Not everyone agrees. The 'naturalistic' philosophers (Nagel, 1944; Quine, 1953) hold that there is no distinction between *a priori* mathematics and *a posteriori* science; mathematics is just a very abstract branch of science and is as empirical as any other branch. I shall discuss the version of the argument in Quine's (1953) *Two Dogmas of Empiricism*, as it is the clearest and most influential statement of this view.

Quine sees knowledge as a system of interconnected statements with a truth value assigned to each. The assignment of truth values over the whole system is likened to a field of force, with experimental experience serving as the boundary conditions. At the periphery of the field are observational statements like 'the reading on this dial is 15.62 volts', which refer more or less directly to particular experiences. In from the periphery are everyday facts and general knowledge, such as 'glass is brittle' and 'copper conducts electricity', which are less specific in their application. Even further in are the laws of physics, then mathematics, and finally, at the very centre, logic. From this system of knowledge we derive experimental predictions, and if a prediction is falsified then we 'reëvaluate' (that is, change the truth values of) some relevant statements.

> But the total field is so underdetermined by its boundary conditions, experience, that there is much latitude of choice as to what statements to reëvaluate in the light of any single contrary experience. (*ibid.*)

So we could say that we have misread a dial (thus reëvaluating a peripheral statement), or that the experiment was conducted wrongly (thus reëvaluating some 'everyday facts'), or that a law of physics has been refuted; we could even revise our mathematics or our logic. Thus any experiment tests our knowledge as a whole: no individual statement can be tested in isolation, even the most down-to-earth observational statement.

We nearly always prefer to revise peripheral parts of our knowledge rather than central parts, and so logic and mathematics are almost immune from revision, being insulated by thick layers of physical theory. However, circumstances could arise in which we would prefer to alter a single axiom of set theory or law of logic in response to an experiment rather than make a large number of *ad hoc* alterations in physics, everyday knowledge and observations. For example, we might adopt a 'quantum logic' in order to simplify the presentation of quantum mechanics. Simplicity and minimal disruption are our guiding principles in revising our knowledge.

> The edge of the system must be kept squared with experience; the rest, with all its elaborate myths or fictions, has as its objective the simplicity of laws. (*ibid.*)

It follows that there is no sharp division between empirical science and *a priori* mathematics and logic.

There is a problem here of how statements make contact with experience. If some statements were purely observational records of 'sense data' then they could be individually refuted when the sense data did not occur as expected. Quine, like most modern empiricists, does not accept such pure observational statements: no individual statement (and hence presumably no finite set of statements) can be tested in isolation. This leaves it unclear how the whole system can ever be said to be incompatible with experience. However, this is a problem for empiricists generally and is irrelevant to Quine's main point, which is to undermine logical necessity. So I shall simply assume, in accordance with Quine's evident intentions, that some theories are compatible with given experiments and others are not, and that this compatibility relation conforms to normal scientific practice for confirming and refuting theories.

There is one very puzzling feature of Quine's account. Logical relations between statements are represented in two ways: as connections between statements and as statements in their own right.

> Reëvaluation of some statements entails reëvaluation of others, because of their logical interconnections – the logical laws being in turn simply certain further statements of the system, certain further elements of the field. (*ibid.*)

Thus, if A implies B then this is represented as an 'interconnection' between A and B, but also as a statement, $A \supset B$, assigned the value *true*. ('Whatever *Logic* is good enough to tell me is worth *writing down*', said the Tortoise (Carroll, 1895), from whom I am borrowing the argument that follows.) Now, are we allowed to assign the truth values *true*, *true*, *false* to the statements A, $A \supset B$, B, respectively? If the answer is yes, then evidently we are at liberty to sprinkle truth values at random, and we can accommodate any experimental results without revising physics, mathematics or logic, contrary to Quine's clear intention. If, however, the answer is no, then *Modus Ponens* is a necessary logical truth, again contrary to intention since he rejects any notion of logical necessity.

Quine may reply that individual statements are meaningless in isolation; only entire theories can be said to be correct or incorrect: 'our statements about the external world face the tribunal of sense experience not individually but only as a corporate body'. So let us restate the Tortoise's argument in holistic terms. Let A and B be two mathematical statements such that, in our present system of knowledge, T, the statements A, $A \supset B$ and B are assigned the values *true*, *false* and *false*, respectively. Let T' be a system just like T except that $A \supset B$ is assigned the value *true*. Now, is T' any worse than T? It seems that T and T' are compatible with the same experiences. It is hard to be sure of this, in the absence of any account of how theories are supposed to be tested against experiment, but assuming it is not too different from normal scientific practice I can appeal to two arguments. The first is that $A \supset B$ can only play a role in experimental prediction via A and B, and hence since T and T' agree on A and B the disagreement on $A \supset B$ can have no empirical consequences. The second is that since $A \supset B$ is a *mathematical* statement it can only affect empirical predictions via physical statements; since T and T' agree on all of these then once again the disagreement on $A \supset B$ is insulated from experience.

If so, then we have no empirical grounds for preferring T to T'. When two theories account for all experimental results equally well then scientists prefer the one that is simplest or most elegant. But this does not help us here. T and T' are equally simple, and in any case the statement that T is more simple or more elegant than T' must lie *outside* science (otherwise T and T' could each contain a statement asserting 'I am the simplest and most elegant system!'). This statement, if true, is also *a priori*, since, as we have seen, isolated statements are not subject to empirical tests.

Is there some coherence or consistency requirement that renders T' immediately unacceptable? The answer is no. The only thing wrong with T' is that it violates the law of *Modus Ponens* (which is contained in both T and T'). But unfortunately this violation is only apparent to us by standing outside both systems and applying ordinary logic. According to T' there is no conflict between asserting *Modus Ponens* and asserting that, for these particular statements, A and $A \supset B$ are true and B is false.

Therefore there are *no* grounds for preferring T to T'. Returning to the force field analogy, the field is not merely *underdetermined* by its boundary conditions but wholly *indeterminate*. This is because there are boundary conditions but *no field equation.*

Quine's theory of knowledge fails because it is a game with no rules. *Some* statements must be necessary if *any* are to be usable in inference. *Modus Ponens* is necessary because it simply recapitulates the purpose for which we introduced the '$\supset$' symbol: we decided to say '$A \supset B$' as a way of expressing the circumstance that we can infer B from A. The necessity is not legislated by God, nor is it a symbol-reciting habit drummed into us at an early age. We are perfectly free to change the usages of the symbols '$\vee$' and '$\supset$', taking them to mean 'apple' and 'orange', perhaps, or to mean the operators of so-called 'quantum logic', and this would lead us to revise statements involving these symbols; but except in this superficial sense these statements are not liable to revision. Their necessity comes from the fact that when we understand a statement our understanding includes a recall of how we intended to *use* the notation in it, and this obliges us to accept certain statements as consequences of our intentions; for example, understanding an axiom schema involves being able to see which formulae are instances of it, even though those formulae were not explicitly considered when the axiom schema was adopted.

Quine spends much of his paper arguing that no informative account can be given of this notion of logical necessity. This is true, but it is not the fault of logical necessity; the desired explanation would be a byproduct of any workable theory of meaning. Unfortunately we do not have a workable theory of meaning. We all accept that some statements are meaningful; we are all reluctant to analyse this fact by saying that there are metaphysical entities called *meanings* that these statements possess; most of us are wary of reducing meaning to the observable behaviour of the language-user; and we have no other analysis to offer. So for the time being we must simply accept meaningfulness, and its corollary necessity, as primitive properties of statements that we recognise on a case-by-case basis.

CONCLUSION

I have stated the central problem in the philosophy of mathematics (or, at least, an instance of it), and used it to reject a number of philosophical positions and to characterise my own position as logicist. The instance of the problem was

- *Explain and justify the shepherd's counting procedure.*

and the problem in general is

- *What does mathematics mean and how can we know for sure that it is true?*

Other questions associated with mathematics are subsidiary or irrelevant. The psychology and sociology of mathematics are irrelevant. A relevant but subsidiary question is that of ontology: do so-called 'mathematical objects' really exist, and what does existence mean anyway? is there a special kind of existence for mathematical objects different from that for physical objects? I do not mind how these questions are answered, nor even whether they are answered at all, provided we end up with an account that tells the shepherd whether he needs to look for lost sheep or not. Counting sheep is an arithmetic problem; and I anticipate that any account of mathematics that is adequate for arithmetic will extend comparatively easily to other branches of mathematics, or will reveal those branches to be philosophically unsound.

CHAPTER 2

WHAT'S WRONG WITH SET THEORY?

I have posed the fundamental question in the philosophy of mathematics as 'what does mathematics mean and how can we know for sure that it is true?' Before developing my answer to this question, I must explain why I do not accept the orthodox answer, namely that mathematics is the study of sets and that our knowledge of mathematics is derived from our set-theoretic intuition using classical logic. In this chapter I shall argue that 'set-theoretic intuition', as formalised in the Zermelo-Fraenkel axioms with the axiom of choice (ZFC), is conceptually incoherent. In the following chapter I shall argue that infinite quantifiers, the distinctive feature of classical logic, are meaningless.

The naive notion of 'set', as it arose in the nineteenth century, contained a mixture of informal ideas; but principally sets were seen as being like properties except that they were extensional. Thus, to say that an element belonged to a set was simply to say that it satisfied the corresponding property, and to say that two sets were equal was to say that two properties were equivalent.

This was why the Zermelo-Russell paradox came as a shock. It was not merely an unexpected discovery about sets; it cast doubt on the whole notion of set. The founders of modern set theory, however, missed this point. They merely concluded that evidently not all properties have sets as their extension, and they launched an inquiry into the question 'What sets are there?'. But this question has no meaning until we establish what the word 'set' means; we might just as well ask 'What xlgjphts are there?'. If sets are no longer to be understood in terms of properties then we need to find an alternative meaning before we can meaningfully ask what sets exist. It is no use appealing to 'set-theoretic intuition' before the word 'set' has been given a meaning; we might just as well appeal to 'xlgjpht-theoretic intuition'.

An example of what I am complaining about is Quine (1969), who (on page 1) proclaims that 'Sets are classes', explains a class as the extension of an open sentence, rejects any notion that sets are collections (in the sense I explain below), then on page 2 reveals that classes aren't extensions of open sentences after all, and then gets distracted into a purely terminological discussion of the words 'set' and 'class'. So now we know two things that sets/classes are *not*, but have no inkling of what they *are*. Thus Quine is in no position to pose the problem (page 3) of 'what open sentences to view as determining classes; or, if I may venture the realistic idiom, what classes there are', the question that dominates the rest of the book.

Quine is here following the tradition in set theory that ignores meaning altogether and simply posits axioms that it is hoped sets might satisfy. This has led to the ZFC system of axioms which is widely accepted today. This attitude of 'postulate first and ask questions later' is perfectly sound as a heuristic method, *provided one remembers to ask the questions*. As with any heuristic procedure the results have to be justified case-by-case on their own merits. In this case the question to ask is: do we arrive at a unique, intuitively adequate list of axioms that sharpens our understanding of the notion being axiomatised, or does the exercise peter out in an arbitrary and complicated mess? The Peano axioms are an example of successful axiomatisation: the axioms articulate all the aspects of natural numbers implicit in the informal notion of applying successor repeatedly to 0. The ZFC axioms, on the other hand, are an example of the opposite; the clumsiness, arbitrariness and lack of intuitive completeness in the axioms indicate what a mixture of conflicting and poorly articulated ideas underlie the use of the word 'set'. Set theorists do not see it this way, because they assume that 'set' has a pre-ordained meaning and regard themselves as engaged on a purely factual enquiry into the shape and size of the set-theoretic universe.

The primary question, following the paradoxes, should be 'what do we mean by "set"?', and the answer should be not a list of axioms about what sets exist but an explanation of the informal ideas underlying our various uses of the word. There are, in fact, several incompatible notions of set. The following list attempts to summarise all the ways in which sets have been construed or justified since Cantor:

(1) sets as *consistent multiplicities* or *multiplicities considered as unities*;
(2) sets as *collections*;
(3) sets as *classes*, in the sense of extensionalised properties;
(4) the *limitation of size* view;
(5) the iterative conception of sets;
(6) sets as an *extrapolation* from finite sets of physical objects;
(7) sets obtained by a transition from potential to actual infinity.

I shall discuss each of these in turn, indicate how each motivates some but not all of the ZFC axioms, and state objections to the use of each as a foundation for ZFC. Finally I shall comment on their incompatibility.

ZFC is, of course, not the only axiomatic system for set theory; however, all the axiomatisations draw on much the same mix of ideas, and nearly all of what I have to say about ZFC applies to nearly all the other systems. Hence I shall only discuss ZFC and shall regard a refutation of ZFC as a refutation of platonist set theory in general.

(1) MULTIPLICITIES CONSIDERED AS UNITIES.

One of Cantor's later definitions of 'set' was as a *consistent multiplicity*.

> For a multiplicity can be such that the assumption that *all* of its elements 'are together' leads to a contradiction, so that it is impossible to conceive of the multiplicity as a unity, as one 'finished thing'. Such multiplicities I call ***absolutely infinite*** or ***inconsistent multiplicities***. ... If on the other hand the totality of the elements of a multiplicity can be thought of without contradiction as 'being together', so that they can be gathered together into '*one* thing', I call it a *consistent multiplicity* or a 'set'. (Cantor, 1899)

This view has been widely accepted by later set theorists. It motivates a restriction of the comprehension principle to avoid the paradoxes, but does not explain the ZFC axioms in detail. Any other set theory that successfully avoids the paradoxes has equal claim to be considered an axiomatisation of consistent multiplicities.

There is an obvious objection to this notion. When we speak of a 'multiplicity' or 'totality of elements' we are using a singular noun phrase and so we are *already* thinking of it as one thing. Thus we cannot coherently say that a multiplicity cannot be thought of as a single thing, because when we say so we are proving ourselves wrong.

A second objection is that the definition assumes a pre-existing notion of multiplicity, and so it fails to prevent someone from saying, since sets are so tricky, let's found mathematics on multiplicity theory instead of set theory! 'Multiplicity theory' leads immediately to the Zermelo-Russell paradox, so the notion of multiplicity must be incoherent. If you define something in terms of an incoherent notion then the definition is unsuccessful. Inserting the qualification 'consistent' does *not* patch it up; qualified nonsense is still nonsense.

The whole formulation of the definition seems clumsy. Multiplicities cannot be 'consistent' or 'inconsistent': only *theories* can be. If positing a multiplicity leads to a contradiction then that does not mean that the multiplicity has the exotic property of inconsistency – it means that there is *no such multiplicity*. One wants to reformulate the definition as follows: there is a set satisfying a certain specification iff it is consistent to assume so.

Defining existence in terms of consistency is an odd position for platonists to take. Platonists think of the mathematical universe as independent of our apprehension of it; they trust that their modes of reasoning are sound but they do not usually expect them to be complete. If the positing of a set leads to a contradiction then we can all agree that the set does not exist, but platonists should not expect that we can refute the existence of every set that doesn't exist, any more than we can discover every set that does. Quite apart from platonism, the proposal that existence equals consistency is logically unsafe. Suppose we have two multiplicities, A and B, whose definitions are so related that it is consistent to assume that A exists or that B exists, but not consistent to assume that both exist (analogous to an irresistible force and an immovable object). Then each set is a consistent multiplicity, so both sets exist, which leads to a contradiction. Quine (1969, p. 37) gives an example of something along these lines.

(2) SETS AS COLLECTIONS

By a *collection* I mean the result of collecting things, for example a heap of sand or a stamp collection. It is immediately clear that collections differ in many important respects from ZFC sets. First, a collection is not usually distinguished from its members: for example to fetch a pair of slippers is to fetch the slippers themselves, not to fetch a third object, the 'pair'. As Frege says, 'If we burn down all the trees of a wood, we thereby burn down the wood' (Geach & Black, 1970, p. 89). A singleton collection is the same as its single member (if it is accepted as a collection at all). Secondly, there is no empty collection: a stamp collection with no stamps is no collection. Thirdly, a collection is created by gathering its members into one place and destroyed by dispersing its members; what does this mean when the members are not located in space and time? (Chihara (1989) argues similarly about heaps.)

Collections are so unlike sets that it is surprising to find set theorists attempting to construe sets as collections. One of Cantor's definitions reads:

> By a 'set' we mean any collection M into a whole of definite, distinct objects m (which are called the 'elements' of M) of our perception or of our thought. (Cantor, 1895; translation quoted in Dauben, 1979, p. 170)

There is clearly a notion here of *collecting* objects, but it is not explained how one goes about collecting objects of perception or thought. How do we locate, grab hold of, carry to one place, and deposit in a heap, such 'objects'? Must we fetch them one at a time, or can we bring them in mental handfuls? What aspects of the 'collection' notion are metaphorical, and how does the metaphor serve to illuminate set-formation? In a similar vein, Tharp (1989) derives the notion of set from 'heaps' and 'agglomerations', and Maddy (1990) regards set theory as based on collections of physical objects; thus a set is located where its elements are, and is created or destroyed when its elements are. Pressed by Chihara (1982) to explain the distinction between a singleton set and its element Maddy suggests identifying the two when the element is a physical object (considered as a single thing), and also dispensing with the empty set and pure sets. This doesn't really answer Chihara, as he could still ask what the difference is between $\{a,b\}$, $\{\{a,b\}\}$, $\{\{\{a,b\}\}\}$, ..., where a and b are two physical objects. However, this does show that Maddy's sets are more like collections than like ZFC sets.

Gödel (1944) is also prepared to treat the empty set and the distinction between $\{x\}$ and x as 'fictions' if necessary, 'introduced to simplify the calculus like points at infinity in geometry'. This seems a questionable analogy. We accept points at infinity without qualms, not due to any syntactic consistency proof but due to a representation of the projective plane as the system of rays in three-dimensional space: on this view points at infinity are simply rays like any other and their presence is natural and inevitable. To apply a similar device to set theory would require interpreting it in a more fundamental system (or in itself in a different way) and so would deprive it of its position as the foundation of all mathematics.

This half-heartedness about the empty set goes back to Zermelo (1908), whose second axiom says: 'There exists a (fictitious) set, the *null* set, 0, that contains no element at all'. To say that the empty set is fictitious is to say that there is no empty set; then why propose as an axiom that there is? I suspect that the intention is that the formal expression 'there exists' or '$\exists$' is not to be interpreted literally: thus $\exists x \forall y\, y \notin x$ is taken as true even though there is not really any such x. If this reading is correct then set theorists owe us an explanation of the real meaning of 'there exists'; if not then they owe us a clear ruling on whether there is an empty set or not, and an axiom system that reflects this decision.

The confusion on this point indicates how strongly set theorists are attracted to the idea of sets as collections, despite the consequences for their axioms.

(3) SETS AS CLASSES

By a *class* I mean the extension of a property. This may be interpreted in two ways: either that for every property there is an associated object called its extension, or that when talking about classes we are really talking about properties but we are restricting ourselves to use them only extensionally. Either way, classes are explained in terms of properties and every property automatically defines a class.

Viewing sets as classes motivates the existence of the empty set, the distinction between a singleton set and its member, the axiom of foundation (corresponding to the well-foundedness of definitions), and unrestricted comprehension. The position as regards the axiom of choice is debatable. Cantor's first definition of sets took them to be classes.

> I call an aggregate (a collection, a set) of elements which belong to any domain of concepts *well-defined*, if it must be regarded as *internally determined* on the basis of its definition and in consequence of the logical principle of the excluded middle. It must also be internally determined whether any object belonging to the same domain of concepts belongs to the aggregate in question as an element or not, and whether two objects belonging to the set, despite formal differences, are equal to one another or not. (Cantor, 1882; translation quoted in Dauben, 1979, p. 83)

The class interpretation is nowadays usually rejected on account of the Zermelo-Russell paradox. This seems to me a strange reaction. The paradox does not depend on extensionality; an intensional version ('the property of a property's not satisfying itself') may be formulated equally easily. The natural response is not 'here is a property with no corresponding set' but rather 'here is an apparent property that turns out on examination to be incoherent'. Similar remarks apply to the Burali-Forti paradox. This *strengthens* the case for interpreting sets as classes: sets and properties suffer from the same diseases, so it is natural to suppose that one caught the infection from the other; and the natural cure is to argue that 'not satisfying oneself' is not a legitimate property. However, what is truly paradoxical about the paradoxes is that they were universally taken as proving the opposite, that sets could not be classes. The official position now is that it is a *factual question* whether for a given property there is a set containing all and only the things possessing the property. This implies that sets are understood in some way independent of properties, and hence not as classes.

Yet the class interpretation maintains a ghostly presence. It is the historical origin of the concept of set and it accounts for more of the ZFC axioms than any other interpretation. When teaching set theory we find it impossible to convey to students the point of the empty set and the distinction between x and $\{x\}$ except using the class interpretation. If the students go on to become set theorists they are told that the class interpretation is mistaken, but they are expected to go on believing by inertia that $\emptyset$ exists and that $x \neq \{x\}$. Ontogeny recapitulates phylogeny in set theory, if not in biology.

(4) LIMITATION OF SIZE

The most distinctive feature of Cantor's view of the mathematical universe (compared, say, with Frege's, Russell's or Brouwer's) is his notion of *limitation of size*. On this view, what distinguishes a legitimate mathematical object from an illegitimate one is that the latter has *too many elements*, as measured by one-to-one correspondences. This distinction between transfinite (not too big) and absolutely infinite (too big) is analogous to a constructivist's distinction between finite and infinite.

This view clearly motivates the axioms of pairing, separation and replacement, and perhaps also the axiom of choice, but not, for example, the axiom of foundation.

I have three objections. First, why is size the important factor? If someone develops an alternative set theory that admits a universal set but excludes other sets in such a way as to avoid the paradoxes while being adequate for ordinary mathematics (Quine's (1937) 'New Foundations', for example), why is this a philosophical mistake? No explanation has been provided on this point (leaving aside Cantor's association of absolute infinity with God (Hallett,

1984, §1.1)).

The second objection is that we are not told how big is too big. In particular, it would be possible to adhere to this view and deny the power-set axiom (Hallett, 1984, §5.1). Imagine two set theorists arguing about the admissibility of a large-cardinal axiom, with one taking a larger view of the set universe than the other. What rational grounds are available for settling the dispute, assuming the axiom is independent of ZFC? Would it be acceptable to say that they have different concepts of 'set', or must we say that one is right and the other wrong? The problem is complicated by the fact that size is measured by one-to-one correspondences, which are themselves sets, and hence their existence is also under dispute. My complaint here is not merely that there is no decision procedure but rather that there are no truth conditions. If there is *one* legitimate concept of set, on this interpretation, then there must be *many*, obtained by restricting the size of sets in various ways. So the limitation of size idea cannot pinpoint a unique meaning for 'set'.

A third objection is that the only reason we have for believing in more than one infinite cardinality is Cantor's diagonal argument; without this we would probably assume that all infinite sets are countable. It bothers me that some diagonal arguments are classified as paradoxes and others are accepted as genuine theorems, simply according to whether they lead immediately to a contradiction. This is a little like classifying one's burglaries as criminal or non-criminal according to whether one is arrested at the scene. If we had decided that the mathematical universe is countable (a plausible view on the class interpretation) *before* discovering Cantor's diagonal argument, we might well stick to our opinion and dismiss the diagonal argument as a paradox.

(5) THE ITERATIVE CONCEPTION OF SETS

A picture of sets that many find attractive is that they are formed by transfinite iteration of the power-set operation. The set universe is arranged in layers, indexed by the ordinals; each layer consists of all subsets of the union of all previous layers. Gödel (1964), for example, explains sets in this way: he says that a set is something obtainable by transfinite iteration of the operation 'set of', starting from basic objects such as the integers.

As a descriptive account this is quite illuminating: it combines the pairing, union, power-set, separation and foundation axioms into a single picture. However, it is little use as an explanation of what sets are. The notion of transfinite iteration is mysterious: as Brouwer (1907, pp. 145–146) complains, the transfinite notion of 'and so on' is not repetition of the *same* thing, as the finite 'and so on' is, and thus the transfinite notion is ill-defined. The iterative picture presupposes the ordinals; it explains nothing about the impredicative notion of '*all* subsets'; and its quasi-constructive style is hypocritical. Maddy (1990, p. 102) comments:

Of course, the temporal and constructive imagery is only metaphorical; sets are understood as objective entities, existing in their own right.

Yet it is just this imagery that accounts for the appeal of the iterative conception. If constructive imagery is attractive, why not become a constructivist? Hallett (1984, §6.2b) argues that on any constructivist view, even one that grants us infinitistic powers of construction, the impredicativity in the separation and replacement axioms cannot be justified. But if constructive imagery is firmly rejected then we lose our picture of the *shape* of the universe of sets: why, for example, should the set universe be populated in such a way that for every set there is another set that is its power set?

(6) EXTRAPOLATION FROM FINITE SETS OF PHYSICAL OBJECTS

Tharp (1989) and Maddy (1990) both propose that set theory is obtained by extrapolation from our everyday experience with small finite sets of physical objects (such as pairs of shoes and collections of pairs of shoes). The everyday sets can all be specified by enumeration or by a rule of membership, but we extrapolate from these to infinite sets not necessarily defined by a rule. The properties of sets in general are obtained by extrapolation from those of finite sets. For example, we know in the case of finite sets that the union of a finite set of finite sets is finite and that any finite set has finitely many subsets; if one replaces the word 'finite' by 'transfinite' one obtains the union and power-set axioms. This seems to be the main motivation for these two axioms.

I have problems with this notion of 'extrapolation' of meaning. Unlike the notions of abstraction and idealisation, which I shall expound in Chapter 4, extrapolation is an obscure technique. Tharp and Maddy both introduce the word 'set' using enumerations or rules of membership but go on to say that neither is essential to the concept: 'any way of picking out elements gives a set', but sets need not be 'given individually by particular specifications' (Tharp, 1989). Is there a difference between a way of picking out elements and a particular specification? Or is he saying that ways of picking out elements only give *some* of the sets? If so then how are the other sets to be understood? Maddy is equally puzzling.

Carrying this notion [iterative formation of finite sets] into the infinite, subcollections are 'combinatorially' determined, one for each possible way of selecting elements, regardless of whether there is a specifiable rule for these selections. (Maddy, 1990, p. 102)

What is the difference between a way and a rule? Surely a rule is simply anything that *determines* which elements to include, and any 'possible way of selecting elements' accomplishes this and so qualifies as a rule.

Perhaps the idea is that rules must be expressed in language, while ways need not be; ways could thus involve quantum systems or free choices. But such a distinction is unconvincing, since we can introduce names for any physical system or free agent and so incorporate their activity into rules. In this way we can formulate undecidable, unpredictable or senseless rules. (E.g., School Rule No. 1 is 'Whatever the Headmaster says, goes', where the Headmaster's decisions may be wholly arbitrary.)

It is surely not intended that rules are things expressible in some fixed formal language while ways are expressible in any terms, since this would rob the distinction of philosophical significance.

The method of extrapolation leaves us with no notion of what sets in general *are*, and thus in extrapolating their properties we are forced to rely on trial and error. Any property that holds for all finite physical sets we posit for all sets, except where this leads to a contradiction. Some properties have to be accepted as being peculiar to finite sets, such as not being in one-to-one correspondence with a proper subset of oneself, or only being well-orderable with one order-type.

This way of introducing infinite sets is really no better than saying, 'Consider *cats*: we meet them in our everyday lives and are familiar with their properties. Now, all the everyday ones are *feline*, but we can extrapolate away from their felineness to obtain a realm of higher, non-feline cats. Higher cats share many properties with their familiar feline brethren. For example, you can still stroke them – or at least, most cat-lovers think so, on the grounds that we like stroking cats and it seems to lead to no harm.'

My objections, then, are to the vagueness of the technique of extrapolation (leading to doubts about which properties of everyday sets carry over to all sets), and the fact that we seem to be extrapolating away from the only thing that gave us a handle on the concept in the first place (namely, the enumeration or rule of membership).

(7) SETS FROM POTENTIAL INFINITIES

One argument used by Cantor to justify the existence of infinite sets went as follows. Everyone, even a constructivist such as Kronecker who claims to reject infinite sets, finds it necessary to speak of arbitrary finite numbers (for example, 'for any arbitrarily large number N there exists a number $n > N$'). This presupposes the existence of *all* numbers $n > N$, taken as a whole, and this is just what is meant by an infinite set.

> Thus every potential infinity (the wandering limit) leads to a *Transfinitum* (the sure path for wandering), and cannot be thought of without the latter. (Cantor, 1887–8; translation quoted in Dauben, 1979, p. 127)

This claim that potential infinity presupposes actual infinity is a popular ploy amongst defenders of classical logic. Hallett (1984, Part I, Introduction) refers to it as Cantor's 'domain principle' and identifies it as one of Cantor's three key principles. I shall argue in the next chapter that, on the contrary, actual infinity is ill-defined and that there is an independently understood notion of potential infinity. However, in the present context there is a much easier way of defeating the argument. This is to point out that set theorists regard the set universe in much the same way that Kronecker regards the natural numbers. Set theorists do not accept the universe of sets as a legitimate mathematical object, and yet they allow themselves to quantify over all sets, assuming without question that classical logic applies to unbounded set quantifiers (see Mayberry (1977) for a dissenting view, however). They are therefore in a worse position to reject the set of sets than Kronecker is to reject the set of natural numbers.

This argument is essentially Dummett's (1991, Chapter 24), who states it in the following form: if we can overcome our initial prejudice against speaking of the 'number of all natural numbers', by telling ourselves that anything has a number even if we cannot count it, how can we be prevented from asking how many transfinite cardinals there are? Merely to cite the paradoxes is 'to wield the big stick, not to offer an explanation'.

SUMMING UP

This completes my list of the various ways of construing sets. Most of them are due to Cantor. For example, the definition

> By an 'aggregate' or 'set' I mean generally any multitude which can be thought of as a whole, i.e., any collection of definite elements which can be united by a law into a whole. (Cantor, 1883, p. 204; translation quoted in Dauben, 1979, p. 170.)

seems to combine the multiplicity-as-unity view, the collection view and the class view. ZFC is a strange hybrid of quasi-constructive and impredicative features. The quasi-constructive features show up in the axioms that say 'for any x there is a y such that ...': it is hard to resist the notion that y is *constructed from* x, and if we succeeded in purging our minds of this notion the intuitive persuasiveness of the axioms would be lost (why *should* the x's and y's be related in this way?). The impredicative features are the unbounded quantifiers (assumed to satisfy classical logic) and the replacement and separation axiom schemata, which can be instantiated with arbitrary formulae. It is hard to imagine a philosopher who could believe in all the axioms at once.

ZFC also seems most ill-adapted to serving as a basis for the rest of mathematics. In one sense the ZFC set universe is disconcertingly big: all traditional branches of mathematics are concerned exclusively with one minute corner of it. Admittedly, twentieth-century platonists have on rare occasions found that axioms about large cardinals have consequences for the highly impredicative properties of sets of reals, but they have been singularly unsuccessful in developing a distinctive mathematical theory for, say, the $\aleph_{37}$'th level of the cumulative hierarchy, to place alongside the existing theories of the natural numbers, the real numbers, and the real functions. One cannot avoid the impression that nearly all of the set universe is mathematically useless.

In another sense the set universe is *too small*. Many modern branches of mathematics are concerned with arbitrary structures, not necessarily sets, satisfying certain axioms. For example, group theory applies to any realisation of the group axioms, even one involving proper classes. To interpret 'for any group' as 'for any set that satisfies the group axioms' is therefore to limit the application of group theory artificially.

It is depressing to read Frege's (1893, §0) criticism of Dedekind's and Shröder's notions of set and reflect on how little conceptual clarification has taken place in the hundred years since. All my complaints against set theory may be summed up in a single charge, one of *axiom hacking*, that is, tinkering with formal systems without proper attention to the underlying informal ideas.

CHAPTER 3

WHAT'S WRONG WITH INFINITE QUANTIFIERS?

POTENTIAL AND ACTUAL INFINITY

The foundation stones of platonist mathematics are the infinite quantifiers, $\forall$ and $\exists$, regarded straightforwardly as infinite conjunction and infinite disjunction. For example, Goldbach's conjecture (that every even number greater than 2 is the sum of two primes) is understood as an infinite conjunction '4 is the sum of two primes and 6 is the sum of two primes and 8 is the sum of two primes and ... '. Wright (1980, I, §7) quotes this example and asks, since we understand the decidable predicate 'is the sum of two primes', the binary operation 'and', and the sequence of even numbers (which we can effectively enumerate), how can we possibly doubt that Goldbach's conjecture is a meaningful proposition? The answer is obvious: the dubious part is the three dots '...'. I shall spend this chapter examining these strange punctuation marks. They can be understood in two ways, traditionally called 'potential infinity' and 'actual infinity'. 'Actual infinity' means viewing the quantifiers as infinitary analogues of conjunction and disjunction: I shall argue that this is meaningless. 'Potential infinity' means viewing the quantifiers, and all other apparent references to infinities, in terms of the counting algorithm and other computational procedures: I shall defend this basic notion of computation against Wittgenstein.

QUANTIFIERS AS INFINITE CONJUNCTION AND DISJUNCTION

The case for construing universal quantification as infinite conjunction seems to be something like the following.

If we have two propositions, A and B, it is clear what it means to say that *both* are true; we write this as $A \wedge B$. This generalises effortlessly to the case of more than two propositions, where we write $A_1 \wedge A_2 \wedge \ldots \wedge A_n$. In fact, for any collection of propositions it makes sense to ask whether *all* of them hold; it is irrelevant whether there are finitely or infinitely many propositions. Imagine, for example, one truth value associated with every point on a line segment; imagine that each point holds up a flag if the value there is *false*, and not if the value is *true*; then one can imagine looking at the line segment and seeing if any flag is raised. This notion makes sense even though one cannot run through all the truth values one by one at a bounded speed in finite time. A similar argument applies to existential quantification. Other visualisations are possible: see for example Benacerraf & Putnam (1964, p. 17).

To help gauge the strength of this argument, compare it with the following one, obtained by substituting sums for conjunctions.

If we have two quantities, x and y, it is clear what it means to talk of the *sum* of the two; we write this as $x + y$. This generalises effortlessly to the case of more than two quantities, where we write $x_1 + x_2 + \ldots + x_n$. In fact, for any collection of quantities it makes sense to speak of the total or sum of them all; it is irrelevant whether there are finitely or infinitely many quantities. The sum is either a finite number or infinity. Imagine, for example, each quantity represented as a line segment; imagine lining up all the segments end to end; then the result must either stretch to infinity or stop at a certain point; in either case this gives us the sum of the quantities. This notion makes sense even though one cannot line up all the segments one by one at a bounded speed in finite time.

Infinite sums, of course, are not generally accepted in this straightforward way by mathematicians, on account of paradoxical sums like $1 - \frac{1}{2} + \frac{1}{3} - \frac{1}{4} + \frac{1}{5} - \cdots$, which can be made to sum to any real number or to diverge by rearranging the terms suitably. This, however, is an accident of history. Suppose we had invented infinite series before negative numbers. Then we would probably have found the above argument for infinite totals highly persuasive. No obvious contradiction would arise from accepting it, and when negative numbers were finally proposed we would regard *them*, not infinite totals, as paradoxical. Only a few philosophical sceptics would insist on explaining infinite sums in terms of the sequence of finite partial sums.

From this comparison between infinite conjunctions and infinite sums (first suggested by Hilbert (1925, p. 378)) I draw three conclusions.

- A commutative and associative binary operation may be extended uniquely to any finite number of unordered arguments, but it is a fallacy to infer by analogy that there is a well-defined infinitary notion.
- A simple visualisation of the infinitary notion is not sufficient justification, either.
- Nor is it sufficient that the assumption of the infinitary notion does not lead to a contradiction.

Even all three conditions together are not sufficient. Thus infinite conjunctions need a special argument, showing why the step from finite to infinite works for conjunctions but doesn't work for sums. As far as I know, no such argument has ever been proposed. It certainly will not do to appeal to the everyday use of the word 'all'; one might just as well appeal to the everyday use of the word 'total' as a justification for infinite sums.

THE ARGUMENT FROM NONSTANDARD MODELS

I hope I have succeeded in casting some doubt on the legitimacy of infinite conjunctions and disjunctions. My principal argument, however, will be against the notion of infinite set on which the infinite quantifiers rely; it resembles Putnam's (1980) argument, though without his general anti-realist perspective. If we are to understand universal quantification over, say, the natural numbers as infinite conjunction then the domain of quantification, the set of natural numbers **N**, must be 'grasped' as a well-defined totality of some sort. Exactly what an infinite totality is, whether it exists, and how we can grasp it given acquaintance with only finitely many of its members, are puzzling questions: it is not even clear what is at stake. However, so much of our cognitive life is mysterious that I do not dismiss infinite sets on that account.

Instead of asking whether **N** exists I shall state the question as whether we can grasp the concept **N**, that is, whether the symbol '**N**' is meaningful when we use it. The reason for formulating the question in this way is that a symbol only has the meaning that *we* give it; it wouldn't do to say that it has a meaning that passes our understanding. If we mean nothing when we use '**N**' then it is meaningless and we should stop using it.

My talk of 'grasping concepts' is not intended as a serious semantic theory; that is, it is not intended as the first step in an analysis of what is involved in understanding a word or a sentence. It simply follows ordinary English usage, which I hope is not too misleading, because I have no better way of talking about meaning. I shall assume that 'grasping' is a cognitive relation between a thinker and a 'concept' and that it is necessary and sufficient to justify the thinker in introducing a name for the concept. It follows that we must be able to grasp concepts *one at a time*, since if we could not grasp A without simultaneously grasping B we would be in no position to introduce a name denoting A (and not denoting B). However, in the case of **N** I shall assume it is only necessary to grasp it up to isomorphism.

Arguing by contradiction, suppose we have a clear grasp of **N** sufficient to support our use of the expression 'for all elements of **N**'. I shall infer a contradiction from this assumption.

The argument turns on the possibility of nonstandard models. It is generally accepted by classical mathematicians that in addition to **N** there are many other sets, called *nonstandard models of arithmetic*, that satisfy just the same propositions of elementary number theory as **N** but are not isomorphic to **N**. (Note that this is *not* an application of Gödel's incompleteness theorems: it is actually a corollary of his *completeness* theorem.) These nonstandard models are usually obtained using ultrafilters, and hence depend on the axiom of choice. However, my argument does not depend on whether my platonist opponent accepts the axiom of choice; it is sufficient that the platonist is unable to rule out the possibility of nonstandard models. Platonists believe that they have a grasp of 'the one true **N**' as opposed to the nonstandard imposters. They regard the non-categoricity of axiom systems as a model-theoretic curiosity posing no threat to their philosophical position

(Fraenkel, Bar-Hillel & Levy, 1973, V, §5).

Sceptics have asked, how do platonists know that they grasp the correct **N**? How do they even know that two platonists are talking about the *same* **N**, or that one platonist is talking about the same **N** on two occasions? Everything they can say to try to distinguish one candidate **N** from another is true or false for them both equally.

The answer one finds in the textbooks on nonstandard analysis is that different models of elementary arithmetic differ in which propositions of *second-order* arithmetic they satisfy; this allows one to distinguish 'the one true **N**' as the one that satisfies induction with respect to all second-order formulae. This, however, is no help, for second-order logic depends on a 'for all subsets of *S*' quantifier. All the second-order characterisation of **N** shows is that two platonists who have different models of arithmetic also have different second-order quantifiers. Second-order quantifiers are even more philosophically problematic than **N**; so to found **N** on them is to do it no favour. Similarly it is no help to define **N** within the universe of set theory: two platonists may have different models of set theory with different power-set operations.

Thus it seems that a platonist's grasp of the correct **N** cannot be demonstrated by any proposition they affirm about it. Hence it cannot be taught to young platonists: it must be innate in the human mind and merely *brought out* by the usual verbal explanations ('0, 1, 2, 3, and so on') which, strictly speaking, apply to all models equally. Any mathematical behaviour (counting, sums, proving and asserting number-theoretic propositions) is carried out identically by a platonist with the correct **N** as by one with a wrong **N**. Even if, as a matter of fact, we all have the right **N**, there may someday arise a strange mental disease that causes its sufferer to learn the wrong **N** from normal early life-experiences. This distressing condition would remain forever undiagnosed as it would lead to no disability; indeed (I claim) it would make no difference to what is going on in the sufferer's head. But in fact evolution is unlikely to have endowed us with any predisposition to grasp the correct **N** since it has no survival advantage over any other **N**. Thus it is no use postulating a special mental faculty that enables us to grasp the correct **N** (as Putnam (1980) suggests the platonist might). If it is legitimate to imagine a person who possesses normal human cognitive abilities plus this special faculty then it is also legitimate to imagine someone who possesses normal human cognitive abilities plus a pathological version of the special faculty that leads them to the wrong **N**; we have no way of defining what it is for the special faculty to be in a healthy or pathological state other than by reference to the standard **N**.

Now, to some philosophers (such as Putnam), who explain mathematical knowledge in terms of linguistic behaviour, social consensus or successful application, this would be conclusive evidence that the distinction between the one true **N** and the others is meaningless; hence the platonist position, that '**N**' denotes a distinct set, cannot be accepted literally. The foundation for 'for all numbers' on infinite conjunction over **N** thereby vanishes. I am not one of those philosophers, however. If I thought that I had a well-defined notion of **N** then it would not bother me particularly if I could not demonstrate its correctness, or even communicate it, to anyone else; moreover, if other people have the wrong **N** then that is their problem.

The trouble is that I *don't* think I have such a notion. I *do* have a well-defined notion of *counting*; I am confident that however long I count I shall always know what comes next, and that when I repeat the counting the sequence of numbers generated is always the same, barring errors. This does not seem to depend on the distinction between the one true **N** and the nonstandard models. There comes a point where we have to dismiss an alleged distinction as meaningless; the point comes when (i) we cannot explain the distinction in terms of anything more basic and (ii) we can understand the subject perfectly well without it. That point has been reached here: there is no notion of standardness for elementarily equivalent models of arithmetic. But, as remarked above, to grasp a concept we must be able to grasp it *in isolation from other concepts*. Hence we cannot grasp **N**. This contradicts the initial assumption that we could grasp **N**. The contradiction establishes that we cannot grasp **N**.

Note that this argument only applies to mathematical infinities and does not rule out the possibility of physical infinities. The reason for the difference is that the purposes of mathematical and physical theories are very different. A physical theory is intended to account for our sensory experience and is judged successful if it provides a simple explanation of a large body of sensory data and correct predictions of future experiences. We would be justified in believing in an infinity of stars if that led to the best available explanation of the observed value of Hubble's constant, say. The notion of 'the infinite collection of stars' could play a well-defined role in a successful physical theory, which is the most that one can ask from any physical concept. Mathematical concepts cannot be justified in the same way: we demand more from **N** than that it play a well-defined role in Peano Arithmetic. Thus Tait's (1986, §3) objection that arguments against mathematical infinity such as mine can be applied equally well to the physical world is mistaken.

There are several weak counter-arguments the platonist may resort to. One is to say that number theory refers indifferently to *all* models, so there is no need to distinguish one as 'the one true **N**'. Goldbach's conjecture, for example, is true in all models or false in all models. However, on this view, what reason is there to think that there are *any* models of the appropriate sort? And how are we to distinguish these models from similar **N**-like structures in which Goldbach's conjecture has the opposite truth value? In any case, the very notion of infinite set is undermined if it is conceded that we are incapable of grasping any single set.

A second possible counter-argument is that, although admittedly we cannot distinguish the one true **N** by formal propositions we may still do so by informal explanation. But this will carry no weight unless the informal explanation is actually given. It certainly won't do to say '**N** contains everything you can count up to and nothing else'.

The effect of my nonstandard models argument is to reinforce the traditional distinction between 'potential infinity' (counting and other iterative processes) and 'actual infinity' (the one true **N**). Platonists typically reject this distinction. Their argument (Dedekind, 1890; Tait, 1983) is that to understand counting or any other algorithm it is not sufficient to understand the basic steps (starting with 0, applying successor): it is necessary also to understand that the steps are to be applied *any finite number* of times until the halting condition is satisfied. Hence counting presupposes number; a grasp of **N** is integral to any meaningful use of numbers. We cannot explain in words or mathematical symbols the difference between the one true **N** and the others because it is such a basic notion that there is no way of discussing numbers without presupposing it – in just the same way that we cannot explain the difference between a sensation of red and a sensation of green because such distinctions are basic to all our talk of colour.

This amounts to a refusal to allow the intuitionist to take iteration as a primitive notion. I have four rejoinders.

- If a concept A is conventionally explained in terms of a concept B, this is merely a fact about natural language usage and does not imply that B is philosophically more fundamental than A. A philosopher who wants to argue that A is primitive cannot be refuted merely by showing that A is synonymous to a phrase involving B. For example, Frege held that 'having the same number as' was more fundamental than 'number', despite surface syntax. Many pairs of concepts are interdefinable (for example, 'open set' and 'closed set'), but it cannot be maintained that each member of the pair is more fundamental than the other. I am free to argue, therefore, that counting is primitive and is not understood in terms of number.
- The notion of continuing a computation indefinitely ('Follow the arrows of this flowchart and keep going') seems to arise unproblematically from our basic understanding of time experience, without dependence on any notion of infinity. This is most of the point of Brouwer's First Act of Intuitionism.
- The fact that two platonists with different models of **N** would still count the same way undermines the claim that counting depends on **N**.
- The platonists' view, that **N** is defined as the intersection of all sets containing 0 and closed under successor, is philosophically perverse:

> *this* is a clear-cut case of defining something comparatively clear in terms of something obscure and secondary.

A further platonist counter-argument is that, although *we* cannot grasp **N**, due to our finite nature, perhaps there are infinite beings who can, by simply inspecting each natural number. **N** therefore would exist as a concept of such a being's thinking, and since we may reason about anything that might exist it follows that we are entitled to reason about **N**. (This resembles Cantor's view that infinite sets exist in the mind of God (Dauben, 1979, p. 229); see also Russell's remark that our inability to run through the whole expansion of π is not '*logically* impossible' but merely '*medically* impossible' (Russell, 1935–6).) The fallacy of this is that if infinite beings are possible then so are nonstandard infinite beings, and we have no way of distinguishing the standard from the nonstandard ones. Each infinite being might use a different **N**, and we are incapable of singling out the standard being and hence the standard **N**.

My conclusions from this discussion are that the platonists' **N** and the associated infinite quantifiers go beyond our fundamental understanding of iterated processes (counting and other computations), and that they are meaningless.

It follows that number theory needs to be reformulated in terms of computations rather than **N**. My proposal is not to *define* **N** as everything generated by counting, thus putting **N** on a firm foundation, but rather to replace all talk of **N** by talk of counting, thus *bypassing* **N**. Our understanding of numbers in general is founded on our ability to count and reflect on the process of counting, but it does not proceed via any notion of 'everything generated by counting'. This amounts to adopting an intuitionistic view of arithmetic. Intuitionists understand number in terms of the counting process rather than in terms of an intended model. One can of course subject intuitionistic formal systems such as Heyting Arithmetic to model-theoretic analysis, but the problem of identifying the 'standard' model does not arise for intuitionists since their grasp of arithmetic does not depend on models at all. Hence the nonstandard models argument cannot be used against intuitionism.

This leads us to the platonist's final defence: that such a view is too weak to support conventional mathematical practice (Peano Arithmetic), and is certainly incapable of extension to analysis and higher mathematics. If so then perhaps the arguments against **N** can be dismissed as simply another variety of philosophical scepticism. The philosopher's job is to explain why mathematics works, or appears to work; if the explanation offered is simply that mathematics *doesn't* work then it is reasonable to reject the explanation rather than mathematics. Therefore to complete the argument against **N** I must show that intuitionism *can* provide an interpretation of arithmetic that explains why it works. Intuitionistic arithmetic and platonist arithmetic are *rival* accounts, since if the intuitionistic account succeeds then **N** is *not* essential to our understanding of number and the final argument for **N** falls.

DEFENCE OF POTENTIAL INFINITY AGAINST WITTGENSTEIN

My basic notion of computation can be attacked from the opposite direction, from a feasibilist or a Wittgensteinian perspective. A feasibilist (Yessenin-Volpin, 1970; Parikh, 1971) holds that even large finite numbers like 10^{1000} don't exist since we cannot in practice count up to them. A proper response to this is in terms of the notion of idealisation (see the next chapter).

Wittgenstein, on the other hand, in his *Philosophical Investigations* (Wittgenstein, 1953), attacks my assumption that an infinite sequence of actions can be *determined* by a rule, at least in the way conventionally assumed when we speak of computations. What he objects to is the common belief that when we learn a rule a special mental process called 'understanding' takes place within us that determines how we will apply the rule in all possible future cases. This is part of his scepticism about mental states and processes in general. *Genuine* examples of mental states are depression, excitement and pain (*op. cit.*, I, §151(a)), and genuine examples of mental processes are 'a pain's growing more and less', and 'the hearing of a tune or a sentence' (I, §154); he also seems to accept visualising a mental image as a mental process or event. But he does not accept *cognitive* mental processes (*being directly aware that*, as opposed to *being directly aware of*); the examples he discusses are

- understanding a word (I, §138–142),
- understanding a rule, having being taught by examples (I, §143–149, 185–242),
- reading a written passage aloud (I, §156–178),
- dreaming (II, §vii),
- seeing an ambiguous picture in a certain way (he gives an example of a picture that can be seen either as a duck's head or a rabbit's head) (II, §xi).

In each case his purpose is to deny that any characteristic mental process is taking place. In the case of words and rules, there is no inner process of understanding: the only criterion of understanding is correct usage. In the case of reading aloud, there is no mental process that distinguishes this from reciting a memorised passage or speaking spontaneously. In the case of dreaming, it makes no sense to ask whether the dream really took place as it seems or whether it was instantaneously fabricated at the moment of waking. In the case of seeing an ambiguous picture, there is no difference between the experience of seeing it as a duck's head and seeing it as a rabbit's head: the image in the mind's eye is the same. In every case it is the person's subsequent behaviour, not their present mental state, that is the criterion for understanding, reading aloud, dreaming, or interpreting a picture in a certain way.

Most of us would dispute this, arguing that there *is* a characteristic mental process in each case and that we are directly aware of it by introspection. We know what it feels like to understand a rule, and we know that we understand the rule even if we never subsequently apply it. Correct usage is a *symptom* of understanding, not a *criterion*.

Wittgenstein's answer, I think, is that when we claim to be introspecting we are really *theorising* about the psychological processes that we think must be going on just below the surface of consciousness: we believe in these psychological processes so firmly that we delude ourselves into thinking that we actually *experience* them. What really happens in our minds when we understand something is a sudden feeling of relief, enlightenment or confidence, and perhaps a mental picture. But these are merely accidental accompaniments of understanding: one could understand perfectly well without having these sensations, and moreover there is nothing *else* going on simultaneously with these sensations that explains why we now know how to apply the rule.

Most of us would nevertheless insist that we *are* directly aware of these mental processes, as directly aware as we are of our physical sensations and mental images.

Wittgenstein defends his position sometimes by taking a behaviourist attitude ('An "inner process" stands in need of outward criteria.' I, §580), and sometimes by what might be called the Argument from Insanity ('Always get rid of the idea of the private object in this way: assume that it constantly changes, but that you do not notice the change because your memory constantly deceives you.' II, §xi). The former argument will not convince anyone who is not already a behaviourist. The latter is useless because it is so indiscriminate: *any* belief can be undermined in this way, including the belief that the Argument from Insanity is unanswerable. All we can do is get on with thinking, in the hope that our thoughts have a sufficient minimal coherence.

He also resorts repeatedly to the argument that we cannot convey these alleged mental processes (our understanding of a rule or the meaning of word, say) to another person unambiguously, either by examples or by verbal explanation, since a finite set of examples is always compatible with more than one rule and since verbal explanations depend on a correct understanding of the words used. Hence we cannot give an account of what it means to understand a rule or word correctly. This much is true, but it does nothing to shake our naive view that we each understand the rule or word individually, even if we cannot fully articulate that understanding and even if, through misinterpretation, we each understand it differently.

Mostly, however, Wittgenstein does not attempt to argue for his position but simply states it in a variety of examples in the hope that eventually the scales will fall from our eyes and we shall see that he is right. He adopts this method because he sees himself as *describing* rather than *explaining* or *theorising* (the distinction is dubious). His examples have the opposite effect on me: they convince me of the absolute necessity of recognising these mental processes if our mental life is not to become even more of a mystery than it already is.

His views have drastic consequences for the way that logical and mathematical necessity are to be understood. Mathematical necessity depends on the notion that an algorithm has a unique correct outcome. Logical necessity depends on the notion that some statements may be seen to be true simply by inspecting the meanings of the words and grammatical constructions used in them. Both these assume that once a rule has been adopted it is now determined what is to count as correct application of the rule in every possible case.

Wittgenstein, of course, denies this view of how rules work. His account of rule-governed behaviour is as follows. Suppose we adopt a general rule (say, rules of arithmetic) and then proceed to apply it to a sequence of cases (say, a series of sums). In calling it a *rule* we mean that each application of it is to be treated as *necessarily correct*: that is, we say not merely 'This is how the sum worked out this time' but 'This is how it *must* work out'. Of course, in practice, we do not always get the same answer when we apply a rule to a case repeatedly, so we need to introduce a notion of 'error' and hence a distinction between what seems to be a faithful application of a rule and what is a genuinely correct application. Wittgenstein argues that this distinction makes no sense in purely introspective terms, since a seemingly correct application of the rule is introspectively just like a genuinely correct application, except in its outcome. This is true whether the rule is being applied by a single person or by a whole society.

In practice, when we speak of errors we merely mean that an individual's application deviates from common usage; there is no ulterior notion of correctness against which either the individual's usage or common usage can be compared. Hence the notion of logical or mathematical necessity arises not from any abstract notion of correct application of rules but from the contrast between an individual and society. A rule is simply a social practice in which we treat individual variations as errors.

It follows that an individual cannot set up and practise a rule in isolation – even though it seems to us that we often do exactly that.

And hence also 'obeying a rule' is a practice. And to *think* one is obeying a rule is not to obey a rule. Hence it is not possible to obey a rule 'privately': otherwise thinking one was obeying a rule would be the same thing as obeying it. (I, §202; see also the 'private languages' argument,

I, §243–280.)

It also follows that the application of a rule to a particular case (the outcome of a sum, say) is not determined in advance. *After* we have applied it to the case we shall say that the outcome we agree on was the only possible correct one; but there was nothing in the rule (as written down or as understood by anyone beforehand) that determined the correct outcome.

An immediate difficulty with this argument is that, although it is intended only to attack private rule-following, it applies equally to public rule-following and hence appears to undermine any notion of obeying a rule. To benefit from training in a rule and to participate in public rule-following you need to be able to determine when society is approving of your usage and when it is trying to correct you. Society might express approval or disapproval by saying 'Yes, that's the way' or 'No, not like that', or in more diverse and subtle ways, but in any case you need to classify the speech you hear and the gestures you see into two categories, those expressing social approval and those expressing disagreement. This classification is *purely private*: it is precisely of the sort that Wittgenstein says is impossible. If his position is correct then one cannot draw a meaningful distinction between one's *seeming to oneself* to apply a rule correctly and be approved by society and one's *actually* doing so. By a similar argument one cannot draw such a distinction for another person's behaviour. It follows that society cannot draw the distinction either (since society is made up of individuals who cannot draw the distinction). It follows that there is no such distinction. (Imagine a student who thought that grade E was higher than grade A, that the point of arithmetic was never to do the same sum the same way twice, and that the objective in school was to leave it as early as possible. Such a person would go through life thinking themselves a calculating prodigy. In what objective sense are they wrong?)

Wittgenstein's position on rule-following has radical consequences for the notion of mathematical proof, which are developed in his *Remarks on the Foundations of Mathematics*; the citations that follow are to the third edition (Wittgenstein, 1978). I shall follow the interpretation in Wright (1980, XXIII). Wittgenstein's idea of proof seems to be that a proof of $5 + 7 = 12$, say, might be a diagram showing twelve dots grouped into two sets of five and seven. We look at the diagram, count the dots, and make the following observations.

1. I have counted five things and seven things and, on this occasion, obtained twelve.
2. The diagram is so perspicuous that I cannot imagine how I can have made a mistake in counting, nor can I imagine how five plus seven could ever even *seem* to give any result other than twelve.

3. If I count five things and seven things repeatedly I shall nearly always get twelve (Wittgenstein, 1978, I, §31; I, §41).

Now, these observations do not *prove* that $5 + 7 = 12$. (1) is an empirical fact, a report of a counting experiment, (2) is about the limitations of my imagination, and (3) is an empirical generalisation. Together they entitle me to treat '5+7' and '12' as interchangeable with practical confidence that this will rarely generate discrepancies – but they do not make $5 + 7 = 12$ a *necessary* truth. The equation only becomes a necessary truth if we adopt it as a 'grammatical rule': we say that from now on any calculation that contradicts $5 + 7 = 12$ shall be treated as erroneous. This, clearly, alters our concepts of correct counting and number, since it could lead us to reject as erroneous counts that we would previously have accepted as correct. The 'proof' and the observations (1)–(3) provide a strong practical motivation for adopting the equation as necessary but they do not *force* us to do so. One could understand the diagram, believe all three observations, and yet choose not to treat $5+7=12$ as a necessary truth. Thus there is an essential element of *decision* in accepting the conclusion of a proof, and each proof requires us to modify our concepts in some way.

The proof doesn't *explore* the essence of the two figures, but it does express what I am going to count as belonging to the essence of the figures from now on.— I deposit what belongs to the essence among the paradigms of language.

The mathematician creates *essences*. (I, §32)

We say that a proof is a picture. But this picture stands in need of ratification, and that we give it when we work over it. (VII, §9)

Can I say: "A proof induces us to make a certain decision, namely that of accepting a particular concept-formation"?

Do not look at the proof as a procedure that *compels* you, but as one that *guides* you.— And what it guides is your *conception* of a (particular) situation. (IV, §30)

What is proved by a mathematical proof is set up as an internal relation and withdrawn from doubt. (VII, §6)

If this is indeed Wittgenstein's conception of the role of proof then I cannot help thinking that it leaves out the essential ingredient. What is the basis for observation (3)? We have counted five plus seven and obtained twelve (on this occasion); moreover it seems inconceivable how any other result could arise. But how do we know that the next time we count it won't seem equally inevitable that the result is thirteen? How does the present perspicuity of the diagram govern how things will seem in the future? Four possible approaches to an answer are suggested in the text.

First, since observation (3) is intended to be taken as a well-justified empirical prediction, one may simply say that it is supported by our previous experience with computations: perspicuous proofs have in the past turned out to be highly reliable predictors of future computations (a square triangle is inconceivable to us, and, sure enough, square triangles have turned out to be exceedingly rare in nature) so we are empirically justified in asserting (3) for each new proof.

Secondly, recognising a proof may be similar to recognising a colour or recognising that a rule has been applied correctly: all these require an act of interpretation and hence a decision.

> Hitherto we have calculated according to such and such a rule; now someone shews us the proof that it can also be done in another way, and we switch to the other technique – not because we tell ourselves that it will work this way too, but because we feel the new technique as identical with the old one, because we have to give it the same sense, because we recognize it as the same just as we recognize this colour as green. . . . It might be said: the reasons why we now shift to a different technique are of the same kind as those which make us carry out a new multiplication as we do; we accept the technique as the *same* as we have applied in doing other multiplications. (IV, §36; see also IV, §30)

Thus the proof-following mystery is assimilated to the rule-following mystery.

Thirdly, observation (3) may rest on a sort of *psychological intuition*.

> I know with *great* certainty that if I multiply 25 by 25 ten times I shall get 625 every time. That is to say I know the psychological fact that this calculation will keep on seeming correct to me . . . . Now is that an empirical fact? Of course – and yet it would be difficult to mention experiments that would convince me of it. Such a thing might be called an intuitively known *empirical* fact. (IV, §44)

Fourthly, Wittgenstein states repeatedly that a proof shows us not *that* a proposition holds but *how* it holds.

> When the proposition seems not to be right in application, the proof must surely shew me why and how it *must* be right; that is, *how* I must reconcile it with experience.
>
> Thus the proof is a blue-print for the employment of the rule. (VI, §3)

The idea seems to be that the proof tells us how to explain away future calculations that appear to conflict with the proposition. Unfortunately he does not give any examples of how a proof can accomplish this.

All these four approaches seem to miss the point of what is so *convincing* about proofs. To change the example, consider the statement that $P \supset P$ is a theorem of propositional calculus. (Forget the meaning of propositional calculus and consider it for present purposes purely as an uninterpreted axiomatic system.) A proof of this statement would be a formal derivation such as

$$\dfrac{P\supset(P\supset P) \qquad \dfrac{P\supset((P\supset P)\supset P) \quad (P\supset((P\supset P)\supset P))\supset((P\supset(P\supset P))\supset(P\supset P))}{(P\supset(P\supset P))\supset(P\supset P)}}{P\supset P}$$

(the axiom schemata used here are $A \supset (B \supset A)$ and $(A \supset (B \supset C)) \supset ((A \supset B) \supset (A \supset C))$, and the rule of inference used is *Modus Ponens*). According to Wittgenstein it is possible to check each step of the derivation and to find it unimaginable how the derivation could contain a mistake, and yet to refrain from treating the statement that $P \supset P$ is derivable as a necessary truth. If someone actually took this stance I would ask them, which formulae used in the derivation *do* you accept as necessarily theorems of propositional calculus? They would have to point to an axiom that they did not accept was necessarily an instance of the relevant axiom schema, or an inference step that they did not accept was necessarily an instance of *Modus Ponens*. Then I would conclude that they did not *understand* the axiom schema or rule of inference in question. My point is that, whatever it means to 'understand' something, it is certain that understanding a schema involves knowing what is an instance of it. Any uncertainty about whether $P\supset((P\supset P)\supset P)$ is an instance of the axiom schema $A \supset (B \supset A)$ is a *criterion* (and not merely a *symptom*) of a lack of understanding of the schema. Understanding a schema probably involves more than simply being able to judge whether given formulae are instances of it, but it certainly *includes* this ability. Wittgenstein seems to make a similar point himself when he says, 'the meaning of "$(x).fx$" is made clear by our insisting on "fa" 's following from it' (I, §10–11), but the discussion is inconclusive.

Now, 'understanding' is such a contentious issue that I could imagine some philosopher insisting that it *is* possible to understand a schema without being certain whether a given (small) formula is an instance of it. But such a sense of 'understanding' is so weak that it is hard to see what it would amount to.

My conclusion is the very unoriginal one that anyone who understands the propositional calculus (as an uninterpreted formal system) is constrained by their own understanding to accept that the formula $P \supset P$ is derivable in it. It is simply not coherent to understand a proof and not accept the conclusion as a necessary truth.

Similarly Wittgenstein's account of rule-following seems to miss the essential ingredient of rules. Suppose, to take up one of his examples, that, after adopting the equation $5+7=12$, we count a group of children and find that there are five boys, seven girls, and thirteen children. Then we say that 'we must have made an error'. The equation has been adopted as a *criterion* of correct counting; adopting the equation modifies our concept of twelve (previously twelve was defined purely in terms of direct counting). Thus we conclude that the total of thirteen children must be wrong. So, at least,

Wittgenstein claims. But this is not in fact how we deal with errors. When the results of two calculations (addition of sub-totals and direct counting) disagree we do *not* simply dismiss one as erroneous. We say there must have been an error, and we *search* for it in *both* calculations. As Dummett (1959) says, 'one cannot make some mistake without there having been some particular mistake which one has made'. Thus we look through the steps of both calculations until we find a step that, when viewed in isolation, is patently wrong. Then we correct the step and repeat the rest of that calculation. If the results still disagree then we search for further errors. Eventually, the discrepancy *always* disappears (unless we give up because we are not sufficiently motivated). This procedure differs from Wittgenstein's account in three important respects. First, we cannot tell, given simply a discrepancy, *which* calculation is erroneous. Secondly, if a calculation is erroneous it must contain a specific error, that is, an individual step that is unmistakably wrong. Thirdly, the error, once found, is enough by itself to invalidate the calculation in which it occurs. The calculation is erroneous *in its own right* because it contains this error: it is not simply *overruled* by the other calculation. The discrepancy with the other calculation is therefore a *symptom*, not a *criterion*, of error.

If we repeatedly found ourselves unable to locate errors in conflicting pairs of (short) calculations, or if we were unsure about whether a particular step was a mistake, then (as Wittgenstein says in another context (I, §37)), 'that would be the end of all sums'.

This provides the key to the *objectivity* of rules, the distinction between applying a rule correctly and merely seeming to do so. If two mathematicians apply the same rule to the same case but obtain different outcomes they do not have to appeal to the community or abandon mathematical necessity. They can compare their working step by step until they locate the first step at which they disagree; then it will be obvious to both of them who has botched this step. Whenever we apply a rule the fear of being corrected in this way is at the back of our minds; we are always thinking, 'This calculation *seems* correct, but would it stand comparison with someone else's?', and this gives us the desired distinction between apparent correctness and actual correctness. (Moreover, it is not strictly necessary to have more than one mathematician; an individual can simply carry out the calculation repeatedly.)

My conclusion on Wittgenstein's philosophy of mathematics is that it fails by his own standards. His objective is to describe mathematical practice, without trying to reform it or justify it, in such a way as to dispel any sense of philosophical mystery. But his account seriously misrepresents the way mathematicians handle errors, an issue that he correctly treats as central to an understanding of mathematical objectivity. And, far from dispelling philosophical mystery, it exacerbates it unbearably. It leaves us completely mystified as to how proofs *prove* and how rules *regulate*. The purpose of stating a rule is precisely to constrain future behaviour, to allow certain actions and disallow others, not to license people to regard whatever they do as a necessary consequence of the rule. If the Ten Commandments had been worded in the style 'Thou shalt not kill; but justifiable homicide is okay, at your own discretion' then Moses would scarcely have bothered to carry them down the mountain.

More importantly, it fails to address the question I posed at the beginning, that of how mathematics can be meaningful. How, for example, is the shepherd in Chapter 1 to know that his sheep-counting procedure is accurate? Wittgenstein deliberately avoids such questions. All he would be willing to say is that the shepherd does what he does and that another shepherd might do something different. Which shepherd would run out of sheep first (given certain reasonable physical assumptions about the world)? To answer this, we need to go beyond Wittgenstein's account and accept that we understand rules, such as counting, in advance of, and *independently* of, our subsequent application of them. The outcome of a computation is pre-determined, not because we never make mistakes but because we can always correct our mistakes.

It is useful to compare this conclusion with the previous discussion on infinite quantifiers. There, I rejected classical quantifiers because we cannot characterise them in language; here I accept the conventional view of rule-understanding *even though* we cannot characterise it in language. Why the difference? I said that an expression may be rejected as meaningless if it cannot be explained in terms of anything more basic and if we can get by perfectly well without it. (Both these criteria are imprecise but still usable.) The point is that we need some realist notion of rule-determination if we are to make sense of the everyday facts of rule-using behaviour; we do *not* need a platonist notion of infinite quantifiers to make sense of arithmetic. The evidence for the former is Wittgenstein's threadbare and mystifying account of mathematics, obtained by trying to do without conventional rule-determination. The evidence for the latter is the rest of this book, where I show how to understand significant parts of mathematics without recourse to classical quantifiers.

It follows that if anyone can give a credible of account of mathematical practice that does not rely on conventional rule-determination then I must abandon everything in this book after this point as philosophically misconceived.

CONCLUSION

The conclusion from this discussion of potential and actual infinity is that we grasp how to count and can reflect on the way we count, but we have no grasp of 'everything that can be generated by counting'. Hence number theory should be developed *directly* from our algorithmic knowledge of counting and not proceed via any notion of infinite set; if such an account succeeds then, as explained above, this completes the refutation of actual infinity.

CHAPTER 4

ABSTRACTION AND IDEALISATION

I have argued that mathematics needs to be reconstructed or reinterpreted in terms of potential infinity. The first step in this task is to examine the notion of a *mathematical object* and its relation to physical objects. This is the subject of the present chapter.

WHY WE NEED TO ABSTRACT AND IDEALISE

A popular and time-honoured theory on the nature of mathematical objects is that they are obtained by *abstraction* from physical objects. Cantor explained sets, and Dedekind explained numbers, in this way. The idea is that we start with a physical object and selectively disregard some of its properties, thus creating an object that is like the original one but without the disregarded properties. For example, if we disregard the colour of a tabby cat we create a colourless cat; if we disregard the number of branches on a tree we obtain a tree with no number of branches; if we disregard the identity and interrelationships of a collection of three objects we obtain the number three. Or possibly we do not create a colourless cat but merely the *idea* of one, which for mathematical purposes is, it seems, just as good.

This remarkable mental faculty fell into disrepute following Frege's hostility (Frege, 1906; Geach & Black, 1970, pp. 84–5), but it is still often appealed to (for example, Tharp, 1991). Related to abstraction is *idealisation*, the process by which we dismiss inconvenient obstacles by pure will power. Thus we say that we can count up to any natural number, even 10^{1000}, 'in principle', idealising away the obvious reasons why we cannot. Idealisation is particularly important for us intuitionists: we wish to restrict mathematics to mental constructions, but we do not wish to restrict ourselves to practically feasible ones. As George (1993) points out, intuitionism depends on a certain distinctive conception of 'what is an appropriate idealization of the actual human condition'.

Abstraction and idealisation are such murky notions that it is sometimes suggested that we should develop a *feasible* mathematics based on constructions that can actually be performed (Wright, 1982). Thus, a natural number would be defined as anything we can actually count up to. The consequences for arithmetic would be severe; for example, the theorem that if m and n are natural numbers then m^n is a natural number would clearly have to go

(Parikh, 1971). Of course, what can be counted up to depends on who is doing the counting, with what material aids, in what circumstances, and in what notation. For simplicity, let us confine our attention to stroke numerals: thus counting up to 5 means writing 'IIIII'.

We might begin with some observations of the following sort.

- Anne can't count up to 10^{1000} because, taking account of the estimated lifetime of the universe, the quantum nature of matter, and the finiteness of the speed of light, one cannot carry out 10^{1000} actions of any kind in this universe.
- Brian can't count up to 10^{100} because the human lifespan is too short.
- Cilla can't count up to 10^5 on a blackboard because she hasn't a large enough supply of chalk (although she might be able to do so in ink on paper).
- David can't count up to 1000 on this blackboard because it is too small.
- Edward can't count up to 13 because he is too superstitious.
- Fiona can't count up to the first even prime number greater than two.

It is clear that to develop a workable theory of what arithmetic problems of this sort are practically soluble we must begin by classifying the multifarious reasons why some are insoluble, in particular the six above. Anne's problem is to do with physics, Brian's is due to human biology, Cilla's is due to her choice of writing material, David's is due to his choice of blackboard, Edward's is due to his psychological inhibitions, while Fiona's seems somehow 'intrinsic' to the task. The importance of this classification is that it indicates what we need to do to overcome the various obstacles. Anne needs to do some physical research in the hope of overturning current physical theories; Brian needs to find some other intelligent lifeform to do the counting; Cilla needs to switch to ink and paper; David merely needs to find a bigger board; in Edward's case the simplest remedy is to find a less superstitious writer; whereas Fiona's problem cannot be solved by any change of writer or writing material and indeed seems to resist any sort of solution.

A second advantage of such a classification is that it achieves what computer scientists call a *separation of concerns*. It is hard to concentrate on the psychology of superstition while also keeping in mind the Earth's finite chalk reserves and the latest discoveries in cosmology. These matters each have their own peculiar laws and techniques and are best studied separately by specialists. Thus we assign Anne's problem to physicists, Brian's problem to gerontologists or exobiologists, and so on.

This leaves us with poor Fiona. Her problem is in fact insoluble. To explain insolubilities of this sort is the special task of *mathematics*. We need a way of discussing Fiona's problem without being distracted by the heated debate between the advocates of whiteboards and blackboards (which is why we need *abstraction*) and without being distracted by controversy on the possible existence of long-lived intelligent Martians (which is why we need *idealisation*).

I am sorry if I seem to be belabouring the very obvious, but many people have looked askance at abstraction and idealisation; intuitionism in particular has been criticised for relying on them (see the discussion of 'extrapolation' below). I am trying to show that both abstraction and idealisation are integral to mathematics and that we could not develop a theory of feasible arithmetic without first developing an abstract, idealised arithmetic.

A similar story could be told for any branch of science, each of which is defined by a characteristic set of abstractions and idealisations.

Philosophers, therefore, would be better employed putting abstraction and idealisation on a firm foundation than trying to do without them. In the next section I shall describe how abstraction and idealisation work; in the final section I shall distinguish them from mere 'extrapolation', as practised by Tharp and Maddy.

HOW ABSTRACTION AND IDEALISATION WORK

Let T be our repertoire of knowledge and arguments concerning iterative physical processes. One example of an iterative physical process is writing strokes on a blackboard or paper, so the problems of Anne and her friends in the previous section fall within the scope of T. But T also applies to other repeatable processes, such as demolishing houses and crossing the Atlantic. Examples of demonstrable propositions in T are:

(1) writing a pair of strokes and then writing another pair is tantamount to writing a quadruple of strokes;
(2) demolishing a pair of houses and then demolishing another pair is tantamount to demolishing a quadruple of houses;
(3) a double-crossing of the Atlantic followed by another double-crossing is tantamount to a quadruple-crossing of the Atlantic;
(4) regardless of what strokes have already been written it is always possible to write another one, provided the supply of chalk has not run out, *etc.*;
(5) regardless of what houses have already been demolished it is always possible to demolish another, provided there are some left, *etc.*;
(6) regardless of what Atlantic crossings have already been made it is always possible to cross again, provided one's boat hasn't sprung a leak, *etc.*.

In the last three examples the '*etc.*' denotes a long list of other obstacles that might prevent the next step from being carried out. In general, an *obstacle* is anything that prevents us from carrying out a precisely specified task; whereas an *error* is any unwitting deviation from the task that allows us to arrive at a wrong result. More precisely, an error is defined as a step that, if we were to examine it critically in isolation, we would see to be unequivocally a violation of the intended procedure. Errors are more insidious than obstacles, because we do not know when we are making one; but they can always be corrected by repeating the task more carefully, whereas obstacles cannot always be overcome. We cannot do the task so carefully as to exclude altogether the possibility of error. Nevertheless, we can make the risk of error arbitrarily low by being sufficiently careful; and in the event of a discrepancy between two repetitions of a process we can always resolve the matter by comparing the processes step by step. As argued in the discussion of Wittgenstein in the previous chapter, it is our ability to repeat computations and correct errors that gives mathematics its objectivity, its distinction between a seemingly correct computation and a genuinely correct one.

Even worse than errors is *imprecision*, a lack of determinacy about whether the task has been carried out correctly. All empirical concepts admit borderline cases, while all mathematical concepts are exact; this creates a difficulty in relating mathematics to the physical world. For example, imagine an apple tree containing two apples that are grown together so that they may be considered as a single apple: then there is no well-defined number of apples on the tree. Or imagine a sequence of stroke marks on a piece of paper in which two strokes are so close together as to be virtually one: then the sequence of strokes is not a well-defined numeral. Körner (1960) regards examples such as these as a decisive argument against logicism and formalism.

The solution to the difficulty rests on the fact that even an imprecise concept has clear-cut cases. If we accidently write a bad symbol token we do not agonise endlessly over whether it is an 'a', an 'α', or an ambiguous case: we simply cross it out and write another one, trying repeatedly until we succeed. The concept of having succeeded is itself imprecise, but this does not matter; we simply continue trying until we have a token that seems to us a clear-cut case. It is always possible to do this, barring obstacles, and having done so it is unlikely that a reader will misinterpret the token.

Similar remarks apply to repeatable processes other than token-writing, such as demolishing houses or crossing oceans. If, for example, we cross the Atlantic from Spain to a Caribbean island and then return home it may be disputed whether we have crossed the entire ocean or not. So we write off the attempt and do it again, this time taking care to land on the mainland of America.

Thus the fact that any process has ambiguous instances does not undermine the whole exercise; it simply introduces another source of unreliability, alongside errors. Misinterpretations can be made arbitrarily rare by insisting on sufficiently clear-cut steps, just as errors can be made arbitrarily rare by being sufficiently careful with each step.

It is usual in the philosophy of mathematics to confine attention to stroke-writing rather than demolition or sea-faring, but the making of marks on flat surfaces has no special philosophical status. Strokes are simply more convenient: it is easy to make a large number of them, and the results of doing so are on display afterwards.

Let W be the physical world, to which the arguments of T apply. Now, *abstraction* is an operation acting on T that produces a simpler system of arguments, A. A deals with the steps of an *unspecified* iteration. Examples of demonstrable propositions in A are:

(7) two steps followed by two steps constitute four steps;
(8) at any stage of iteration it is always possible to take another step, barring obstacles;
(9) if the agent takes two steps and then another two, and then counts the result, then barring obstacles, errors and imprecision the answer will be four.

Here, (7) comes by abstraction from the propositions (1)–(3) of T; (8) comes by abstraction from (4)–(6); and (9) is a consequence of (7). In A the iterated process and the controlling agent (if any) are unspecified. A applies to the physical world W just as much as T does; to apply A we choose a process, an agent and an occasion, and then any sound argument in A can be instantiated to a sound argument in T. The advantages of carrying out arguments in A instead of T are that a single argument in A corresponds to many arguments in T and that in A we are undistracted by the nature of the iterated process.

In A, the phrase 'barring obstacles' means 'if no obstacles are encountered on this occasion'; the phrase 'barring errors' means 'if no errors are made on this occasion'. A single 'step' in A really means a *repeated attempt* to carry out the step until the result is accepted as clear-cut. The phrase 'barring imprecision' then means 'if the step, accepted as clear-cut, is not subsequently misinterpreted'.

Equality, for expressions in A, is defined as follows. We say '$X = Y$' in A iff any instantiation of X in W could also be an instantiation of Y, and vice versa. (Here, X and Y are viewed in isolation, not as part of some larger expression or argument. Thus, in (7), the two occurrences of 'two steps' have the same possible instantiations and so are equal, even though when embedded in the expression 'two steps followed by two steps' they cannot both denote the same pair of steps.) If '$X = Y$' holds in A then X and Y are interchangeable in all statements of A. This follows because all statements of A correspond by instantiation to statements of T, and replacing X by Y does not alter the instantiated statements.

The disadvantage of A is that it refers to unspecified obstacles, errors and imprecision, which merely complicate the arguments. So the next step is *idealisation*, which produces an even simpler system of arguments, I, by omitting all phrases like 'barring obstacles, errors and imprecision'. Our three propositions of A become:

(10) two steps followed by two steps constitute four steps;
(11) at any stage of iteration it is always possible to take another step;
(12) if the agent takes two steps and then another two, and then counts the result, then the answer will be four.

I applies to the physical world, *W*, but not in a straightforward literal way. For any sound argument in *I* there is a corresponding sound argument in *A*, obtained by replacing each atomic proposition (asserting that a certain process has a certain outcome) with the proposition that the process has that outcome provided no obstacles prevent it from being completed, no errors are made, and all steps are clear-cut. The argument in *A* may then be instantiated to a sound argument in *T* by choosing a particular iterative process, a particular agent and a particular occasion.

Arguments of *I* tell us what happens on those occasions in which no obstacles are encountered and no erroneous or imprecise steps are taken. *I* is therefore a useful system in which to derive truths about *W*, at least for iterative processes in which obstacles, errors and imprecision are rare.

A *number* may be defined within *I* as a stage of iteration; 0 is defined as the stage of being about to start; the successor operation is defined as the taking of an additional step. With this terminology the propositions (10)–(12) may be rewritten as:

(10′) two plus two is four;
(11′) for any number we can form the successor number;
(12′) if the agent adds two and two, the answer will be four.

This approach to arithmetic via idealisation should be compared with the 'If-Thenist' approach, which understands any proposition *P* of arithmetic as a conditional statement, 'If there is a model of Peano Arithmetic then *P* holds in it'. If-Thenism is a different way of dealing with obstacles. It amounts to prefixing all propositions of *T* with 'If there is an infinite supply of chalk (or whatever) then ...', while the propositions of *A* would begin with 'If an unlimited number of steps is possible, then ...', and idealisation would simply remove this qualification. The drawback of this method is that there *isn't* an infinite supply of chalk or anything else, so we would be deducing consequences from a false premise. We could rephrase it as 'If there *were* an infinite supply of chalk then it would necessarily follow that ...'; this interpretation is meaningful, but it is hard to see how it can be applied to the real world of finite resources. Think back to the example of the shepherd counting sheep in Chapter 1. We need an assurance that if the rock is big enough to hold the tally then the finiteness of the rock does not affect the calculation. My account provides this, by prefixing a qualification 'if the rock is big enough for this calculation' to *each atomic proposition* in *T*; the If-Thenist account, which prefixes 'if the rock is infinitely big' to the *whole* proposition, does not, since of course the rock is *not* infinitely big.

This way of introducing numbers may be generalised to other classes of abstractions, for example, trees. A *tree* is either an atom or composite; each atom is an instance of one of a specified list of atom types; each composite tree may be decomposed in a unique way into subtrees. Numbers are a special sort of tree with a single atom type, 0, and one subtree (the *predecessor*) for every composite tree. Whereas numbers were abstracted from iterative processes, trees in general are abstracted from any class of things, events or processes described recursively by atoms and combination rules. Syntactic examples spring to mind: bracketed expressions, sentences in context-free languages, and tree diagrams consisting of circles connected by line segments. But again I stress that there is nothing special about marks on flat surfaces. Recursive structures occur widely in nature and human artifacts, as the following examples show.

- Sequences of crossings of the Atlantic and the Pacific, giving rise by abstraction to lists on a two-element set.
- A (botanical) tree, where the subtrees are defined by the branching and the atoms are leaves. (A bare twig is a composite tree with no subtrees.)
- An astronomical system consisting, for example, of a star, orbited by planets, orbited by moons, orbited by meteoroids, and so on.
- The history of cell division of an embryo, starting with the fertilised egg.
- A machine consisting of components, which themselves consist of components, and so on – provided that any component may be analysed into sub-components in at most one way.
- Christmas presents, defined as follows. A supply of atomic presents of various types is provided (book tokens, handkerchiefs, bottles of sherry ...). A composite present is formed by taking several presents and wrapping them in paper. The wrapping paper ensures that there is only one way of decomposing a composite present into sub-presents.

We should perhaps restrict attention to those examples where it makes sense to think of analysing a structure into its components and combining several structures into a single structure; thus the botanical tree and the history of cell division should perhaps be excluded. Trees are then obtained from these recursive systems by abstraction and idealisation in the same way that numbers were obtained from iterated processes; in fact, iterated processes are simply a special kind of recursive system.

For a precise and formal account of the idealised system I based on trees, see the Theory of Constructions, in Part II.

This completes my account of abstraction and idealisation, which are perfectly innocuous procedures. The reason they have attracted suspicion in the past is that they have been radically misunderstood. Idealisation has been construed as a process of imagining that our capacities were extended by a finite amount – a circular explanation, since we need idealisation to define finiteness (Wright, 1982). The abstract system of arguments A has been seen as applying not to W but to an 'abstract world' W_A (containing numbers and trees instead of blackboards, houses and oceans) while I has been seen as applying to an ideal world W_I (containing an indefatigable agent who can count arbitrarily high). Thus abstraction and idealisation have been seen as operations that create new worlds out of old ones (colourless cats out of tabby cats, for example). W_I is a rather strange place to live in: it is hard to imagine what it would be like to be able to count up to 10^{1000}, never to forget anything, to have a brain the size of a galaxy. Even worse, it is sometimes supposed that abstraction does not *create* W_A but *transforms* W *into* W_A (thus we do not create a colourless cat alongside the tabby one but rob the cat of its colour); Frege seems to have understood abstraction in this way. These absurdities are avoided if one keeps firmly in mind that abstraction and idealisation are operations on systems of arguments not on worlds. W_A and W_I are unnecessary. The position is illustrated in Figure 1.

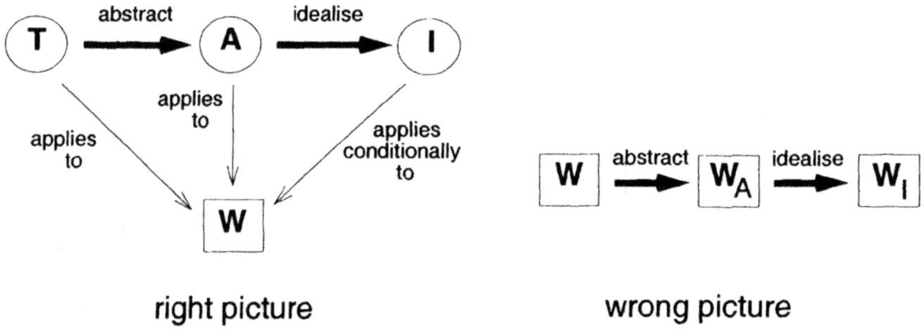

Figure 1: how abstraction and idealisation work.

COMPARISON OF IDEALISATION WITH EXTRAPOLATION

It is important to distinguish the process of idealisation from that of *extrapolation*, as used, for example, by Maddy (1988, 1990), Tharp (1989) and Lavine (1994, VIII) to obtain the general concept of set from finite sets.

Lavine begins with a finite theory in which all quantifiers range over indefinitely large finite domains of various sizes. By omitting the bounds on the quantifiers he obtains a theory in which all quantifiers range over a single infinite domain. Extrapolation, then, appears as a purely syntactic operation of omitting quantifier bounds; given a formula in the infinite theory it would be possible to restore the quantifier bounds and interpret it in a purely finite way. This resembles what I have called idealisation. Lavine, however, resists such an eliminative interpretation of infinity (pp. 264–265); he regards extrapolation as a substantial semantic step that a constructivist would be unlikely to accept (p. 307). Thus his extrapolation is fundamentally different from idealisation.

George (1988) does not distinguish between idealisation and extrapolation, and this leads him to argue against intuitionism. He begins by pointing out that intuitionists commonly accuse platonists of 'allowing mathematical truth to outrun all possible practice'; intuitionists claim that their own view of mathematics is more intelligible since it is expressed wholly in terms of operations that mathematicians can effectively carry out. George argues that the intuitionist is squeezed between two extreme positions: feasibilism and transfinitism. *Feasibilism* rejects idealisation and demands that we restrict mathematics to constructions that can actually be carried out in practice. (George call this *actualism*; it is also known as *strict finitism* and *ultra-intuitionism.*) *Transfinitism*, or *Cantorian finitism*, is the position of Maddy (1988) 'that infinite sets are not so different from finite ones, that the most basic properties are ones they share'. Now, George argues that the feasibilist can bring the same charge against the intuitionist as the intuitionist brings against the transfinitist, that of 'appealing to fictitious extrapolatory abilities'. Thus intuitionism is merely the result of a half-hearted attempt to explain mathematics in terms of actual mathematical practice, a project that in any case George considers ill-conceived. This argument originated with Bernays:

> What does it mean to claim the existence of an Arabic numeral for the foregoing number [$67^{257^{729}}$], since in practice we are not in a position to obtain it?
>
> Brouwer appeals to intuition, but one can doubt that the evidence for it really is intuitive. Isn't this rather an application of the general method of analogy . . . ? . . . In short, the point of view of intuitive evidence does not decide uniquely in favor of intuitionism. (Bernays, 1935)

Tait (1986, §12) makes the same point.

Contrary to this, I shall argue that there are important differences between the feasible-to-finite step and the finite-to-transfinite step, and that the differences all work in the intuitionist's favour.

First, the transfinitist has a problem of deciding *which* properties of finite sets transfer to transfinite sets. Even the most ardent transfinitist has to accept that there is a genuine distinction between finite and infinite, and that there are some properties peculiar to finite sets and some peculiar to infinite sets. The intuitionist may ask, how do we know about the peculiarly infinite properties of infinite sets, given acquaintance only with finite sets? How do we know that infinite sets resemble finite sets even remotely? The positing of properties for infinite sets is bound to rely on trial and error: desirable properties are postulated, and withdrawn if they lead to a contradiction. This does not bother Maddy, who sees mathematics as ultimately an empirical science, but it is uncomfortable for the rest of us who want mathematical theorems to be necessarily true. There is no analogous problem for the intuitionist, who does not recognise a precise distinction between feasible and infeasible constructions and consequently never has a problem of whether any properties are peculiar to feasible constructions.

The second difference is the mirror image of the first. The intuitionist may try to embarrass the feasibilist by asking, which of my numbers do you accept as feasible? 10^4? 10^8? 10^{16}? 10^{32}? 10^{64}? The feasibilist dare not give a clear answer, for fear of being trapped into saying something like '34219163 is feasible but 34219164 isn't' and because in any case what is feasible depends on the non-mathematical circumstances. It is futile for the transfinitist to try the same ploy against the intuitionist ('Which of my transfinite ordinals do you accept?') because the answer is obvious: 'All the ones below ω'.

The cause of these differences is that the feasible-to-finite step is an idealisation, a *simplification* of argument by deleting qualifying clauses ('provided no obstacles occur'), whereas the finite-to-transfinite step is something very different, a *complication* of argument by introducing new methods (Cantor's (1883) 'second principle of generation'). In idealisation there is a clear distinction between what is idealised away (the obstacles) and what remains; in the finite-to-transfinite extrapolation we extrapolate away our entire understanding of sets (based on enumeration and rules of membership). Idealised arguments apply (unreliably) to the same world as the unidealised arguments, whereas extrapolated arguments purport to apply to a much larger world.

Idealisation, then, is quite unlike extrapolation; and there is no inconsistency in accepting the feasible-to-finite idealisation but not the finite-to-transfinite extrapolation.

CONCLUSION

Mathematical arguments are abstract and idealised. They appear to deal with abstract and ideal objects but in fact deal (unreliably) with iterative processes and recursively structured objects in the physical world. The word 'unreliably' here indicates that their results are only guaranteed to be correct on the occasions when no obstacles, errors and ambiguities occur. This unreliability is tolerable in practice because:

- when an obstacle occurs we always know it;
- we can always reduce the risk of errors by being more careful, and we can always resolve discrepancies between two processes by comparing them step by step;
- we can always reduce the risk of imprecision by performing steps more carefully, and we can always resolve ambiguities by discarding them and repeating the step.

CHAPTER 5

WHAT ARE CONSTRUCTIONS?

CONSTRUCTIONS AS RECURSIVE DATA STRUCTURES

Constructivism is the thesis that mathematics is an activity of 'mental construction'. Discussions of constructivism tend to concentrate on the question of constructive reasoning and neglect the more fundamental question: what are constructions? In this chapter I shall attempt to answer this and compare my answer with Brouwer's views.

The word 'construction' is a metaphor. Clearly it is supposed to make us think of building houses or machines by connecting components together. For any sort of construction, components come in certain basic types (bricks, wheels, pistons, ...); we take as many instances as we like of each type and connect them together with certain combination procedures (cementing, gluing, soldering, screwing, ...), subject to certain constraints (only two things can be glued together at once, bricks have to be cemented not soldered, ...) to form an arbitrarily large construction. The word 'construction' is used both for the activity of constructing and for the constructed result. Sundholm (1983) distinguishes a third sense of the word: the process of construction considered as a completed object.

The traditional mathematical example of a construction occurs in Euclidean geometry. A Euclidean construction is formed from any number of instances of the basic building blocks (points, lines and circles) by applying certain ruler-and-compass combination rules.

If one had to define constructions in general, one would surely say that a type of construction is specified by some *atoms* and some *combination rules* of the form 'Given constructions $x_1, \ldots x_k$ one may form the construction $C(x_1, \ldots x_k)$, subject to certain conditions on $x_1, \ldots x_k$'. A construction, then, is defined recursively as either an atom or $C(x_1, \ldots x_k)$, where C is a combination rule and $x_1, \ldots x_k$ are constructions satisfying the conditions for applying C.

To state it more shortly, a construction is a recursive structure (see Chapter 4 for examples). A *mathematical* construction is an *abstract* recursive structure (or a *tree*, as I called it in Chapter 4). Each abstract atom corresponds to a *type* of physical atom: for example the abstract atom *brick* might correspond to physical bricks. In an abstract construction an atom may be used several times over (e.g., *house*(*brick*, *brick*, ... *brick*)), but in a physical construction each atom may only be used once (thus a house consists of many different bricks of the same type, not the same brick used many times). Equality of abstract constructions is defined by: $x = y$ iff x and y could be instantiated by the same physical constructions. Equivalently, $x = y$ iff x and y are built out of the same atoms using the combination rules in the same way.

This definition of constructions accords well with the way in which they seem traditionally to have been viewed by constructivists. The very simplest type of construction allows just a single atom (call it '0') and a single combination rule (given a construction x we may construct $S(x)$) with no associated conditions: these constructions are called *natural numbers*. This explains the special elementary status that constructivists have always accorded the natural numbers. Other types of construction, obtained by allowing more atoms, combination rules and associated conditions, include lists, stacks, trees, algebraic expressions, logical formulae, sentences and proof trees.

This view of constructions is certainly adequate to provide a foundation for arithmetic (see Part III); it even suffices for analysis (in Part IV I expand the interpretation of some constructions but do not enlarge the constructive universe). Further extensions of mathematics, for example to a general theory of species, may require a broader notion of construction; it is not possible to circumscribe in advance all the 'Acts of Intuitionism' we may make in future. Hence the thesis I shall defend in this book is that the constructions involved in arithmetic and analysis are precisely the recursive structures.

The natural *operations* on constructions are functions that are built out of the following primitive steps: making atomic constructions, combining constructions into a composite construction, taking constructions apart into their components, and branching according to the structure of given constructions. Such functions are precisely the *recursive functions*.

A construction may itself be interpreted as a recursive function: then it is called a *program*. Under such an interpretation, constructions act on other constructions, thus forming a computational system. This gives constructions an operational semantics as well as a syntax.

All this will be very familiar to computer scientists. All I am saying is that constructions are expressions in a pure functional programming language. Bishop (1970) speaks of constructive mathematics as concerning itself with 'the precise description of finitely performable abstract operations', and comments that all such operations may be reduced (by coding, presumably) to operations on integers. This view is essentially equivalent to mine.

CHURCH'S THESIS

If this view is correct it seems natural to close the constructive universe by asserting Church's Thesis, that *any* constructive operation can be implemented as a recursive function and thus represented as a construction. Two forms of Church's thesis need to be distinguished. The *weak* form states that all 'mechanically computable' functions are recursive; most people, including intuitionists, are persuaded by Turing's (1937) arguments in favour of this. The *strong* form states that all 'humanly computable' functions are recursive, and it is this version that has attracted scepticism from intuitionists. Kreisel (1970) treats the strong thesis as an open question, and Dummett (1977, p. 264) describes it as 'not particularly plausible from an intuitionistic standpoint'. Others besides intuitionists, such as Gödel (Wang, 1974, X, §7), have also expressed scepticism about the strong thesis. I have no rigid opinion on this, but I am inclined to share Turing's (1950) views on artificial intelligence, which imply that the strong and weak theses are equivalent. For simplicity I shall stick to recursive functions, thus implicitly treating the strong thesis as true. If anyone produces a convincing counter-example I expect it will be possible to enlarge our notion of recursive function to include it.

CONSTRUCTIONS AND LANGUAGE

Although I think it will be generally agreed that the above account represents a reasonable formulation of constructivism, it may be disputed whether it deserves to be called *intuitionistic*. There are three alleged distinctions that I may be accused of ignoring:

(1) the distinction between finitism (based on mere 'combinatorial' considerations) and intuitionism (which includes non-combinatorial notions such as proof (Gödel, 1958) and human judgement (Bishop, 1967, Appendix B));
(2) the distinction in subject matter between logic and mathematics (mathematical constructions having a specifically mathematical content);
(3) the distinction, so important to Brouwer, between linguistic reasoning and languageless mathematical activity.

Thus it may be claimed that my notion of construction is too broad, since it includes 'linguistic constructions' as well as 'mathematical constructions', or that it is too narrow, since it excludes non-combinatorial constructions and anything not expressible in language. In either case it may be considered more appropriate to some variety of constructivism other than intuitionism, such as finitism, formalism or Markov-style constructivism.

In the case of (1), I doubt that there is a distinct and viable finitist position in the sense intended; I defer discussion of this to Chapter 9. As for (2), I have already argued in Chapter 1 that logic and mathematics have equal claim to having a special subject matter. I shall discuss (3) here, in the light of Brouwer's own writings and interpretations by van Stigt (1990) and Detlefsen (1990a). It is necessary for me to discuss Brouwer's views on this point in some detail, for if he is right then my view of the constructive universe is fatally distorted by my 'syntactic' perspective and I have missed the essentially open-ended nature of mathematical construction. I shall try to show in this chapter that there is no philosophical mistake in construing recursive syntactic structures as mathematical constructions. Of course it does not follow from this that all mathematical constructions are recursive structures, but I can at least show that recursive structures are adequate for arithmetic and analysis, and I have already conceded that the notion of construction may need to be augmented to go beyond analysis.

BROUWER'S VIEW OF MATHEMATICS AND LANGUAGE

Brouwer regarded mathematics as obtained by abstraction from our awareness of the flow of time. The story starts with the *Primordial Happening*. A sensation occurs in the mind of the Subject. No sooner has it occurred than it begins to recede into the past, so there are now two sensations: the original one, now retained in memory, and its continuation in the present. The two may however be considered as a unity (the second sensation experienced in the light of the memory of the first). Next the second sensation also recedes into the past, giving three sensations, and the process repeats indefinitely.

> The becoming-aware-of-time is the fundamental happening of the intellect: a moment in life falls apart into two qualitatively different things of which the one gives way to the other but is retained in memory. . . .
>
> Of this temporal two-ity, born out of time-awareness, or this two-membered time sequence of phenomena, one of the elements can in turn and in the same way fall apart into two parts; in this way the temporal three-ity or three-element time sequence is born. Proceeding this process, the self-unfolding of the fundamental happening of the intellect creates the time sequence of phenomena of arbitrary multiplicity. (Brouwer, 1933b, pp. 45–6)

If we abstract from the content of the sensations we get the 'empty forms' of the *Primordial Intuition*: one thing splitting into two, two united into an ordered pair (a 'two-oneness' or 'two-ity'), and indefinite iteration thereof. Brouwer (1907, p. 180) lists the primitive elements of the Primordial Intuition as including the notions of 'continuous', 'entity', 'once more' and 'and so on'. From the Primordial Intuition mathematics is constructed: the natural numbers are nested ordered pairs, the ultimate elements of which are not particular sensations but the empty forms of 'first event', 'second event', and so on. This is the *First Act of Intuitionism*.

> However, only at the highest levels of civilization does mathematical activity reach full maturity; this is achieved through the mathematical abstraction, which divests two-ity of all content leaving only its empty form as the common substratum of all two-ities.
>
> This common substratum of all two-ities forms the primordial intuition of mathematics, which through self-unfolding introduces the infinite as a perceptual form and produces first

of all the collection of natural numbers, then the real numbers and finally the whole of pure mathematics or simply of mathematics. (Brouwer, 1933b, pp. 47–8)

The First Act of Intuitionism ... recognizes that intuitionistic mathematics is an essentially languageless activity of the mind having its origin in the perception of a move of time, i.e. of the falling apart of a life moment into two distinct things, one of which gives way to the other, but is retained by memory. If the two-ity thus born is divested of all quality, there remains the empty form of the common substratum of all two-ities. It is this common substratum, this empty form, which is the basic intuition of mathematics. (Brouwer, 1952)

Heyting (1974) gives a similar account in very different language. Thus arithmetic arises from abstract awareness of the past and present. The *Second Act of Intuitionism* is awareness of the future as an open-ended field of free choices, giving rise to the notion of an infinitely proceeding sequence and hence to the intuitionistic continuum (van Stigt, 1990, §4.6.3).

(Pure) mathematics is an activity of mental construction using only the Primordial Intuition. It is independent of any particular sense experience, even though the Primordial Intuition was itself originally obtained from sensation. Mathematics is perfectly precise and rigorous, and needs no philosophical justification or foundation.

Mathematics may also applied to the task of organising our sensations into causal sequences, physical objects, and scientific theories. Applied mathematics uses *hybrid* constructions built out of particular sensations using the Primordial Intuition ('iterative complexes of sensations' (Brouwer, 1948)); these are to be distinguished sharply from the constructions of pure mathematics, built out of the empty forms.

Language is one such impure construction, and consequently it can never be wholly precise. This applies equally to 'natural' languages, such as English, and to formal languages, such as predicate logic, algebraic expressions and equations. Brouwer took 'logic' to be any general reasoning expressed in language, and I shall follows this usage in this chapter.

For more details on all these points see van Stigt (1990, §§3.8–3.10).

Brouwer regarded the laws of logic as inductive generalisations from the linguistic records of our past constructive experience. The tautology $A \supset (B \supset (A \wedge B))$, for example, arises as follows. When carrying out mathematical constructions we often take two constructions and combine them. If the two constructions are represented symbolically as P and Q then we represent the combined construction by concatenating the linguistic representations P and Q and inserting punctuation marks, thus: (P, Q). If the original constructions are of types expressed as A and B respectively, the combined construction is said to be of a composite type, which we represent as $A \times B$ or $A \wedge B$. Suppose we do this many times, with different constructions of various types. Then, looking back over the linguistic record of our constructions, we notice the common pattern and postulate, as an empirical law, that any two constructions can always be paired and thus we can always go from constructions of types A and B to a construction of the type $A \wedge B$. We express this using the '$\supset$' symbol as $A \supset (B \supset (A \wedge B))$. This law is well-supported by experience; therefore, in future, given two constructions, we may be highly confident that we shall be able to make a pair of them. But this is only an *empirical prediction*: we cannot be *mathematically certain* until we actually try it and succeed.

> [Speaking of syllogistic reasoning:] However, looking at the *words* that accompany this primitive form of mathematics, we notice in them a surprising mechanism with a regularity which is not clear a priori. (Brouwer, 1907, p. 131)

> Moreover, the function of the logical principles is not to guide arguments concerning experience subtended by mathematical systems, but to describe regularities which are subsequently observed in the language of the arguments. (Brouwer, 1908)

Detlefsen (1990a) supports this view of $A \supset (B \supset (A \wedge B))$ and insists that the prediction is falsifiable. How do we know, he asks, that any two constructions can always be paired in the way expressed linguistically by concatenation?

> 'Pairing' is thus a mere tag for an undescribed and rather dubious ... mental operation that is supposed to allow us to take any two separate constructional acts and turn them into a single (complex) act whose content is the conjunction of the contents of the separate experiences. (p. 528)

He accepts that two linguistic expressions can always be concatenated, but

> though 'concatenation' may be clear as an operation on syntactical entities, it gives no indication of what its counterpart at the level of mental proof might be. (p. 529)

Thus $A \supset (B \supset (A \wedge B))$ is a record of past success but is not binding on the future. Similarly the principle of excluded middle, $A \vee \neg A$, is a record of success in *decidable* contexts, but is falsified by unsolved problems.

Detlefsen's argument is a highly unconventional one, as pairing is normally accepted as an uncontroversial constructive operation. I am not aware that Brouwer ever questioned the legitimacy of pairing; however, Detlefsen is certainly reflecting accurately Brouwer's general views on the relations between mathematics, mathematical language, and theoretical logic.

It is clear from all this that mathematics is philosophically prior to logic, that logic is merely a record of past mathematics and an imperfect guide to future mathematics, and that to attempt to found mathematics on formal logical systems, in the manner of Frege, Russell or Hilbert, is to eat the menu instead of the dinner.

One quick answer to Detlefsen is that, even if he is right about $A \supset (B \supset (A \wedge B))$, other logical laws such as $(A \wedge B) \supset A$ and $(A \wedge B) \supset B$ are still sound, since we would not accept a construction as being of type $A \wedge B$ unless we could decompose it into components of type A and B. Thus $A \supset (B \supset (A \wedge B))$

may have to be abandoned along with the Excluded Middle, but that is no argument against logic as a whole. A second quick answer is that pairing is simply Brouwer's abstract two-ity and is therefore at the heart of the Primordial Intuition.

More importantly, however, this account is a serious misrepresentation of the way language is used in mathematics. The laws of logic are *not* obtained from empirical studies of the way people think. The tautology $A \supset (B \supset (A \wedge B))$ is accepted on the basis of our understanding of $\supset$ and $\wedge$, not as an inductive generalisation from a large number of instances that we have accepted in the past.

Brouwer was hostile to any essential reliance on language in mathematics. His objections were complex and can be disentangled into three theses.

- *The separation thesis:* there is a genuine philosophical distinction between mathematical constructions and linguistic expressions. A linguistic expression may at best *accompany* a mathematical construction; it cannot itself *be* a mathematical construction.
- *Syntactic instability:* linguistic expressions, considered purely as combinations of symbols, are inherently imprecise and unstable.
- *Semantic instability:* the correspondence between linguistic expressions and the thoughts they represent is inherently imprecise and unstable. ('The essential weakness of language is the tenuous, unstable link between thought and physical symbol' in van Stigt's (1990, p. 281) words.)

These theses are related as follows. Syntactic and semantic instability lean heavily on the separation thesis (assuming that mathematical constructions and mappings between them are precise and stable). The strongest thesis is syntactic instability: it implies semantic instability. One could, however, accept semantic instability (perhaps on the grounds that semantics is philosophically obscure in comparison with syntax) but believe that when linguistic expressions are divested of their meaning and considered purely as symbolic structures they become precise and amenable to rigorous metalinguistic reasoning. This seems to have been Heyting's view, at least some of the time. He affirms 'the fundamental ambiguousness of language' and then continues:

> However, let us take another point of view. We may consider the formal system itself as an extremely simple mathematical structure; its entities (the signs of the system) are associated with other, often very complicated, mathematical structures. In this way formalizations may be carried out inside mathematics, and it becomes a powerful mathematical tool. (Heyting, 1956, p. 5)

This leads him to the position (p. 6) that 'every logical theorem ... is but a mathematical theorem of extreme generality; that is to say, logic is a part of mathematics'; but see also p. 102, where he seems to say that the rigour of this correspondence is undermined by semantic instability.

Brouwer asserted the separation thesis in all his publications:

[Replying to an imaginary logician:] The words of your mathematical demonstration merely accompany a mathematical *construction* that is effected without words. (Brouwer, 1907, p. 127)

The first act of intuitionism completely separates mathematics from mathematical language, in particular from the phenomena of language which are described by theoretical logic. (Brouwer, 1954)

One apparent exception to this is in his definition of spreads, where he insists that the elements of an infinitely proceeding sequence are *symbols*:

First of all we determine an indefinitely proceeding sequence of symbols by a first symbol and a law which derives from each of these symbols the next one. We could e.g. choose the sequence ζ formed by the successive 'numerals' 1, 2, 3, ... (Brouwer manuscript, in van Stigt (1990), Appendix 12)

See also van Stigt (1990, p. 374). On the other hand, Brouwer sometimes says that the terms of the sequence are chosen from 'mathematical entities previously acquired' rather then symbols (Brouwer, 1952). In an apparent attempt to reconcile these alternatives he says:

Because mathematics is independent of language, the word *symbol* (*Zeichen*) and in particular the words *complex of digits* (*Ziffernkomplex*) must be understood in this definition in the sense of *mental symbols*, consisting in previously obtained mathematical concepts. (Brouwer, 1947b)

Here he appears to concede that symbols can sometimes be mental and mathematical, which amounts to an abandonment of the separation thesis. This concession, is, however, highly atypical of his published views. Not too much significance should be read into this anomaly, particularly since, in the *Cambridge Lectures on Intuitionism* (Brouwer, 1981), delivered between 1947 and 1951, he described infinitely proceeding sequences as composed of natural numbers rather than numerals.

As for the two instability theses, he asserts the impossibility of reliable communication, whether verbal or non-verbal, in his early work *Life, Art and Mysticism* (1905, Chapter 5). However, he does not rely heavily on instability arguments in his early philosophy of mathematics (Brouwer, 1907, 1908), where indeed he seems to accept that logical arguments can reliably accompany mathematical constructions. His complaint against logicians in (1907), (1908) and (1912) is that they neglect the question of whether there are any mathematical constructions accompanying their logical systems, that they rely on logical principles that are merely empirical regularities, and that this leads them to contradictions and uninterpretable notions such as Cantor's second number class. He refers to language as 'imperfect' and 'defective' (Brouwer, 1907, pp. 141, 169), but these references are incidental to his argument. His later works, however, make clear that any use of language is inherently imprecise, even when manipulated metamathematically; this indicates a commitment to syntactic as well as semantic instability, although he never clearly distinguished the two.

It follows that the language of daily intercourse between people with a limited memory, being necessarily imperfect, limited and of insecure effect, even if it is organised with the utmost practically attainable refinement and precision, . . . (Brouwer, 1933a; see also Brouwer, 1947b)

There are indications that he regarded language as unstable because it relies upon human memory to associate a construction with a word: since memory is fallible we cannot be sure that we always recall the same construction when we use a given word. On the question of how to secure the exactness of mathematics he says:

The answer is that the languageless constructions which arise from the self-unfolding of the basic intuition, are exact and true, by virtue of their very presence in the memory, but that the human faculty of memory which must survey these constructions, is by its nature limited and liable to error, even when it seeks the support of linguistic signs. (Brouwer, 1933a)

It is clear from this that the private use of language does not *introduce* unreliability into mathematics: it *reduces* (but does not completely eliminate) the unreliability caused by fallible memory. Hence the fallibility of memory does not explain the unreliability of linguistic reasoning in comparison with mathematical construction.

It seems that the real reason for the imprecision of language is the diversity of its everyday uses: 'the stability and exactness which language in its grammar and vocabulary seems to possess formally, are to a great extent lost again, since far more elementary notions are needed in every-day practice than there are words and modes of connecting words in language' (Brouwer, 1933b, p. 51; see also Brouwer, 1948). This is clearly an assertion of semantic, though not syntactic, instability.

CRITICISM OF BROUWER'S VIEW OF LANGUAGE

I shall oppose all three theses: I shall argue that linguistic expressions have just as much right to be considered 'constructions' as numbers have (Brouwer concedes this, though he usually prefers to call them 'verbal edifices'), and further that amongst constructions there is no genuine distinction to be drawn between the 'linguistic' and the 'mathematical' ones.

What seems absurd about Detlefsen's pairing argument (see above) is the assumption that when forming pairs we are doing two things in parallel: pairing mathematical constructions and concatenating linguistic representations. It seems to me that we are only doing one thing, which may be called mathematical or linguistic indifferently.

Some further examples will help. Consider the following list of (what I would call) constructions, ranging from the unmistakably linguistic to the unmistakably mathematical:

(1) a sentence in English;
(2) a sentence in a formal language, such as a formula or a formal derivation in an axiomatic system;
(3) the parse tree of a sentence in a language;
(4) a tree as an abstract data structure;
(5) a list as an abstract data structure;
(6) a natural number.

Now, which of these are linguistic and which are mathematical? (1) is clearly linguistic and (6) is clearly mathematical. (2) is also surely linguistic: intuitionists do not distinguish between 'natural' languages like English and formal notations like predicate calculus. (3) is merely an alternative representation of (1) or (2); (3) is also a special case, or rather an application, of (4). (5) is a special case of (4): a list is a tree in which each node has at most one subtree. (6) is a special case of (5): a number is a list on a one-element set, with 0 being the empty list and successor being the operation of adding an element. It is very hard to see where in this sequence of examples there is room for a linguistic/mathematical distinction.

In identifying numbers with a special type of list I have assumed that the system of natural numbers is only specified up to isomorphism, so that we may take any infinite progression as our number system. I think that anyone who wanted to maintain the linguistic/mathematical distinction would have to dispute this 'structuralist' view: they would need to argue that numbers have some special *mathematical* character not possessed by lists of any type, and hence not captured in Peano's axioms. Indeed, some people have claimed exactly this: that the natural number system is determined in some way beyond isomorphism by its everyday use in counting. Thus, Russell (1919, §I) says

> We want our numbers not merely to verify mathematical formulae, but to apply in the right way to common objects. We want to have ten fingers and two eyes and one nose. A system in which "1" meant 100, and "2" meant 101, and so on, might be all right for pure mathematics, but would not suit daily life.

Hempel (1945, §7) agrees:

> The Peano system permits of many different interpretations, whereas in everyday as well as in scientific language, we attach one specific meaning to the concepts of arithmetic.

Now it is certainly true that we need to explain how formal arithmetic can be used for everyday counting purposes, and to verify that, for example, we say 'ten' when asked to count the fingers on two hands. However, the conclusion drawn by Russell and Hempel is stronger: that we must understand zero and successor as specific things, not merely as the generators of an arbitrary infinite progression. Quine (1969, p. 81) denies this, correctly in my view. Imagine two people counting somebody's fingers using two different infinite progressions as number systems. They will both say 'one, two, three, four, five, six, seven, eight, nine, ten: there are *ten* fingers', regardless of the fact that they mean different things by the numerals. The difference in number systems does not affect their use of numerals for everyday counting and arithmetic purposes; it only shows up if one asks questions like 'Is Julius Caesar equal to seven?', and in everyday terms such questions would be dismissed as neither true nor false but senseless. Thus the structuralist account accords *better* with everyday usage than does Russell's and Hempel's view. So even the system of natural numbers is only known up to isomorphism; the same is certainly true for the real numbers (who cares whether a real is a Dedekind cut or a class of Cauchy sequences?) and for other mathematical systems.

In order to argue against the mathematics/language distinction I shall have to assume its existence temporarily; I shall take natural numbers as prototypical examples of mathematical objects, and logical symbols and English words as prototypical examples of linguistic objects. I shall try to show that, despite the different historical connotations of the words 'logic' and 'mathematics' there is no genuine distinction to be drawn that would justify Brouwer's view that a logical proof is not a mathematical construction. To see the alleged distinction clearly it is necessary to consider briefly several other distinctions with which it is liable to be confused.

First, there is the distinction between formality and informality. It is sometimes assumed that 'logic' is a particular axiomatic system (first-order classical predicate calculus, typically), while 'mathematics' is simply whatever mathematicians do. Thus Gödel's incompleteness theorems are taken to show that mathematics transcends logical reasoning (Bernays, 1935, p. 283). This is an arbitrary and groundless interpretation: it makes perfect sense to speak of formal logic and informal logic, formal mathematics and informal mathematics.

A second distinction is between types and tokens. If the letter 'A' is written twice on a blackboard then we have two *tokens*, both of the same *type*, namely 'A'. Likewise if we have two collections of five objects then we could call them two 'tokens' of the 'type' 5. A number and a linguistic symbol are both types rather than tokens. As such they are abstractions and do not occur in a particular mathematician's brain at a particular time or on a particular piece of paper (contrary to much of Brouwer's rhetoric).

Thirdly, one can draw a distinction between, for example, the abstract notion of pairing and the particular punctuation scheme by which we represent it (thus we could represent a pair as (A,B) or $\langle A|B\rangle$ or in a variety of other ways). This may conceivably be what Detlefsen has in mind when distinguishing between mental pairing and syntactic concatenation. However, this distinction is not between mathematics and syntax but between *abstract syntax* and *concrete syntax* (see Lalement (1993, §2.2.5) for more information on this distinction). A similar distinction arises within formal logic, between, say, the abstract syntax of an existentially quantified formula (which simply stipulates that for every variable x and formula A there is a formula that is the existential quantification of A with respect to x) and the concrete syntax by which it is represented (such as $\exists x A$ or $(\exists x{:}\,A)$ or $(\exists x)A$ or $(\mathrm{E}x)A$). Thus this distinction is no use for the purpose of distinguishing formal logic from mathematics: when we discuss predicate calculus philosophically we are thinking of a single system (the abstract syntax) not the various concrete-syntax realisations of it.

A fourth distinction is between the private and the public. Brouwer thought of language as essentially a social mechanism by which we try to control other people and conquer nature; he refers to it as part of 'the power-grid of will-transmission' (Brouwer, 1933b, p. 50). 'All verbal utterances are more-or-less developed verbal imperatives, i.e. speaking can always be reduced to commands or threats, and understanding to obeying' (Brouwer, quoted in van Stigt (1990, p. 197)). Thus any use of language is tainted by impure motives and imprecision. Mathematics, in contrast, is an essentially individual activity carried out for its own sake; this accounts for its unique rigour and precision.

The contrast is contrived. Both mathematics and language are, as a matter of historical fact, social creations, and are learned by the individual from society; both have social and technological uses; both however can be practised and extended by an isolated individual for their own sake. It is about as easy to imagine mathematics being created from scratch by an isolated genius as it is to imagine language being so created. If one is tainted by its origin then so is the other.

A fifth distinction is between a concept as an object of thought and the same concept expressed in a physical representation (as a sequence of sounds or ink marks, say). Brouwer seems to have regarded mathematical objects as existing in a mathematician's mind and as becoming linguistic when they are written down or spoken, and Heyting (1956, p. 15) shows a similar tendency. Yet surely it is possible to think in words and symbols without writing or speaking anything. For example, we can carry out proofs in propositional calculus purely mentally; this is surely still 'logic' even though it is not written down. Conversely, could we not carry out mathematical arguments using physical representations that are not linguistic? (Think of doing sums by counting apples, for example.) Brouwer may well be right that the use of physical representations introduces unreliability into our arguments, but this is no help in separating mathematics from language. A number and an

English word are both 'mental objects' in the same sense.

A sixth distinction is that language typically has a meaning beyond itself whereas mathematics does not. When we reason linguistically about numbers we have algebraic equations and numbers in mind at once; the equations 'accompany' the numbers, as Brouwer puts it. Whereas if we think about numbers non-linguistically we just have the numbers. This might suggest that non-linguistic thinking is more direct and rigorous than linguistic thinking. But this cannot be what Brouwer is getting at. For it would imply that if we ignore the meaning of the linguistic expressions and study them simply as pieces of syntax we restore full rigour. Thus this distinction can only be used to support the semantic instability thesis, not syntactic instability. In any case, the distinction merely seems to be between thinking with two sorts of mental entity and thinking with only one sort; it does not show that the equations are any less mathematical than the numbers.

None of the above distinctions can plausibly be identified with the mathematics/language distinction; at least, none will do the job that Brouwer wants, that of establishing that a proof in symbolic logic is not a mathematical construction. All of these distinctions are *independent* of the alleged mathematics/language distinction (at least, if it is insisted that numbers are 'mathematical' and words are 'linguistic') and they are therefore no help in substantiating the separation thesis. It seems to me that *all* the arguments used by Brouwer to separate mathematics from language rely on a confusion with one of the above distinctions. Language requires four elements:

- a finite alphabet of character types;
- the ability to recognise and generate an endless supply of tokens of each character type, with each token being an instance of a unique type;
- the ability to combine tokens into arbitrarily large structures, subject to self-imposed grammatical restrictions;
- the ability to parse a linguistic structure uniquely into substructures and ultimately into tokens.

These look to me suspiciously like applications of the Primordial Intuition. Brouwer does indeed admit that language is a construction using the Primordial Intuition, but he believes that it is an *impure* one incorporating particular sensations, or perhaps incompletely abstracted patterns of sensations ('iterative complexes of sensations'). However, when one considers the variety of physical representations that have been used for language (ink marks, sounds, gestures, flag signals, flashing lights, smoke signals, electric pulses) one sees that language is a *complete abstraction* from the physical nature of the signal. The Primordial Intuition is likewise a complete abstraction from the Primordial Happening. So the difference between them disappears.

In my account of abstraction and idealisation (Chapter 4) I argued that numbers were abstracted from repeatable operations (cf Brouwer's (1908) characterisation of pure mathematics, 'mathematical systems exempt of living sensation, i.e. systems constructed out of the abstraction of repeatable phenomena, out of the intuition of time', and also Heyting (1956, p. 13)). Linguistic expressions are abstracted in the same way from such things as sequences of ink marks. Since the making of ink marks is one example of a repeatable operation, and since any repeatable operation could in principle be used to express any language (particularly if it admits a number of varieties that can be taken as characters in an alphabet), numbers and linguistic expressions are essentially equivalent abstractions.

Looking back at the examples of constructions (1)–(6), it is hard to see how (6) has a rigorous precision that is lost as one moves up the list, at least as far as (2). Is the linguistic numeral 'four' any less precise than the mathematical number four? On this basis I reject the separation and syntactic instability theses.

To assess the semantic instability thesis, consider the following three examples of mappings between constructions.

(7) A recursive mapping from expressions of the type of example (2) to other expressions of the same type is called a *syntax-directed translation scheme*. As a simple, well-known example, consider Gödel's (1933a) mapping from classical number theory into intuitionistic number theory, which he used to show their proof-theoretic equivalence:

$$\begin{aligned} A &\mapsto A \quad \text{if } A \text{ is atomic} \\ A \wedge B &\mapsto A' \wedge B' \\ A \vee B &\mapsto \neg(\neg A' \wedge \neg B') \\ A \supset B &\mapsto \neg(A' \wedge \neg B') \\ \neg A &\mapsto \neg A' \\ \forall x\, A &\mapsto \forall x\, A' \end{aligned}$$

where $A \mapsto A'$ and $B \mapsto B'$.

(8) An example of a recursive mapping from expressions of the type of example (6) to other expressions of the same type is the addition function

$$\begin{aligned} (0, n) &\mapsto n \\ (Sm, n) &\mapsto Sx \quad \text{where } (m, n) \mapsto x. \end{aligned}$$

This works in a similar way to (7), using a recursive clause for each kind of composite expression (Sm) and a clause for atomic expressions (0).

(9) Finally (the most important example), consider a semantic mapping from expressions in an axiomatic system (say, Heyting Arithmetic, the standard formalisation of intuitionistic number theory) to mathematical

constructions (whatever they are). Imagine a mapping Pr that maps a formal Heyting Arithmetic derivation D of a theorem A to a construction $Pr(D)$ that is an intuitionistic proof of A (whatever that means). Such a mapping would be analogous to the Curry-Howard correspondence, which maps natural deduction derivations to proof terms. The mapping Pr is defined by recursion on D, in the style of examples (7) and (8): for each axiom A of Heyting Arithmetic a proof of A is specified, and for each rule of inference, $\frac{B \ldots C}{A}$, it is shown how to obtain a proof of A from proofs of $B, \ldots C$. It follows that any theorem of Heyting Arithmetic, derived by a derivation D, has a proof, $Pr(D)$. Indeed, D may be regarded *as* the proof, since as soon as we know D the conversion to $Pr(D)$ is routine and since, from an intuitionistic point of view, knowing how to carry out a construction is as good as having done it (see the Brouwer manuscripts in van Stigt (1990), Appendix 8, p. 450, and Appendix 9, p. 454; also Heyting (1956), p. 99). D is a high-level specification of how to build the construction $Pr(D)$, or equivalently it *is* the construction, expressed in high-level terms. The existence of such a Pr would show that formal reasoning in Heyting Arithmetic is a sound way of arriving at intuitionistic proofs. (I shall actually construct such Pr functions for various axiomatic systems later in this book.)

What view would Brouwer take of such a Pr function? The issue is whether logical arguments can be *reliably* accompanied by mathematical constructions. In his early period Brouwer accepted that one can apply logical principles purely formally, in full confidence that the corresponding mathematical constructions could be carried out, provided one avoids use of the Excluded Middle (Brouwer, 1907, pp. 159–160; Brouwer, 1908). In his later publications, however, he inserted an explicit qualification implying semantic, and probably also syntactic, instability.

Suppose that, in customary mathematical language trying to deal with an intuitionist mathematical system, the figure of an application of one of the principles of classical logic occurs, does then this figure of language accompany an actual languageless mathematical procedure in the actual mathematical system concerned?

A careful examination reveals that the answer is, in general, in the affirmative, as far as the principles of identity, contradiction and syllogism are concerned, if one allows for the inevitable inadequacy of language as a mode of description and communication. But with regard to the principle of the excluded third it cannot, except in special cases, be affirmative (Brouwer, 1947a, p. 1; see also Brouwer, 1952)

It seems that, for the later Brouwer, possessing D is *not* tantamount to possessing $Pr(D)$ since D and Pr are inherently imprecise while $Pr(D)$, as a mathematical construction, is precise. However, looking back at the examples (1)–(9), the kinship between (2) and (6) is so strong that I cannot accept that formal derivations are imprecise while numbers are precise, and the analogy between the 'syntactic' mapping (7), the 'mathematical' mapping (8) and the 'semantic' mapping (9) is so strong that I cannot accept that addition is precise while *Pr* is imprecise. A 'semantic' mapping is simply a function between two classes of constructions, and seems to me no more unstable than any other function. It follows that formal logic is a rigorous predictor of what mathematical constructions are possible (provided we drop the Excluded Middle, of course), and furthermore that constructing a formal derivation *D* is tantamount to constructing the intuitionistic proof *Pr*(*D*).

In this respect I would seem to be deviating significantly from Brouwerian orthodoxy (at least, that of the later Brouwer) and Detlefsen's interpretation, and perhaps missing the central point of intuitionism. This is not entirely my fault. Brouwer's classification seems arbitrary, and Detlefsen tells us a lot about what constructive activity is *not*, but gives no examples or insight into what it *is*. So I must persist in my misunderstanding, if misunderstanding it is.

INTUITION

I have spoken of Brouwer's 'Primordial Intuition' and 'intuitionism', but apart from these special expressions I have so far carefully avoided any use of the term 'intuition'. This reflects three misgivings about the word.

First, there is the vagueness and variety of its uses. Mathematicians use the word indiscriminately for any ideas that are not expressed formally – thus lumping together notions that are informal because they *underlie* one's chosen formal foundational system with notions that are simply too confused or poorly developed to be formulated rigorously.

My second misgiving about 'intuition' concerns its connotation of mystic insight. The word suggests a special mental faculty by which we derive infallible knowledge not obtainable by ordinary means. Thus Körner (1960, VII), for example, rejects intuitionism on the sole ground that intuitionists have disagreed on supposedly 'self-evident' constructions. He cites the principle of *ex falso quodlibet*, *false* $\supset A$, which, construed intuitionistically, states that we could transform a given proof of an absurdity, if such a thing were possible, into a proof of an arbitrary formula *A*. This principle is an accepted part of intuitionistic logic, although Heyting (1956, p. 102) is hesitant about it, saying that it requires a wider interpretation of implication, while Dummett (1977, §1.2) suggests that it may require a special stipulation, Kolmogorov (1925) rejects it as lacking intuitive foundation, and van Dalen (1973, §2.1) defends it as an immediate consequence of the decidability of proof. Since intuitionists can disagree, Körner argues, intuitionistic logic cannot follow from a self-evident and rigorous intuition.

Now, Körner is surely being unfair here. He is applying a standard that nothing could satisfy. However lucid and rigorous a theory is, there is no defence against someone's disputing its principles through misunderstanding them, nor against someone's hi-jacking its terminology to advance a rival theory (e.g., Griss's views on negation, perhaps). Intuitionism is no *more* vulnerable to this than any other philosophical position; intuitionism does not rely on a faculty of self-evident intuition any more than any other position. Every philosophy is based on informal ideas that are supposed to be accepted on their own evidence. Brouwer was given to proclaiming his founding ideas as Acts of Intuitionism, while other authors are more inclined to play them down and present their theories as 'presuppositionless', but this is simply a difference of expository style. The specific point about *ex falso quodlibet* is easily dealt with. I shall show in Chapter 19 that this principle arises naturally and inevitably out of a general theory of constructions. It does not need to be specifically postulated and it raises no special interpretational difficulties.

My third reservation about 'intuition', as used by Brouwer to describe the structural invariants of time awareness, is that these invariants are intended to be known *a priori* (Brouwer, 1907, p. 99; 1912). That is, Brouwer claims (following Kant) that we know the structure of time independently of any experience; the reason for this is that we are so constructed that we can only take in sensations in a certain way, and this imposes a temporal structure on events. Now, as I see it, the structure of time is given by two binary relations on events: $A = B$ (read as 'A is simultaneous with B') iff A and B are experienced together; $A \prec B$ (read as 'A is before B') iff A could be remembered while B is being experienced (where 'could be' allows for what we would normally call forgetting or simply not recalling). It is essential to our notions of present experience, memory, forgetting and free will that $=$ be an equivalence relation and $\prec$ be a partial order. In addition it is conventionally assumed that $\prec$ is a total, dense order with no last element.

However, it seems to me that the properties of $\prec$ are contingent: we can imagine future time not satisfying them. Cosmologists consider the possibility that there may be a literal end to the universe, a last moment of time. The many-worlds interpretation of quantum mechanics can be read as saying that time is continually branching – nobody notices the branching ordinarily because any *initial segment* of time is totally ordered. Many physicists doubt that time is infinitely divisible. I can even, just about, imagine time suddenly turning two-dimensional. In short, it seems perfectly coherent to imagine the required properties of $\prec$ failing.

Does this matter for mathematics? I think that intuitionist mathematics would survive the failure of these properties, just as Euclidean geometry survived Einstein. We would simply have to speak of 'Brouwerian' time, by analogy with 'Euclidean' space. And this demonstrates that mathematics is not founded *epistemologically* on the properties of time.

It is, however, so founded *cognitively*. Our ability to understand infinity is founded on our understanding of a process repeated in time. The point is best made by a comparison with Frege. Consider the problem of defining Frege's ancestral relation, 'y follows in the ϕ-series after x', where ϕ is a given binary relation. In modern terminology the ancestral relation is simply the transitive closure, ϕ^+. Frege (1884, §79) defines it by

$$x \, \phi^+ \, y \text{ iff } \forall F\,((\forall c\,(x \, \phi \, c \supset F(c)) \wedge \forall d, e\,((F(d) \wedge d \, \phi \, e) \supset F(e))) \supset F(y))$$

(where $F(u)$ means that u falls under the concept F); whereas anyone who regards mathematics as abstracted from time experience would define it by

$$x \, \phi^+ \, y \text{ iff } y \text{ may be obtained from } x \text{ by iterating } \phi\text{: } x \, \phi \, a \, \phi \, b \, \phi \, c \, \phi \, \ldots \, \phi \, y$$

(where iteration is understood by abstraction from sequences of events). The question is not which definition is mathematically more elegant or convenient but which is philosophically more elementary. Frege dismisses the iterative definition as follows.

Next, there may be those who will prefer some other definition as being more natural, as for example the following: if starting from x we transfer our attention continually from one object to another to which it stands in the relation ϕ, and if by this procedure we can finally reach y, then we say that y follows in the ϕ-series after x.

Now this describes a way of discovering that y follows, it does not define what is meant by y's following. Whether, as our attention shifts, we reach y may depend on all sorts of subjective contributory factors, for example on the amount of time at our disposal or on the extent of our familiarity with the things concerned. Whether y follows in the ϕ-series after x has in general absolutely nothing to do with our attention and the circumstances in which we transfer it; on the contrary, it is a question of fact . . .

My definition lifts the matter onto a new plane; it is no longer a question of what is subjectively possible but of what is objectively definite. (Frege, 1884, §80)

I have two counter-arguments. First, we can remove the subjective factors from the notion of 'shifting attention' by abstraction and idealisation, as explained in Chapter 4, so producing an objective mathematical notion of iteration. Secondly, Frege's account depends on quantification over concepts, a notion that he simply takes for granted. The prospects for founding iteration on our time awareness seem much better than the prospects for founding second-order quantifiers on our supposed grasp of a surveyable platonist universe of concepts.

It is on this vital point that I think Brouwer is right: he is starting mathematics *at the right place*, and this leads him to an essentially sound view of the mathematical world. This is why, despite disagreeing over the status of language and the *a-priority* of time, I persist in calling myself an intuitionist. The fundamental notions required for intuitionistic mathematics (namely, the abstract and idealised notions of iteration, recursive construction and recursive transformations) are in practice accepted by nearly everyone, especially by formalists, who could not even define a formula without them. Brouwer has provided the only explanation we have of *why* and *how* they are fundamental. On Frege's view, in contrast, arithmetic would be just another branch of second-order logic.

CHAPTER 6

TRUTH AND PROOF OF LOGICAL FORMULAE

So far I have argued that abstract and idealised talk about numbers, other constructions, and recursive functions is meaningful. Thus, for example, it is meaningful to assert equations and inequalities such as

$$7^7 > 100 \times 7! \tag{1}$$

and compound statements involving equations and inequalities such as

$$\text{if } 231 \times 886 = 204666 \text{ and } 886 \neq 0 \text{ then } 204666 \div 886 = 231, \tag{2}$$

and even statements involving letters, such as

$$\text{if } m \times n = p \text{ and } n \neq 0 \text{ then } p \div n = m,$$

provided the letters denote specific numbers or other constructions. I have, however, claimed that it is meaningless to say

$$\text{for all numbers, } m, n, p, \text{ if } m \times n = p \text{ and } n \neq 0 \text{ then } p \div n = m, \tag{3}$$

or

$$\text{there is a number } n \text{ such that } n^n > 100 \times n! \tag{4}$$

since quantifiers over infinite sets are meaningless. Hence the possibilities for arithmetic would seem rather limited.

All the same, it is hard to dismiss (3) and (4) as meaningless with complete sincerity. General considerations about multiplication and division would seem to show that (3) is true by definition *regardless* of the values of m, n and p; it is hard to contemplate (2) without feeling that it would hold whatever numbers were chosen, and how else can we say this but by asserting that (3) is true? Likewise, (1) would appear in some sense to justify (4).

There is a solution to the dilemma. Intuitionists would understand (4) not as an assertion that somewhere in the mathematical world there lives an n such that $n^n > 100 \times n!$ but as a *demand* for an n such that $n^n > 100 \times n!$, or as the *problem* of finding such an n. The demand is satisfied, or the problem is solved, by 7 – or, as we say technically, 7 is a 'proof' of (4). Clearly (4) must be meaningful, otherwise we would be in no position to say that 7 is a 'proof' of it while 6 is not; but it gets its meaning by a strange route, via this special notion of 'proof'.

This approach may be generalised to other logical formulae, involving logical connectives and quantifiers nested without restriction, provided the atomic formulae are decidable. Formulae may be understood in terms of what would count as a 'proof' of them. In Kreisel's (1962) words, 'The *sense* of a mathematical assertion denoted by a linguistic object A is intuitionistically determined (or understood) if we have laid down what constructions constitute a *proof* of A.'

This is an unusual use of the word 'proof', and shows how intuitionistic semantics diverges from classical semantics. Classically, a formula is understood as a *proposition*. That is to say, it is understood in terms of what is required for it to be true (given an assignment of values to its free variables). The classical semantics of a formal system (say, first-order number theory) may be thought of as a box with two inputs: a formula and an assignment of values to variables. The output is the truth value of the formula. A *proof*, then, is any argument showing that a formula is true for all values of the variables. The notion of proof does not enter into the semantics at all: it is purely derivative.

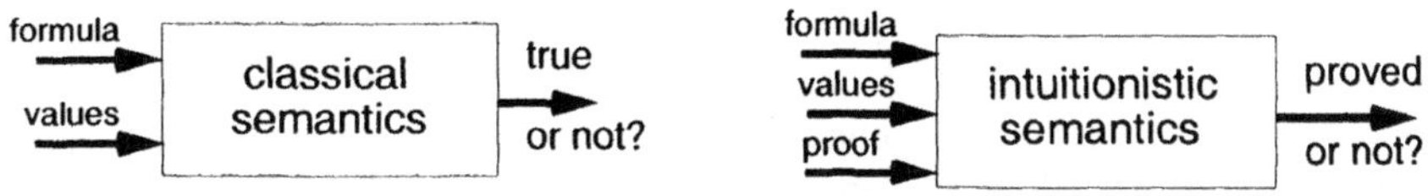

Figure 2: classical and intuitionistic semantics.

Intuitionistic semantics is fundamentally different (see Figure 2). The box has three inputs: the formula, an assignment of values to variables, and a construction. The output is a truth value, depending on all three inputs; it is interpreted as saying whether the construction is a proof of the formula (with those values for the variables) or not. Thus if you input the formula (4), no values, and the construction 7, the output is *true*; if you input (4), no values, and 6, the output is *false*.

Note that the intuitionistic semantics does not mention truth. So where does truth come in and how is it related to proof? There seem to be four views on this.

(1) It is often said that once a formula has been proved then it *becomes* true, and remains so forever. Before it was proved it had no truth value.
(2) Alternatively one may say that when a formula is proved this shows that it was true all along; if its negation is proved then it was false all along.
(3) A more austere view is that an assertion 'A is true' (where A is a formula containing infinite quantifiers) is simply an abbreviation for a certain

assertion of the form 'P is a proof of A'. 'A is true' is to be read as '*mumble, mumble* is a proof of A'. We say 'A is true' when we have a proof P but do not wish to specify it at the moment because it is irrelevant to our present purpose.

(4) The simplest view of all is to drop the concept of truth altogether for formulae of this kind and simply speak of proof. Thus, $\exists n\,(n+3=5)$ is proved by 2 but is nevertheless not *true*: it is grammatically incapable of possessing a truth value.

Interpretation (1) is often linked to a thoroughgoing anti-realism in which truth for indicative sentences generally (not just mathematical ones containing infinite quantifiers) is explained in terms of what is required to prove them. Dummett (1975), for example, suggests two grounds for rejecting classical logic in favour of intuitionistic logic. The first is the doctrine that 'meaning is use'; the second is a radical scepticism about subjunctive conditionals, under which even a proposition such as '$10^{10^{10}}+1$ is prime or composite' does not become true until we compute whether $10^{10^{10}}+1$ is prime. Goodman (1979a), Wright (1982) and George (1988) point out that arguments of this sort lead not to intuitionism but to a feasibilist or Wittgensteinian position. Although there is some overlap of attitudes between Wittgenstein and intuitionism (his most nearly intuitionist remark is 'Generality in mathematics does not stand to particularity in mathematics in the same way as the general to the particular elsewhere.' (Wittgenstein, 1978, V, §25)) it is generally recognised that Wittgenstein's doctrines are just as destructive of intuitionist philosophy as of platonism (Gonzalez, 1991).

Intuitionism takes a robustly realist view of constructions and decidable operations on them, as Weinstein (1983) emphasises, while simultaneously holding that there is something suspect about the notion of truth when applied specifically to mathematical formulae containing infinite quantifiers. This is not part of a general scepticism about truth; radical anti-realist assumptions are not necessary here (as interpretation (4) shows) and they are too powerful for the job. I sympathise with Heyting's position, that intuitionism should be based on minimal philosophical presuppositions. ('The only philosophical thesis of mathematical intuitionism is that no philosophy is needed to understand mathematics.' (Heyting, 1974).)

Interpretation (2) is more appropriate to a classical mathematician interested in constructivity than to a constructivist. On this view it makes sense to ask whether Goldbach's conjecture is true, even though we have not yet proved it. We cannot assert that Goldbach's conjecture is either true or false but perhaps we can assert that it is either true or not true. In addition a question arises of whether everything that is true is destined to be proved, or merely *could* be proved in some sense requiring an infinite quantifier over future time or possible futures. Hodes (1982) advocates interpreting 'A is true' as 'there exists a proof of A', where existence is understood intuitionistically, just as in 'there exists a natural number'. But this does not accomplish much: the point of the exercise is to explain intuitionistic truth and existence claims, not to presuppose them.

Dummett (1977, §1.2) tries to steer a middle course between interpretations (1) and (2), which I don't entirely understand.

Interpretation (3) has the great merit of short-cutting all these difficulties while not impeding constructive mathematics in any way. A clear statement of this position is in Nordström *et al.* (1990, Chapter 4). The disadvantage is that it tempts a careless reader to lapse into the previous two interpretations. Thus, many authors on intuitionism seem to start with interpretation (3) and then slip into (1) or (2). An example is Martin-Löf (1987), who says that 'to know that a proposition is true is to know a proof of it', and then he cancels the two 'know' s to conclude that to be true is to be provable.

For these reasons I prefer (4). It avoids all philosophical difficulties associated with the relation between truth and proof. The only cost is that we must remember to refrain from describing a formula as true even after we have found a proof of it. Think of a formula as an *incomplete proposition*, a proposition with a gap. It has no truth value on its own; its meaning consists in the fact that when a construction is inserted in the gap it becomes a complete proposition and so becomes true or false. A construction whose insertion yields a true proposition is called a *proof*. Interpretation (4) is recommended by Dummett (1982, p. 91); it is also consistent with Hilbert's notion of a 'partial proposition' (Hilbert, 1925, p. 378; Bernays, 1935, p. 278) and with Kolmogorov's (1932) view of formulae as statements of problems.

Which of the interpretations is orthodox intuitionism? Heyting (1956, p. 3) banishes the question as metaphysical. Goodman's (1970, §1) account is typical:

> For the intuitionistic mathematician, on the other hand, it does not make sense to talk about the truth of a mathematical proposition independently of the question of whether or not one has a proof of it. Classically, in order to explain the meaning of a proposition it suffices to give the conditions under which it is true. Intuitionistically, one must rather explain what it means to have a proof of the proposition.

He goes on to explain how the proof of a formula is defined in terms of proof of its components. If there is a notion of truth at all it would appear to be wholly derived from the notion of proof, and to be *theoretically idle*, since it plays no role in building up the meaning of a formula. We have the choice therefore of dispensing with it or retaining it as a parasite. Goodman, however, does not draw this conclusion, and it is not clear which of the four above interpretations he took in 1970. Subsequently (Goodman, 1979b) he settled for interpretation (1) and abandoned intuitionism on the grounds that it 'denies the objective reality of mathematical truth' and 'without the practical reality of mathematical truth there would be no such thing as mathematical rigour.' This mistake could be avoided on interpretation (4), under which the proof relation is perfectly objective.

THE CREATIVE SUBJECT

Let's try another way of approaching the question. Consider a typical event in the constructive life of a Brouwerian Mathematical Subject: at eight o'clock one evening, while taking a bath, Maud suddenly realises that 420232437 is an odd perfect number. (It probably isn't really, but never mind.) As philosophers of mathematics, how are we to analyse this event? It seems to me that the event has three components:

(1) the construction produced (420232437);
(2) the property that makes the construction interesting (being an odd perfect number);
(3) the mathematically irrelevant circumstances (the person, the time, the place, and the reasons that made her consider the particular number 420232437).

In intuitionistic terminology (at least, my version of it), (1) is called a 'proof', (2) is called a 'formula', and (3) is not called anything. The event is summed up as

$$420232437 \vdash \text{being an odd perfect number.} \qquad (*)$$

(I shall always use the '$\vdash$' symbol for the intuitionistic proof relation.) I should explain how we distinguish the three components. (2) is distinguished from (1) and (3) by the fact that we could formulate (2) separately and use it in two ways: before the discovery of a proof, as a *statement of a problem* (imagine Maud sitting in the bath, musing 'An odd perfect number?'), and afterwards as an *exclamation* (imagine Maud proclaiming from the bathroom window, 'An odd perfect number!'). The distinction between (1) and (3) arises because constructions must be repeatable. When Maud reports her discovery so that she or someone else can duplicate it, she does not say 'Get into the bath at 8 o'clock, ...'; she says 'Consider the number 420232437, ...'. Constructive events *must* be repeatable, otherwise there would be no notion of *correcting errors*, and indeed no notion of error and no objectivity in mathematics (see Chapters 3 & 4). (Brouwer is often loosely described as a 'subjectivist' because he believed that mathematics is all in the mind, but he was committed to objectivity in my sense just as much as any other mathematician.) We therefore must be able to distinguish the aspects of a constructive experience that need to be recreated when repeating the construction from those that do not, that is, to distinguish (1) from (3).

It is (2) that is the closest intuitionistic analogue to a classical *proposition* ('There is an odd perfect number'); the exclamation usage of (2) is analogous to a classical *assertion* or *judgement*. The exclamation may be regarded as *justified* (by the possession of 420232437) or as *misleading* (if uttered by someone who has not found an odd perfect number). The notions of 'justified' and 'misleading', however, are quite unlike the classical notions of 'true' and 'false', in that classical truth depends only on the proposition, while justified exclamation depends obviously and essentially on the possession of a proof. In an attempt to appease classical logicians, intuitionists have traditionally referred to these exclamations as 'judgements' and the formulae exclaimed as 'propositions', even speaking of them as 'true' and 'false'. This terminology seems to me completely misguided. An intuitionist view of mathematics is solely concerned with recording (the mathematically relevant aspects of) the constructions produced by various mathematicians at various times. This statement needs to be qualified in two ways if it is not to be misleading. First, when we speak of mental constructions we are concerned with mental types rather than mental tokens, since it is essential that constructions be repeatable. Secondly, some mathematical constructions entail general statements about infinite classes of constructions: if we construct a general method for transforming Xs into Ys then we have discovered that if the method is applied to any X it will produce a Y. Thus we are doing more than reporting individual constructive episodes in our mental life. The fact remains, however, that occurrences of constructions are the whole story; there are no *additional* facts about truth or provability. That is why the primitive assertion of intuitionism is of the form $(*)$ above.

A failure to analyse Maud's constructive experience into its natural components leads to spurious philosophical problems and diversions – most notoriously, the Theory of the Creative Subject, in which the basic assertion is not $(*)$ but

$$\vdash_{(8\text{ o'clock})} \text{being an odd perfect number.}$$

Proof, on this view, is a relation between a time and a formula. Axioms about this relation are posited that involve quantifying over all times. Confusion ensues about what it means to have a proof at a particular time (if $\vdash_t A$ and $\vdash_t A \supset B$ does it follow that $\vdash_t B$ or that $\exists t' > t\ (\vdash_{t'} B)$?) and also between modal and actual time (does $\exists t\,(\vdash_t A)$ mean that A is destined to be proved or that A may be proved if we follow the right constructive path?); see Dummett (1977, §6.3) for more on these questions.

All this seems to me a pointless excursion into epistemic logic. The relation '$\vdash_t A$' really means 'the Subject has a construction P at time t and knows that $P \vdash A$', and it is only the '$P \vdash A$' part that is relevant to intuitionistic semantics, while the epistemic difficulties are all in the rest of the phrase. If we are going to include time in the proof relation, why not go all the way and say

$$\vdash_{(\text{Maud, in bath, 8 o'clock})} \text{being an odd perfect number} \quad ?$$

CONCLUSIONS

To give an intuitionistic account of a branch of mathematics it is necessary and sufficient to define the relation $P \vdash A$ (meaning that the construction P proves the formula A) and to explain and justify any necessary relationships of the form 'if $P \vdash A$ then $Q \vdash B$'. The notion of truth applies to expressions of the form $P \vdash A$ but not to the formula A in isolation.

In this book I shall use the word 'proof' in two senses: in the sense of intuitionistic proof, as expounded in this chapter, and in the usual mathematical sense of the (formal or informal) argument in support of a proposition. This ambiguity is undesirable, but it is an attempt to conform to common mathematical usage, and it will always be clear which sense is intended.

CHAPTER 7

THE NEED FOR A THEORY OF CONSTRUCTIONS

The principal task of this book is to provide a *theory of constructions*, explaining what constructions are, what we can do with them, and how they provide a foundation for predicate calculus, arithmetic and analysis. Before doing so I should explain why we need such a theory, why the informal remarks on the subject by Brouwer and Heyting are not sufficient.

THE MEANINGS OF THE LOGICAL CONSTANTS, (I): HOW TO PROVE A FORMULA

Consider the standard intuitionistic explanations of the meanings of the logical constants. The meaning of a formula containing logical constants is given by what you have to do to prove it.

- To prove an atomic formula, expressing the result of a computation, you simply need to carry out the computation.
- To prove a conjunction $A \wedge B$ you must prove A and prove B.
- To prove a disjunction $A \vee B$ you must either prove A or prove B (also, some would say, you must indicate which you have done).
- To prove an implication $A \supset B$ you must show how to prove B given that you can prove A, that is, how to transform any state of affairs in which you have proved A into one where you have proved B.
- To prove a universal quantification $\forall x A$ you must show how, given any construction x (of the correct type), to prove A for that value of x.
- To prove an existential quantification $\exists x A$ you must specify an instance x (of the correct type) and prove A for that value of x.

THE MEANINGS OF THE LOGICAL CONSTANTS, (II): WHAT IS A PROOF?

The above definitions are consistent with the views of constructivists generally (see for example Heyting, 1956, VII; Bishop, 1967, §1.3; van Dalen, 1973, §2.1; Dummett, 1977, p. 12; Beeson, 1985, §II.6; Troelstra & van Dalen, 1988, Chapter 1, §3.1). Indeed, they could be taken as constitutive of constructivism in the wider sense (including intuitionism, Markov's and Bishop's positions, and various forms of 'finitism', but not including predicativism). Differences of opinion, however, emerge in the attempt to express them more concretely. Intuitionists (and apparently most other constructivists) believe that proving means finding a construction of a certain sort, called a *proof* (Heyting, 1956, p. 98); so we ought to ask, for each of the above clauses, how the proving activity is represented as a construction. (Some such concrete representation seems required if the $\supset$-clause is to become manageable.) This construction should encapsulate the difference in epistemic state between someone who has proved the formula and someone who understands it but has not yet proved it.

What, then, is a proof? One view (Sundholm, 1983, proposition III) is that proofs are sequences or trees of formulae, like the proof trees in classical formal logic, in which case the $\wedge$-clause means that the only way to prove $A \wedge B$ is by $\wedge$-introduction from A and B, while the $\vee$-clause says that the only way to prove $A \vee B$ is by $\vee$-introduction from A or from B. But what then does the $\supset$-clause mean? The only possible answer, on this reading, is that a proof of $A \supset B$ is a proof of B from the premise A; likewise a proof of $\forall x A$ would have to be a free-variable proof of A. However, as Dummett (1977, §1.2) points out, this is far too restrictive, as we would be unable to justify induction or the inference

$$\frac{\forall x\,(A \supset B)}{\exists x A \supset \exists x B}$$

on this basis.

A second view is that a proof of $A \wedge B$ is a pair (P, Q), where P is a proof of A and Q is a proof of B, a proof of $\forall x A$ is a function transforming any given x to a proof of A, and similarly for the other clauses. Thus a proof is not a tree of formulae but a pair, a function of a certain type, or whatever other construction is required by the proof clauses. I shall assume this formulation since it seems to involve no loss of generality. Possessing a proof of $A \wedge B$ is tantamount to possessing P and Q, which in turn is tantamount to possessing the pair (P, Q). So, even if the proof is not really a pair, the proof and pair are interconvertible and possession of one represents the same state of knowledge as possession of the other. As Dummett (1977, p. 344) remarks, it makes no difference from the intuitive point of view precisely how a proof is coded as a construction.

Having abandoned the first view we must distinguish carefully between a *derivation*, which is a tree of formulae conforming to the axioms and rules of some axiomatic system, and an *intuitionistic proof*, which is a pair, a function, or some other construction as required by the proof clauses.

This reveals a legitimate sense in which intuitionistic mathematics really is prior to logic (taking 'logic' to mean predicate calculus, the theory of the logical constants $\wedge$, $\vee$, $\supset$, $\forall$ and $\exists$): the logical constants are to be *defined*, via the above proof clauses, in terms of an underlying theory of constructive reasoning. Hence the logical constants are not available for *use* in constructive reasoning, so we need a theory to spell out what *is* available.

This is a murky area, and constructivists are not in agreement even on the broadest issues. We all agree that to prove $\forall x A$ we must know how to transform any given x to a proof of A, but what does this mean? Here are three readings:

(1) we possess, somewhere in our repertoire of functions, one that, unbeknownst to us, converts any x into a proof of A; or
(2) we possess such a function and we know by divine revelation that it converts any x into a proof of A; or
(3) we possess such a function and know by constructive reasoning that it converts any x into a proof of A.

Heyting (1974), for example, says 'A proof of $[\forall x A]$ consists in a general method which converts the construction of a natural number x into a proof of $[A]$.' This suggests interpretation (1); but it could plausibly be argued that, constructively speaking, to possess a general method entails knowing that it *is* a general method. Interpretations (1) and (2) seem to me rather perverse; only interpretation (3) is faithful to the intention that a proof should encapsulate the difference in knowledge between someone who has made a certain constructive discovery and someone who has not.

If interpretation (3) is adopted then the proof of $\forall x A$ must be not just a function F but a pair (E, F), where E is constructive evidence that F maps any x to a proof of A; likewise a proof of $A \supset B$ is a pair (E, F), where E is evidence that F maps proofs of A to proofs of B. These references to 'evidence' are known as *second clauses*, and are formalised in Kreisel's and Goodman's systems (see the next chapter). A variation on this is to require that F be presented in such a way as to make it manifest that it does what it is supposed to. This amounts to incorporating E in the description of F; this approach is adopted in Martin-Löf's system.

But what is this notion of 'evidence'? According to Beeson (1985, §II.6) it is simply constructive proof; thus the definitions of the logical constants are circular, like Tarski's classical truth definitions. This is, I think, a constructivist but not an intuitionist position, since it removes any sense in which constructive activity is prior to predicate calculus. It also overlooks the fact that the logical constants (if that is what they are) on the right-hand sides of the proof clauses are only used in a very restricted way; therefore to understand the right-hand sides we do not already need to grasp the full sense of the logical constants. This raises the possibility that the clauses for the logical constants can serve as genuine *explanations* of their meaning, as clearly intended by Heyting (1956, VII) and reaffirmed by Dummett (1977, p. 409). Then 'evidence' must refer to some primitive system of constructive reasoning not using the logical constants. Items of evidence need to be representable as constructions so that they can be incorporated in the proofs of $A \supset B$ and $\forall x A$.

USES OF A THEORY OF CONSTRUCTIONS

The goal of a theory of constructions is to provide a coherent and self-contained system of constructive reasoning that is philosophically prior to predicate calculus and that can be used to give an interpretation of formal systems such as Heyting Arithmetic in a way that is faithful to the intended meanings of the logical constants. A successful theory of constructions could be used to shed light on the murky areas of intuitionistic logic, such as:

- why the principle of excluded middle is constructively unsound (the answer may seem obvious from the informal definitions of the logical constants, but Tait (1983) and Troelstra & van Dalen (1988, exercise 1.3.4) have pointed out that these definitions, together with the principle that truth equals provability, are consistent with classical logic);
- the status of Markov's principle, $\neg\forall n\, P(n) \supset \exists n \neg P(n)$ for decidable predicates P, which is accepted by some constructivists but not by intuitionists;
- whether, and in what sense, proof is decidable (see Chapter 10);
- whether 'impossible constructions' are admissible, as used in the logical principle of *ex falso quodlibet*, *false* $\supset A$ (see Chapter 19);
- why intuitionists believe that having a method for producing a construction is tantamount to having the construction (Heyting, 1956, p. 99; Dummett, 1977, p. 20);
- to investigate the completeness of intuitionistic predicate calculus and other formal systems with respect to the intended meanings of the logical constants;
- to determine the extent to which constructive reasoning can be formalised (see Chapter 10);
- to explicate, or exorcise, Brouwer's notion of a 'fully analysed proof', as used in his proof of the principle of Bar Induction;
- to compare intuitionistic reasoning with Hilbert's finitary reasoning (see Chapter 9);
- to compare different constructivists' notions of constructive proof (Beeson (1985, p. 410) suggests that these may have little in common beyond adherence to the laws of Heyting's predicate calculus and arithmetic);
- to assess Sundholm's (1983) distinction between a proof as a 'process of construction', a proof as the 'object obtained as the result of a process of construction', and a proof as a 'construction-process as object';

- to compare intuitionistic proof in the full sense with apparently weaker notions such as free-variable proof and proof from hypotheses (Dummett, 1977, p. 15);
- to understand the impredicativity involved in quantifying over all proofs, in the definition of implication (it is very hard to capture the full intended meaning of intuitionistic implication in a theory of constructions, as the next chapter shows).

Discussing the last point, Dummett (1977, §7.2) draws a distinction between *canonical proofs* (the proofs mentioned in the definitions of the logical constants) and *demonstrations* (the 'proofs' that appear in textbooks, which are merely means for constructing a canonical proof). He argues that the universe of canonical proofs must be stratified according to the logical complexity of the formulae involved, otherwise the definitions of $\supset$ and $\forall$ would become vacuous. (This should be compared with Goodman's theory of constructions (Chapter 8), in which a similar stratification of proofs is proposed to avert a self-referential paradox.) On the other hand, when Dummett considers the need for 'second clauses' he concludes that there is no way of circumscribing the complexity of the evidence E: 'it is at just this point that any confines within which we seek to enclose the possible complexity of a proof of a given conclusion will be burst.'

Dummett arrives at the drastic conclusion that the notion of canonical proof will never be completely stable and that theorems of the form $A \supset B$ can never be regarded as permanently proved, since we may one day invent exotic proofs of A that cannot be transformed into proofs of B.

It is to save intuitionism from this fate that a theory of constructions is necessary. I shall spend the rest of this book setting up this theory and using it to interpret predicate calculus, arithmetic and analysis. In the process, some of the questions listed above will be solved; the others will become more amenable to rigorous treatment.

In summary, Weinstein (1983) is surely correct in saying of the intuitionist philosophy that 'The success of this account depends upon our ability to give an explanation of the basic proof relation. Problems of both a technical and philosophical nature have frustrated attempts to explicate this relation. If intuitionism is to remain of philosophical interest, these problems must be overcome.'

CHAPTER 8

THEORIES OF CONSTRUCTIONS

The questions for this chapter are: what does the intuitionistic proof relation mean? and what system of constructive reasoning is required for defining and manipulating this relation? I shall discuss attempts to answer these questions by Kleene, Gödel, Kreisel, Goodman, Scott and Martin-Löf; I shall explain why I find none of their systems entirely satisfactory; and I shall sketch informally the ingredients that an adequate theory answering these questions should contain.

PROOF, EVIDENCE AND PROTOLOGIC

Constructive reasoning seems to involve three notions:

- the notion of *proof* of a formula, represented here by the symbol '$\vdash$'; for example one might define

$$N \vdash \exists n\,(n^n > 100 \times n!) \quad \text{iff} \quad N^N > 100 \times N!$$

- a notion of *evidence*, as used in the second clauses; for example one might define

$$(E, F) \vdash A \supset B \quad \text{iff} \quad E \text{ is evidence that } F \text{ maps proofs of } A \text{ to proofs of } B$$

  (the term 'evidence' is due to Kleene (1945), who seems to have been the first to point out the need for it)
- a reasoning system, typically a sequent calculus, for deriving statements of the form $P \vdash A$ or 'E is evidence that ...' from other such statements; I shall call a system of this sort *protologic*. Protologic is typically used for establishing that all theorems of intuitionistic predicate calculus or arithmetic have proofs in the sense of $\vdash$.

Note that I am not *advocating* such a three-fold division of labour; I simply introduce it as a framework encompassing all the systems discussed below. There is no standard terminology for these three components; not all systems have all three, and some authors use the word 'proof' for two or more of them. I shall use the terms 'proof', 'evidence' and 'protologic' strictly as above. I shall reserve the word 'formula' for expressions of predicate calculus and arithmetic (unlike many of the authors cited below). I shall also consistently use the symbol '$\vdash$' for the proof relation, the symbol '$\rightarrow$' for the sequent arrow in protologic, and the symbols '$\wedge$', '$\vee$', '$\supset$', '$\exists$' and '$\forall$' for the logical constants in formulae.

KLEENE'S REALISABILITY

The first theory of constructions that I am aware of is due to Kleene (1945). He begins with Hilbert's and Bernays' notion of an intuitionistic formula as a 'partial judgement' or 'incomplete communication'. A *proof* is a construction that completes the communication. The proof relation $P \vdash A$ (which Kleene expresses as *P realises A*) has the following clauses for the logical constants. (Note that constructions are coded as natural numbers, 2^a3^b codes the pair of a and b, and $\{e\}$ is the e'th partial recursive function.)

$e \vdash A$ (for atomic A)	iff A is true and e is 0
$2^a3^b \vdash A \wedge B$	iff $a \vdash A$ and $b \vdash B$
$2^03^a \vdash A \vee B$	iff $a \vdash A$
$2^13^b \vdash A \vee B$	iff $b \vdash B$
$e \vdash A \supset B$	iff, for all q, if $q \vdash A$ then $\{e\}(q) \vdash B$
$2^x3^a \vdash \exists x A$	iff $a \vdash A$
$e \vdash \forall x A$	iff, for all x, $\{e\}(x) \vdash A$

No protologic is provided, but Kleene reports several results, using conventional mathematical arguments, relating realisability to intuitionistic and classical truth; in particular, every theorem of intuitionistic number theory is realisable.

Kleene points out how realisability deviates from intuitionistic proof. An intuitionistic proof is intended to encapsulate a mathematical discovery so that it can be communicated to someone else: 'A complete communication of the discovery to another person would have to provoke in the latter the same discovery' (§2). Thus, in the case of $\forall x A$, to complete the communication we need not just the function $\{e\}$ but also *evidence* that it does indeed map any number x to a proof of A. In other words, we need second clauses.

Kleene's notion of realisability has since been developed by other authors in directions that take it further away from intuitionistic proof; these developments are outside the scope of this survey.

GODEL'S DIALECTICA INTERPRETATION

Gödel (1958) provides a formal interpretation of intuitionistic arithmetic into a system of finite-type primitive recursive arithmetic T (see also Weyl (1921) and Gödel (1941)). T is a many-sorted theory with variables for (natural) numbers, functions from numbers to numbers, functionals acting on those functions, and so on up to all finite types. The intended model of T is the computable functions of all finite types (Gödel takes this as an informally understood primitive notion).

The formulae of T are built up from equalities $A = B$ (where A and B are terms of the same type) using the propositional connectives. Equality is intensional and decidable at all types, so the propositional connectives may be construed truth-functionally. The axioms are those of propositional logic with equality, the Peano axioms, and definition by λ-abstraction and primitive recursion.

Gödel maps each formula A of arithmetic to a formula $\exists \underline{x} \forall \underline{y} A_D(\underline{x}, \underline{y})$, where $\underline{x}$ and $\underline{y}$ are sequences of variables of various types and $A_D(\underline{x}, \underline{y})$ is a formula of T. His interpretation theorem is that if A is derivable in Heyting Arithmetic then, for some sequence of terms $\underline{t}$, $A_D(\underline{t}, \underline{y})$ is derivable in T.

This can be construed as a theory of constructions in the following way. The constructions are the computable functions of finite type. A proof of A is the sequence of terms $\underline{t}$ in the interpretation theorem. There are no second clauses and hence there is no evidence. The protologic is the axioms and rules of T.

The interpretation of the logical constants is essentially equivalent to Kleene's except for the $\supset$-clause

$$A \supset B \quad \text{maps to} \quad \exists \underline{V}, \underline{Z} \forall \underline{y}, \underline{w} \, (A_D(\underline{y}, \underline{Z}(\underline{y}, \underline{w})) \supset B_D(\underline{V}(\underline{y}), \underline{w}))$$

where A and B map to $\exists \underline{y} \forall \underline{z} A_D(\underline{y}, \underline{z})$ and $\exists \underline{v} \forall \underline{w} B_D(\underline{v}, \underline{w})$ respectively. This clause is significantly stronger than the usual intuitionistic interpretation: according to Gödel's clause, to prove $A \supset B$ one must produce functions $\underline{V}, \underline{Z}$ such that an instance $B_D(\underline{V}(\underline{y}), \underline{w})$ of $B_D(\underline{v}, \underline{w})$ follows from a *single* instance $A_D(\underline{y}, \underline{Z}(\underline{y}, \underline{w}))$ of $A_D(\underline{y}, \underline{z})$. Intuitionistically, one would allow $\underline{v}$ to depend not just on $\underline{y}$ but on the proof of $\forall \underline{z} A_D(\underline{y}, \underline{z})$, and one would allow the proof of an instance of $B_D(\underline{v}, \underline{w})$ to depend on more than a single instance of $A_D(\underline{y}, \underline{z})$.

This difference is deliberate. Gödel intends T as the minimal extension of 'finitary' reasoning in which intuitionistic logic can be interpreted; he does not advocate adopting his interpretation as the intended meaning of the logical constants. Bishop (1970), however, does. He claims it provides a purely 'numerical' interpretation of proof that is adequate for mathematics. He regards $\forall \underline{y} A_D(\underline{t}, \underline{y})$ as a *complete mathematical statement* and A or $\exists \underline{x} \forall \underline{y} A_D(\underline{x}, \underline{y})$ as an *incomplete mathematical statement*. To prove A we must produce $\underline{t}$ and recognise somehow that $\forall \underline{y} A_D(\underline{t}, \underline{y})$ holds.

KREISEL'S ABSTRACT THEORY OF CONSTRUCTIONS

The first theory of proof to incorporate evidence seems to be Kreisel's (1962; 1965, §§2.21, 2.31). This theory is intended primarily for technical purposes since Kreisel (1962, p. 199) regards Heyting's informal explanations of the logical constants as already adequately clear.

Kreisel's notion of construction includes basic constructions such as 0 and 1 (representing *true* and *false*) and various truth functions acting on 0 and 1, such as negation, conjunction and implication; but it also includes non-finitary 'abstract' constructions (in Gödel's (1958) sense). All functions are total (Kreisel, 1962, p. 206), and hence all terms denote constructions. Apart from this, Kreisel's theories are very general and assume very little about the shape and nature of the constructive universe.

The proof relation $P \vdash A$ is expressed by $r_A(P, a_1, \ldots a_n) = 0$ or by $\Pi[P, a_1, \ldots a_n; A]$, where $a_1, \ldots a_n$ are values for the free variables of A; the construction r_A or Π is defined by recursion on the structure of A.

Evidence is provided by a primitive undefined predicate π: if t is a term (denoting a construction) and α is a notion (a decidable property of constructions) then $\pi(t; x \cdot \alpha[x])$ means that t is evidence that any construction x satisfies α. This predicate is assumed to be decidable: 'we recognize a proof of an assertion of this form when we see one' (Kreisel, 1965, §2.14). (Note that Kreisel uses the word 'proof' for both proof (Π) and evidence (π).) This implies that proof (Π) will be decidable.

Evidence is used in his proof clauses for $\supset$ and $\forall$, which may be paraphrased as follows.

$(c_1, c_2) \vdash A \supset B$ iff c_1 is evidence that,

for any construction q, $q \vdash A \Rightarrow c_2(q) \vdash B$

$(c_1, c_2) \vdash \forall x A$ iff c_1 is evidence that, for any construction x, $c_2(x) \vdash A$

where $\Rightarrow$ is a truth-functional operator acting on the truth values 0 and 1.

Kreisel's (1965) protologic is an axiomatic system involving truth-functional combinations of expressions such as $a = b$ and $\pi(a; x \cdot \alpha[x])$, where a and b are terms and α is a notion. (The 1962 system was restricted to equalities $a = b$, but the effect is the same.) Since the terms a and b may contain free variables the meaning of $a = b$ is that, for all values of the free variables, a and b denote the same construction.

The axioms of the protologic, in the 1965 system, include a reflection principle

$$\pi(a; x \cdot \alpha[x]) \Rightarrow \alpha[t].$$

This allows evidence to be transferred into protologic: if we have evidence a for $\alpha[x]$ then we can derive protologically any instance $\alpha[t]$. Both systems also have a converse principle, allowing for transfer from protologic to evidence: if p is a protological derivation of $\alpha \Rightarrow \beta$ and x does not occur in α then

$$\alpha \Rightarrow \pi(a_p; x \cdot \beta[x])$$

is an axiom, for some term a_p.

Then for any theorem A of intuitionistic predicate calculus a term p can be obtained such that $p \vdash A$ is provable in protologic.

Kreisel (1962) gives a finitist consistency proof for a weakened version of his system. However, a strong version of the system is inconsistent (Goodman, 1970, §9).

GOODMAN'S THEORY OF CONSTRUCTIONS

Goodman (1970) defines a theory of constructions as 'a type-free and logic-free theory directly about the rules and proofs which underlie constructive mathematics'. Goodman's system is a development of Kreisel's, modified to avoid the inconsistency and to admit partial functions.

His language of terms is extremely simple, consisting of constants (which include combinators and pairing functions), variables and functions applications. λ-abstractions are available as metanotation, translatable into combinator expressions in the usual way. (The semantics of this part of the system is discussed in Goodman (1972).)

Goodman assumes a primitive evidence constant π (he calls it a 'proof predicate') such that

πuv has the value *true* iff v is evidence that, for all z, uz has the value *true*.

π is assumed to be decidable but is otherwise unexplained: 'We shall take such proofs as basic and will not attempt to analyze their internal structure.' However, in Goodman (1973a, §6) he provides a hint of the intended meaning: 'We can think of the insight [represented by v] as the visualization, or grasping, of the totality of the computations of the values of the function [u].'

For each formula A he defines a term $|A|$ (like Kreisel's r_A) such that $|A|xy$ has the value *true* iff $y \vdash A$ (with x representing a sequence of values for A's free variables). His clauses for the logical constants resemble Kreisel's. Most importantly, in the clause for $\supset$ he interprets the conditional operation $\Rightarrow$ truth-functionally, as Kreisel does, although with non-strict evaluation in the second argument. The justification he gives for this is that the clause is intended as a *definition* of implication and therefore the $\Rightarrow$ operator used in it must be something *simpler* than implication if the definition is to escape circularity. The truth-functional expression '$X \Rightarrow Y$' is of course only applicable if X and Y are *decidable*; hence intuitionistic proof must be decidable.

Goodman's protologic differs substantially from Kreisel's; it is a calculus of sequents of the form

$$A = B, \ldots C = D \rightarrow E = F$$

where $A, B, \ldots C, D, E, F$ are terms and $X = Y$ means X and Y have the same construction as their values. There are structural rules, axioms and rules for equality, axioms for the combinators and other constants, and an induction rule.

If one assumes in addition a reflection principle that everything for which there is evidence is true then Goodman (1970, §9) shows informally that this leads to a self-referential paradox. He blames this on the impredicativity of $\supset$, and recovers by stratifying his universe of constructions according to the principle that *proofs must be about objects already constructed.* He defines a sequence of *grasped domains* as follows. The basic domain B contains 'purely finitistic' things (the partial recursive functions, I think); for any domain x, the *extended domain* Ex includes x, all proofs about members of x, and certain other constructions. Every construction must belong to one of the grasped domains $B, EB, E(EB), E(E(EB))\ldots$. This sequence does not continue into the transfinite: we cannot define $E^{\omega}(B)=\bigcup_n E^n(B)$ because 'The rule which leads from the nth level to the $(n+1)$st level is not a rule which we can understand' (Goodman, 1973a, §6). This is a puzzling statement because in the previous paragraph he undertook to explain this incomprehensible rule, finishing with the sentence 'Continuing in this way, we can construct the nth level for arbitrary n.' Continuing in *what* way? Why explain something that cannot be understood? My puzzlement increases on reading (Goodman, 1970, §10) that there are constants that do not belong to any grasped domain. The evidence predicate π is modified to take a third argument, the grasped domain over which the given proof applies; consequently π itself belongs to no grasped domain and apparently is not a construction.

This is a problem that often arises with stratified systems. First it is argued on general grounds that every well-defined object must belong to some level; then we are shown some special objects that don't.

To make matters worse, Goodman is forced to posit two reducibility operators to cope with the impredicativity in the interpretation of $\supset$. (Kreisel (1965, §2.215) also mentions, but does not adopt, a reducibility hypothesis.) Weinstein (1983) argues that Goodman's reducibility principle is implausible.

Axioms are added to protologic for grasped domains, the reducibility operators and the evidence predicate π; a reflection principle (1970, VIII7) is also added, without paradox this time, and a 'rule of proof' (1970, VIII8) saying that protological derivations can be turned into evidence, as in Kreisel's system.

The resulting theory is equivalent to Heyting Arithmetic: a formula is a theorem of Heyting Arithmetic iff there is a term that proves it (Goodman, 1973a,b).

Kreisel (1970) endorses Goodman's theory with one reservation, that although most 'notions' have a limited range of significance (which can be identified with a grasped domain) there are some notions that apply to all constructions.

SCOTT'S CONSTRUCTIVE VALIDITY

Scott (1970) complains (justly) that Kreisel and Goodman provide no analysis of their notion of evidence, which enters into their systems in a 'mysterious way simply to allow certain properties to be decidable'; he dismisses the notion as unnecessary.

His system includes constructions and species, with a single sort of variables ranging over both. The constructions include certain basic objects and total functions mapping constructions to constructions; apart from this, what he means by a 'construction' is far from clear, but certainly differs greatly from the notion I describe in Chapter 5. In some ways his universe of constructions is very large, including within it a model of higher-order arithmetic. On the other hand it is does not appear to include all the recursive functions: he deliberately omits an iteration operator, in order to keep the functions total. Constructions are described by a language of terms containing constants, variables, function applications, pairs and pair projections, infinite products (which are identified with λ-abstractions), infinite trees, infinite disjoint sums, and primitive recursion on trees.

The logical constants are defined as usual but without second clauses ($\supset$ is defined in terms of $\forall$ and $\vee$ is defined in terms of $\exists$). Like many constructivists, Scott is squeamish about the principle of *ex falso quodlibet*, so in his interpretation of a logical formula he inserts instances of the principle as a side condition where required. Every formula of arithmetic and analysis is interpreted as a species, so that the proof relation, $P \vdash A$, is represented by species membership, $P \in A$.

His protologic is a calculus of sequents $A, \ldots B \rightarrow C$, where the expressions $A, \ldots B, C$ are of the form $\sigma = \tau$ or $\sigma \in \tau$ (σ and τ being terms). The protologic consists of structural rules and axioms for equality (which constitute what he calls 'urlogic'), together with axioms and rules for the various constructs in the term language.

Scott shows how, given any theorem of intuitionistic predicate calculus A, to find a term P and a protological derivation of the sequent $\rightarrow P \in A$; and he suggests that the converse may hold.

Since there is no system of evidence his proof relation $\in$ is apparently not decidable. As he acknowledges in a postscript, this is unsatisfactory: 'the desired reduction of logical complexity has not been obtained'. He suggests that reflection principles could be used to produce higher-order constructions and reinstate a theory like Kreisel's, a proposal I shall adopt in my own theory of constructions (see Chapters 9 & 10).

MARTIN-LOF'S TYPE THEORY

Martin-Löf's system (1975) is a development of Scott's. It is based on the Curry-Howard correspondence (Lalement, 1993, §4.2.5), according to which a logical formula is interpreted as a type and the proofs of the formula are objects of that type. Consequently the proof relation $P \vdash A$ is written as $P \in A$ ('P is of type A').

The *closed normal terms* (listed by Martin-Löf on p. 100) form a model of the theory and hence may be regarded as the constructions; Martin-Löf does not speak of an *intended* model, so the possibility of other constructions, not expressible as closed normal terms, remains. All functions representable as closed terms are total (by Theorem 3.3) and recursive (by Theorem 3.15).

A sequence of *universes*, $V_0, V_1, V_2 \ldots$, is assumed, each closed under the term-forming constructs, such that $V_0 \in V_1 \in V_2 \in \cdots$.

The logical constants are explained according to the Curry-Howard correspondence: thus, $\forall$ is the infinite product Π, $\supset$ is a special case of Π, $\exists$ is the infinite disjoint union Σ, $\wedge$ is a special case of Σ, and $\vee$ is a disjoint union $+$. Equality, numbers and recursion are also dealt with, thus giving an interpretation of the language of first-order arithmetic.

It follows from this that there are no second clauses and hence no evidence.

The protologic is a natural deduction system involving judgements of the form $P \in A$ and P conv Q (the latter meaning that P and Q are definitionally equal). The system has rules for term formation and 'conv', and four rules for every term-forming construct: a reflection (or formation) rule, an introduction rule, an elimination rule, and a conversion rule.

The key result (Theorem 3.3 and the following paragraph) is that every closed term reduces to a unique closed normal term. An application of this is that the proof relation $P \in A$ is decidable (for closed terms P and A). For we can reduce P and A to unique closed normal terms P' and A', and then extract from P' its type and compare it with A' (Theorem 3.13). This relies on the fact that every closed normal term has a unique type, which is evident from its syntactic form. This is in accordance with the principle that every construction is presented to us as an element of a specified type (§1.1).

Every formula in the language of Heyting Arithmetic is derivable in Heyting Arithmetic if and only if it has a proof (that is, an element of the corresponding type) in a version of Martin-Löf's system without universes (Thompson, 1991, Theorem 8.8).

The system is elaborated in Martin-Löf (1982, 1984), in which we are told that the protologic deals with judgements of the following four forms:

$$A \text{ type}, \qquad A = B, \qquad a \in A, \qquad a = b \in A$$

meaning, respectively, that A is a type, A and B are the same type, a is an object of type A, and a and b are equal objects of type A (note that every type now has an associated equality relation; this replaces the syntactic 'conv' relation in the 1975 system). The 1982 and 1984 systems reveal a shift in motivation: Martin-Löf's purpose is now not to represent the full notion of intuitionistic proof but only the 'computational' part of proof (Beeson, 1985, §XI.23). Functions are now treated extensionally and proof becomes an abstract version of realisability (see Diller & Troelstra, 1984). I shall therefore restrict attention to the 1975 system.

The decidability of the proof relation is bought at a price. Under the Curry-Howard correspondence, the terms of the system correspond to formal derivations in a natural deduction formulation of Heyting Arithmetic; reducing a closed term to closed normal form corresponds to normalising a natural deduction derivation (Girard *et al.*, 1989, §§3.5–3.6). In effect, Martin-Löf's system takes canonical proofs, in Dummett's (1977, §7.2) sense, to be normalised natural deduction derivations. Dummett himself rejects such an identification (pp. 398–399) for the following reason. A natural deduction derivation for $A \supset B$ corresponds to a function that we can *immediately recognise* as transforming proofs of A to proofs of B. By contrast, a proof of $A \supset B$ in the full intuitionistic sense is a function that can be recognised, *perhaps with the aid of a further proof*, as transforming proofs of A to proofs of B; the 'further proof' (which I would call the 'evidence') may be arbitrarily complex.

Dummett's point seems to me sound. In Martin-Löf's system the only functions representable are those definable by primitive recursion: these include the primitive recursive functions themselves and also functions defined by higher-order primitive recursion such as Ackermann's function, but this is still only a recursively enumerable subclass of all the recursive functions (Thompson, 1991, §4.8). Now, suppose we had a function that we knew transformed proofs of A to proofs of B, but the function was not representable in the system. Then we would have a proof of $A \supset B$ in the intuitionistic sense but not in Martin-Löf's sense. The system cannot be extended to include all recursive functions because this would also admit partial recursive functions and hence the normalisation theorem would fail and proof would no longer be decidable.

The conclusion is that there is a conflict between the decidability of the proof relation and the desire to represent the full meaning of the intuitionistic logical constants, at least in systems of this sort with no second clauses.

Sundholm (1983) suggests a new way of looking at the decidability question. He distinguishes between the *process of constructing* a proof and the *constructed proof object*; he proposes that it should be decidable whether a given constructive process produces a proof of a given formula A, but it need not be decidable whether a given constructed object is a proof of A. If P is the constructed object, the process of constructing P seems to correspond to the protological derivation D of the judgement $P \vdash A$. Sundholm's proposal seems to be that protologic is decidable but $\vdash$ need not be, and hence there is no need for second clauses; the evidence is provided by the construction process rather than being part of the constructed proof object.

It is not clear to me what this new point of view accomplishes. Recall from Chapter 7 that the intended meaning of $A \supset B$ is that we imagine we have proved A and we show how to obtain a proof of B. That is, we imagine that we not only possess a proof object P for A but also remember the process D by which we obtained it; we can use both P and D to build a proof of B. Hence we must regard the pair (D, P), rather than P, as the real proof of A (in the canonical sense used in the definition of $\supset$). Then the proof relation $(D, P) \vdash A$ becomes decidable once more, and second clauses have been reinstated, with D as the evidence. Sundholm's apparent intention is to avoid second clauses by insisting that the proof of B depend only on P and not on D. In that case, circumstances could arise in which we know how to obtain a proof of B provided we first succeed in proving A, but yet we do not have a proof of $A \supset B$. This seems to be an unjustifiable divergence from the intended meaning of implication. (The question discussed in this paragraph was first raised by Kreisel (1970, pp. 145–146, note 11).)

UNRESOLVED QUESTIONS

The above theories, of course, raise more questions than they answer.

First, what are constructions? It is not enough to be given a language of terms denoting (some of the) constructions; we need a much more explicit idea of the shape and size of the constructive universe than anyone has given so far. This is particularly so for those systems that distinguish between 'finitary' and higher-order constructions; Goodman's and Martin-Löf's stratified systems provide some idea of how higher-order constructions arise but they cannot be the whole story: having constructed a sequence of levels it always makes constructive sense to take the union of them. The issue here is the *open-endedness* of constructive mathematics, which is related to the problem of how to handle the impredicativity in the interpretation of $\supset$. It would be absurd to demand a formal theory of open-endedness, but we do need an explanation of how open-endedness is *accommodated* within a formal theory of all constructions.

Secondly, we need an informative theory of evidence. Kleene, Gödel, Scott and Martin-Löf try to do without evidence, and lose contact with the intended notion of intuitionistic proof as a consequence. Kreisel and Goodman simply posit evidence as an undefined primitive notion. We need to know the general form of the statements to which evidence applies, which is the general form of a *complete mathematical communication* in Kleene's sense; Kreisel, Goodman and Bishop take it to be of the form 'for any x (in a domain), $A(x)$', where A is decidable, but this needs an argument. We also need to know about the nature of the constructions that are used as evidence. The assumption that the evidence predicate is decidable (and hence proof is decidable) ought to be justified (or refuted) from a theory of evidence, not simply stipulated.

Thirdly, the nature and status of protologic are obscure. Since evidence is apparently intended to deal with the general form of a constructive statement it is puzzling that we find ourselves introducing a separate system of protologic. What is its relation to evidence? What is the general form of a protological expression: a sequent (Goodman, Scott), or an equality (Kreisel, 1962), or a universally quantified propositional function of equalities (Gödel), or a universally quantified decidable statement (Kreisel, 1965), or one of Martin-Löf's four judgement forms, and why? What axioms and rules should be included in protologic?

Most importantly, what philosophical significance does an interpretation of arithmetic in a theory of constructions have? The intention is to reduce intuitionistic predicate logic and arithmetic to a more elementary system of constructive reasoning. If this is to succeed we need to characterise both evidence and protologic as coherent systems (not just opportunistic collections of axioms and rules) that are philosophically more primitive than predicate logic and that are themselves philosophically justifiable. This would be easier if we had one basic system instead of two, as we could then argue that it formalised constructive reasoning in general.

MY ANSWERS TO THE UNRESOLVED QUESTIONS

Briefly, my answers to the above questions are as follows.

Constructions are recursive structures interpretable as partial recursive functions, as explained in Chapter 5. There are no 'higher-order' constructions. I shall represent constructions by a term language that is like Goodman's but without the stratification and the mysterious undefined operators. *Every* construction will be expressed by a term; or rather, every construction will *be* a term: there is no need for a distinction between mathematics and syntax (see Chapter 5). The open-endedness of constructivity is represented via well-foundedness of trees (see Chapter 10), taking up Scott's suggestion of using reflection principles.

Evidence is *coded protologic*. That is, a piece of evidence (as used in the second clauses) is a protological derivation encoded as a construction. Thus there is only one basic system of constructive reasoning, protologic.

Protologic is a calculus of sequents as explained in the next section. I shall argue that the sequents represent the general form of a constructive statement and that the axioms and rules embody all of the ingredients of our understanding of constructions.

INFORMAL INGREDIENTS OF PROTOLOGIC

Protologic is the system of reasoning that underlies predicate calculus and has a direct constructive meaning. The name 'protologic' is intended to suggest its relation to 'logic' (in two senses of the word). First, if we take 'logic' to mean predicate calculus then protologic is *prior to* logic. Secondly, if we take 'logic' to be the general theory of deductive reasoning then protologic is the *most primitive form* of logic, at least for mathematical purposes.

Consider this second sense of 'logic'. In my discussion of Quine (Chapter 1) I asserted that logical necessity arises from the fact that when we understand the meaning of a word this understanding is inextricably tied up with seeing that certain general statements hold for it. Thus a byproduct of defining a construction is an ability to reason constructively with it. Constructions are essentially programs (Chapter 5), so we may take as an example the factorial function, defined by

$$fact(0) \triangleq 1, \qquad fact(n+1) \triangleq (n+1) \times fact(n).$$

Now, one could not be said to understand this definition unless one could see the correctness of arguments such as

$$\begin{aligned}\text{for any } n, \quad fact(n+3) &= (n+3) \times fact(n+2) \\ &= (n+3) \times (n+2) \times fact(n+1) \\ &= (n+3) \times (n+2) \times (n+1) \times fact(n)\end{aligned}$$

(assuming the usual properties of $+$ and $\times$). The equations are true 'by definition', as we say. A second example is the function f defined on positive integers by

$$f(1) \triangleq 1, \quad f(2n) \triangleq f(n) + 1, \quad f(2n+1) \triangleq f(6n+4) + 1 \quad (n \geq 1).$$

It is not known whether f is total, so we need to phrase our arguments in the style

$$\begin{aligned}&\text{for any } n, v \geq 1, \text{ if } f(2n+1) = v \text{ then } f(6n+4) + 1 = v \\ &\qquad\qquad\qquad \text{and so } f(3n+2) + 2 = v\end{aligned}$$

or possibly

$$\begin{aligned}&\text{for any } m, v \geq 1, \text{ if } [f(m) = v \text{ and } odd(m)] \\ &\qquad\qquad \text{then } [f(3m+1) + 1 = v \text{ and } even(3m+1)] \\ &\qquad\qquad \text{therefore } f((3m+1)/2) + 2 = v.\end{aligned}$$

The typical form for the statements that make up these arguments seems to be

for any constructions $x, \ldots y$, if A and $\ldots B$ have the value *true* then so does C

where the terms $A, \ldots B, C$ represent computations involving $x, \ldots y$.

But what do 'for any', 'if ... then' and 'and' mean here? They cannot be understood as $\forall$, $\supset$ and $\wedge$, since we are supposed to be working at a level below predicate calculus. Nor can 'if ... then' be truth functional, since it is applied to computations such as $f(m) = v$ that may not halt. Let us therefore invent a new notation for them:

$$(x, \ldots y)\, A,\ \ldots\ B \rightarrow C$$

where 'for any', 'if ... then' and 'and' are represented by brackets, an arrow, and a comma, respectively. The apparent universal quantification 'for any $x, \ldots y$' is to be understood in the sense of Russell (1908, II). If P is a constructive argument, 'for any x, P' means that P works *regardless* of x; that is, we can see the correctness of P without needing to enquire into the meaning of x. P is really an *argument schema*; it becomes a complete argument if a value is chosen for x, but even in advance of choosing x the schema is *already sound*, so that we are guaranteed that whatever instance of P we form it will be sound. Russell distinguishes this from the classical 'for all x', which is an infinite conjunction of all instances; it is also different from the constructivists' '$\forall x$', where the proofs of the instances may take different forms for different values of x. Russell traces the distinction between 'for all' and 'for any' to Frege (1893, §17) and ultimately to Euclid's general and particular enunciations; the notion of 'for any' is also taken as primitive by Herbrand (1931, footnote 3).

The three phrases 'for any', 'if ... then' and 'and' are not three 'protological constants' but are inseparable parts of a single construct $(x, \ldots y)\, A,\ \ldots\ B \rightarrow C$, which I shall call a *protological sequent*.

The above examples of constructive reasoning require a protological axiom schema

$$(x, \ldots y)\, A \rightarrow B$$

where B is obtained by rewriting subterms of A according to the defining clauses of the constructions used; in addition we need various structural rules to give a workable sequent calculus.

What else do we need? We need to capture all the knowledge about a construction that is implicit in an understanding of its definition. The functions *fact* and f above are defined by recursion; a recursive definition says not only that the function defined satisfies its recursion equations but that it *only* has a value when the recursion equations require it to. This requires a principle of induction. All constructivists agree that induction is a legitimate element of constructive reasoning, at least if it is only applied to decidable predicates; they either take this for granted or give a circular justification (Heyting, 1956, p. 14; Troelstra & van Dalen, 1988, Chapter 3, §1.1). I shall attempt to say something more informative here about why induction is intrinsic to constructive reasoning.

Recall from Chapter 4 how mathematical constructions (such as numbers) are obtained by abstraction from physical constructions (such as counting processes or other repetitive events). Consider two *intensionally isomorphic* physical computation processes A and B. By 'intensionally isomorphic' I mean that there is a one-to-one correspondence between the *ways* in which they are carried out, such that they may be regarded as instantiations of the same abstract computation. For example, A and B may be the 'same' computation repeated by one person at different times, or they may be the 'same' computation carried out in different languages (English and French, say). The use of abstraction requires that A and B should deliver the same (or corresponding) results: if the English computation halts at step 72 with the result 'five' then the French computation should halt at step 72 with the result 'cinq'. This requirement may be stated thus: if there is a mapping (such as English-to-French translation) that maps any step of A to a correct step of B then the whole computation A will map to the whole computation B. The *local* correspondence between individual steps implies a *global* correspondence between the entire computations.

Some principle of this form is required if we are to abstract from computation episodes; if we could not do this then we would have no notion of *repeating* a computation, and therefore no notion of *error* and *correctness* in computations (see Chapters 3 & 4). Computation episodes would not be correct or erroneous; they would just *happen*. Hence the result of a computation episode would have no consequences beyond itself.

If such a principle *is* granted, however, it gives us a powerful supply of constructive arguments in sequent form

$$(x, \ldots y)\, A \rightarrow B;$$

this is to be read as 'the computation A is related step by step to the computation B (regardless of $x, \ldots y$) in such a way that if the former results in the value *true* then so must the latter'. The principle can be generalised, since what counts as a 'step' is somewhat arbitrary. For a precise formulation see the Fxpt Rules in the formal version of protologic (Chapter 17) and the commentary in Chapter 18.

A sequent $(x, \ldots y)\, A, \ \ldots\ B \rightarrow C$ is an incomplete proposition, since it refers to an intensional relationship between the computations $A, \ldots B$ and the computation C without specifying it. The protological derivation by which the sequent is derived in the sequent calculus of Chapter 17 indicates how this intensional relationship is built up; thus the protological derivation completes the proposition, showing *how* the computations $A, \dots B$ are related to the computation C. Because a sequent is an incomplete proposition, sequents cannot be nested. That is, it would be meaningless to form complex expressions such as

$$(x)\ [(y)\, A,\ [B \to C] \to D],\ E,\ F \to [(z)\, G \to H],\ I.$$

To summarise, protologic expresses all that we know about constructions when we understand their definitions. Protologic is a coherent whole: it would make no sense, for example, to study the proof-theoretic effects of weakening or omitting the induction principle, since induction is founded solidly in the abstraction operation that divides the world into mathematical and non-mathematical aspects, without which mathematics would be impossible. Protologic is expressed in terms of sequents $(x, \dots y)\, A,\ \dots\ B \to C$, which express intensional relationships between computations $A, \dots B$ and a computation C and which entail that, for any $x, \dots y$, if $A, \dots B$ result in *true* then so must C. Since this means that we can know the result of C before evaluating it, this gives protologic a direct 'predictive' meaning, in Bishop's (1970) terminology.

The story is not quite complete. For simplicity I have omitted *reflection principles*, which, as I shall argue in Chapter 10, also belong in protologic.

EVIDENCE AS ENCODED PROTOLOGIC

Evidence is whatever we use to establish statements of the form

for any q, if $q \vdash A$ then $F(q) \vdash B$ (in the definition of $\supset$),
for any x, $F(x) \vdash A$ (in the definition of $\forall$).

Note that both of these are in the sequent form $(x, \dots y)\, A, \dots B \to C$ used by protologic (if we allow the number of terms $A, \dots B$ to be zero). Since protologic represents directly meaningful constructive reasoning in general it seems compelling to identify evidence with protologic and say

$(E, F) \vdash A \supset B$ iff E is the code of a protological derivation of $(q)\, q \vdash A \to F(q) \vdash B$

$(E, F) \vdash \forall x A$ iff E is the code of a protological derivation of $(x)\ \to F(x) \vdash A$.

(The $\supset$-clause will need further modification for technical reasons; see Chapter 27.) This identification also has the advantage that it raises the possibility of justifying Kreisel's and Goodman's axioms stating that protological derivations can be transferred into evidence and vice versa.

Note that on this view the 'if ... then ...' in the definition of $\supset$ is not truth-functional (as claimed by Kreisel and Goodman) and the general form of the statements to which evidence applies is a sequent rather than the 'for any x (in a domain), $A(x)$' form assumed by Kreisel, Goodman and Bishop.

CHAPTER 9

HILBERT'S FORMALISM

In this chapter I attempt to disentangle the complex relationship between intuitionism and Hilbert's formalism. I do this for two reasons: to dispel the widespread impression that Hilbert's philosophy is a *rival* to intuitionism, and to advance the formulation of constructive reasoning begun in the previous chapter. In this chapter, the word 'intuitionism' will refer only to mathematics based on the First Act of Intuitionism: Brouwerian choice sequences and species are excluded from consideration here since there is nothing in formalism with which to compare them.

Interpreting Hilbert is a difficult task, partly because even his most comprehensive philosophical expositions (Hilbert, 1925, 1927) are very sketchy, and partly because it is far from clear how Hilbert responded to developments in logic of the 1930s (Gödel's incompleteness theorems and recursive function theory). My purpose in discussing Hilbert is ultimately philosophical rather than exegetic; I am concerned with the question of what options are available to someone who accepts Hilbert's general outlook and objectives.

It is usual to interpret Hilbert's formalism in such a way as to bring it into immediate conflict with Gödel's second incompleteness theorem. Such an interpretation is not forced upon us by his writings. I shall present a more sympathetic exposition.

HILBERT'S PROGRAMME

According to most accounts (for example, Körner, 1960, IV), Hilbert set out to put classical mathematics on a firm philosophical foundation by showing that it could be expressed in a single formal system and that this system is consistent. The formal system itself was to be treated as a meaningless symbol structure, and its consistency was to be proved using meaningful ('contentual') finitary metamathematical reasoning.

This account leaves one puzzled as to why a consistency proof should be considered to remove all philosophical doubts from mathematics. It would, to be sure, guarantee us freedom from paradoxes; we could continue to live in 'Cantor's paradise'. But the cost would be high: we would have to give up the common mathematical belief that terms such as 'real number' and 'set' have a meaning and that mathematical theorems express their properties. Mathematicians take the apparent consistency of mathematics as evidence for the soundness of its underlying ideas; but a consistency proof would remove this evidence by explaining away consistency as a purely syntactic property of the formal system. What one wants from a philosophy of mathematics is an *interpretation* of mathematics, according to which the concepts and theorems mean something (even if not what we naively thought). A mere consistency proof provides no such interpretation, and so mathematics would lose much of its appeal.

This is, in any case, a misleading description of Hilbert's programme. It is true that Hilbert sought to express common mathematical practice in a formal system and to prove the system consistent. But the above account overlooks the special reason why Hilbert wanted a consistency proof.

Hilbert (1925) took it for granted that certain simple mathematical statements, including numerical equations such as $5^3 - 5 = 5!$, were meaningful and sound in just the way that mathematicians commonly understand them. Call this subsystem of mathematics *D*. Since there are no variables or logical constants in *D*, every formula is decidable: it has a truth value that can be discovered simply by evaluating it. Proofs in *D* consist simply of step-by-step evaluations of equations. Hilbert continues as follows:

> But the science of mathematics is by no means exhausted by numerical equations and it cannot be reduced to these alone. One can claim, however, that it is an apparatus that must always yield correct numerical equations when applied to integers. But then we are obliged to investigate the structure of the apparatus sufficiently to make this fact apparent. (Hilbert, 1925, p. 376)

Thus, let *M* be a formal system adequate to express the whole of classical mathematics. *D* is a subsystem of *M*, weaker both in what it can express and in what it can prove. Hilbert's contention is that *M* is a *conservative extension* of *D*: that is, any formula of *D* can be proved in *M* iff it can be proved in *D*. If so, then *M* may be regarded as a system for deriving true numerical equations in *D* without having to evaluate them. Suppose that, using some sophisticated argument involving transfinite set theory, we succeed in deriving in *M* an equation such as $123456789 \div 9 = 13717421$. Then the conservativeness property tells us that the equation is true, that is, that a direct evaluation would verify it. Using *M* in this way we can derive a large number of true equations that it would be tedious or impracticable to verify directly and individually. This is the proper role of mathematics: as a sound tool for deriving directly checkable equations.

Note that this is a *holistic* view of mathematics. The system *M* has a meaningful interpretation (or at least a purposeful role), but the individual formulae in *M* do not and therefore should not strictly speaking be regarded as expressing propositions. Hilbert calls them *ideal propositions*, by analogy with ideals in number theory, points at infinity in projective geometry, and imaginary numbers in analysis: the ideal propositions are added to the real propositions in order to produce a more mathematically tractable system.

Since every formula in *D* can be proved or refuted in *D* by a computation, and since the computation may be expressed as a proof in *M*, it follows that the conservativeness of *M* over *D* is a consequence of the consistency of *M*. Hence the problem of justifying the use of *M* reduces to the problem of proving it consistent:

> for, extension by the addition of ideals is legitimate only if no contradiction is thereby brought about in the old, narrower domain, that is, if the relations that result for the old objects whenever the ideal objects are eliminated are valid in the old domain. (p. 383)

As this quotation illustrates, Hilbert did not distinguish clearly between consistency and conservativeness, and sometimes spoke as if consistency were an end in itself. However, the example with Fermat's last theorem in Hilbert (1927, p. 474) makes it clear that conservativeness is the real goal.

In summary, Hilbert's programme was to codify mathematics in a formal system *M*, prove the consistency of *M*, infer that *M* is a conservative extension of *D*, and thereby justify *M* as a tool for deriving correct formulae of *D*.

FINITARY REASONING

How do we set about showing that *M* is consistent? To say that a formal system is consistent is to say that there is no derivation tree conforming to the rules of the system whose conclusion (root node) is a standard absurdity such as $1 \neq 1$. Since formulae and derivation trees can be coded as numbers, with the syntax, axioms and rules being represented as recursive functions, this is equivalent to saying that there is no natural number with a certain decidable property. Thus the metamathematical reasoning we use in proving consistency is essentially a form of number theory (Hilbert, 1925, p. 383), known as *finitary* or *finitist* reasoning. This number-theoretic reasoning must be accepted as directly meaningful and sound – there is no question of formalising it and proving its consistency.

Hilbert did not specify precisely what was included in the term 'finitary reasoning', probably because he expected to produce a consistency proof that his colleagues would accept as finitary on its own merits, without recourse to a general definition. It is generally considered today that no known consistency proof for arithmetic deserves to be called finitary and that probably no finitary consistency proof is possible. To make sense of this claim it is essential to delimit as precisely as possible the class of finitary arguments. There are three issues to be settled: the subject matter of finitary reasoning, the general form of a finitary proposition, and the finitarily acceptable modes of inference. I shall consider Hilbert's (1925) account in detail.

In the first place (p. 376), finitary reasoning is exclusively concerned with discrete symbol structures ('extralogical concrete objects'):

If logical inference is to be reliable, it must be possible to survey these objects completely in all their parts, and the fact that they occur, that they differ from one another, and that they follow each other, or are concatenated, is immediately given intuitively

Examples of such objects are numerals, formulae and equations.

The first sort of finitary proposition to be introduced are the numerical equations $A = B$, where A and B are terms built up from numerals using arithmetic operations; note that no variables have been introduced yet (p. 376). In addition, truth-functional combinations of such equations and statements involving decidable predicates (such as '... is a prime number') are finitary (p. 377). These make up the system I have called *D*. *Infinite* truth-functional combinations (i.e., quantified equations) are not finitary:

such a passage to the infinite is no more permitted without special investigation and perhaps certain precautionary measures than the passage from a finite to an infinite product in analysis, and initially it has no meaning at all. (p. 378)

An existentially quantified decidable statement can only be interpreted as a 'partial proposition' – not a real proposition but merely an incomplete sketch of one (p. 378).

On page 378 Hilbert considers for the first time *general* statements, such as 'if a is a numeral, we must always have $a + 1 = 1 + a$'. This is not a numerical equation, nor can it be understood as an infinite conjunction of numerical equations, 'but only as a hypothetical judgement that comes to assert something when a numeral is given.' Such hypothetical judgements are finitarily meaningful, but they cannot meaningfully be negated or used as subformulae of composite propositions.

On page 380 variables are introduced into the equation language; Hilbert distinguishes carefully between $a + b = b + a$ as a meaningful hypothetical judgement (conditional on the construction of arbitrary numerals a and b), and the meaningless ideal proposition $a + b = b + a$ (as used in ordinary mathematical practice).

In summary, Hilbert says (p. 380) that there are three kinds of formula:

(1) 'finitary propositions that contain only numerals, like $3 > 2$, $2+3=3+2$, $2 = 3$, and $1 \neq 1$', to which classical logic applies in an unproblematic way;
(2) 'finitary propositions of problematic character';
(3) 'ideal propositions'.

What is in the second category? Hilbert's remarks are cryptic, but it is clear from the context that what is 'problematic' about them is that classical logic does not apply to them. Hence, to identify the problematic finitary propositions we need to look back at Hilbert's examples to see which ones are finitarily meaningful and yet not subject to classical logic. There is only one plausible candidate: the hypothetical judgement $a + 1 = 1 + a$ on page

378. Detlefsen (1990b, p. 353) claims that bounded existential quantifications over decidable propositions are also 'problematic finitary', but I find this an implausible reading.

Two examples of finitary argument are given on page 383: the standard proof of the irrationality of $\sqrt{2}$ and a consistency proof for M. So clearly we must admit arguments of the following forms as finitary:

for any numerals a, b, if $a^2 = 2b^2$, and a and b are co-prime,
then a is even so b is even; hence contradiction

and

for any formal derivation d in M, if the conclusion of d is $1 \neq 1$
then ... contradiction.

It appears from this that the general form for a (problematic) finitary proposition is

$$\text{for any } a, b, \ldots c, \text{ if } P(a, b, \ldots c) \text{ then } Q(a, b, \ldots c) \qquad (*)$$

where $a, b, \ldots c$ represent numbers, formal derivations, or elements of other symbol systems, and P and Q represent computations to be applied to $a, b, \ldots c$.

This reading of Hilbert is not quite the orthodox one. Kreisel (1965, §3.411) takes the general form of a finitary proposition to be a free-variable equation $a(n) = b(n)$ (tacitly quantified over n), and similarly Weinstein (1983) takes it to be $\forall n\, A(n)$, where A is decidable. On this reading, in order to represent the $\sqrt{2}$ argument we must interpret 'if ... then ...' as a truth-functional operator. Hence I think Kreisel and Weinstein would rewrite my general finitary proposition $(*)$ as

$$\text{for all } a, \ldots c,\ P(a, \ldots c) = \textit{false} \text{ or } Q(a, \ldots c) = \textit{true}.$$

This is somewhat unnatural. Hilbert's notion of hypothetical judgement suggests a more natural way of reading the 'if ... then' in $(*)$:

$Q(a, \ldots c)$ comes to assert a truth when $a, \ldots c$ are given
satisfying $P(a, \ldots c)$.

This interpretation is essentially different from the truth-functional one, as becomes clear if one allows P and Q to be partial functions.

I conclude that finitary propositions are either computations or general hypothetical statements of the above form.

The next question is what forms of reasoning are finitarily sound. I propose (uncontroversially) that they include, among other things, simple rewriting of terms (replacing *definiens* by *definiendum*) and also induction using decidable predicates (otherwise there would be no hope of consistency proofs!).

Notice that, on this view, finitary reasoning turns out to resemble closely the protologic introduced in Chapter 8. It is my intention eventually to identify finitary reasoning with protologic, but I am not yet ready to do so because I have not yet finished describing either protologic or finitary reasoning.

THE SIGNIFICANCE OF GODEL'S INCOMPLETENESS THEOREMS

My interpretation of Hilbert's programme is unorthodox in another respect. Let us call the system of finitary reasoning *F*. It is commonly said (Giaquinto, 1983) that Hilbert wanted to prove that *M* was a conservative extension of *F*, rather than a conservative extension of *D* as I have claimed. A variation of this, proposed by Detlefsen (1990b), is that Hilbert wanted to show that *M* is *consistent with F*.

In defence of my interpretation I appeal to the first passage from Hilbert (1925, p. 376) quoted above. *M* is required to 'yield correct numerical equations when applied to integers', where by 'numerical equations' Hilbert means equations involving numerals *but not variables*: equations with variables are not considered until p. 378 and are not counted as 'numerical equations'. Thus Hilbert's requirement is that *M* should be a conservative extension of *D*.

It may be argued that conservativeness over *D* is tantamount to conservativeness over *F*: the example with Fermat's last theorem in Hilbert (1927, p. 474) shows how the former extends to the latter. However, I still object to the formulation '*M* is a conservative extension of *F*', not because of the word 'conservative' but because of the word 'extension', which implies that every argument in *F* also belongs to (or can be formalised as) a proof in *M*. This interpretation would bring Hilbert into direct collision with Gödel's second incompleteness theorem, for if *F* is a subtheory of *M* then a finitary consistency proof for *M* is (or can be formalised as) a consistency proof for *M* in *M*, which is impossible if *M* is consistent.

Most commentators on Hilbert's programme do indeed seem to believe that *F* is required to be a subsystem of *M*, and hence that his philosophy has been refuted by Gödel's theorem.

The second theorem implies the impossibility of proving the consistency of formalized classical mathematics by finitist methods. For in spite of a certain vagueness in demarcating the notion of finitist proofs, any such proof can be arithmetized and incorporated into [formal first-order arithmetic]. (Körner 1960, IV)

See also Kreisel (1965, p. 98), Dummett (1977, p. 2) and Giaquinto (1983). Hilbert's successors have responded by relativising his programme, that is, by identifying useful weakened versions of classical theories and proving their consistency relative to other, more constructive, predicative or finitary theories (Sieg, 1988; Simpson, 1988; Feferman, 1988). Sieg says of this approach that 'It does not claim to solve foundational problems once and for all, but rather tries to provide material for an informed reflection on the epistemology of (parts of) mathematics'.

The assumption that F is a subtheory of M seems to be motivated partly by the misunderstanding mentioned above, that Hilbert intended M to be a conservative extension of F rather than of D, and partly by a vague feeling that finitary reasoning ought to be weaker than classical reasoning (Shapiro, 1992). Bernays (1935) comments that no finitary argument has yet been suggested that cannot be formalised in Peano Arithmetic but that lacking a precise delimitation of finitary argument it is hard to demonstrate the point conclusively.

Detlefsen (1979), however, points out that Hilbert's programme does not require F to be a subtheory of M. The nearest Hilbert himself comes to a direct statement on the matter is:

For, if to the real propositions we adjoin the ideal ones, we obtain a system of propositions in which all the simple rules of Aristotelian logic hold and all the usual methods of mathematical inference are valid. (Hilbert, 1927, p. 471)

This sentence says two things: that the *language* of F is contained in the *language* of M, and that M contains classical logic; it does not state any relation between the real and the ideal *modes of reasoning*.

How then do Gödel's incompleteness theorems apply to Hilbert? The theorems may be paraphrased as follows: informal contentual reasoning concerning the natural numbers has the ability to transcend any recognisably consistent formal system, however strong the ingredients we put in the formal system. Therefore an informal system F, intended to encompass all our meaningful arguments about numbers, cannot be contained within any formal system M, even if M is adequate for normal present-day mathematical practice. But as Detlefsen (1979) also points out, Hilbert does not require F to be formalisable. Therefore there is no threat to Hilbert's programme, as Gödel himself pointed out at the end of his original paper (Gödel, 1931), though he changed his mind shortly afterwards (Gödel, 1933b). (See also Detlefsen (1990b) and Detlefsen (1986) for further objections to the use of Gödel's first and second theorems, respectively, against Hilbert's programme.)

Finitary reasoning, then, cannot be encompassed by any formal system. The next section will develop the mathematical consequences of this fact.

FINITARY AND CONSTRUCTIVE REASONING

In the early days of Hilbert's programme it was assumed by many that finitary reasoning was simply constructive reasoning with numbers (Herbrand, 1931). However, it is now generally believed that finitary reasoning is a weaker notion than constructive reasoning, at least as understood by intuitionists. The distinction appears to have originated in the 1920s and to have been first stated in print by Bernays (1935) and Gödel (1933b, 1958), who attributed it to Hilbert (1925), though Hilbert's remarks do not seem to justify this; subsequently it was taken up by researchers into intuitionism (Kreisel, 1962, 1965, 1970; Weinstein, 1983; Troelstra & van Dalen, 1988, Chapter 1, §4.5) and formalism (Giaquinto, 1983; Feferman, 1988). The usual account of the distinction is as follows.

Finitary (or *finitist*) reasoning is concerned exclusively with *concrete, combinatorial* symbols (*spatio-temporal configurations*) and operations on them. We are allowed to reason about infinitely many concrete symbols at once, but all the properties, operations and arguments we use must be *intuitive, visualisable* or *surveyable.* For example, a function on symbols is only finitarily acceptable if it can be computed mechanically and if we can see, by surveying all possible executions, that it is total (Kreisel, 1965, §3.412; Tait, 1981).

Constructive (or *intuitionistic*) reasoning, however, goes beyond the finitary viewpoint by considering *abstract* objects such as proofs, functions from concrete symbols to concrete symbols, functionals of higher type, rules, and 'constructions in general'. It admits properties, such as being a proof of a formula and being a meaningful rule, that are decidable (in the sense that we can decide them using intelligent judgement) but are not necessarily recursive. This looks like a denial of Church's Thesis, at least in its strong form, but it is not quite. One may accept this view and still believe that for every 'abstract' function on numbers there is a recursive function extensionally equivalent to it; thus Church's Thesis becomes a (rather implausible) axiom of reducibility for intuitionistic mathematics (Kreisel, 1970).

Giaquinto's (1983) characterisation is typical: 'intuitionism was based on an abstract phenomenology of mental constructions yet to be made precise, whereas Hilbert's formalism was based on mechanical (primitive recursive) reasoning about finite arrays of symbols.'

The words in italics above are being used in a special sense that is never really explained; instead of an explanation we get a long list of synonyms. Some of the terminology is most unfortunate. 'Spatio-temporal', 'concrete' and 'abstract' suggest the token-type distinction, which is *not* what is intended; McCarty (1983) is surely correct to speak of concrete mathematics as concerned with *physically realisable items* rather than *physically realised items*. Moreover, the word 'intuitive' invites confusion with 'intuitionistic';

for example, Weinstein (1983) says that for intuitionists, but not for Hilbert, 'decidable properties of proofs may involve considerations about the intuitive content of these mathematical constructions', which is the opposite of the usage of 'intuitive' above. The terms *combinatorial* and *non-combinatorial* give the best expression of what seems to be intended. The idea, I think, is that, although proofs, meaningful rules, and so on, may individually be expressed in symbols, there is no systematic notation that expresses *all* the valid proofs or rules and *only* valid ones, so we cannot be said to have expressed what makes a proof correct or a rule meaningful; therefore these notions always elude spatio-temporal representation.

As the Giaquinto quotation indicates, the concrete-abstract distinction leans heavily on the apparent *obscurity* of the intuitionistic notion of proof. Any explicit account of intuitionistic proof, such as I am giving in Chapters 8 and 10, is bound to diminish the apparent distance between intuitionism and finitism.

Attempts to express the finitary viewpoint formally have produced two characterisations: (i) finitary reasoning admits induction with decidable properties over any ordinal less than ε_0, but not over ε_0 itself (Gödel, 1958); and (ii) finitary reasoning is Primitive Recursive Arithmetic (Tait, 1981).

Gödel explains his ordinal characterisation as follows.

> the validity of inference by recursion up to ε_0 surely cannot be made immediately intuitive, as it can up to, say, ω^2. More precisely, we can no longer survey the various structural possibilities that obtain for descending sequences, and therefore we cannot intuitively recognize that every such sequence will necessarily terminate. (Gödel, 1958, p. 243)

Gödel is unwilling to commit himself on whether our inability to survey up to ε_0 is merely a practical matter of feasibility, like our inability to count up to 10^{100}, or whether it is intrinsic, like our inability to count up to ω. If the former then the 'finitary standpoint' he is describing would have no special philosophical significance (it would vary from person to person and from time to time), so I shall take him to be tentatively asserting the latter.

ε_0 seems a very odd choice for the first unsurveyable ordinal. If we can survey the lesser ordinals ω, ω^ω, ω^{ω^ω}, and so on, why can we not survey the *sequence* of these ordinals (which is only of order-type ω), and is this not tantamount to surveying the *limit*, ε_0, of the sequence? After all, any descending sequence in ε_0 is a descending sequence in one of ω, ω^ω, ω^{ω^ω}, ...

A *regular* ordinal would seem a better choice. However, whatever ordinal we choose, it remains the case that if you can survey all the ordinals less than α then you can survey α; to believe in induction over all ordinals less than α is tantamount to believing in induction over α. The only way out appears to be to say that a finitist can survey each ordinal below a certain α, when presented individually, but cannot *know* that they can survey all of them and hence cannot make the transition to α. In other words, no one can knowingly be a finitist.

Let's try the other characterisation and see if it is any less puzzling. Primitive Recursive Arithmetic, first formulated by Skolem (1923), is Heyting Arithmetic without the logical constants. The formulae are of the form $M = N$, where M and N are terms built up from 0 and numeric variables using primitive recursive functions. (Skolem also allows truth-functional combinations of such equations, but this involves no gain in generality.) The logic may be expressed as a natural deduction system with equality, Peano's axioms, and an induction rule.

A consequence of identifying finitary reasoning with Primitive Recursive Arithmetic is that the only functions known to be total are the primitive recursive ones; this accords with Gödel's (1972) statement that 'while any primitive recursive definition is finitary, the general principle of primitive recursive definition is not a finitary proposition, because it contains the abstract concept of function'. Let us examine this doctrine more closely. There are two possible reasons for rejecting the general concept of primitive recursive function. One might accept any particular primitive recursive definition but be unable to grasp the class of primitive recursive definitions in its entirety and hence be unable to diagonalise out of it; or, more radically, one might reject the general concept of function, as suggested by Gödel's statement. I shall discuss the two possibilities separately.

First, is it really possible to accept each instance of primitive recursion while denying the general concept? The primitive recursive functions can all be expressed as expressions built up from 0, successor, projection functions, composition, and a recursion operator; for example, in the notation of Chapter 16, multiplication is

$$rec_1(rec_0(0, proj_2^2), comp_2(rec_1(proj_1^1, comp_1(S, proj_3^3)), proj_2^3, proj_3^3)).$$

We can enumerate all such expressions primitive recursively; let R_m be the m'th such expression. (Assume R_m is represented as a unary function, using a coding of tuples in the usual way.) We can also express the evaluation of $R_m(n)$ as a process of applying simple primitive recursive rewriting rules repeatedly until a numeral is obtained. Hence we can define the universal function over the primitive recursive functions, U, by

$$U(m,n) = R_m(n).$$

U itself is not primitive recursive. It would appear to be the simplest example of a finitarily inadmissible function: the finitist can execute the function, for any given m and n, but cannot survey all possible executions and therefore cannot see that execution halts for all m and n.

Now, if I ever met a flesh-and-blood finitist, in this sense, I would point out to them the general notation for primitive recursive functions, the enumeration R_m, and the universal function U, and ask them why, since they accept each instance R_m they do not accept U. There are, as far as I can see, only four ways in which they might reply.

(1) 'I don't grasp the common pattern you are trying to point out. To me, each primitive recursive function is unique; they don't fit into any systematic scheme that I can see.'
(2) 'I see the pattern but I don't accept its general validity. Admittedly, all the functions I have ever defined can be expressed in your notation, and admittedly I cannot produce a single counter-example of an R_m that is unacceptable to me, but still the notation is not *generally* sound: it is just an amazingly successful heuristic device.'
(3) 'I accept that each instance R_m is admissible, but U is not.'
(4) 'Now I see that U is admissible.'

Of these responses, (1) seems to me disingenuous: the finitist can certainly grasp the primitive recursive mapping that turns m into R_m and can certainly execute U on any given m and n, since U works by iterating a primitive recursive function until a primitive recursive halting condition is satisfied. Response (2) is a possible position, though an extremely uncomfortable one: the finitist simply lives with the miracle that non-finitary methods produce meaningful results without seeking a rational explanation. Response (3) is incoherent: if you accept that each R_m is total then you have conceded that U is total. Response (4) seems to me the only viable one, but it amounts to abandoning finitism.

I concede that it is a perfectly mathematically well-defined policy to admit each primitive recursive function in one's reasoning while excluding U. I also concede that it is psychologically possible that a person could understand each particular instance of primitive recursion and yet fail to understand a general notation for it. What I question is whether this constitutes a significant philosophical position that is more secure than accepting primitive recursion in general. It is equally psychologically possible that a person could understand addition and subtraction but balk at multiplication and division. One could imagine explaining to them that doubling means adding a number to itself and tripling means adding three copies of a number together, and in general every multiplication is a repeated addition. It is perfectly possible that they could still fail to understand, no matter how many examples one goes through. But that does not justify treating their limited arithmetic as a highly sceptical philosophical position that is more secure than arithmetic involving multiplication and division. The same applies to the finitist who accepts particular primitive recursions but not primitive recursion in general: to present this as scepticism (and not merely stupidity) it is necessary to show that there is some philosophical advantage in taking this position. What is needed is a way of justifying primitive recursive functions on a case-by-case basis that does not immediately provide a general justification for primitive recursion, and it is hard to see how this could be possible. In a formal system it is certainly the case that accepting a general principle is strictly stronger than accepting all its instances, but things are different for informal philosophical argument since any grounds for accepting all the instances are also grounds for accepting the general principle.

This brings us to the second possible reason for rejecting the concept of primitive recursive function. The trouble might not be in the concept of primitive recursion but in the concept of function. Tait (1981) says: 'For the finitist to recognize the validity of primitive recursive arithmetic, he must recognize the *general* validity of definition of functions by primitive recursion. But he cannot even formulate this since it involves the notion of function.' Gödel says something very similar (see above). This claim that a finitist doesn't understand the concept of function is obscure to me. The finitist needs to understand *something* about functions in general in order to: (i) know what to do with them (that is, understand that one is supposed to present them with arguments and obtain values by a process of execution), and (ii) understand questions of the form 'Do you accept ... as defining a function?'. If one asked a finitist 'What sort of thing is multiplication?', they would surely say something like 'It is a symbol expression that I can use to turn two numbers into a number'. This minimal understanding of the function concept, apparently presupposed by Tait (see p. 532 in particular) and Gödel, is quite adequate to understand U: one simply grasps the (primitive recursively specifiable) syntax and operational semantics of the class of expressions R_m, and (tries to) survey the structural possibilities for executions of expressions conforming to this syntax. Everything, then, depends not on a grasp of a general concept of function but on this metaphorical notion of 'surveyability'.

Unfortunately, all the rhetoric about 'spatio-temporal configurations', 'surveyability' and 'the finite standpoint' is very little help when it comes to drawing the boundary of the finitary. All recursive definitions can be expressed in discrete symbolic notation; yet to see that the function defined is total always requires an act of understanding. To visualise infinitely many computation sequences necessarily requires some grasp of infinity; this visualisation must be intensional, since no one can imagine infinitely many individual things at once. U does not involve an abstract concept of function any more than individual primitive recursive functions do. Obviously, recursive definitions get harder to visualise or survey as they get more complicated, and everyone will have a limit to what they can visualise in practice, but the functions do not suddenly become unvisualisable-in-principle at any point. I do not object in principle to the notion of surveyability; indeed, I shall make use of something similar myself in the next chapter. I do, however, object to Gödel's and Tait's strange and arbitrary stipulations about what can and cannot be surveyed.

My conclusion is that what Gödel and Tait call 'finitism' is a perfectly well-defined mathematical system but is not a philosophical position that anyone could consciously subscribe to (a conclusion that I think Tait would endorse). To realise that you are a finitist is to become an intuitionist.

It follows, then, that if we are to turn Hilbert's sketchy remarks about finitary reasoning into a stable philosophical standpoint we must admit steps such as the one that takes us from the primitive recursive functions R_m to the universal function U. This is a *reflection principle*; it takes us from a formal system, each expression in which is grasped individually, to a grasp of the system as a whole. Of course, the story does not stop at U. Given U we can form the system of all functions defined using primitive recursion and U, and then define the universal function U_1 over this system. Repeating this procedure gives a sequence of successively more powerful functions $U, U_1, U_2, \ldots$, leading to a limit function U_ω defined by

$$U_\omega(m,n) = U_m(n)$$

(where I am assuming each U_m function is expressed as a unary function using a coding of pairs). Naturally, U_ω is followed by $U_{\omega+1}, U_{\omega+2}, \ldots$. There is no way we can call a halt to this progression once we have admitted reflection principles. The sequence continues to all U_α for which we can recognise α as well-founded (by surveying all the descending sequences, in Gödel's terminology). As we saw in the discussion of ε_0, there is a difficulty in determining the extent of our ability to recognise ordinals as well-founded. I shall return to this problem in the next chapter.

My proposal to admit reflection principles is analogous to Detlefsen's (1979) argument that a form of ω-rule should be included in finitary reasoning. Detlefsen remarks that 'the dubitability of $T \cup Con(T)$ is no greater than that of T'; that is, any philosophical grounds adequate for accepting a theory T as sound are also adequate for accepting the soundness of its formalised consistency statement, $Con(T)$. He further argues that, for theories with finitely many axioms and axiom schemata, T is no more dubitable than some finitely axiomatised part of T ('epistemic compactness'). It follows that the finitist who accepts a particular formal system as contentually sound is inexorably led to accept reflection principles, indefinitely iterated; this generates a transfinite sequence of ordinal logics (Turing, 1939). Detlefsen's ω-rule needs to be formulated with care if it is to escape triviality (Ignjatović, 1994), but his point is nevertheless sound.

The preceding discussion shows that the finitist is obliged to accept an open-ended system of recursive functions; this echoes Herbrand's (1931,

p. 628) statement that 'it is impossible to describe all intuitionistic procedures of reasoning (since it is impossible to describe all procedures for the construction of functions)', where Herbrand identifies 'intuitionistic' with 'finitist'. For the sake of a uniform notation for recursive functions it is tempting also to admit *partial* recursive functions. These are certainly finitarily meaningful as *descriptions of algorithmic processes*, since each step in execution is primitive recursive; it ought to be possible for a finitist to discuss such processes without prejudging whether they will halt. For this reason I propose to allow the computations P and Q in the general Hilbertian finitary hypothetical judgement $(*)$ discussed above,

$$\text{for any } a, b, \dots c, \text{ if } P(a, b, \dots c) \text{ then } Q(a, b, \dots c),$$

to be partial recursive functions. Hilbert gives little guidance on what class of recursive functions he would admit – unsurprisingly, since recursive function theory did not exist at the time of writing. There can certainly be no finitary objection to allowing the P function in $(*)$ to be partial recursive, since the phrase 'for any $a, \dots c$, if $P(a, \dots c)$' may be rewritten using standard theorems of recursive function theory in the form 'for any $a, \dots c, u$, if $P'(a, \dots c, u)$', where P' is primitive recursive and u codes a computation of $P(a, \dots c)$. However, when one tries to rewrite '$Q(a, \dots c)$' as '$Q'(a, \dots c, v)$' in the same way, there is a difficulty of where the value of v comes from. If we supplied information about how v is obtained from $a, \dots c, u$ then the hypothetical judgement $(*)$, with partial recursive P and Q, could be understood in terms of primitive recursive functions, P' and Q', and so would certainly be finitarily intelligible.

This hypothetical judgement should be compared with the general sequent form in protologic

$$(x, \dots y)\, A, \ \dots \ B \rightarrow C$$

(see Chapter 8). The fact that sequents may have several terms $(A, \dots B)$ on the left-hand side while hypothetical judgements have only one (P) is obviously inessential since several terms can be combined into one using truth-functional conjunction; thus protological sequents and hypothetical judgements agree in syntax. They also agree in intended meaning. In both cases we are not simply making an extensional statement about the results of the computations $P(a, \dots c)$ and $Q(a, \dots c)$: we are also relating the computation processes u and v. Hypothetical judgements with partial recursive functions are finitarily acceptable provided they are interpreted intensionally, as relations between (primitive recursively specifiable) computation processes – exactly as for protological sequents.

Furthermore, the case for including reflection principles in finitary reasoning applies with equal force to protologic. Hence, as I shall argue in detail in Chapter 10, protologic consists of the informal ingredients listed in Chapter 8 plus reflection principles. This makes protologic the same as finitary reasoning. I conclude, then, that formalists and intuitionists both have the same underlying system of constructive reasoning. This may be a controversial conclusion. It is commonplace to point out that finitary reasoning is weaker than intuitionistic predicate calculus (since the former does not allow nested quantifiers), and it is frequently inferred that finitism is weaker than intuitionism; but this is a misleading comparison. Intuitionistic predicate calculus does not play an analogous role in intuitionism to the role of finitary reasoning in formalism. In intuitionism, intuitionistic predicate calculus is not taken as primitive but is explained in terms of an underlying theory of constructions (see Chapter 7), which is the intuitionists' version of 'contentual reasoning'. Intuitionists do not think in predicate calculus; they think in constructions, and predicate calculus emerges as a set of high-level regularities in the constructive activity. So the proper comparison should be between finitary reasoning and protologic, and these are identical because they are both based on the same informal set of ideas.

INFORMAL AND FORMAL PROOFS

Hilbert's programme involved two steps: expressing mathematical practice in a formal system, and proving the system consistent. I have already discussed an objection to the second step, that it is impossible on account of Gödel's second incompleteness theorem. The first step has also come under attack. All varieties of formalism rest on the assumption that mathematics can, and indeed should, be formalised. The normal historical process of clarifying mathematical concepts and making proofs more rigorous naturally leads to formal systems. Informal mathematics is mathematics that has not yet been fully formalised and is therefore not yet wholly rigorous. I shall call this the *Central Dogma of Formalism.* There is no assumption here that a single formal system will suffice for all mathematics for all time, simply that any particular body of ideas and techniques can be formalised.

Davis & Hersh (1981) oppose this with the thesis that mathematics is inherently informal. What is generally called a proof in mathematics, they point out, is actually a prose argument, interspersed with mathematical notation, that is intended to convince a knowledgeable reader of the truth of the theorem. Such an informal proof will be organised into steps of reasoning, like a formal proof, but the steps are much larger than those permitted in any formal system and rely essentially on intuition and routine arguments that the reader is expected to fill in. Indeed, an informal proof is often *more* convincing than a formal one as it is much shorter and gives us a better understanding of what is going on. Past generations of mathematicians such as Newton and Euler used the concepts of real number, convergence, and so on, without knowing what are today regarded as their precise definitions; on a formalist account it is a mystery how they made any progress. Tragessor (1992) argues in a similar vein.

Now, formalism is perfectly willing to acknowledge that informal concepts and arguments often pre-date the corresponding formalisation and have a separate heuristic role. However the Central Dogma insists that for *epistemological* purposes informal proofs have no life of their own: an informal proof is only convincing to the extent that it hints at a formal proof.

Davis & Hersh (*ibid.*, pp. 363–369) illustrate their notion of informal proof by describing a heuristic argument, due to Good and Churchhouse, for the Riemann hypothesis, which they seem to find completely convincing, more so even than a very complicated formal proof would be. Personally I am unconvinced by the heuristic argument. How are they to persuade me? If it were an ordinary mathematical proof, of the sort that is published in books and journals, Davis & Hersh could break the argument into smaller and smaller steps until I am forced to accept it or an error comes to light. At each stage they could ask me 'Which step of the proof do you challenge?'; I would have to choose one such step, and they would break it down into even smaller steps. By the time we had reached the level of detail of a completely formal proof the dispute would be settled. This ability to formalise any part of a proof on demand is what makes it convincing. Davis & Hersh's heuristic argument cannot be so formalised, and that is why it is unconvincing to anyone who is not immediately swayed by the rhetoric.

This reveals why the Central Dogma of Formalism is correct. Formal proofs, meaningfully interpreted, are the most convincing examples of reasoning known to us; anything else is, to some degree, unconvincing. Ideally, therefore, all proofs would be formal; the second-best is that all proofs should be written out in sufficient detail as to convince the reader that they can be formalised. Davis & Hersh manage to give the impression that complete formality is a remote, superhuman goal, but this is not so, as is shown by the fact that we can write and read lengthy computer programs, which are completely formal. Programmers, forced to deal with complex formal objects, have developed appropriate notations (high-level programming languages) to make them tractable. Analogous techniques for mathematics have not yet been developed because mathematicians (unlike computers) are prepared to tolerate informality.

FORMALISM AND INTUITIONISM

So far I have been defending Hilbert against all objections. It may be asked, why do I, as an intuitionist, defend Hilbert? Aren't intuitionism and formalism supposed to be deadly enemies?

It is true that Brouwer and Hilbert saw each other as opponents. Their true relationship, however, is more complex. Brouwer did not see the opposition as fundamental: he argued (1927) that the choice between formalism and intuitionism would 'be reduced to a matter of taste, as soon as the following insights ... are generally accepted', that is, as soon as the formalists also become intuitionists.

Two of these four insights, Brouwer noted, were already accepted by Hilbert. These were: (i) the distinction, invented by Frege (1885) and Brouwer (1907, pp. 173–174), between formal mathematics as a meaningless symbol system and meaningful ('contentual') metamathematics; and (ii) the invalidity of the principle of excluded middle in contentual reasoning. The other two insights were not yet accepted.

Third insight. The identification of the principle of excluded middle with the principle of the solvability of every mathematical problem.

Fourth insight. The recognition of the fact that the (contentual) justification of formalistic mathematics by means of the proof of its consistency contains a vicious circle, since this justification rests upon the (contentual) correctness of the proposition that from the consistency of a proposition the correctness of the proposition follows, that is, upon the (contentual) correctness of the principle of excluded middle. (Brouwer, 1927)

These two 'insights' are misunderstandings on Brouwer's part. First, Hilbert's belief that all mathematical problems are solvable plays *no* part in his formalist philosophy of mathematics. Moreover, Hilbert never claimed to possess a general method for solving all problems, as would be required from a constructive point of view to justify the principle of excluded middle in contentual reasoning. The other misunderstanding is that Brouwer did not see why Hilbert should regard a proof of consistency as establishing the *correctness* of the formal system; so he guessed that Hilbert was relying on a contentual principle of excluded middle somehow to go from consistency to correctness. Brouwer seems not to have been aware of the role of *conservativeness* in Hilbert's programme. This is borne out by the following later criticism of 'New Formalism' (Hilbert).

But no attention was paid by New Formalism to the circumstance that, between the perfection of mathematical language and the perfection of mathematics proper, no clear connection can be seen. (Brouwer, 1952)

Heyting (1956, p. 4; 1974) displays similar attitudes. Brouwer's view of Hilbert, then, was based on a clear perception of where Hilbert borrowed from Brouwer but simple misunderstandings of the distinctive features of Hilbert's programme. (See Giaquinto (1983) for more on this.)

Hilbert's view of Brouwer was more accurate. He understood that the real point of difference between them was that Brouwer insisted that every formula should have a meaningful interpretation, whereas Hilbert required only that the formal system as a whole should have a meaningful interpretation with respect to its decidable part (Hilbert, 1927, p. 475).

This insight can be developed as follows. Suppose we want to carry out Hilbert's programme for a given formal mathematical system. A particularly simple form that the consistency proof might take is to associate a finitarily meaningful predicate P_A with every formula A in such a way that

- for each axiom A we can find an object that satisfies P_A,
- for each rule of inference $\dfrac{A \cdots B}{C}$ we can obtain an object satisfying P_C given objects satisfying $P_A, \ldots P_B$,
- for each variable-free atomic formula ('numerical equation') A there is an object satisfying P_A iff A is true.

Then it would follow that a numerical equation is only derivable in the system if it is true, as required. But this consistency proof is also an *intuitionistic interpretation* of the formal system, if we read P_A as '... is a proof of A'. Hence the intuitionist's problem of making sense of a formal system is simply the formalist's problem of proving its consistency – with one difference, that the intuitionist insists on an interpretation (P_A) for every formula A, whereas the formalist would be satisfied with an interpretation of the system as a whole and hence has a wider choice of consistency proofs.

Here I rely on two controversial points of interpretation: that the formalist and the intuitionist have the same underlying system of contentual constructive reasoning (as argued above), and that Brouwer's view of linguistic entities as strictly external to mathematics is mistaken (as argued in Chapter 5).

CONCLUSIONS ON FORMALISM

Hilbert's formalism can best be described as a variety of constructivism that accepts the use of formal systems as a reliable aid in mathematical constructive activity. My interpretation of Hilbert is unorthodox on the following points:

- finitary reasoning consists of decidable propositions and hypothetical judgements (rather than decidable propositions and universal generalisations of them, as is usually supposed);
- finitary reasoning is not a subtheory of ideal mathematics, and hence Gödel's second incompleteness theorem poses no threat to Hilbert's programme;
- finitary reasoning includes reflection principles;
- finitary propositions may involve partial recursive functions and not merely primitive recursive functions;
- finitary reasoning equals protologic, the fundamental system of constructive reasoning underlying intuitionistic arithmetic (which, of course, is not to say that finitary reasoning equals intuitionistic arithmetic);

- formalism is compatible with intuitionism, in that a successful intuitionistic interpretation of a branch of mathematics would accomplish Hilbert's programme for that branch.

When I call myself a formalist I mean two things: that I accept the Central Dogma that all proofs must be formalisable (though not in a single system), and that my interpretation of arithmetic and analysis in this book may be regarded as carrying out Hilbert's programme (see Chapter 35 especially).

CHAPTER 10

OPEN-ENDEDNESS

FORMALITY AND INFORMALITY

The discussion of 'finitist' functions in the previous chapter drew attention to an essential aspect of mathematics, which I shall call *open-endedness* – its ability to transcend any particular formal system. A mathematical domain is *open-ended* when, given any (sound) formal theory supposedly characterising it, we have a general method for producing an element of the domain outside the theory. Open-endedness (also called *essential incompleteness* and *incompletability*) implies that a foundation for mathematics can never be entirely formal. The question therefore arises of what use formality is in founding mathematics and how the formal part of the foundation is related to the informal part.

In a typical axiomatic foundation, say ZFC set theory, the formal part is the axiom system, where all syntactic devices are decidable and so is the notion of a correct ZFC proof. The informal part consists of seeing that the axioms are true and the rules of inference preserve truth; this depends on an informal understanding of the logical constants, the universe of sets, and set membership. The answer to the question 'what use is formality?' is obvious in this context. Formality separates, in any set-theoretic argument, the philosophical concerns (about the legitimacy of the general concept of set) from the purely mathematical concerns (about whether standard set-theoretic modes of reasoning are being applied correctly). If set theory were not formalised every new theorem would provoke a fresh philosophical dispute; long proofs would be subject to greater suspicion than short proofs since the dubious set-theoretic assumptions were being used a greater number of times. The benefit of formalisation is that the philosophical principles can be summed up in a finite list of schemata; then philosophical criticism of the principles can proceed independently from the mathematical elaboration of their consequences. The axioms mark the boundary between the philosophy and the mathematics.

This separation of qualitatively different concerns is the essential means by which any difficult intellectual problem is solved. We may expect formality to play a similarly helpful role in any foundation of mathematics.

The unsatisfactory aspect of this, at least in the case of ZFC set theory, is that we are unable to give a rational account of how the informal part, the choice of axioms, is carried out. Set theorists have indeed developed some heuristic principles for choosing axioms (Maddy, 1988), but they have little to say on their rational justification and there is often disagreement on how they are to be applied in a particular case. If two set theorists disagree on whether to admit a strong axiom of infinity, it is not clear whether there is an objective matter of fact about who is right. (See Skolem (1922) for further discussion of the use of a single axiomatic system as a foundation for mathematics.)

We require, then, for any foundation of mathematics, that it sharply delimit the division of labour between the formal and informal parts, that it explain how the two parts fit together, and that it say something informative about what the *correctness* of the informal part consists in. This is the task of the present chapter.

OPEN-ENDED DOMAINS

The phenomenon of open-endedness arises in all philosophies of mathematics. Here are some examples of open-ended domains.

1. The total recursive functions. Given any recursively enumerable class of total recursive functions it is always possible to generate by diagonalisation a total recursive function outside the class.
2. The class of well-defined rules. The semantic paradoxes arise from the fact that we cannot pin down the meaning of 'well-defined' definitively.
3. Large cardinals in set theory. Each axiom of infinity suggests a stronger one.
4. Ordinal logics. An attempt to characterise informal arithmetic proof might start with Heyting Arithmetic or Peano Arithmetic, add a formalised statement of its consistency to give an extended theory, and repeat the extension operation transfinitely (Turing, 1939). The question of how far the iteration can proceed was raised but not answered in the previous chapter.
5. Well-founded trees. Mirimanoff's paradox (see below) shows that the universe of all well-founded trees can never be circumscribed.
6. Intuitionistic proof. Goodman's paradox (see Chapter 8) shows that proof eludes characterisation in any decidable form.

These examples are of course related. Clearly, the same phenomenon is at work in all the cases.

One of the examples stands out as being particularly tractable: example (1) is the only case where we have a very explicit, universally agreed, philosophically uncontentious theory of what the objects in question are like. I shall therefore take it as the model for understanding open-endedness in general.

Imagine that we were discussing the general notion of an 'effective' function in the 1920s, before the development of recursive function theory. We might identify simple classes of effective functions, such as the primitive recursive functions, and observe that it is always possible to extend the class by repeated diagonalisation. We might plausibly infer that there is a vast uncountable hierarchy of effective functions. This would seem paradoxical, since surely every effective function is defined by a rule expressible in a finite number of characters from a fixed alphabet (note the link between examples (1) and (2)). We might conclude that the notion of effectiveness is ill-defined or ineffable.

This is, however, not the way effective functions are understood today. We broaden our point of view to include partial functions, and then all functions become expressible in a single formal notation (such as Turing Machine codes). The universe of effective functions is not enormously big; it is just an undecidable subset of a countable domain. A total recursive function has two aspects: a decidable aspect (its syntax as a Turing Machine and its operational semantics) and an undecidable aspect (its totality).

The same shift in perspective can be applied to the other examples. In the case of set theory, Gödel (1946) has suggested that, by analogy with the absolute notion of computability, there may be absolute notions of demonstrability and definability, not relative to a formal system, which are closed under diagonalisation. Thus we might some day be able to characterise axioms of infinity: 'An axiom of infinity is a proposition which has a certain (decidable) formal structure and which in addition is true.' Further, Kreisel (1965, p. 113) suggests that for every set-theoretic formula A there might be an axiom of infinity A' such that $A \iff A'$ is provable in ZFC. If so, this would split set-theoretic truth into two components: a decidable component (proof in ZFC) and an undecidable component (the truth of the axiom of infinity). I propose that a similar approach be taken to the other examples of open-ended domains.

PROOF AND WELL-FOUNDEDNESS

Of all the examples of open-ended domains given above, the one that is most important for my purposes is (6), intuitionistic proof. Intuitionists have traditionally been very evasive about what a proof actually *is* (although they have had a lot to say about what it *isn't*), and this has created the impression that intuitionistic proof is something exotic and transcendental, forever eluding rational description (hence the spurious concrete/abstract distinction discussed in Chapter 9). I find this situation unsatisfactory and intend to be completely explicit about what a proof is.

As argued in Chapter 9, contentual constructive reasoning must include the ability to formalise one's past constructive reasoning and, by diagonalisation, transcend it. Thus, once we have grasped some methods of constructive argument, say the protologic outlined informally in Chapter 8, it is possible to code such arguments in a formal axiomatic system and define a construction (that is, a recursive function) DT_0 such that

$$DT_0(D,S) = \mathit{true} \quad \text{iff} \quad D \text{ is the code of a formal derivation tree for the sequent } S$$

where the formal derivations are formalised applications of the protological methods of argument previously recognised as sound. Having defined DT_0 we immediately recognise that the *reflection principle*

$$(x,\ldots y)\; DT_0(D, [\![\rightarrow C]\!]) \rightarrow C$$

is generally sound, that is, if the sequent $\rightarrow C$ is derivable then the value of C is *true*. If we add this reflection principle as an axiom schema to protologic then we obtain a larger system of constructive argument, coded by a construction DT_1 instead of DT_0. Immediately we see that the reflection principle

$$(x,\ldots y)\; DT_1(D, [\![\rightarrow C]\!]) \rightarrow C$$

is generally sound; and if we add it as an axiom schema to protologic we obtain a system represented formally by a new construction DT_2. If we iterate this procedure to give $DT_3, DT_4, \ldots$, we may next recognise that all reflection principles of the form

$$(x,\ldots y)\; DT_n(D, [\![\rightarrow C]\!]) \rightarrow C,$$

where n=0, 1, 2, ..., are sound, and if we add them all to protologic we obtain a still larger system, represented by a new construction DT_ω. The procedure may be continued to give DT_α for all ordinals α that we can recognise as well-founded.

Thus we have an open-ended system of protological theories (ordinal logics). However, as I suggested in the previous section, this is not the correct way to look at it. Instead of thinking of protologic as indefinitely extensible we should think of it as a subclass of a decidable system. That is, define a construction DT representing derivations that include the basic ingredients of protologic (as described in Chapter 8) plus any reflection principle of the form

$$(x,\ldots y)\; DT(D, [\![\rightarrow C]\!]) \rightarrow C.$$

(See Chapter 21 for the formal definition of DT.) Then this system is closed under reflection. The price we pay, of course, is that it is no longer sound; the sound derivations are an undecidable subsystem of this formal system (just as the total recursive functions are an undecidable subclass of the partial recursive functions).

The soundness of a derivation D may be investigated as follows. D is a finite tree of sequents, including in general some reflection principles of the form $(x, \ldots y)\ DT(U, [\![\rightarrow C]\!]) \rightarrow C$. The soundness of D depends on the soundness of all instances of all the U's (that is, for all values of the constructions $x, \ldots y$). Each of these U instances depends for its soundness on all instances of the derivation trees occurring in *its* reflection principles, and so on. Let us form a new tree, consisting of D as the root node, all the instances of the derivations U as the child nodes of D, and so on: in general, X is a child of Y iff X is an instance of a derivation tree occurring in one of Y's reflection principles. Call this tree the *reflection tree* of D. Then D is sound provided D's reflection tree is well-founded.

The conclusion is that all protological arguments can be represented in a single formal notation (coded by DT), and that the soundness of a protological argument (coded formally by D) for a sequent S depends on two factors: whether $DT(D, S)$ holds and whether D's reflection tree is well-founded. Thus protological reasoning turns out to have two aspects: a decidable aspect, which can be specified once-and-for-all by a construction DT, and an undecidable aspect, well-foundedness. (The programme sketched here is carried out explicitly in Chapter 21.)

This accomplishes the task set at the beginning of this chapter, at least in the case of protological argument: it explains how protologic consists of a formal and an informal component, and it identifies what the correctness of the informal component consists in. This account requires that well-foundedness be treated as a primitive notion, grasped informally prior to all formalisation. This seems a plausible position in view of the way in which a perception of well-foundedness seems to be required whenever we make intelligent use of any formal system: we understand a definition, or are convinced by a proof, or grasp that a function is total, by following the train of logical dependencies backwards until they cease. This also seems to be the idea behind the notion of 'grounded truth', as used in semantic theories with 'truth gaps'. Feferman (1991) describes two ways of extending Peano Arithmetic to make it closed under reflection and shows that they are equivalent to two ordinal logics based on Peano Arithmetic. In these theories 'truth' is a partial predicate applying only to grounded formulae, that is, formulae whose truth is determined ultimately by the truth values of the atomic formulae; an ungrounded formula would be, for example, one whose truth depends in a circular way on itself. Groundedness, then, is a form of well-foundedness, existing outside the formal system.

The special extra-formal role of well-foundedness is brought out vividly in proofs of program correctness. A program is said to be *correct* iff given any input satisfying a certain precondition it produces output satisfying a certain postcondition. A proof of correctness consists of two components: a proof of *partial correctness* (which says that the program will produce correct output if it halts) and a proof of halting. Partial correctness can be completely axiomatised in predicate calculus. Proof of halting proceeds by mapping each possible state of a program during an iteration to an element of a well-founded tree and showing that the element decreases at each iteration step, or, to put it more picturesquely, by surveying the structural possibilities for the execution of the program. (See Loeckx & Sieber, 1984.) Thus proof of correctness consists of a decidable component and an informal perception of well-foundedness. Program correctness is especially relevant to my concerns, since protologic is simply a program correctness calculus. I cannot, however, make use of the usual correctness calculi as they stand since they rely on predicate calculus. My point is merely that they correctly identify well-foundedness as an essential element of a correctness proof standing outside the formal system.

However, we are not yet finished. Recall from Chapter 8 the distinction between evidence and proof. Protologic is only a theory of evidence. The account just given of evidence (as consisting of a decidable part plus well-foundedness) needs to be extended to proof.

At first sight this may appear difficult. It is clear from Chapter 8 that proofs consist of such things as numbers, protological derivations, pairs of proofs, and functions mapping proofs of one sort to proofs of another sort. However, I shall show that all these may be regarded as trees of various higher types. A logical formula specifies a type of tree, and a proof of the formula is a tree of that type which satisfies a certain decidable condition and is well-founded. Thus proof, like evidence, has a decidable part and a well-foundedness part.

TREE CODINGS

One reason for caution in taking well-foundedness as a primitive, informally understood notion is Mirimanoff's paradox, which goes as follows. Take all the well-founded trees (that is, one representative of each isomorphism class) and connect them together into a super-tree T that has all the well-founded trees as subtrees. Observe that T is well-founded, since any descending sequence of nodes starting at the root passes immediately into a well-founded tree. Therefore T is (isomorphic to) one of its own subtrees. This implies that there is an infinite descending sequence of nodes in T, contradicting the well-foundedness of T.

This paradox might suggest that there is no general notion of well-foundedness, only limited notions relating to particular classes of trees. I prefer, however, to interpret it in a different way. In reality there are no trees in the abstract, only trees *coded as constructions.* Indeed, it is even misleading to speak of coding trees as constructions, as if trees and constructions were two different sorts of mathematical object. Constructions are the only mathematical objects there are. What we should really say is that some constructions are *interpreted as trees.* To explain what this means we need the following terminology.

A tree consists of *nodes*, one of which is the *root* node. Each node has zero or more *child* nodes; a node with no children is called a *leaf*. The *descendants* of a node are its children, its children's children, and so on. Every node except for the root is a child of exactly one other node and is a descendant of the root; the root itself is a child of no node. A *subtree* is a tree consisting of one of the root's children and all the child's descendants.

To interpret a construction as a tree means to specify which constructions are (interpreted as) its subtrees and to specify how to interpret those constructions as trees. The simplest way to do this is to define a *class* of trees by a relation R between constructions such that if $R(X, Y)$ holds then Y is considered a subtree of X; by applying R repeatedly we can develop the entire structure of the tree (that is, its subtrees, its subtrees' subtrees, and so on) given the construction representing the whole tree. Alternatively we may apply different relations at different points in the tree; but the appropriate relation to use at any stage must be specified when we explain how to interpret a construction as a tree.

I shall call a *tree coding* any scheme for interpreting a class of constructions as trees, whether by a single relation R, by several relations, or in any other way.

An example of a tree coding is coded protologic. As indicated above, a construction D may represent a formal protological derivation in the form of a finite tree of sequents. But D also has an associated reflection tree, which is typically infinitely branching. Thus the one construction D can be interpreted in two ways, as a derivation tree or as a reflection tree, according to the tree coding used. (See Chapter 21, where the reflection tree coding is called *rt*.)

A trivial example of a tree coding is one that says: let any construction X represent a tree consisting of a root node with no children. In Chapter 21 this coding is called *leaf* because in such a tree the root is a leaf.

Two tree codings, Π and Σ, can be combined to give more complex codings. Define a coding, called *product*(Π, Σ) in Chapter 21, by saying that a construction is interpreted as a tree iff it is a pair (X, Y), where X is interpreted as a tree according to Π and Y is interpreted as a tree according to Σ; X and Y are the subtrees of (X, Y). Note that (X, Y) is well-founded iff both X and Y are.

Also, define a coding *map*(Π, Σ) by saying that, to interpret a construction F as a tree, we apply F to all constructions representing *well-founded* trees according to the coding Π, and we regard the values obtained as the subtrees of F. That is, T counts as a subtree of F iff $F(X)$ has the value T for some X

representing a well-founded tree of type Π. The subtree T is to be interpreted as a tree according to the coding Σ. The point of this complicated definition is that it implies that F is well-founded iff all its subtrees are, that is, iff F maps well-founded trees (of type Π) to well-founded trees (of type Σ). This shows that the concept of a mapping between well-founded trees of various types can be construed as a higher form of well-foundedness.

The *map*(Π, Σ) coding may be considered unsatisfactory in that to determine whether T is a subtree of F we need to search through infinitely many constructions X to find one that is mapped to T by F. This defect could be remedied by complicating the interpretation of F slightly, as follows. A subtree of F is now a pair (X, T) such that X is a well-founded tree according to Π and F maps X to T; moreover, (X, T) has a single subtree T, interpreted according to Σ as before. However, in this book I shall stick to the first formulation for simplicity.

The above devices allow us to define a large supply of tree codings by starting with *rt* and *leaf* and using *product* and *map* repeatedly. Statements of the form 'X is well-founded and Y is well-founded' or 'F maps well-founded trees to well-founded trees' can be construed as single statements of well-foundedness, using the *product* and *map* devices. This will be useful in the interpretation of the logical constants (Chapter 25): it will allow me to say that

- any logical formula is semantically analysed as having two components, a recursive function A and a tree coding Π;
- a construction P is a proof of the formula iff $A(P)$ has the value *true* and P represents a well-founded tree according to Π.

Thus intuitionistic proof, like evidence, consists of a decidable aspect and an undecidable well-foundedness aspect.

My account of interpreting constructions as trees amounts to saying that to define a tree coding one must specify (a) which constructions represent trees, (b) for those that do represent trees, which constructions represent their subtrees, and (c) the tree coding for interpreting the subtree constructions. An equivalent but more elegant way of saying this is that to specify a tree coding it is necessary and sufficient to specify which sequences of constructions count as branches; where a *branch* is a finite sequence of constructions, of which the first represents the whole tree and each subsequent construction represents a subtree of the previous one.

Notice how Mirimanoff's paradox is defused. It is not that there is no general notion of well-foundedness but rather that there is no universal coding of trees. Every so-called tree is actually a construction interpreted as a tree according to some particular tree coding, and one cannot define in a non-circular way a tree coding that allows the paradox to work.

I assume that once we have defined a tree coding we grasp the notion of well-foundedness for trees of this type. More precisely, once we have defined what it means for a sequence of constructions to be a branch we grasp what it means for a construction to represent a well-founded tree. This assumption is necessary to make sense of the $map(\Pi, \Sigma)$ coding. In a tree coding such as $map(map(rt, leaf), rt)$, the meanings of *rt* and *leaf* are grasped directly; from that we grasp the meaning of a branch of type $map(rt, leaf)$, hence a well-founded tree of type $map(rt, leaf)$, hence a branch of type $map(map(rt, leaf), rt)$, and finally a well-founded tree of type $map(map(rt, leaf), rt)$. In short, we have a general grasp of well-foundedness.

I cannot defend this assumption, should it come under philosophical attack. I merely assert it. If I were given to Brouwerian rhetoric I would call the grasp of well-foundedness the *Third Act of Intuitionism.* It may be objected that well-foundedness can be understood in several ways: when saying that a tree is well-founded do I mean (a) that induction up the tree is generally sound, or (b) that every descending sequence of nodes reaches a leaf? Clearly (a) implies (b); the converse is Brouwer's principle of Bar Induction. I understand well-foundedness as (a) and so do not need Bar Induction.

DECIDABILITY OF PROOF

I have claimed that intuitionistic proof consists of a decidable component and an undecidable component. This appears to conflict with the orthodox view, originating with Kreisel, that intuitionistic proof must be decidable. On closer inspection it turns out that Kreisel and others mean the word 'decidable' in a special sense: they mean that in any given case we can decide whether a construction proves a formula, not that we can encapsulate our decision ability in an algorithm. Kreisel (1962) says 'we are adopting the basic intuitionistic idealization that we can recognize a proof when we see one', and it is hard to quarrel with this. Intuitionism rejects the platonist notion of an external mathematical reality, according to which statements are true or false independently of our knowledge, and replaces it with a semantics based on proof. Equally, the intuitionist cannot accept that a construction is or is not a proof independently of our knowledge; it is inherent in the very notion of intuitionistic proof that we can effectively recognise one when we see one. A proof is a construction that encapsulates a mathematical discovery (see Chapter 7); if we could not recognise whether a given construction were a proof then it would thereby fail to be a proof.

There is a second reason for wanting proof to be decidable. Goodman (1970, §6) and Dummett (1977, p. 409) point out that Heyting's accounts of the meanings of the logical constants are intended as 'genuine *explanations*', and hence the right-hand sides must not presuppose the logical constants. Therefore the words 'and', 'or' and 'if' occurring on the right-hand sides can only be understood as truth-functional operators; this means that the proof judgements they operate on must have truth values in the classical way and hence must be decidable.

So far I agree (except that I do not regard the 'if' in the implication clause as truth-functional, as explained in Chapter 8, but that does not affect the argument). However, I think I disagree on the reason why we cannot express our capacity to decide proofhood in an algorithm. According to the orthodox intuitionistic story, proof is effectively decidable but probably not algorithmically decidable, since proof depends on an intuitive understanding of the meaning of the constructions involved (Dummett, 1977, pp. 264, 357). This stress on meaning and understanding seems also to be behind the general agnosticism on Church's Thesis (Bishop, 1967, Appendix B; Kreisel, 1970; Troelstra, 1977, §1.3). It is also central to the debate on finitary reasoning (see Chapter 9), where it is often claimed that intuitionism is concerned with the meaning of constructions while finitism relies purely on formal combinatorial properties of symbol patterns.

I have severe reservations about this line of thought. It assumes that human beings are not equivalent to Turing Machines. Neuroscience suggests that the human mind is a function of the human brain and that the brain is a complex system of cells whose individual behaviour is explicable by well-understood chemical and physical laws; if so then it appears likely that a human mind could be simulated by a Turing Machine. Whether such a Turing Machine would be genuinely conscious is irrelevant for present purposes, since all that concerns us is that it would duplicate our decisions on which constructions are proofs of which formulae. Hence the distinction between machine-computable and person-computable would vanish.

Now, the possibility of such a Turing Machine is, of course, a controversial issue in the philosophy of artificial intelligence. I do not mean to settle it here, merely to point out that it cannot be banished by casual introspection or by brandishing the words 'meaning' and 'understanding'. Philosophers of mathematics would be well advised to remain neutral on the question.

Let us consider the possibility that a particular mathematician, Andrew, is equivalent to some Turing Machine, T_A. Then we can enumerate the constructions that Andrew would recognise as proofs and from the enumeration we can produce a proof that Andrew would not recognise. (If one accepts my view that proofs are essentially well-founded trees then we simply hook all of Andrew's trees together to form a single super-tree.) Clearly Andrew cannot carry out this procedure himself, because if he did he would recognise the new proof as sound; so what is stopping him? He can certainly understand the general procedure that produces a new proof from an enumeration of proofs; the one thing he cannot do is recognise T_A as a correct simulation of himself.

Another mathematician, Barbara, may be able to recognise all the proofs that Andrew can and also recognise that T_A simulates Andrew. Hence Barbara can produce and recognise the proof that is beyond Andrew. Andrew's proofs are 'mechanical' from Barbara's point of view since she can encompass them all in a single algorithm T_A, but from Andrew's point of view they rely essentially on individual acts of understanding. Barbara, in turn, is described by another Turing Machine, T_B, but this fact can only be recognised by a higher mathematician, Charles. And so on.

Two conclusions emerge from this. First, 'mechanicalness' is a relative matter: each Turing-Machine-equivalent mathematician has his or her own conceptual horizon, limiting the proofs that he or she can recognise. Secondly, 'mechanicalness' is a collective property of a system of proofs rather than a property of individual proofs. After all, any individual proof P, however sophisticated, can always be recognised by an algorithm, namely the algorithm that says 'Accept the input string iff it is P'.

A special case of this 'conceptual horizon' phenomenon is Gödel's and Tait's portrait of the finitist (Chapter 9), who can recognise each primitive recursive function but cannot recognise primitive recursion in general. This positioning of the conceptual horizon is not particularly plausible or worthy of special attention.

Now consider the alternative possibility, that Andrew is *not* equivalent to any Turing Machine. Then one can still imagine a higher mathematician, Barbara, who can recognise that all the proofs Andrew recognises are sound and who can enumerate all such proofs and hence generate a proof that Andrew cannot recognise. Andrew himself cannot recognise that all his proofs are sound. A similar picture emerges as in the previous case.

In conclusion, I reject the absolute distinction between mechanical and meaningful, and replace it with a relativistic picture of conceptual horizons. *Relativistic*, however, does not mean *subjectivist*: the correctness of proofs is still an objective matter since it is tied to well-foundedness, even though we cannot capture well-foundedness formally.

The task set at the beginning of this chapter was to explain how mathematics splits into formal and informal aspects and to describe the nature and objective correctness of the informal part (without falling into the trap of attempting to formalise it). My answer is that the informal aspect is well-foundedness. An intuitionistic proof is a tree, whose correctness consists in satisfying a decidable (that is, recursive) condition and in being well-founded. This is a much more explicit account of what a proof is than has been given by any intuitionist hitherto.

I have not yet explained in detail how the formal and informal aspects work together to produce a notion of proof consistent with the intuitionists' traditional understanding of the logical constants. This will become clear later (Chapter 21 onwards).

CHAPTER 11

ANALYSIS

Previous chapters have been exclusively concerned with logic and arithmetic, based on what Brouwer called the *First Act of Intuitionism*. To extend the treatment to the theory of real numbers requires, in Brouwer's view, a new mental ingredient, the *Second Act of Intuitionism*. Further extensions would require additional new insights and new sorts of construction. This process is never completed: there is no definitive inventory of ideas on which the whole of mathematics is based.

I share this view. In this chapter I shall consider how the step from arithmetic to analysis should be accomplished. I define *analysis* as a system of first-order predicate logic containing variables for real numbers and the axioms for a complete ordered field. Variables for sets of reals or functions defined on the reals are not included. The completeness property is expressed in an axiom schema

$$(\exists x A \wedge \exists y \forall x (A \supset x \leq y)) \supset \exists z \forall y (z \leq y \iff \forall x (A \supset x \leq y))$$

for any formula A in which y and z do not occur free. Analysis may be interpreted in second-order arithmetic, if we regard a real number as a Dedekind cut of rationals, a rational as a pair of integers, and an integer as a pair of natural numbers, in the usual way. Hence the task for this chapter is to justify the introduction of second-order quantifiers into arithmetic.

Some constructivists would approach this problem by setting up a weak predicative system of analysis that is a conservative extension of arithmetic but is still adequate for most of conventional mathematical practice (see Sieg (1988) for information on such systems). My aim, however, is to interpret the second-order quantifiers in their full impredicative sense. The theory I produce in the end will be formally identical to classical analysis, without the special continuity principles usually associated with intuitionistic analysis. Nevertheless, my theory will be based on genuinely intuitionistic philosophical principles.

CHOICE SEQUENCES

Brouwer and all later intuitionists have interpreted analysis in terms of the notion of a *choice sequence* or *infinitely proceeding sequence.* A choice sequence is a sequence $\alpha = (\alpha_0, \alpha_1, \alpha_2, \ldots)$ 'whose terms are chosen more or less freely from mathematical entities previously acquired' (Brouwer, 1952; see also the Brouwer manuscripts in van Stigt, 1990, pp. 445, 457). In another manuscript Brouwer adds,

> *Such freedom is to be understood in its widest sense: it may entail the absence of any law, or it may entail a set of restrictions, which fail to determine the sequence uniquely; these restrictions may even be allowed to change during the development of the sequence.* (van Stigt, 1990, p. 434)

The meanings of 'chosen' and 'freely' are not entirely clear. 'Chosen' implies that some agency is determining the terms, while 'freely' implies that some other agency is *not* determining the terms; this is not very informative. Heyting's formulation is more illuminating:

> ... a sequence that can be continued ad infinitum. The question how the components of the sequence are successively determined, whether by a law, by free choices, by throwing a die, or by some other means, is entirely irrelevant. (Heyting, 1956, p. 32)

The key word here is 'irrelevant'. When treating a sequence as a choice sequence we pay no attention to the process by which the terms are generated but simply consider the terms themselves. Choice sequence theory applies to all sequence-generating processes, whether deterministic, probabilistic or completely arbitrary. The intention is that choice sequences be treated purely extensionally: any operation carried out with a choice sequence should only depend on a finite number of the terms of the sequence.

CRITIQUE OF CHOICE SEQUENCE THEORY

The basic idea of a choice sequence seems clear and simple, but puzzling complications arise when we attempt to develop it into a formal theory adequate for an interpretation of analysis.

Troelstra (1977) and Dummett (1977, §7.4) provide good surveys of the development of the subject. For simplicity let us confine our attention to sequences whose terms are natural numbers. We wish choice sequences to satisfy various principles such as

Equality: $\alpha = \beta \iff \alpha \equiv \beta$

Density: $\forall \mathbf{u}\, \exists \alpha\, \alpha \in \mathbf{u}$

Open Data: $\forall \alpha\, \forall \beta\, ((A(\alpha,\beta) \wedge \alpha \neq \beta) \supset \exists n\, \forall \gamma \cong_n \alpha\, (\gamma \neq \beta \supset A(\gamma,\beta)))$

$\forall\alpha\exists n$-continuity: $\forall \alpha\, \exists n\, B(\alpha, n) \supset \forall \alpha\, \exists m\, \exists n\, \forall \beta \cong_m \alpha\, B(\beta, n)$

where 'α', 'β' and 'γ' are choice sequence variables, 'm' and 'n' are natural number variables, '$\mathbf{u}$' is a variable for finite sequences, $\equiv$ and $=$ are intensional and extensional equality of choice sequences respectively, $A(\alpha, \beta)$ depends extensionally on α and β and contains no other choice sequence variables, $B(\alpha, n)$ depends extensionally on α and contains no other choice sequence variables, $\alpha \in \mathbf{u}$ holds iff $\mathbf{u}$ is an initial segment of α, and $\gamma \cong_n \alpha$ holds iff γ and α agree on their first n terms.

We also wish the universe of choice sequences to satisfy various versions of the axiom of choice and to be closed under continuous operations: that is, the result of applying a continuous operation to a choice sequence should be a choice sequence. (A *continuous operation* is a function f mapping a choice sequence α to a choice sequence β, where each term β_n is calculated by f from only some initial segment $(\alpha_0, \alpha_1, \alpha_2, \ldots \alpha_{k(n)})$ of α.) Unfortunately, these *desiderata* are not compatible. If β is defined as $f(\alpha)$ then this typically implies some constraints on the terms of β. For example, β may be defined by

$$\forall n \quad \beta_n = \alpha_n + 1.$$

Then we have some intensional information about β: we know that $\forall n\, \beta_n > 0$ and $\forall n\, \beta_n = \alpha_n + 1$. This violates the Open Data Principle, which says in effect that all our knowledge of choice sequences is extensional. The usual response to this is to distinguish various classes of choice sequence, each class satisfying a different subset of the above principles: *lawless* sequences are given purely extensionally and satisfy Open Data, while other, *restricted* choice sequences such as β have an intensional component. A special case of a restricted sequence is a *lawlike* sequence, whose terms are fully determined by a rule.

This distinction leads to a proliferation of notions of choice sequence and creates difficulties that have never been satisfactorily resolved. I shall summarise the various classes of choice sequence listed by Troelstra and Dummett.

(1) The simplest notion is that of a sequence given by a generating process subject to no constraints (other than that the terms be of a certain type, usually natural numbers); 'at any stage of the process, *all* the information regarding α consists of a finite initial segment' (Troelstra, 1985, §2.2). Troelstra (1983, p. 208) calls these *proto-lawless* sequences; they are believed to satisfy Equality, Open Data and $\forall\alpha\exists n$-continuity, but not Density.

(2) Proto-lawless sequences can be made to satisfy Density if we permit an initial segment of the sequence to be set 'by hand', thus ensuring that there is always a sequence starting with any given finite number of terms. These modified sequences are known as *lawless* sequences and are believed still to satisfy the Equality Principle (Kreisel, 1968). Troelstra

(1983, p. 212) points out that Open Data may fail if the formula $A(\alpha, \beta)$ involves the function Φ_I defined by the rule: for any lawless sequence γ, $\Phi_I(\gamma)$ is the initial segment of γ that is set 'by hand'. However Troelstra asserts that Open Data holds for all formulae $A(\alpha, \beta)$ not involving Φ_I.

(3) A notion of choice sequence adequate for analysis should include restricted, and even lawlike, sequences as well as lawless ones; ideally, it should be closed under continuous operations. A simple way of including restricted sequences is to consider as a choice sequence any sequence that is constrained by a spread law but is otherwise lawless; this includes lawless and lawlike sequences as special cases. However, this class of choice sequences is still not closed under continuous operations (Dummett, 1977, p. 426).

(4) Brouwer (1952) viewed a choice sequence as given by a sequence of choices of pairs (α_n, R_n), where α_n is the nth term of the sequence and R_n is a finite set of restrictions to be applied to subsequent choices $\alpha_{n+1}, \alpha_{n+2}, \ldots$ (where $R_m \subseteq R_n$ for $m < n$). The restrictions in R_n are usually taken to be expressible as a single spread law: that is, as n increases α is required to belong to successively narrower spreads. Unfortunately, as Troelstra (1977, p. 131) shows, this class of choice sequences is again not closed under continuous operations.

(5) Closure under continuous operations can be achieved if we define a choice sequence as any sequence obtained by applying a continuous operation to a tuple of lawless sequences. (A variation on this is to choose a single lawless sequence α and consider all sequences obtainable from α by applying continuous operations.) Unfortunately this class of choice sequences fails to satisfy $\forall\alpha\exists n$-continuity (Troelstra, 1977, p. 65).

(6) In an attempt to satisfy both closure under continuous operations and $\forall\alpha\exists n$-continuity, Troelstra (1977, Appendix C; 1983, §8) describes a modified and much more complicated notion, *GC sequences*, which are generated freely subject to a sequence of restrictions that involve defining the sequence in terms of subordinate choice sequences using continuous operations.

(7) Dummett (1977, pp. 435–451) finds Troelstra's account of GC sequences ambiguous and argues that the proofs of $\forall\alpha\exists n$-continuity and closure under continuous operations fail. Dummett describes an even more complicated version of this idea for which, he argues, Troelstra's proofs work.

Although notion (7) appears to satisfy the formal requirements for intuitionistic analysis, it is hard to accept it as a satisfactory explication of the informal notion of 'a sequence that can be continued ad infinitum'. As Goodman (1979a) remarks, Dummett 'finishes with a notion built up from lawless sequences and continuous functionals in a manner so complex that it is hard to believe that he intends this to be the fundamental notion on which analysis is to be founded.' Note that Troelstra (1977, pp. 77, 138, 158) expresses a similar caution about the not-quite-so-complicated GC sequences. The development of the concept of choice sequence through a series of ever more complicated formulations, each of which tries to patch up the surface defects of the previous, until the final version collapses under its own weight, is characteristic of a formal theory that has lost touch with its underlying informal ideas (cf axiomatic set theory).

Vesley (1979) and Gielen *et al.* (1981) express doubts about the wisdom of attempting to reduce Brouwer's concept of choice sequence to a supposedly more primitive concept of lawless sequence. I am inclined to agree that classifying the choice sequences into proto-lawless, lawless and non-lawless is a mistake. Consider the following examples.

Example 1. Consider a sequence α that is lawless except for the constraint that each term is 0 or 1; now define a 'mirror-image' sequence β by applying the following continuous operation to α:

$$\forall n \ \ \beta_n = 1 - \alpha_n.$$

The relationship between the sequences α and β is symmetric, so it seems arbitrary to say that α is a lawless sequence of 0s and 1s while β is not. The two sequences cannot *both* be lawless, because that would contradict the Open Data Principle (taking $A(\alpha, \beta)$ as $\forall n \ \alpha_n + \beta_n = 1$). As mentioned above, the lawless sequences are believed to satisfy Open Data, provided the formula $A(\alpha, \beta)$ does not involve Troelstra's Φ_I function.

Troelstra's (1977, pp. 16, 48) explanation, in a situation similar to this mirror-image example, is that we have a choice of which sequence to consider as lawless; 'we cannot refer to [α] and [β] as both being lawless within the same context'.

Dummett (1977, p. 421), however, denies this interpretation.

> But, again, it is not a matter of there being two choice sequences, either of which we may, if we will, take as being a lawless sequence, provided that we do not so take the other, but of its being impossible that we should know [the mirror-image relation] to hold of any two lawless sequences.

His point seems to be that, for any two mirror-image choice sequences, one must have been defined first and the other defined in terms of the first; thus the second cannot be lawless.

It is curious that Dummett and Troelstra disagree on this fundamental matter affecting the very nature of lawlessness and yet their disagreement has no consequences for the subsequent development of the theory. This suggests that, even at this early stage, the formal theory is not closely tied to the informal ideas.

The mirror-image example throws serious doubt on the notion of a lawless sequence (and the notion of a proto-lawless sequence, to which the same considerations apply). Suppose we have generated α by repeatedly tossing a coin and counting Heads as 0 and Tails as 1. Suppose another person observes this process and also uses it to generate a choice sequence, but they interpret Heads as 1 and Tails as 0. A third person agrees with us that Heads is 0 and Tails is 1, but thinks that the proper way to read a coin is to crawl underneath the table (which is made of glass) and look at the *underside* of the coin. Clearly the second and third persons will regard β as 'the proto-lawless sequence generated by the coin-tossing' and α as its mirror image. It seems to me that there is no objective matter of fact about which of α and β is directly obtained from the coin-tossing and which is defined in terms of the other. It follows that α and β must be treated symmetrically. The same considerations apply to any other sequence-generating process: there are always many equally good ways of interpreting the process as the generation of a sequence of numbers, for we have a choice of which aspects of the process to regard as relevant and how to encode those aspects as a number. If β is obtained from α by a continuous operation then any process that may be construed as generating α may be re-construed as generating β instead. Perhaps we should take up Troelstra's suggestion (see the quotation above) that lawlessness depends on the context; this means, I take it, that a sequence may be described as lawless provided we are not at the same time considering other sequences with respect to which it is constrained. But this is tantamount to abandoning the notion of lawlessness altogether.

Example 2. This example also concerns the application of Open Data to lawless sequences. Let α be any lawless sequence and define β by prefixing the sequence α by 0. Consequently we have

$$\beta_0 = 0 \qquad \forall n \; \beta_{n+1} = \alpha_n.$$

Now, α consists of finitely many pre-specified terms followed by infinitely many free choices; therefore so does β (it just has one more pre-specified term). Thus β is also lawless. But this again contradicts Open Data (taking $A(\alpha, \beta)$ as $\forall n \; \beta_{n+1} = \alpha_n$).

Example 3. The lawless sequences are generally believed to satisfy the Equality Principle, $\alpha = \beta \iff \alpha \equiv \beta$ (Kreisel, 1968, §2; Troelstra, 1977, §2.5; Dummett, 1977, p. 420). Troelstra's (1983, p. 208) justification reads, in part, 'if $\alpha \not\equiv \beta$, then at no stage will the available data (always involving only finitely many values of α and β) ever permit us to assert that α and β will forever coincide, hence $\alpha \neq \beta$'. (Cf also Troelstra (1985, p. 221): 'the *only* way in which we can know that α, β have the same values for all arguments is to know that α, β are a priori given to us as the *same* source (process).') Now, let m be any natural number and let α be any proto-lawless sequence.

Define a lawless sequence β whose first term β_0 is set 'by hand' to m and whose subsequent terms are chosen by the same free generating process as α's terms. Thus we have

$$\beta_0 = m \qquad \forall n > 0\ \beta_n = \alpha_n. \tag{$*$}$$

Now, we have $\alpha \not\equiv \beta$, since α and β are given to us in different ways. In case this is not quite convincing, consider again Troelstra's Φ_I function, defined above; $\Phi_I(\alpha)$ is the empty sequence, while $\Phi_I(\beta)$ is a one-term sequence. Thus $\alpha \not\equiv \beta$.

Hence, assuming the Equality Principle, $\alpha \neq \beta$. But from $(*)$ we have $\alpha = \beta \iff \forall n\, \alpha_n = \beta_n \iff (\alpha_0 = \beta_0 \wedge \forall n > 0\, \alpha_n = \beta_n) \iff \alpha_0 = m$. Therefore, $\alpha_0 \neq m$.

Since this holds for arbitrary m and α we have $\forall m \forall \alpha\ \alpha_0 \neq m$ (where α ranges over the proto-lawless sequences), and hence by intuitionistic logic $\neg \exists \alpha\, \exists m\ \alpha_0 = m$ – in words, no proto-lawless sequence has a first term!

Example 4. Let α be a proto-lawless sequence and let β be a sequence given by the same generating process as α, but in the case of β we discount the first ten terms as 'practice runs' and regard the sequence proper as starting at α_{10}. Consequently, $\forall n\ \beta_n = \alpha_{n+10}$. Clearly β is related to α by a continuous operation but $\beta \not\equiv \alpha$. Now, is β proto-lawless? It surely must be, since the process generating the terms of β is just as free as the process generating the terms of α (it is, after all, the same process). However, admitting both α and β as proto-lawless is clearly contrary to the principle of Open Data, which is conventionally assumed to hold for proto-lawless sequences (Troelstra, 1983, p. 211). Now, Open Data is the most fundamental of all continuity principles, and the proto-lawless sequences are the most extensional of all types of choice sequence; if Open Data fails even for proto-lawless sequences then the prospects for intuitionistic analysis are bleak. The most that could be salvaged would be a weak form of Open Data

$$\forall \alpha\, (A(\alpha) \supset \exists n\, \forall \gamma \cong_n \alpha\, A(\gamma))$$

where $A(\alpha)$ has no free choice sequence variables other than α.

It is for this reason that Troelstra (1983, p. 215) says (in connection with a different example) 'we cannot select a subsequence [of a lawless sequence] and call it lawless as well'. And yet if we consult Troelstra's own criterion for proto-lawlessness in the same paper, 'at any moment only an initial segment is known, and no restrictions are imposed on future choices' (p. 208), it is hard to see how it can apply to α and not to β. The choice of β_{257} is the same act as the choice of α_{267}, but it would be absurd to regard this as a *restriction* on the choice of β_{257} (unless it is a restriction for a thing to be the same as itself). Note that β is *not* defined in terms of α: rather, β and α are read from the same generating process in different ways.

What all four examples show is that the notions of proto-lawless sequence and lawless sequence are trickier than they look. For example it is not clear whether a lawless sequence is a proto-lawless sequence *prefixed* by finitely many terms (as assumed in Example 2) or whether it is a proto-lawless sequence with finitely many terms at the beginning *overwritten* by new values (as assumed in Example 3): the two conceptions are not necessarily equivalent in view of Example 4. The conventional explanations of these concepts do not adequately justify the principles that are attributed to them and the theory that is based on them. The difficulty is in doing justice simultaneously to the intensional and the extensional aspects of choice sequences.

GETTING RID OF TIME: BLACK BOXES

I have a more fundamental misgiving about choice sequences as conventionally depicted, concerning their temporal aspect. Brouwer (1955) says that time enters into mathematics in two ways: first, propositions acquire truth values as we discover proofs of them; secondly, mathematical objects such as choice sequences may actually *grow* in time.

> in intuitionistic mathematics a mathematical entity is not necessarily predeterminate, and may, in its state of free growth, at some time acquire a property which it did not possess before. (Brouwer, 1955)

I am extremely uncomfortable with this intrusion of time into mathematics. How, for example, is the time-scale of the choice sequence related to the time-scale of the intuitionist analyst who reasons about it? Is it possible that $\alpha = \alpha$ may be false because α has grown in the interval between writing the left-hand side of the equation and writing the right-hand side? Posy (1974) argues that Brouwer's introduction of choice sequences required a fundamental alteration in his ontology of mathematical constructions; he further proposes (Posy, 1976, 1977) that choice sequences are best interpreted in epistemic terms, using the theory of the creative subject, thus in effect linking Brouwer's two sources of time variation.

I have already argued in Chapter 6 that the first kind of time variation (propositions' acquiring truth values as we prove them) is unnecessary and misconceived. I shall now argue the same for the other kind.

I propose to replace the picture of a choice sequence as a process generating values one after another with the picture of a *black box*, that is, a device that generates a value when supplied with a natural number argument. The output of the black box is determined by its input: presenting the same input on different occasions always produces the same output regardless of any other values previously input. When reasoning about black boxes we consider only the output produced for particular inputs, taking no account of what is going on inside the box.

Black boxes are equivalent to choice sequences, but have the advantage of avoiding the temporal element. By 'equivalent' I mean that any black box can be simulated by a choice sequence and vice versa. To see this, consider any choice sequence α, generating values $\alpha_0, \alpha_1, \alpha_2, \ldots$ one after the other. This can be simulated by a black box that outputs α_n when the input is n: if we feed the inputs $0, 1, 2, \ldots$ in succession to such a black box then it will produce the outputs $\alpha_0, \alpha_1, \alpha_2, \ldots$, thus simulating the choice sequence. Conversely, consider any black box; let α_n be its output when the input is n. This black box can be simulated by the choice sequence $\alpha_0, \alpha_1, \alpha_2, \ldots$, in the following way. When an input of n is provided we generate the first $n + 1$ terms $\alpha_0, \ldots \alpha_n$ of the choice sequence (unless we have already done so, in which case we simply remember the terms generated) and output α_n. Thus the choice sequence, used in this way, behaves like the black box.

This argument shows that choice sequences and black boxes are equally useful as a foundation for intuitionistic analysis. A disadvantage of taking black boxes as the basic notion is that we need to make the assumption that the behaviour of the box is deterministic and history-independent: that is, it will always produce the same output for a certain input. This may seem a slightly artificial assumption given that we are pretending not to know anything about what is going on inside the box. No such assumption is required for choice sequences. However, the overriding advantage of black boxes over choice sequences is that a black box is not in a 'state of free growth' and cannot 'acquire a property which it did not possess before'. Thus the intrusion of time is revealed to be inessential. I shall accordingly take black boxes rather than choice sequences to be the fundamental notion from now on. I shall extend the idea slightly by allowing as input any construction rather than merely natural numbers, and I shall assume that the output is a construction. The continuity issues discussed in the previous section arise for black boxes just as they do for choice sequences, but in view of the difficulties described I shall not adopt any continuity, equality or choice principles.

Since a construction is essentially a program, the Theory of Constructions may be viewed as a theory of digital computation. Adding black boxes turns it into a theory of hybrid computation, that is, a theory of a digital computer embedded in a system of devices of various sorts (black boxes) that pass constructions to and from the digital computer. The digital computer is plugged into this system at several *sockets* by means of which it can communicate with the black boxes. The digital computer's programming language contains two new primitive commands: (i) connect a new black box to a certain socket; (ii) send a construction to a certain socket and receive a construction back. Once a box has been connected it remains plugged in for the rest of the computation. Every socket has a unique name, say 's_1', 's_2', 's_3', ..., but black boxes are anonymous; in both of the above commands the computer names the socket in question, but in the first command it has no control over which box will be connected to the socket. This introduces an element of indeterminism: if the same computation is executed repeatedly, different boxes may be chosen and different results may be obtained.

The notion of sockets is useful in explaining how we distinguish one black box from another. It is usually assumed in choice sequence theory that we have two kinds of knowledge about a (proto-lawless) choice sequence: *extensional* knowledge (finitely many terms of the sequence) and the *identity* of the sequence. The latter is necessary if we are to talk about more than one choice sequence at once without confusing them, and yet it may seem mysterious how we can have just this isolated piece of intensional information about a choice sequence. The notion of sockets makes this clearer. The digital computer is aware of all its sockets but has no knowledge of which black boxes are plugged into them. Sockets are finite, completely known objects with unique names. The digital computer can handle sockets at will in computations (their names could be constants in the programming language); in particular it can compare one socket with another.

PUTTING BLACK BOXES TO WORK

On the basis of the above explanation it would be possible to introduce symbols for black box sockets into protologic. If s_1 is a socket name then s_1X would be a term that is evaluated by evaluating X, sending the resulting construction to the socket s_1, and receiving a construction back.

However, this is not quite what is needed for the interpretation of analysis. I shall use black boxes in the informal part of the interpretation, as tree codings. Recall from Chapter 10 that a tree coding is defined by specifying which finite sequences of constructions $X_1, \ldots X_k$ count as branches of a tree. A black box may be used for this purpose. If we supply the list $[X_1, \ldots X_k]$ (which is a single construction) to a black box, we can say that the sequence $X_1, \ldots X_k$ is a branch iff the black box returns the construction *true*. On second thoughts, if $X_1, \ldots X_k$ is a branch then any initial segment $X_1, \ldots X_i$ (for $i < k$) must also be a branch, so we need to modify our procedure to incorporate this constraint. Let us feed the sequence of lists $[X_1], [X_1, X_2], [X_1, X_2, X_3], \ldots [X_1, \ldots X_k]$ one at a time to the black box, and regard $X_1, \ldots X_k$ as a branch iff *true* is returned in response to each list. In this way any black box defines a tree coding, which may be represented formally by the corresponding socket name. We do not of course know precisely which sequences of constructions the box will accept as branches, but we can still carry out well-foundedness arguments involving this tree coding that work *regardless* of the input-output behaviour of the black box (or, as I shall usually say, regardless of the input-output behaviour of the socket). That is, we can say things like 'for any input-output behaviour of the socket s_1, T is a well-founded tree according to the coding Π', where Π may involve the tree coding expressed by s_1. From such a statement we may infer that T is a well-founded tree according to Π', where Π' is the tree coding obtained from Π by replacing s_1 by a tree coding Σ; the justification for this inference is that we may imagine that plugged into the socket s_1 is a black box that simulates Σ.

This amounts to a sort of universal quantification over tree codings. It is analogous to the 'for any construction' quantifier used in protologic (see Chapter 8). It does not assume that we have any grasp of the 'universe of all tree codings', any more than the 'for any construction' quantifier requires a grasp of the 'universe of all constructions'. It simply requires that we grasp what it means for a given device to serve as a black box, that is, to receive and return constructions in a deterministic and history-independent way, and that we can form general well-foundedness arguments concerning the inputs and outputs of such a device that do not involve consideration of its inner workings. It seems clear that we can do this.

This brings out a difference between the 'for any input-output behaviour of a socket' and the 'for any construction' quantifiers. The latter allows one to dissect the given construction and consider the various possibilities for its components. The former does not: we are speaking only about the *behaviour* of the socket. Also, the former as yet only applies to statements of well-foundedness.

This new 'for any input-output behaviour of a socket' quantifier is powerful enough to serve as the foundation stone for an interpretation of analysis, as Part IV shows. It is usual for constructivists to object to analysis on account of the impredicativity of the 'for all real numbers' or 'for all subsets of **N**' quantifier, but I believe the above argument shows that this objection is misconceived. Suppose we produce a general argument of the form 'for any input-output behaviour of s_1, ...', and then we somehow use this to define a tree coding Σ. We may next imagine that plugged into s_1 is a black box simulating the tree coding Σ, and thus, as pointed out above, we may instantiate the general argument by removing the 'for any input-output behaviour of s_1' prefix and replacing 's_1' by 'Σ' throughout. The usual predicativist's objection to this is that to understand 'for any behaviour of s_1' we must first grasp the range of the variable 's_1', and this range cannot include Σ because Σ is only defined *after* grasping 'for any behaviour of s_1'. The objection is fallacious since it harks back to the view of a universal quantifier as an infinite conjunction, according to which a grasp of the quantifier depends on a grasp of every possible conjunct. I have already argued against this view of quantifiers in Chapter 3. Even apart from predicativity concerns, if a grasp of 'for all natural numbers' depended on a grasp of each natural number individually then it would be forever out of our reach, since we can only grasp finitely many things at a time.

My intensional 'for any construction' and 'for any behaviour of a black box' quantifiers do not presuppose a grasp of *any* construction or black box, let alone all of them. (As Frege says, 'It is surely clear that when anyone uses the sentence "all men are mortal" he does not want to assert something about some Chief Akpanya, of whom perhaps he has never heard' (Geach & Black, 1970, p. 83).) The predicativist objection therefore represents a kind of half-hearted constructivism, in which some constructive scruples are adopted but quantifiers are still viewed extensionally.

So much for the universal black box quantifier; what about the corresponding existential quantifier? This is much harder to justify. It may seem that 'for some input-output behaviour of a socket s_1, ...' can be asserted if we can specify precisely how the black box plugged into s_1 is to behave to make '...' true. But it is not clear what is required *in general* for such an existentially quantified argument to be sound. (By contrast, a universally quantified argument 'for any behaviour of s_1, ...' is really an *argument schema*, which works regardless of the meaning of s_1.) Hence I shall take a cautious view and not introduce such an existential quantifier. This means that in the interpretation of analysis in Part IV we shall lack a second-order existential quantifier. However, this defect can be remedied in a purely formal way by the device of defining the second-order existential quantifier, $\exists^2$, as $\neg\forall^2\neg$, just like the interpretation of the existential quantifier of Peano Arithmetic in Heyting Arithmetic (see Chapters 33 and 47). This gives a theory whose logic is formally classical, that is, an interpretation of Second-Order Peano Arithmetic.

CONCLUSION

I propose to interpret analysis (or rather second-order number theory) in terms of the notion of a black box, which is an atemporal variant of the usual intuitionistic notion of a choice sequence. Black boxes will be used to reason about 'arbitrary' tree codings, leading ultimately to a second-order universal quantifier and hence to Second-Order Peano Arithmetic. This programme is carried out in Part IV.

I make no use of the various continuity and choice principles that other intuitionists use in analysis. This is not because I reject them but because, as explained above, it is not yet clear how to incorporate them without damaging the integrity of the choice sequence or black box idea. In any case, they are not required for my interpretation of Second-Order Peano Arithmetic. (They probably would be involved in a third-order version, involving quantification over real functions.) One should not however jump to the conclusion that my interpretation of analysis is 'not intuitionistic'. What makes an interpretation of mathematics intuitionistic or platonistic is not any detail of the formal systems it deals with (such as the presence or absence of the axiom $A \vee \neg A$) but rather the philosophical starting point.

PART II: THE THEORY OF CONSTRUCTIONS

CHAPTER 12

INTRODUCTION TO PART II

Following the informal principles in Part I (particularly Chapters 5 & 8), we are now in a position to set up a formal theory of constructions. This consists of two main parts: a *Term Language* for expressing constructions, and a *Protologic* expressing the relations between constructions.

The theory of constructions will subsequently be used to interpret arithmetic and analysis.

MODULAR ORGANISATION

I shall organise the mathematical chapters in Parts II, III and IV as a sequence of *theories* and *intermediate chapters.* Each theory is a chapter consisting of a list of definitions and theorems; a theorem is either a derived expression in a formal system or a metamathematical assertion about a formal system. Between each theory and the next is an intermediate chapter in which the concepts of the next theory are defined in terms of the concepts of the previous theory (and those of earlier theories), and the theorems of the next theory are proved using those of the previous theory (and earlier theories). An intermediate chapter may in addition contain *local* definitions and theorems, used only within that chapter.

Thus, each theory is the interface between the intermediate chapter that precedes it and everything that follows it. Each theory summarises the preceding intermediate chapter, omitting local definitions and theorems, many details of definitions, and all proofs.

There are four reasons for adopting this modular organisation.

- It makes explicit the logical dependencies between the various definitions and theorems that make up the theories.
- The theories may be of intrinsic mathematical or philosophical interest, independently of their use here as stepping-stones on the way to an interpretation of arithmetic and analysis. Each theory is a self-contained module giving a description of constructive mathematics at a particular level of detail and abstraction.

- It makes the arguments easier to follow by controlling cross-references. When reading a proof in an intermediate chapter one only needs to keep in mind the definitions and theorems stated in the previous theory (with occasional references back to earlier theories) plus the ones introduced so far in the present chapter, not everything in all the chapters so far.
- It allows the reader to skip chapters and resume the story at the next theory.

SUMMARY OF THE FIVE THEORIES IN PART II

- The *Term Language* (T) is a simple notation for expressing constructions: it consists of some constants representing primitive constructions, variables, and function applications. The semantics of terms is given by a *reduction* relation.
- The *Expanded Term Language* (ET) consists of the Term Language plus generalised λ-abstractions and instantiation notation. A *compilation* relation, $\mapsto$, defines a mapping from ET into T. Some metanotation is introduced, including a powerful mechanism for defining functions by pattern matching, and is characterised by metamathematical theorems.
- *Protologic* (P) is a sequent calculus for deriving intensional relationships between computations.
- *Expanded Protologic* (EP) consists of Protologic plus some higher-level derived sequents and rules of inference.
- *The Coding of Trees* (CT) deals with the undecidable component of the theory of constructions, the well-foundedness of trees (see Chapter 10). This theory is concerned with setting up coding schemes for various classes of tree, most importantly protological derivation trees and their associated reflection trees. This requires that protological derivations be coded as constructions; a function DT is defined such that $DT(D, X)$ reduces to *true* iff D is the code of a protological derivation for the sequent X.

Between T and ET is an intermediate chapter interpreting all the ET constructs and properties in T. Similarly, there are intermediate chapters between P and EP, and between EP and CT. There is, however, no need for an intermediate chapter between ET and P. In addition, there are occasional chapters of commentary to explain how the theories implement the philosophical requirements of Part I. The final chapter of Part II discusses the use of ET as a functional programming language.

CHAPTER 13

DESIGN OF THE TERM LANGUAGE

The first of the theories listed above is the *Term Language*: this is a simple notation for expressing constructions. Since any construction may be regarded as an algorithm, the term language is a programming language. Any programming language would do, but to allow easy mathematical manipulation I use a pure functional language.

Unfortunately, existing functional languages are unsuitable as a basis for intuitionistic mathematics because they are inspired by various versions of the λ-calculus.

WHAT'S WRONG WITH λ-CALCULUS?

I quarrel with two features of λ-calculus: extensionality and normal order reduction. The λ-calculus takes an extensional view of functions: f is regarded as the same function as g if fx and gx have the same value for all x (this is shown by the fact that terms inside λ-abstractions can be reduced). Since this applies recursively, f is also equal to g if $fxy = gxy$ for all x, y, and so on. The result is a system of functions in which:

- the notion of equality is obscure, since equality of functions depends on equality of values, where the values are objects of the same kind as the functions (and any obscurity about when 'two' functions are equal is an obscurity about what a function is);
- it is hard to 'survey' the whole universe of functions;
- the existence of a normal form, and hence the meaning, of a term depends on what reduction orders are allowed;
- it is unclear what, if anything, a term with no normal form means;
- there is no intended model of the system (various models are known (Barendregt, 1984, Chapter 18), but none can be plausibly construed as a formulation of what Schönfinkel, Curry and Church had in mind when they founded combinatory logic and λ-calculus).

It is generally accepted (by friend and foe alike) that intuitionism requires an intensional notion of function (Dummett, 1977, §1.2; Tait, 1983).

My second objection is to normal order reduction, that is, the policy of always reducing the leftmost subterm first (for example, reducing $(\lambda x.y)A$ to

y before reducing A). The case for normal order is that if *any* reduction order reduces a term to an irreducible term then so does normal order reduction. Hence, although many reduction orders are permitted in λ-calculus, normal order is generally considered the most important. The case against it is that it loses contact with the origin of λ-calculus as a general theory of function application. Consider for example the functions *plus*, *pred*, *sub*, *mult* and *div* defined on natural numbers by

$$\begin{aligned} plus(0,n) &= n \\ plus(Sm,n) &= S(plus(m,n)) \\ pred(0) &= 0 \\ pred(Sn) &= n \\ mult(0,n) &= 0 \\ mult(Sm,n) &= plus(mult(m,n),n) \end{aligned} \qquad \begin{aligned} sub(m,0) &= m \\ sub(m,Sn) &= pred(sub(m,n)) \\ div(0,n) &= 0 \\ div(Sm,n) &= S(div(sub(Sm,n),n)) \end{aligned}$$

where S is the successor function. (*div* is integer division, rounding upwards.) Then what is the value of $mult(0, div(5,0))$? A supporter of normal order reduction would argue, 'If you reduced $div(5,0)$ first you would get an infinite loop, so it is better to apply the *mult* rule first, giving an answer 0; this is the only value you could get by any reduction route, so clearly we should accept it as the correct value'. Any other mathematician, however, would say, 'The *mult* rule cannot be applied until we know that the argument $div(5,0)$ succeeds in denoting a number: that is, we must reduce $div(5,0)$ first, and since this gives no result we must conclude that the entire expression has no value'. The latter view has been adopted as standard after centuries of experience with mathematical notation, as it leads to a clear and simple semantics; the former view seems to me an attempt to frustrate understanding. None of the usual reduction strategies in λ-calculus quite corresponds to the standard mathematical notion of function evaluation. If reduction is *not* intended to represent function application, as mathematicians in general understand it, then what is it for?

The literature on λ-calculus and functional programming does not seem to address the question of intended meaning. When researchers in this field speak of 'semantics' they mean a mapping from functional programs into λ-terms (Revesz, 1988, §5.4; Hannan, 1993), which in turn are mapped into some non-constructive system (Meyer, 1982), or the functional program may be mapped into a non-constructive system directly (Cardelli & Longo, 1991); from the point of view of explicating meaning this would seem to be a step backwards, since the functional program, as an algorithm, is philosophically much better understood than the non-constructive object it maps to. A further step back is typically taken in category-theoretic semantics where one abandons the idea that λ-terms denote *functions* acting on *sets of objects*; as

Hyland (1991) points out, a purely algebraic characterisation of 'application' and 'λ-abstraction' severs all contact with the original informal notion of λ-calculus as 'a theory of a countable world of intensional functions'. Brookes & Geva (1992) use category-theoretic notions (comonads and the associated Kleisli category) to model lazy, demand-driven evaluation, in which partial information about a function's value is determined by partial information about the argument. Their semantics is intensional in that it represents the sequence of steps by which the argument is evaluated. However, this is not the sort of intensionality one needs for intuitionistic mathematics, in which the argument of a function is a construction (not a term that needs evaluating) and the aim is to model the steps in the application of the function to the argument. Other approaches to 'non-extensional semantics' (Martini, 1992) do not seem to address the question of what non-extensional information is required.

For these reasons I cannot sympathise with Barendregt's (1984, p. vii) statement that 'Constructions in the lambda calculus give the right intuition for constructions in, for example, the semantics of programming languages'. There seems to be no clear semantics for λ-calculus as a language of algorithms acting on other algorithms; λ-calculus therefore cannot be used as it stands as a foundation for intuitionistic mathematics.

The 'λ-calculus–like' principles I wish to retain are as follows.

- The objects of interest (call them *constructions*) are functions from constructions to constructions.
- A *term* is a specification of a sequence of function applications involving constructions; the steps in carrying out the function applications are called *reductions*.
- In addition, terms may contain *variables*, interpreted as representing arbitrary constructions.
- Consequently, the following are terms: a constant (denoting a basic construction), a variable, and a function application, AB, where A and B are terms; terms of this form will be called *simple* terms.
- For every term A and variable x there is a term $(\lambda x.A)$ denoting a function that maps x to A (in some sense).

AN INTENSIONAL AND COMPOSITIONAL REPLACEMENT FOR λ-CALCULUS

Contrary to λ-calculus, I require the following three features.

- Intensional functions: $(\lambda x.A)$ is not equal to $(\lambda x.B)$ unless A is the same term as B. $(\lambda x.A)$ cannot be reduced.

- Compositional semantics: the value of a term is obtained from (and only from) the values of its subterms; thus if a subterm has no value then the whole term has no value.
- Variables stand for arbitrary constructions: a true statement such as 'A reduces to B' or 'A cannot be reduced', involving terms with variables, should continue to hold when the variables are replaced by arbitrary constructions. (This condition is essential if variables are really to stand for arbitrary constructions, yet it is not satisfied by any of the common varieties of λ-calculus.)

The intended semantics of terms is roughly as follows (thinking only of terms without variables for the moment).

- The *reference* or *value* of a term is the term obtained by reducing the original term repeatedly until it cannot be reduced any more. If reduction never halts then the term has no value. The *sense* of a term is the reduction process starting at that term. Every term has a sense. The reference and sense of a term together constitute its meaning. (This terminology is modelled on Frege's (1892).)

(This semantics will be refined in the next chapter.) These stipulations have the following consequences.

- Terms are reduced from the inside out. In reducing AB we must reduce A and B completely and then apply the value of A to the value of B. It doesn't matter in which order we reduce A and B as the two reductions are independent (hence this is not quite applicative order reduction). Thus, the term AB obtains its value only via the values of A and B. (Here I follow Goodman (1972), although Goodman doesn't explicitly define a reduction relation; his semantics is given by a valuation function.)
- If two terms reduce to the same irreducible term then they have different senses but the same reference (like the phrases 'the morning star' and 'the evening star', which both refer to Venus), while if a term has no irreducible form then it has a sense but no reference (like the phrase 'the present king of France').
- The conversion rules of λ-calculus fail. Let $A\binom{X}{x}$ mean the result of replacing each free occurrence of the variable x in the term A by the term X, assuming no variable clashes occur. Then $(\lambda x.A)X$ and $A\binom{X}{x}$ need not have the same value: for example, $(\lambda x.(\lambda f.fx))(plus(2,2))$ has the value $(\lambda f.f4)$, which is different from $(\lambda f.f(plus(2,2)))$, since $(\lambda f.f(plus(2,2)))$ is a function that, when applied to f, adds 2 and 2, whereas $(\lambda f.f4)$ does no addition but simply uses the result, 4. (However, $(\lambda x.A)x$ will reduce to A, thus ensuring that $(\lambda x.A)$ is in some sense a function mapping x to A.) Furthermore, $(\lambda x.Ax)$ does not have the same value as A – the latter may have no value, while the former is irreducible

and is its own value. Whether $(\lambda x.A)$ has the same value as $(\lambda y.A(^y_x))$ is somewhat arbitrary: we are free to set up the system so as to distinguish them or not, as we choose.

- There is a conflict between the above requirements: clearly, the value of $(\lambda x.A)$ depends on the sense of A, not merely its value. The conflict is resolved by distinguishing two languages:
  - the *Term Language*, consisting only of simple terms (constants, variables and function applications);
  - the *Expanded Term Language*, containing simple terms and additional constructs such as $(\lambda x.A)$.

  Terms in the Expanded Term Language may be compiled into simple terms in the usual way using *s*, *id* and *k* combinators. The requirement that the value of a term depend only on the values of its subterms applies only to simple terms. The use of two languages combines the advantages of both languages: the $(\lambda x.A)$ notation is much more readable than combinators, while in proofs by structural induction on terms one need only consider simple terms.
- There are two notions of substitution. *Textual substitution*, $A(^X_x)$ (mentioned above), is a cumbersome concept due to the problem of variable clashes and the distinction between free and bound occurrences of variables. Also it has no simple semantic significance, due to the failure of β-conversion. More useful is *semantic substitution* or *instantiation*, $A\begin{bmatrix}X\\x\end{bmatrix}$, in which A is compiled into a simple term before x is replaced by X; since there are no bound variables in simple terms this circumvents the problems of textual substitution. Another way of reading $A\begin{bmatrix}X\\x\end{bmatrix}$ is as 'A, evaluated in an environment in which x has the same value as X'. Some people express this as '**let** $x = X$ **in** A' or 'A **where** $x = X$'.

The Expanded Term Language consists of the Term Language plus the $A\begin{bmatrix}X\\x\end{bmatrix}$ instantiation construct and a generalisation of the $(\lambda x.A)$ construct. Compiling expanded terms into simple terms amounts to replacing λ-abstractions by combinators and then carrying out the instantiations.

In addition to these constructs there is a variety of metanotation necessary to make the language usable, including a mechanism for defining functions by pattern matching. Each new notational device could be treated as an extension to the language, with a compilation operation mapping terms containing the new notation into terms not containing it, just as was done with the $(\lambda x.A)$ and $A\begin{bmatrix}X\\x\end{bmatrix}$ constructs. However, it would be profligate to do this every time, and in any case the compilation operation is usually straightforward and obvious, so in such cases I shall treat the notation as metanotation rather than as an official part of the Expanded Term Language.

SYNTACTIC OBJECTS

As argued in Chapters 4 and 5, mathematical constructions are abstract recursive structures built out of a given supply of atoms. From now on I shall call the atoms *characters*, I shall specify a fixed set of atoms called an *alphabet*, and I shall assume that the atoms are combined into sequences (rather than more complicated structures such as binary trees). Thus the constructions may be thought of as character strings. The advantage of thinking of them in this way is that syntactic techniques such as formal grammars, formal translation schemes, and metanotation may be applied conveniently to them. Thus all my 'objects' are expressions in various languages; if they denote something they denote other expressions, or possibly themselves, not some other non-syntactic kind of entity.

Here, of course, I am relying heavily on my main conclusion from Chapter 5, that there is no valid distinction between 'mathematical' recursive structures and 'linguistic' recursive structures, and hence that a syntactic perspective may be applied to mathematical constructions without any change of subject matter or loss of rigour.

USE-MENTION CONVENTIONS

One of the peculiarities of formal logic is that having defined a formal language to work with one hardly ever writes down a term in the language. Usually the expressions one actually writes down differ from terms in two respects:

- they contain some metanotation;
- they contain metavariables referring to arbitrary variables or terms, and metaconstants referring to particular terms.

For example, one may write '$(\lambda x.(A,B) = nil)$'. The pair construct here, '(A,B)', is metanotation, so one really means '$(\lambda x.sAB = nil)$' (using s as the pairing operation). The '=' notation is also metanotation and so needs rewriting. Further, the letter 'x' is not a variable but is a metavariable denoting a variable, while the letters 'A' and 'B' denote terms. The term one really means is the one obtained by replacing 'x', 'A' and 'B' by what they denote. All this can be stated systematically as follows.

An *identifier* is a sequence of one or more identifier characters, possibly with more such characters attached as subscripts or superscripts, where an *identifier character* is a roman or greek letter (upper or lower case, possibly underlined), a digit, a prime, or one of the special characters '$\mathcal{E}$', '$*$', '$\wedge$', '$\exists$', '$\Box$', 'Δ', '$\bullet$', '$\forall$', '$\bigwedge$', '$>$' or '$\supset$'. An identifier may be used in the following four ways.

- As a constant. There are eleven constants, which are atoms of the Term Language.
- As a metavariable denoting a variable. In this case the identifier will begin with a lower-case letter. Examples: 'x', 'y_1', '*first*', '*rest*', 'π', 'x''.
- As a metavariable denoting a term. In this case the identifier will begin with an upper-case letter (unless otherwise specified). Examples: 'A', 'P', '$Tree_{11}$'.
- As a metaconstant, introduced to denote a particular term. Examples: '*DT*', '$\wedge_w$', '$proj_i^k$', '$\forall$', '$\Box_d$'.

A *metaterm* of the Expanded Term Language is an expression that is like a term except that it may contain metavariables, metaconstants and metanotation. A metaterm *denotes* the term obtained by replacing all metavariables and metaconstants by what they denote and rewriting all metanotation. (The term is said to be an *instance* of the metaterm.)

For example, consider applying this procedure to the metaterm '$(\lambda x.(A, B) = nil)$'.

1. Assume 'x' denotes the variable '$\mathcal{X}$', 'A' denotes the term '$((\mathcal{X})(\mathcal{Y}))$', and '$B$' denotes the term '$(kk)$'; then replacing gives '$(\lambda \mathcal{X}.(((\mathcal{X})(\mathcal{Y})), (kk)) = nil)$'.
2. The pair metanotation '$(((\mathcal{X})(\mathcal{Y})), (kk))$' is rewritten as '$s((\mathcal{X})(\mathcal{Y}))(kk)$'. The equality metanotation '$s((\mathcal{X})(\mathcal{Y}))(kk) = nil$' is rewritten as '$equal\ (s((\mathcal{X})(\mathcal{Y}))(kk))\ nil$'. Brackets, usually omitted for readability, are restored, giving finally the term '$(\lambda(\mathcal{X}).((equal((s((\mathcal{X})(\mathcal{Y})))(kk)))nil))$'.

Note that the λ-abstraction notation does not need rewriting since it is an official part of the Expanded Term Language rather than a metanotation. Thus, in this example,

$$\text{`}(\lambda x.(A, B) = nil)\text{'} \quad \text{denotes} \quad \text{`}(\lambda(\mathcal{X}).((equal((s((\mathcal{X})(\mathcal{Y})))(kk)))nil))\text{'}$$

or, to state it more simply,

$$(\lambda x.(A, B) = nil) \quad \text{is} \quad \text{`}(\lambda(\mathcal{X}).((equal((s((\mathcal{X})(\mathcal{Y})))(kk)))nil))\text{'}.$$

This shows that terms may be denoted by metaterms without using inverted commas or Quinean quasi-quotation. Note that, under these conventions, every term is a metaterm denoting itself. (This *self-denotation* does not interfere with the *reference* and *sense* of the term, introduced above. I shall reserve the words 'denote' and 'denotation' for the relation between metanotation and formal notation; it is entirely unrelated to the semantics of the formal notation itself.)

These use-mention conventions will be applied to all the term languages, sequent languages and formula languages used in this book; they are, I think, in accord with common mathematical practice but they are rarely spelt out, even in works of mathematical logic. (For some alternative ways of dealing with use and mention see Frege (1893), Carnap (1937) and Quine (1951).)

CHAPTER 14

THE TERM LANGUAGE

The Term Language is a simple functional programming language for expressing constructions. A construction on its own cannot get up to any mischief, but if you put two of them together a *computation* generally ensues; this is expressed by a reduction relation, $\triangleright$.

A *term* represents a stage in the execution of a computation: thus if a construction F is applied as a function to a construction X to give as value a construction Y this computation is expressed as a sequence of reductions $FX \triangleright T_1 \triangleright T_2 \triangleright \cdots \triangleright Y$. Terms may also contain *variables*, which stand for unspecified constructions. Thus a term represents an incompletely specified stage in a computation, and a construction is simply a term that has no variables and cannot be further reduced. A reduction sequence in which the terms contain variables represents a 'schematic computation' – a computation that works regardless of what constructions are substituted for the variables.

My list of constants is similar to that of Goodman's (1972) combinatory logic (except for *fxpt*); my reduction relation defines the evaluation process that I think is intended in his semantics, or something very much like it.

SYNTAX OF THE TERM LANGUAGE

- The *constants* are '*s*', '*id*', '*k*', '*equal*', '*if*', '*true*', '*false*', '*former*', '*latter*', '*fxpt*' and '*nil*'.
- A *variable* is a finite sequence of one or more characters from the alphabet '$\mathcal{A}$' – '$\mathcal{Z}$'. The variables may be enumerated in an infinite sequence in order of increasing length, variables of the same length being ordered lexicographically; this ordering will be called *standard order*.
- An *atom* is a constant or a variable.

DEFINITION. A *term* is a sentence in the following language. (I shall make use of standard techniques for defining formal languages; see Aho, Sethi & Ullman (1986).)

- The alphabet (set of characters) is { '*a*', '*d*', '*e*', '*f*', '*i*', '*k*' – '*u*', '*x*', '$\mathcal{A}$' – '$\mathcal{Z}$', '(', ')' }.

- The lexicon (set of tokens) is { '*con*', '*vbl*', '(', ')' }.
- Lexical analysis consists of recognising constants and replacing them by the token '*con*', recognising variables and replacing them by the token '*vbl*', and recognising the tokens '(' and ')'.
- The grammar of the language is as follows.
  - The terminals are the tokens.
  - The sole nonterminal is T, which is the start symbol.
  - The production rules are $T \rightarrow con \mid (\ vbl\ ) \mid (\ T\ T\ )$.

EXAMPLE. Consider the term '$((id((s(\mathcal{ABC}))k))((\mathcal{A})nil))$'. Lexical analysis turns it into the sequence of tokens ((*con* ((*con* (*vbl*)) *con*)) ((*vbl*) *con*)), where spaces are used to separate the tokens. This is parsed as a T by matching it against (T T):

$$(\ \underbrace{(\ con\ (\ (\ con\ (\ vbl\ )\ )\ con\ )\ )}_{T}\ \underbrace{(\ (\ vbl\ )\ con\ )}_{T}\ )$$

Let's look at how the second subterm, ((*vbl*) *con*), is parsed: it is recognised as a T by matching it against (T T):

$$(\ \underbrace{(\ vbl\ )}_{T}\ \underbrace{con}_{T}\ )$$

where the two subterms (*vbl*) and *con* are recognised as T's directly using the rules $T \rightarrow (\ vbl\ )$ and $T \rightarrow con$. There are in total eleven terms (that is, substrings matched against T) in the original term.

EXERCISE. Apply the same procedure to the term $(s(((\mathcal{XY})nil)(\mathcal{C})))$ and verify that the term contains fourteen tokens and seven terms.

METATERMS

As explained in Chapter 13, identifiers will be used as metaconstants, denoting particular terms, and as metavariables, denoting arbitrary terms or variables. A *metaterm* is an expression that is like a term except that it may contain metaconstants and metavariables and that brackets may be omitted in three contexts: the brackets around the whole metaterm and around variables may be omitted, and so may the brackets in the context $(AB)C$. Optional brackets and spaces may also be added to make the metaterm more readable. To state it more precisely, we add ' ' and the identifier characters to the alphabet, we add three new tokens '*termcon*', '*vblvbl*' and '*termvbl*' to the lexicon, and we modify lexical analysis so that spaces are removed, metaconstants are replaced by '*termcon*', metavariables denoting variables are replaced by '*vblvbl*', and metavariables denoting terms are replaced by '*termvbl*'; the grammar for metaterms is

$$T \rightarrow T\,L \mid L$$
$$L \rightarrow \mathit{con} \mid \mathit{vbl} \mid (\,T\,) \mid \mathit{termcon} \mid \mathit{vblvbl} \mid \mathit{termvbl}$$

where the start symbol is T.

An *instance* of a metaterm is a term obtained as follows.

1. Choose a term or a variable, as required, for each metavariable in the metaterm.
2. Replace all metaconstants and metavariables by the terms or variables they denote.
3. Add and remove brackets as required.
4. Remove spaces.

(This could be specified formally as a syntax-directed translation from metaterms to terms; I shall not do so, for lack of a universally agreed notation for such mappings.)

EXAMPLE. Consider the metaterm '*AB*(*id x*)', where '*A*' is a metavariable denoting a term, '*B*' is a metaconstant denoting the term '(*s*((𝒳)*id*))', and '*x*' is a metavariable denoting a variable. Lexical analysis turns the metaterm into the token sequence *termvbl termcon* (*con vblvbl*), which as parsed as shown:

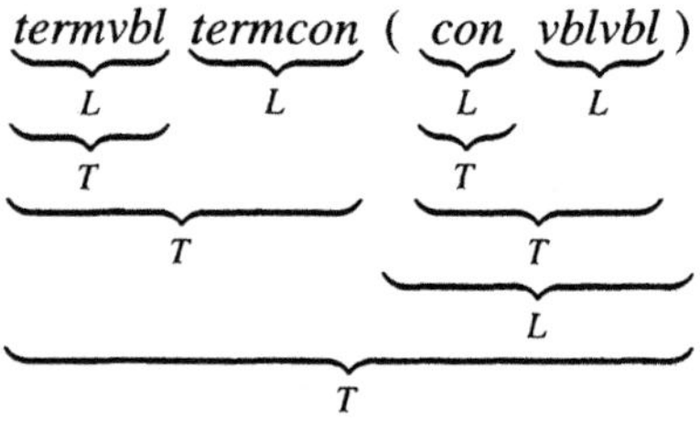

We can form an instance of the metaterm as follows.

1. Let the metavariable '*x*' denote the variable '𝒳' and let the metavariable '*A*' denote the term '(*kk*)'.
2. Replacing metaconstants and metavariables gives '(*kk*)(*s*((𝒳)*id*))(*id* 𝒳)'.
3. Adding brackets gives '(((*kk*)(*s*((𝒳)*id*)))(*id* (𝒳)))'.
4. Removing spaces gives the term '(((*kk*)(*s*((𝒳)*id*)))(*id*(𝒳)))'.

A metaterm may be used in two ways: as a schema, denoting any of its instances, or to denote a particular one of its instances. The latter will be implied when particular terms or variables have already been chosen for the metavariables.

EXERCISE. Verify that '$AB(id\ x)$' has the same instances as '$(AB)(id\ x)$', but not the same instances as '$A(B(id\ x))$'.

It is often convenient to use superfluous brackets to emphasise the structure of a metaterm. My syntax allows such brackets to be added and removed at will, without affecting the term denoted by the metaterm. For example, it is literally correct, as well as highly convenient, to say that $(AB)C$ is the same term as ABC, and $f(x)$ is the same term as fx: of course '$(AB)C$' is a different metaterm from 'ABC', but they denote the same term.

REDUCTION

The semantics of terms is given by the reduction relation, $\rhd$, between terms, defined by the following clauses. The clauses must be applied in order, except for the two on line (1), which may be applied in either order: that is, a clause $X \rhd Y$ is only applicable if no clause higher in the list applies to X. It follows that when reducing a term all proper subterms must be reduced first, and only when they cannot be reduced further may reduction rules be applied at the top level.

$$
\begin{array}{rll}
AB \rhd A'B \ \text{ if } A \rhd A', & AB \rhd AB' \ \text{ if } B \rhd B' & (1) \\
sABC & \rhd\ (AC)(BC) & (2) \\
id\,A & \rhd\ A & (3) \\
kAB & \rhd\ A & (4) \\
equal\ A\ A & \rhd\ true & (5) \\
equal(AB)(CD) & \rhd\ false \quad \text{if } equal\ A\ C \rhd false & \\
 & \qquad\qquad\ \text{or } equal\ B\ D \rhd false & (6) \\
equal\ a\ b & \rhd\ false \quad \text{if } a \text{ and } b \text{ are constants} & (7) \\
equal\ a\ (UV) & \rhd\ false \quad \text{if } a \text{ is a constant} & (8) \\
equal\ (UV)\ a & \rhd\ false \quad \text{if } a \text{ is a constant} & (9) \\
equal\ A\ B & \rhd\ equal\ A\ B & (10) \\
if\ true\ A\ B & \rhd\ A & (11) \\
if\ false\ A\ B & \rhd\ B & (12) \\
if\ C\ A\ B & \rhd\ if\ C\ A\ B & (13) \\
former\,(A\,B) & \rhd\ A & (14) \\
former\ a & \rhd\ nil \quad \text{if } a \text{ is a constant} & (15) \\
former\ x & \rhd\ former\ x \quad \text{if } x \text{ is a variable} & (16) \\
latter\,(A\,B) & \rhd\ B & (17)
\end{array}
$$

$$\begin{aligned} \textit{latter } a &\rhd \textit{nil} \quad \text{if } a \text{ is a constant} && (18)\\ \textit{latter } x &\rhd \textit{latter } x \quad \text{if } x \text{ is a variable} && (19)\\ \textit{fxpt } A\, B &\rhd A(\textit{fxpt } A)B && (20)\\ x\,A &\rhd x\,A \quad \text{if } x \text{ is a variable} && (21) \end{aligned}$$

- Let $\rhd^*$ be the reflexive and transitive closure of $\rhd$. If $A \rhd^* B$ then I shall say that A *reduces to* B. Let $\lhd$ and $\lhd^*$ be the converses of $\rhd$ and $\rhd^*$. Let $\rhd^*\lhd$ be the equivalence relation generated by $\rhd$. Let $A \not\rhd$ mean that A is *irreducible* (that is, there is no B such that $A \rhd B$).
- A *construction* is an irreducible term with no variables. (Equivalently, a construction is a constant or a term of the form *sA*, *sAB*, *kA*, *equal A*, *if A*, *if A B*, *true A B ⋯ C*, *false A B ⋯ C*, *fxpt A*, or *nil A B ⋯ C*, where $A, B, \ldots C$ are constructions.)

EXAMPLES. Using superscripts on '$\rhd$' to indicate which reduction rules are being applied:

(a) *if* (*equal* (*latter*(*sskx*)) (*kx*)) (*fxpt k y*) (*fxpt id*)
$\rhd^{1,2}$ *if* (*equal* (*latter*((*sx*)(*kx*))) (*kx*)) (*fxpt k y*) (*fxpt id*)
$\rhd^{1,17}$ *if* (*equal* (*kx*) (*kx*)) (*fxpt k y*) (*fxpt id*) $\rhd^{1,5}$ *if true* (*fxpt k y*) (*fxpt id*)
$\rhd^{11}$ *fxpt k y* $\rhd^{20}$ *k*(*fxpt k*)*y* $\rhd^{4}$ *fxpt k* $\not\rhd$

(b) *equal* (*k*(*sy*)) (*id*(*k*(*nil z*))) $\rhd^{1,3}$ *equal* (*k*(*sy*)) (*k*(*nil z*)) $\rhd^{6}$ *false* $\not\rhd$, where the second step works since *equal* (*sy*) (*nil z*) $\rhd^{6}$ *false*, which works since *equal s nil* $\rhd^{7}$ *false*

(c) *equal k k* $\rhd^{5}$ *true* $\not\rhd$ (it doesn't also reduce to *false* using rule 7 because rule 7 can only apply when rules 1–6 are inapplicable)

(d) *if true x* (*fxpt id y*) $\rhd^{1,20}$ *if true x* (*id*(*fxpt id*)*y*) $\rhd^{1,3}$ *if true x* (*fxpt id y*) $\rhd^{1,20}$ *if true x* (*id*(*fxpt id*)*y*) $\rhd \cdots$ and so on in an infinite loop (no other reduction rule can apply until *fxpt id y* has finished reducing, which it never does).

THEOREM 1. (Confluence.) If $T \rhd U$ and $T \rhd V$, where U is not V, then there is a term W such that $U \rhd W$ and $V \rhd W$.

Proof. The only indeterminism in the reduction rules is

$$AB \rhd \begin{cases} A'B & \text{where } A \rhd A', \\ AB' & \text{where } B \rhd B'. \end{cases}$$

Then take W as $A'B'$. ∎

THEOREM 2. If a term reduces to an irreducible term then any other sequence of reduction steps starting at the same term reduces in the same number of steps to the same result.

Proof. From the confluence theorem. ∎

EXERCISE. Show that there are 120 ways of reducing the term

$$\mathit{if}\ (\mathit{equal}\ x\,(\mathit{id}\ x))\ (s\ \mathit{id}\ \mathit{id}\ \mathit{true})\ (kkkk)$$

to an irreducible term and that they all give the same result in seven steps. Here are two of the 120 ways:

(i) $\mathit{if}\ (\mathit{equal}\ x\,(\mathit{id}\ x))\ (s\ \mathit{id}\ \mathit{id}\ \mathit{true})\ (kkkk)$
$\triangleright\ \mathit{if}\ (\mathit{equal}\ x\,x)\ (s\ \mathit{id}\ \mathit{id}\ \mathit{true})\ (kkkk)$
$\triangleright\ \mathit{if}\ (\mathit{equal}\ x\,x)\ (s\ \mathit{id}\ \mathit{id}\ \mathit{true})\ (kk)\ \triangleright\ \mathit{if}\ \mathit{true}\ (s\ \mathit{id}\ \mathit{id}\ \mathit{true})\ (kk)$
$\triangleright\ \mathit{if}\ \mathit{true}\ ((\mathit{id}\ \mathit{true})(\mathit{id}\ \mathit{true}))\ (kk)\ \triangleright\ \mathit{if}\ \mathit{true}\ (\mathit{true}(\mathit{id}\ \mathit{true}))\ (kk)$
$\triangleright\ \mathit{if}\ \mathit{true}\ (\mathit{true}\ \mathit{true})\ (kk)\ \triangleright\ \mathit{true}\ \mathit{true}\ \not\triangleright$,

(ii) $\mathit{if}\ (\mathit{equal}\ x\,(\mathit{id}\ x))\ (s\ \mathit{id}\ \mathit{id}\ \mathit{true})\ (kkkk)$
$\triangleright\ \mathit{if}\ (\mathit{equal}\ x\,(\mathit{id}\ x))\ ((\mathit{id}\ \mathit{true})(\mathit{id}\ \mathit{true}))\ (kkkk)$
$\triangleright\ \mathit{if}\ (\mathit{equal}\ x\,(\mathit{id}\ x))\ ((\mathit{id}\ \mathit{true})\mathit{true})\ (kkkk)$
$\triangleright\ \mathit{if}\ (\mathit{equal}\ x\,(\mathit{id}\ x))\ ((\mathit{id}\ \mathit{true})\mathit{true})\ (kk)$
$\triangleright\ \mathit{if}\ (\mathit{equal}\ x\,(\mathit{id}\ x))\ (\mathit{true}\ \mathit{true})\ (kk)\ \triangleright\ \mathit{if}\ (\mathit{equal}\ x\,x)\ (\mathit{true}\ \mathit{true})\ (kk)$
$\triangleright\ \mathit{if}\ \mathit{true}\ (\mathit{true}\ \mathit{true})\ (kk)\ \triangleright\ \mathit{true}\ \mathit{true}\ \not\triangleright$.

THEOREM 3.

- $A \triangleright^{*}\triangleleft B$ iff $A \triangleright^{*} C \triangleleft^{*} B$ for some C.
- If $A \triangleright^{*}\triangleleft B \not\triangleright$ then $A \triangleright^{*} B$.

Proof. From the confluence theorem. ∎

THEOREM 4. If $A \triangleright B$ then any variable occurring in B also occurs in A.

THE INTENDED INTERPRETATION OF CONSTRUCTIONS

A construction F may be regarded as representing an algorithm that receives as input a construction X, concatenates F and X, and reduces FX to a construction, which it outputs. This algorithm is a transformation from constructions to constructions; it is indeterministic in reduction order though not in outcome, and it may not halt. In some cases it is rather trivial: for example, *true* represents an algorithm that converts X to *true* X and *false* represents an algorithm that converts X to *false* X. (Since '*true*' and '*false*' are different identifiers it follows that *true* and *false* represent different algorithms.) Other constructions perform useful jobs: the constant *fxpt* acts as a *least fixed-point* operator which provides for iteration and recursion, *equal* acts as an algorithm that tests two constructions for equality, and *former* and *latter* act as algorithms that dissect a composite construction into its two components.

I say 'algorithm' rather than 'function' to emphasise that its intension matters: there is no attempt to regard all extensionally equivalent algorithms as equal. Any two constructions represent different algorithms.

THE INTENDED INTERPRETATION OF TERMS WITHOUT VARIABLES

A term with no variables represents a computation that, if it halts, will halt at a construction (by Theorem 4). The steps of the computation are specified by the reduction relation. More precisely, a variable-free term has a *sense* and a *reference*, defined as follows.

The *sense* of a variable-free term is the collection of reduction processes starting at that term, and the *reference* or *value* of the term is the construction at which all the reduction processes halt. A process of reduction to irreducible form is called an *evaluation*. Thus:

- a term lacks a value if the reduction processes never halt (it still possesses a sense, however);
- two terms have the same value but different senses if they reduce to the same construction.

EXAMPLE. The sense of the term *s id id id* is the pair of reduction processes

$s\,id\,id\,id \;\triangleright\; (id\,id)(id\,id) \;\triangleright\; id(id\,id) \;\triangleright\; id\,id \;\triangleright\; id \;\not\triangleright$

$s\,id\,id\,id \;\triangleright\; (id\,id)(id\,id) \;\triangleright\; (id\,id)id \;\triangleright\; id\,id \;\triangleright\; id \;\not\triangleright$

and the value of the term is *id*. All the terms *s id id id*, (*id id*)(*id id*), *id*(*id id*), (*id id*)*id*, *id id* and *id* have different senses but the same value.

THE INTENDED INTERPRETATION OF TERMS WITH VARIABLES

Terms with variables have no sense or reference; they get their meaning indirectly from the fact that if one replaces their variables by constructions one obtains terms without variables.

DEFINITION. An *instantiation of a term by constructions* consists of choosing a construction X for each variable v occurring in the term and replacing each occurrence of v by X. A *simultaneous instantiation by constructions* consists of doing this to two or more terms, where each occurrence of v must be replaced by the same construction X in each term.

The idea is that a reduction $A \rhd B \rhd C \rhd \cdots \rhd Z \not\rhd$ involving terms with variables is to be regarded as a *reduction schema* encompassing all the instances $A' \rhd B' \rhd C' \rhd \cdots \rhd Z' \not\rhd$, where $A', \ldots Z'$ are obtained from $A, \ldots Z$ by a simultaneous instantiation of the variables by constructions. For example, the reduction

$$\begin{aligned} &\mathit{latter}(\mathit{if}\ (\mathit{equal}\,(\mathit{id}\,x)\,x)\ (ky)\ z) \\ &\rhd \mathit{latter}(\mathit{if}\ (\mathit{equal}\,x\,x)\ (ky)\ z) \\ &\rhd \mathit{latter}(\mathit{if}\ \mathit{true}\ (ky)\ z) \rhd \mathit{latter}(ky) \rhd y \not\rhd \end{aligned}$$

encompasses such instances as

$$\begin{aligned} &\mathit{latter}(\mathit{if}\ (\mathit{equal}\,(\mathit{id}\,(kk))\,(kk))\ (k(ss))\ (\mathit{true}\,\mathit{false})) \\ &\rhd \mathit{latter}(\mathit{if}\ (\mathit{equal}\,(kk)\,(kk))\ (k(ss))\ (\mathit{true}\,\mathit{false})) \\ &\rhd \mathit{latter}(\mathit{if}\ \mathit{true}\ (k(ss))\ (\mathit{true}\,\mathit{false})) \rhd \mathit{latter}(k(ss)) \rhd ss \not\rhd\ . \end{aligned}$$

This interpretation presupposes that reduction satisfies the following two properties:

(1) if $A \rhd B$ then $A' \rhd B'$, where A' and B' are any simultaneous instantiation of A and B by constructions;
(2) if $A \not\rhd$ then $A' \not\rhd$, where A' is any instantiation of A by constructions.

These properties seem simple and obvious, as well as necessary if variables are to do their job of 'standing for' arbitrary constructions. Unfortunately they are impossible to satisfy: consider

$$\mathit{latter}(\mathit{if}\ (\mathit{equal}\,(\mathit{id}\,x)\,\mathit{nil})\ (ky)\ z) \rhd \mathit{latter}(\mathit{if}\ (\mathit{equal}\,x\,\mathit{nil})\ (ky)\ z) \rhd \cdots ?$$

The instances of this reduction take a different course depending on whether the construction replacing x is *nil* or not; there is no schema covering all the reductions.

What should *equal x nil* reduce to? (Or *if x y z* or *former x* or *xy*, for that matter?) It is not safe to reduce to any other term, because that would violate property (1). It is not safe even to call it irreducible, since that would violate (2). Dropping (2) and retaining (1) is not an option, since (1) and (2) are intimately interdependent: a term can only reduce when all its subterms are irreducible.

The solution in such cases is to make the term reduce to itself (see rules 10, 13, 16, 19 and 21). Then we can salvage properties (1) and (2) in a slightly weakened form, in the following theorem.

THEOREM 5.

- If $A \ \triangleright^* \ B$ then $A' \ \triangleright^* \ B'$, where A' and B' are any simultaneous instantiation of A and B by constructions.
- If $A \not\triangleright$ then $A' \not\triangleright$, where A' is any instantiation of A by constructions.

Proof. At least one of the following three cases must hold.

Case 1: $A \ \triangleright \ T$, where T is a term other than A. Then $A' \ \triangleright \ T'$ by the same reduction rule, where A' and T' are any simultaneous instantiation of A and T by constructions.

Case 2: $A \ \triangleright \ A$. Then $A' \ \triangleright^* \ A'$, in zero reduction steps, where A' is any instantiation of A by constructions.

Case 3: $A \not\triangleright$. Then $A' \not\triangleright$, where A' is any instantiation of A by constructions.

The theorem follows by induction. ∎

Thus the reduction of a term with variables still says something about the reduction of its instantiations, but the steps do not always correspond one-to-one.

SUMMARY

A metaterm *denotes* a term, which may be *instantiated by constructions* to give a term with no variables, whose *value* (if any) is a construction, which *represents* an algorithm acting on constructions.

PARTIAL REDUCTION

Partial reduction is a reduction-like process for obtaining information about the evaluation of a term given incomplete information about it. Typically, given a term F, one wants to know something about the evaluation of FX for an unknown construction X. I shall define two partial reduction relations, $\rightharpoonup$ and $\rightarrow$, for use in setting up the Fxpt Rules in Protologic (see Chapter 17).

DEFINITION. Given two variables, f and v, define a partial reduction relation $\rightharpoonup$ (with respect to f and v) by the following clauses, where the two on the first line must be applied before the others.

$$AB \rightharpoonup A'B \quad \text{if } A \rightharpoonup A', \qquad AB \rightharpoonup AB' \quad \text{if } B \rightharpoonup B'$$

$$sABC \rightharpoonup (AC)(BC)$$

$$id\,A \rightharpoonup A$$

$$kAB \rightharpoonup A$$

$$\begin{array}{rcll}
\mathit{if}\; C\, A\, B X & \rightharpoonup & \mathit{if}\; C\, (AX)\, (BX) & \\
va & \rightharpoonup & v & \text{if } a \text{ is a constant} \\
vv & \rightharpoonup & v & \\
v(AB) & \rightharpoonup & vAB & \\
fa & \rightharpoonup & v & \text{if } a \text{ is a constant} \\
fv & \rightharpoonup & v & \\
f(\mathit{if}\; C\, A\, B) & \rightharpoonup & \mathit{if}\; C\, (fA)\, (fB) & \\
a & \rightharpoonup & v & \text{if } a \text{ is } \mathit{equal}, \mathit{true}, \mathit{false}, \mathit{former}, \\
 & & & \quad \mathit{latter}, \mathit{fxpt} \text{ or } \mathit{nil} \\
x & \rightharpoonup & v & \text{if } x \text{ is a variable other than } f \text{ and } v
\end{array}$$

Let $\overset{*}{\rightharpoonup}$ be the reflexive and transitive closure of $\rightharpoonup$, let $\leftharpoonup$ and $\overset{*}{\leftharpoonup}$ be the converses of $\rightharpoonup$ and $\overset{*}{\rightharpoonup}$, and let $\overset{*}{\leftrightharpoons}$ be the equivalence relation generated by $\overset{*}{\rightharpoonup}$. Let $A \not\rightharpoonup$ mean that $A \rightharpoonup B$ for no B.

(Note: I referred in the above definition to 'two variables', f and v. It is implied that they are different variables: there is no need to say 'two distinct variables', since if they were the same there would not be *two* of them. If I wanted to allow f and v to be the same variable I would say instead 'Let f and v be variables . . . '.)

THEOREM 6. (Confluence) If $T \rightharpoonup U$ and $T \rightharpoonup V$, where U is not V, then there is a term W such that $U \rightharpoonup W$ and $V \rightharpoonup W$. Consequently if $A \overset{*}{\leftrightharpoons} B \not\rightharpoonup$ then $A \overset{*}{\rightharpoonup} B$.

Proof. As in Theorems 1 and 3. ∎

DEFINITION. A term Φ is *continuous* iff $v(\Phi fv) \overset{*}{\rightharpoonup} v$, where $\rightharpoonup$ is defined with respect to f and v, for two variables f, v not occurring in Φ. (This definition is independent of the choice of f and v.)

Informally, to say that Φ is continuous means that Φfv depends on f only in an extensional way; that is, in the evaluation of ΦFX, for any constructions F and X, the only time the evaluation depends on the internal structure of F is when terms of the form $F\,T$, for some T, are reduced. This rules out reductions such as *former* $F \triangleright \cdots$ and *if* $F\,A\,B \triangleright \cdots$, since they depend on the structure of F; but it permits reductions such as $\mathit{id}\, F \triangleright F$ and $sFAB \triangleright (FA)(FB)$, since they work regardless of F's structure. The point of $\rightharpoonup$ is to determine whether a term A depends on f only extensionally; $A \overset{*}{\rightharpoonup} v$ is a sufficient condition for this, and $vA \overset{*}{\rightharpoonup} v$ is a weaker sufficient condition, adequate for most cases – hence the above definition.

The notion of continuity will be used in Chapter 17, where the protological Fxpt Rules will be restricted to continuous functions.

EXAMPLES. Let Φ be $s(s(ks)(s(kk)(s\,id\,(s\,id\,(kk)))))id$. Then

$$\begin{aligned} v(\Phi fv) &\stackrel{*}{\rightharpoonup} v(s(k(f(fk)))fv) \rightharpoonup v(s(k(fv))fv) \rightharpoonup v(s(kv)fv) \\ &\rightharpoonup v((kvv)(fv)) \rightharpoonup v(v(fv)) \rightharpoonup v(vv) \rightharpoonup vv \rightharpoonup v \end{aligned}$$

so Φ is continuous.

Now let Ψ be $s(k(s\,id))(s(k(s\,id))(s(kk)id))$. Then

$$\begin{aligned} v(\Psi fv) &\stackrel{*}{\rightharpoonup} v(s\,id\,(s\,id\,(kf))v) \stackrel{*}{\rightharpoonup} v(v(vf)) \rightharpoonup vv(vf) \rightharpoonup v(vf) \rightharpoonup vvf \\ &\rightharpoonup vf \not\rightharpoonup \end{aligned}$$

so Ψ is not continuous.

DEFINITION. Given two variables, f and v, define another partial reduction relation $\rightharpoondown$ (with respect to f and v) by the following clauses, which must be applied in order (except that there is no ordering between the two rules on the first line).

$$\begin{aligned} AB \rightharpoondown A'B \quad \text{if } A \rightharpoondown A', \qquad & AB \rightharpoondown AB' \quad \text{if } B \rightharpoondown B' \\ fA &\rightharpoondown v \\ vA &\rightharpoondown v \\ Av &\rightharpoondown v \\ sABC &\rightharpoondown (AC)(BC) \\ id\,A &\rightharpoondown A \\ kAB &\rightharpoondown A \end{aligned}$$

Let $\stackrel{*}{\rightharpoondown}$ be the reflexive and transitive closure of $\rightharpoondown$. Let $A \not\rightharpoondown$ mean that $A \rightharpoondown B$ for no B.

DEFINITION. A term Φ is *non-constant* iff $\Phi fx \stackrel{*}{\rightharpoondown} v$, where $\rightharpoondown$ is defined with respect to f and v, for three variables f, v, x not occurring in Φ. (This definition is independent of the choice of f, v and x.)

Informally, $\Phi fx \stackrel{*}{\rightharpoondown} v$ means that the evaluation of ΦFX, for any constructions F and X, involves at least one reduction of the form $F\,T \,\triangleright\, \cdots$, for some term T. Hence if F is an empty function and Φ is non-constant then ΦF will also be an empty function.

Non-constant functions will be used in Chapter 17, where one of the Fxpt Rules will involve a restriction to non-constant continuous functions.

EXERCISE. Let Φ be $s(s(ks)(s(kk)id))(k(s(kk)id))$. Show that Φ is non-constant by completing the partial reduction

$$\Phi fx \rightharpoondown \cdots \rightharpoondown f(kx) \rightharpoondown \cdots$$

EXERCISE. Let Ψ be $s(k(s(s\,if\,k)))k$. Show that Ψ is not non-constant by completing the partial reduction

$$\Psi fx \rightharpoondown \cdots \rightharpoondown s\,if\,k\,x\,f \rightharpoondown \cdots$$

CHAPTER 15

FROM THE TERM LANGUAGE TO THE EXPANDED TERM LANGUAGE

The Term Language (T) is very simple and hence very easy to reason about; however, it is also very hard to use. Thinking of the Term Language as machine code, we need a higher-level language, an *Expanded Term Language* (ET), for day-to-day use. This, together with its accompanying metanotation, will have all the facilities one expects in a functional programming language: λ-abstractions, function definitions using pattern matching, pairs, tuples, lists, numbers, and primitive and general recursive functions. The language is general-purpose, so a version of Church's Thesis may be formulated for it. It is also possible to encode terms within the language.

See Chapter 23 for a comparison of ET with other functional programming languages.

The Expanded Term Language is compiled into the Term Language in a way based on the traditional interpretation of λ-calculus into combinators.

SYNTAX OF THE EXPANDED TERM LANGUAGE (ET)

The notions of constant, variable and atom are understood as in the Term Language.

DEFINITION. A *term* of ET is a sentence in the following language.

- The alphabet is that of T plus { 'λ', '.', '$\Big[$', '$\Big]$' }.
- The lexicon is that of T plus { 'λ', '.', '$\Big[$', '$\Big]$' }.
- Lexical analysis consists of recognising constants and replacing them by the token '*con*', recognising variables and replacing them by the token '*vbl*', and recognising every other token.
- The grammar of the language is as follows.
  - The terminals are the tokens.
  - The sole nonterminal is T, which is the start symbol.
  - The production rules are $T \rightarrow con \mid (\, vbl \,) \mid (\, T \; T \,) \mid (\lambda \, T \, . \, T \,) \mid (\, T \left[{}^{T}_{vbl} \right])$

Note that this grammar contains that of T. From now on, the word 'term' will mean a term of ET; a term of T will be called a *simple* term. A term of the form ($\lambda\ T\ .\ T$) is called a λ-*abstraction*; a term of the form ($T\left[\begin{smallmatrix}T\\ vbl\end{smallmatrix}\right]$) is called an *instantiation*.

EXAMPLE. Applying lexical analysis to the term '$(\lambda(k(\mathcal{X})).((\mathcal{X}\mathcal{Y})\left[\begin{smallmatrix}id\\ A\end{smallmatrix}\right]))$' gives the following sequence of tokens, which is parsed as shown.

$$\underbrace{(\lambda\ (\underbrace{\underbrace{con}_{T}\ \underbrace{(\,vbl\,)}_{T}}_{T})\ .\ (\underbrace{\underbrace{(\,vbl\,)}_{T}\left[\begin{matrix}\overbrace{con}^{T}\\ vbl\end{matrix}\right]}_{T}))}_{T}$$

Note that the parse is unique.

METATERMS

Metaterms are defined as expressions that are like terms except that they may contain metaconstants, metavariables and the metanotation introduced later in this chapter, and that some brackets may be omitted. Optional brackets and spaces may also be added. To state it more precisely, the alphabet is that of metaterms of T plus the characters needed for metanotation; the lexicon is that of metaterms of T plus {'λ', '.', '[', ']' }; lexical analysis is the same as for metaterms of T; and the grammar is

$$T \rightarrow T\,L \mid T\left[\begin{smallmatrix}T\\ vbl\end{smallmatrix}\right] \mid L$$

$$L \rightarrow con \mid vbl \mid (\,T\,) \mid (\,\lambda\,T\,.\,T\,) \mid termcon \mid vblvbl \mid termvbl \mid \ldots$$

where the start symbol is T and '...' represents production rules for all the metanotation introduced below. *Instances* of metaterms are defined and used as in T – to obtain an instance, replace each metaconstant or metavariable by a term or variable, add and remove brackets as required, remove spaces, and rewrite all metanotation.

EXAMPLE. Let 'A' denote '(kk)', 'B' denote '(ss)', 'C' denote '$(\mathcal{X})$', 'x' denote '$\mathcal{X}$', 'y' denote '$\mathcal{Y}$', and 'z' denote '$\mathcal{Z}$'. Then the metaterm '$AB\left[\begin{smallmatrix}A\\ x\end{smallmatrix}\right]C\left[\begin{smallmatrix}B\\ y\end{smallmatrix}\right]\left[\begin{smallmatrix}A\\ z\end{smallmatrix}\right]$' denotes the term '$((((((kk)(ss))\left[\begin{smallmatrix}(kk)\\ \mathcal{X}\end{smallmatrix}\right])(\mathcal{X}))\left[\begin{smallmatrix}(ss)\\ \mathcal{Y}\end{smallmatrix}\right])\left[\begin{smallmatrix}(kk)\\ \mathcal{Z}\end{smallmatrix}\right])$'.

Observe that $AB\begin{bmatrix}C\\x\end{bmatrix}$ is the same term as $(AB)\begin{bmatrix}C\\x\end{bmatrix}$, not $A(B\begin{bmatrix}C\\x\end{bmatrix})$: that is, the metaterms '$AB\begin{bmatrix}C\\x\end{bmatrix}$' and '$(AB)\begin{bmatrix}C\\x\end{bmatrix}$' denote the same term, for any A, B, C and x.

The first piece of metanotation is that a multiple instantiation, $A\begin{bmatrix}X_1\\x_1\end{bmatrix}\cdots\begin{bmatrix}X_k\\x_k\end{bmatrix}$, will often be abbreviated to $A\begin{bmatrix}X_1,\ldots X_k\\x_1,\ldots x_k\end{bmatrix}$ or $A\begin{bmatrix}\underline{X}\\\underline{x}\end{bmatrix}$, where $\underline{x}$ is the (possibly empty) sequence of variables $x_1,\ldots x_k$ and $\underline{X}$ is the sequence of terms $X_1,\ldots X_k$. Underlined lower-case letters generally will be used to denote sequences of variables. Similarly, underlined capital letters will be used to denote sequence of terms; the intended number of terms in the sequence will always be clear from the context (for example, if the notation $A\begin{bmatrix}\underline{X}\\\underline{x}\end{bmatrix}$ is used then $\underline{X}$ is intended to have the same number of terms as $\underline{x}$ has variables). Note that multiple instantiation is not *simultaneous* instantiation: $\begin{bmatrix}\underline{X}\\\underline{x}\end{bmatrix}$ is simply an abbreviation for a sequence of instantiations applied one at a time.

INTERPRETATION OF THE EXPANDED TERM LANGUAGE IN THE TERM LANGUAGE

Every term is compiled into a simple term by applying the compilation relation, $\mapsto$, repeatedly. $\mapsto$ is defined as follows, where the clauses may be applied in any order.

$$
\begin{aligned}
AB &\mapsto A'B \quad \text{if } A \mapsto A' \\
AB &\mapsto AB' \quad \text{if } B \mapsto B' \\
(\lambda A.B) &\mapsto (\lambda A'.B) \quad \text{if } A \mapsto A' \\
(\lambda A.B) &\mapsto (\lambda A.B') \quad \text{if } B \mapsto B' \\
(\lambda PQ.B) &\mapsto s(s(k(\lambda P.(\lambda Q.B)))\mathit{former})\mathit{latter} \\
(\lambda a.UV) &\mapsto s(\lambda a.U)(\lambda a.V) \quad \text{if } a \text{ is an atom} \\
(\lambda x.x) &\mapsto \mathit{id} \quad \text{if } x \text{ is a variable} \\
(\lambda a.b) &\mapsto kb \quad \text{if } a \text{ and } b \text{ are atoms and not both the same variable} \\
A\begin{bmatrix}B\\x\end{bmatrix} &\mapsto A'\begin{bmatrix}B\\x\end{bmatrix} \quad \text{if } A \mapsto A' \\
A\begin{bmatrix}B\\x\end{bmatrix} &\mapsto A\begin{bmatrix}B'\\x\end{bmatrix} \quad \text{if } B \mapsto B' \\
PQ\begin{bmatrix}B\\x\end{bmatrix} &\mapsto (P\begin{bmatrix}B\\x\end{bmatrix})(Q\begin{bmatrix}B\\x\end{bmatrix}) \\
x\begin{bmatrix}B\\x\end{bmatrix} &\mapsto B \\
a\begin{bmatrix}B\\x\end{bmatrix} &\mapsto a \quad \text{if } a \text{ is an atom other than } x
\end{aligned}
$$

Let $\overset{*}{\mapsto}$ (read as 'compiles to') be the reflexive and transitive closure of $\mapsto$. Let $\leftarrow\!\shortmid$ and $\overset{*}{\leftarrow\!\shortmid}$ be the converses of $\mapsto$ and $\overset{*}{\mapsto}$. Let $\overset{*}{\leftrightarrow}$ (read as 'is equivalent to') be the equivalence relation generated by $\mapsto$. Let $A \not\mapsto$ mean that there is no B such that $A \mapsto B$.

EXAMPLE. Let x and y be two variables. Then

$$\begin{aligned}(\lambda x.nil\,x\,y) &\mapsto s(\lambda x.nil\,x)(\lambda x.y) \mapsto s(s(\lambda x.nil)(\lambda x.x))(\lambda x.y)\\ &\mapsto s(s(k\,nil)(\lambda x.x))(\lambda x.y) \mapsto s(s(k\,nil)(\lambda x.x))(k\,y)\\ &\mapsto s(s(k\,nil)id)(k\,y) \not\mapsto .\end{aligned}$$

EXERCISE. What would the result have been in the previous example if x and y had been the same variable?

Observe that the compilation of a term of the form $(\lambda x.B)$ resembles the traditional translation of λ-abstractions into combinators. However, a more general kind of λ-abstraction, $(\lambda A.B)$, is allowed; its properties will be discussed from Theorem 15 onwards.

EXERCISE. Let x and y be two variables. Show that $(\lambda x.yx)\begin{bmatrix}true\\x\end{bmatrix}\begin{bmatrix}false\\y\end{bmatrix} \overset{*}{\mapsto}$ $s(k\,false)id \not\mapsto$.

You should have found in the previous exercise that when a term of the form $A\begin{bmatrix}B\\x\end{bmatrix}$ is compiled the lambda-abstractions and instantiations in A have to be compiled away before B can be substituted for x. This avoids complications arising from substitution into the scope of bound variables.

THEOREM 1. $A \not\mapsto$ iff A is simple.

THEOREM 2. Any compilation sequence $A \mapsto B \mapsto C \mapsto \cdots$ halts. For any term A there is a unique term A^* such that $A \overset{*}{\mapsto} A^* \not\mapsto$.

(I shall call A^* the *compiled form* of A, and shall always use this asterisk notation for it.)

Proof. To see that compilation always halts, consider the following mapping d from terms to positive integers.

$$\begin{aligned} d(a) &\text{ is } 1 \quad \text{if } a \text{ is an atom}\\ d(AB) &\text{ is } d(A)+d(B)+1\\ d((\lambda A.B)) &\text{ is } 12^{d(A)}d(B)\\ d(A\begin{bmatrix}B\\x\end{bmatrix}) &\text{ is } 2d(A)d(B)\end{aligned}$$

It is straightforward to check that if $T \mapsto T'$ then $d(T) > d(T')$. Therefore any compilation sequence halts.

The proof that A^* is unique uses the usual confluence property: if $T \mapsto U$ and $T \mapsto V$ then there is a W such that $U \overset{*}{\mapsto} W$ and $V \overset{*}{\mapsto} W$. (There are ten cases where U and V may differ, and in each case the desired W is obvious.) Therefore, as usual, if $A \overset{*}{\mapsto} B$ and $A \overset{*}{\mapsto} C$ there is a D such that $B \overset{*}{\mapsto} D$ and $C \overset{*}{\mapsto} D$. If in addition $B, C \not\mapsto$ then B, D and C must be the same term. ∎

THEOREM 3. $A \overset{*}{\leftrightarrow} B$ iff A^* is B^*.

THEOREM 4. $A\begin{bmatrix} x \\ x \end{bmatrix} \overset{*}{\leftrightarrow} A$.

Proof. $A\begin{bmatrix} x \\ x \end{bmatrix} \overset{*}{\mapsto} A^*\begin{bmatrix} x \\ x \end{bmatrix} \overset{*}{\mapsto} A^* \overset{*}{\leftarrow\!\shortmid} A$, where the second step is shown by structural induction on the simple term A^*, using the compilation clauses

$$x\begin{bmatrix} x \\ x \end{bmatrix} \mapsto x, \qquad a\begin{bmatrix} x \\ x \end{bmatrix} \mapsto a, \qquad PQ\begin{bmatrix} x \\ x \end{bmatrix} \mapsto (P\begin{bmatrix} x \\ x \end{bmatrix})(Q\begin{bmatrix} x \\ x \end{bmatrix}),$$

where a is any atom other than x. ∎

FREE VARIABLES

It is normal to distinguish between *free* and *bound* occurrences of a variable in a term: for example, in the term $(\lambda x.yx)x$ one would say the first two occurrences of x are bound and the third is free; also, any occurrence of x in A is bound in $A\begin{bmatrix} X \\ x \end{bmatrix}$. However, the notion of an *occurrence* of a variable is rather hard to handle rigorously – though it can be done (Lalement, 1993, §2.2.4). It is much easier to define a binary relation 'occurs-free-in' between variables and terms, and bypass 'occurrences' altogether. The intention is that x occurs-free-in A iff x occurs (in the informal sense) in A^*. The 'occurs-free-in' relation is all that is needed; there is no need to define 'occurrence', 'bound variable' or 'scope'.

DEFINITION. The relation $v \in T$ (read informally as 'the variable v occurs free in the term T') is defined as follows.

$v \in a$	iff v is a, for any atom a
$v \in AB$	iff $v \in A$ or $v \in B$
$v \in (\lambda A.B)$	iff $v \notin A$ and $v \in B$
$v \in A\begin{bmatrix} B \\ x \end{bmatrix}$	iff ($v \in A$ and x is not v) or ($v \in B$ and $x \in A$).

This notation may be extended to sequences of variables and terms: $x, y, \ldots z \in A, B, \ldots C$ means that *each* of the variables $x, y, \ldots z$ occurs free in *each* of the terms $A, B, \ldots C$; while $x, y, \ldots z \notin A, B, \ldots C$ means that *none* of the variables occurs free in *any* of the terms. The *free variables* of T are the variables v such that $v \in T$.

EXAMPLES. Let x, y and z be any three variables.

(1) $x \in (kx)(yz)$ since $x \in kx$ since $x \in x$.

(2) $x \in (\lambda y.xzy)$ since $x \notin y$ and $x \in xzy$, since $x \in xz$ since $x \in x$.

(3) $y \notin (\lambda y.xzy)$ since $y \in y$.

EXERCISES. Again, let x, y and z be any three variables. Show that

$$y \in (ky)\begin{bmatrix} z \\ x \end{bmatrix}, \qquad y \notin (ky)\begin{bmatrix} z \\ y \end{bmatrix}, \qquad z \notin (ky)\begin{bmatrix} z \\ x \end{bmatrix}, \qquad z \in (ky)\begin{bmatrix} z \\ y \end{bmatrix}.$$

THEOREM 5. Any term has finitely many free variables.

THEOREM 6. If $A \mapsto B$ then $x \in A$ iff $x \in B$. Hence $x \in A$ iff $x \in A^*$.

THEOREM 7. $A\begin{bmatrix} X \\ x \end{bmatrix} \overset{*}{\leftrightarrow} A$, if $x \notin A$.

Proof. $A\begin{bmatrix} X \\ x \end{bmatrix} \overset{*}{\mapsto} A^*\begin{bmatrix} X \\ x \end{bmatrix} \overset{*}{\mapsto} A^* \overset{*}{\mapsfrom} A$, where the second step is verified by structural induction on the simple term A^* using the compilation clauses

$$a\begin{bmatrix} X \\ x \end{bmatrix} \mapsto a, \qquad PQ\begin{bmatrix} X \\ x \end{bmatrix} \mapsto (P\begin{bmatrix} X \\ x \end{bmatrix})(Q\begin{bmatrix} X \\ x \end{bmatrix}),$$

where a is any atom other than x. ∎

BINDING INDEPENDENCE

I shall often define a term by saying something like 'let F be $(\lambda x.(\lambda y.Ay(Bx)))$, where x and y are two variables', without saying what the variables x and y are. To justify definitions of this sort it is necessary to show that the term F is the same, up to compilation, regardless of the choice of x and y, as long as they are not the same variable and are not already present in A and B. To state this formally, I introduce the following temporary metanotation.

DEFINITION. The relation $v \prec T$ (informally, *v occurs in T*) is defined by

$$
\begin{array}{ll}
v \prec a & \text{iff } v \text{ is } a, \text{ for any atom } a \\
v \prec AB & \text{iff } v \prec A \text{ or } v \prec B \\
v \prec (\lambda A.B) & \text{iff } v \prec A \text{ or } v \prec B \\
v \prec A\begin{bmatrix} B \\ x \end{bmatrix} & \text{iff } v \prec A \text{ or } v \prec B \text{ or } v \text{ is } x.
\end{array}
$$

THEOREM 8. For any T, there are only finitely many v such that $v \prec T$.

EXAMPLE. x, y and z are the variables occurring in $(\lambda x.xy\begin{bmatrix} y \\ z \end{bmatrix})$.

THEOREM 9. If $A \mapsto B$ and $v \prec B$ then $v \prec A$.

DEFINITION. $T\left\{{y \atop x}\right\}$ (informally, *T with x replaced by y*), is defined by

$$
\begin{array}{rll}
x\left\{{y \atop x}\right\} & \text{is } y & \\
a\left\{{y \atop x}\right\} & \text{is } a & \text{if } a \text{ is an atom other than } x \\
(PQ)\left\{{y \atop x}\right\} & \text{is } (P\left\{{y \atop x}\right\})(Q\left\{{y \atop x}\right\}) & \\
(\lambda A.B)\left\{{y \atop x}\right\} & \text{is } (\lambda A\left\{{y \atop x}\right\}.B\left\{{y \atop x}\right\}) & \\
(T\begin{bmatrix} V \\ v \end{bmatrix})\left\{{y \atop x}\right\} & \text{is } (T\left\{{y \atop x}\right\})\begin{bmatrix} V\left\{{y \atop x}\right\} \\ v\left\{{y \atop x}\right\} \end{bmatrix}. &
\end{array}
$$

(Observe that, in $T\left\{{y \atop x}\right\}$, y is substituted for x indiscriminately, without regard for binding.)

THEOREM 10. If $A \not\mapsto$ then $A\begin{bmatrix} y \\ x \end{bmatrix} \overset{*}{\mapsto} A\left\{{y \atop x}\right\}$.

THEOREM 11. If $A \mapsto B$ and $y \not\prec A$ then $A\left\{{y \atop x}\right\} \mapsto B\left\{{y \atop x}\right\}$.

THEOREM 12. If $x \notin A$ and $y \not\prec A$ then $A\left\{{y \atop x}\right\} \overset{*}{\leftrightarrow} A$.

Proof. Let $A \mapsto A' \mapsto \cdots \mapsto A^* \not\mapsto$. Since $y \not\prec A$ we have $y \not\prec A, A', \ldots A^*$ by Theorem 9. Hence $A\left\{{y \atop x}\right\} \mapsto A'\left\{{y \atop x}\right\} \mapsto \cdots \mapsto A^*\left\{{y \atop x}\right\}$ by Theorem 11. Also, $x \notin A^*$ since $x \notin A$, by Theorem 6. So

$$
A\left\{{y \atop x}\right\} \overset{*}{\mapsto} A^*\left\{{y \atop x}\right\} \overset{*}{\mapsfrom} A^*\begin{bmatrix} y \\ x \end{bmatrix} \overset{*}{\leftrightarrow} A^* \overset{*}{\mapsfrom} A
$$

by Theorems 10 and 7, as required. ∎

DEFINITION. A notation (whether metanotation or an official part of the language) is *binding-independent* iff the choice of variables for its bound metavariables makes no difference, up to compilation, provided different metavariables denote different variables not otherwise occurring in the notation. More precisely, a notation (denoting a term) is binding-independent iff replacing all occurrences of a variable that does not occur free in the term by a variable that does not occur in the notation leaves the term unchanged, up to compilation.

THEOREM 13. All constructs of the Expanded Term Language are binding-independent.

Proof. This simply restates the previous theorem. ▮

Thus the choice of x in $(\lambda x.A)$ or $A\begin{bmatrix}X\\x\end{bmatrix}$ makes no difference, up to compilation, provided $x \notin X$.

EXAMPLE. Let u, v, w, x, y, z be six variables. Then, by binding independence,

$$(\lambda x.ky(xzw\begin{bmatrix}y\\z\end{bmatrix})) \overset{*}{\leftrightarrow} (\lambda u.ky(uvw\begin{bmatrix}y\\v\end{bmatrix})).$$

REDUCTION IN THE EXPANDED TERM LANGUAGE

DEFINITION. The reduction relations $\triangleright$, $\triangleright^*$, $\triangleleft$, $\triangleleft^*$, $\triangleright^*\triangleleft$ and $\not\triangleright$ in the Term Language are extended to the expanded language by

$$\begin{array}{llll} A \triangleright B & \text{iff } A^* \triangleright B^* & A \triangleright^* B & \text{iff } A^* \triangleright^* B^* \\ A \triangleleft B & \text{iff } A^* \triangleleft B^* & A \triangleright^*\triangleleft B & \text{iff } A^* \triangleright^*\triangleleft B^* \\ A \not\triangleright & \text{iff } A^* \not\triangleright \,. & & \end{array}$$

EXAMPLES. Let x and y be two variables.

(1) $(\lambda x.(kx)(yy))z \triangleright ((\lambda x.kx)z)((\lambda x.yy)z)$, since
$(\lambda x.(kx)(yy))z \overset{*}{\mapsto} s(s(kk)id)(s(ky)(ky))z \triangleright ((s(kk)id)z)((s(ky)(ky))z)$
$\overset{*}{\leftarrow\!\dashv} ((\lambda x.kx)z)((\lambda x.yy)z)$.

(2) $equal(former\ x)(latter\ y)\begin{bmatrix}xk,ky\\y,x\end{bmatrix} \triangleright^* equal\ k\,k \triangleright true$, since
$(equal(former\ x)(latter\ y)\begin{bmatrix}xk,ky\\y,x\end{bmatrix})^*$ is $equal(former(ky))(latter((ky)k))$.

EXERCISE. Let x, y be two variables. Show that $equal(former\ x)(latter\ y)\begin{bmatrix}ky,xk\\x,y\end{bmatrix}$ reduces only to itself.

DEFINITION. A *construction* is an irreducible term with no free variables.

That is, a term is a construction in the Expanded Term Language iff it compiles into a construction of the Term Language.

EXERCISE. Let x, y, z be three variables. Verify that $(\lambda y.(\lambda x.xy))$ and $(\lambda x.xyz)\begin{bmatrix} kz \\ y \end{bmatrix}\begin{bmatrix} true \\ z \end{bmatrix}$ are constructions.

THEOREM 14. (The instantiation theorem.)

- If $X \not\triangleright$ and $A \triangleright^* B$ then $A\begin{bmatrix} X \\ x \end{bmatrix} \triangleright^* B\begin{bmatrix} X \\ x \end{bmatrix}$.
- If $X \not\triangleright$ and $A \not\triangleright$ then $A\begin{bmatrix} X \\ x \end{bmatrix} \not\triangleright$.
- If $X \triangleright^* Y$ then $A\begin{bmatrix} X \\ x \end{bmatrix} \triangleright^* A\begin{bmatrix} Y \\ x \end{bmatrix}$.

Proof. For the first part, since $A\begin{bmatrix} X \\ x \end{bmatrix} \overset{*}{\mapsto} A^*\begin{bmatrix} X^* \\ x \end{bmatrix}$ and $B\begin{bmatrix} X \\ x \end{bmatrix} \overset{*}{\mapsto} B^*\begin{bmatrix} X^* \\ x \end{bmatrix}$ it is sufficient to show $A^*\begin{bmatrix} X^* \\ x \end{bmatrix} \triangleright^* B^*\begin{bmatrix} X^* \\ x \end{bmatrix}$. This follows from the $\triangleright$ clauses in the Term Language, as in Theorem 5 of the Term Language. The other two parts are similar. ∎

EXAMPLES.

(1) Consider the reduction

$$\begin{aligned} &\mathit{if}\ (\mathit{equal}\ (\mathit{latter}(sx))\, x)\ (kx)\ \mathit{nil} \\ &\triangleright\ \mathit{if}\ (\mathit{equal}\, x\, x)\ (kx)\ \mathit{nil} \\ &\triangleright\ \mathit{if}\ \mathit{true}\ (kx)\ \mathit{nil}\ \triangleright\ kx \not\triangleright\ . \end{aligned}$$

If we instantiate x by ky throughout and compile we obtain

$$\begin{aligned} &\mathit{if}\ (\mathit{equal}\ (\mathit{latter}(s(ky)))\, (ky))\ (k(ky))\ \mathit{nil} \\ &\triangleright\ \mathit{if}\ (\mathit{equal}\, (ky)\, (ky))\ (k(ky))\ \mathit{nil} \\ &\triangleright\ \mathit{if}\ \mathit{true}\ (k(ky))\ \mathit{nil}\ \triangleright\ k(ky) \not\triangleright\ . \end{aligned}$$

(2) $\mathit{if}\ (\mathit{equal}\, (\mathit{former}\, x)\, s)\ (kx)\ \mathit{nil}\ \triangleright\ \mathit{if}\ (\mathit{equal}\, (\mathit{former}\, x)\, s)\ (kx)\ \mathit{nil}$, so, instantiating x by ky, we obtain $\mathit{if}\ (\mathit{equal}\, (\mathit{former}\, (ky))\, s)\ (k(ky))\ \mathit{nil}\ \triangleright^*$ $\mathit{if}\ (\mathit{equal}\, (\mathit{former}\, (ky))\, s)\ (k(ky))\ \mathit{nil}$, which is trivially true since $\triangleright^*$ is reflexive. This tells us nothing about the reduction of this term, which is as follows:

$$\begin{aligned} &\mathit{if}\ (\mathit{equal}\, (\mathit{former}\, (ky))\, s)\ (k(ky))\ \mathit{nil}\ \triangleright\ \mathit{if}\ (\mathit{equal}\, k\, s)\ (k(ky))\ \mathit{nil} \\ &\triangleright\ \mathit{if}\ \mathit{false}\ (k(ky))\ \mathit{nil}\ \triangleright\ \mathit{nil} \not\triangleright\ . \end{aligned}$$

THEOREM 15. $(\lambda A.B) \not\triangleright$.

Proof. $(\lambda A.B) \overset{*}{\mapsto} (\lambda A^*.B^*)$, which is shown to be irreducible by structural induction on the simple term A^*, as follows.

Case 1: A^* is of the form PQ. Then

$$(\lambda A^*.B^*) \mapsto s(s(k(\lambda P.(\lambda Q.B^*)))\mathit{former})\mathit{latter} \not\rhd,$$

since, by inductive hypothesis, $(\lambda P.(\lambda Q.B^*)) \not\rhd$;

Case 2: A^* is an atom a. Then, using structural induction on the simple term B^*,

- $(\lambda a.UV) \mapsto s(\lambda a.U)(\lambda a.V) \not\rhd$, since, by inductive hypothesis, $(\lambda a.U) \not\rhd$ and $(\lambda a.V) \not\rhd$;
- $(\lambda a.a) \mapsto \mathit{id} \not\rhd$, if a is a variable;
- $(\lambda a.b) \mapsto kb \not\rhd$, if a and b are atoms but not both the same variable. ∎

The $(\lambda x.B)$ notation is intended to suggest a function that maps any x to B, so we require $(\lambda x.B)x \rhd^* B$ to hold. Similarly we would expect $(\lambda A.B)A \rhd^* B$ to hold: that is, $(\lambda A.B)$ should be a function that maps any construction that is an instantiation of A to the corresponding instantiation of B. On second thoughts, this is only feasible in the case where A is irreducible (or reduces to an irreducible term). The following theorem is a generalisation that not only gives $(\lambda A.B)A \rhd^* B$ when $A \not\rhd$ but can also give useful results when A is reducible.

THEOREM 16. $(\lambda A.B)A\left[\frac{\underline{V}}{\underline{v}}\right] \rhd^* B\left[\frac{\underline{V}}{\underline{v}}\right]$, provided $\underline{V} \not\rhd$ and $A\left[\frac{\underline{V}}{\underline{v}}\right] \not\rhd$.

Proof. $(\lambda A.B)A\left[\frac{\underline{V}}{\underline{v}}\right] \overset{*}{\mapsto} (\lambda A^*.B^*)A^*\left[\frac{\underline{V}}{\underline{v}}\right] \rhd^* B^*\left[\frac{\underline{V}}{\underline{v}}\right] \overset{*}{\mapsfrom} B\left[\frac{\underline{V}}{\underline{v}}\right]$, as required, where the second step is verified by structural induction on the simple term A^* as follows.

Case 1: A^* is PQ. Then $P\left[\frac{\underline{V}}{\underline{v}}\right] \not\rhd$ and $Q\left[\frac{\underline{V}}{\underline{v}}\right] \not\rhd$, and hence

$$\begin{aligned}
(\lambda A^*.B^*)A^*\left[\tfrac{\underline{V}}{\underline{v}}\right] &\overset{*}{\mapsto} s(s(k(\lambda P.(\lambda Q.B^*))\left[\tfrac{\underline{V}}{\underline{v}}\right])\mathit{former})\mathit{latter}(PQ\left[\tfrac{\underline{V}}{\underline{v}}\right]) \\
&\rhd^* (\lambda P.(\lambda Q.B^*))\left[\tfrac{\underline{V}}{\underline{v}}\right](\mathit{former}(PQ\left[\tfrac{\underline{V}}{\underline{v}}\right]))(\mathit{latter}(PQ\left[\tfrac{\underline{V}}{\underline{v}}\right])) \\
&\rhd^* (\lambda P.(\lambda Q.B^*))\left[\tfrac{\underline{V}}{\underline{v}}\right](P\left[\tfrac{\underline{V}}{\underline{v}}\right])(Q\left[\tfrac{\underline{V}}{\underline{v}}\right]) \\
&\overset{*}{\mapsfrom} (\lambda P.(\lambda Q.B^*))P\left[\tfrac{\underline{V}}{\underline{v}}\right](Q\left[\tfrac{\underline{V}}{\underline{v}}\right]) \rhd^* (\lambda Q.B^*)\left[\tfrac{\underline{V}}{\underline{v}}\right](Q\left[\tfrac{\underline{V}}{\underline{v}}\right]) \\
&\overset{*}{\mapsfrom} (\lambda Q.B^*)Q\left[\tfrac{\underline{V}}{\underline{v}}\right] \rhd^* B^*\left[\tfrac{\underline{V}}{\underline{v}}\right],
\end{aligned}$$

as required, using the inductive hypothesis twice.

Case 2: A^* is an atom, a. Then the required reduction, $(\lambda a.B^*)a\left[\frac{V}{\underline{v}}\right] \triangleright^* B^*\left[\frac{V}{\underline{v}}\right]$, follows by the instantiation theorem (14) from $(\lambda a.B^*)a \triangleright^* B^*$, which is verified by structural induction on the simple term B^*, as follows:

- if B^* is XY then $(\lambda a.XY)a \mapsto s(\lambda a.X)(\lambda a.Y)a \triangleright ((\lambda a.X)a)((\lambda a.Y)a) \triangleright^* XY$, by inductive hypothesis, as required;
- if B^* is a and a is a variable then $(\lambda a.a)a \mapsto id\, a \triangleright a$, as required;
- if B^* is an atom b, where a and b are not both the same variable, then $(\lambda a.b)a \mapsto kba \triangleright b$, as required. ∎

EXAMPLE. Let w, x, y, z be four variables. Then $(\lambda x.\mathit{if}\ (\mathit{equal}\, x\, y)\ w\ z)x \overset{*}{\mapsto} s(s(s(k\ \mathit{if})(s(s(k\ \mathit{equal})\mathit{id})(ky))(kw))(kz))x \triangleright^* \mathit{if}\ (\mathit{equal}\, x\, y)\ w\ z$.

EXERCISE. Show that $(\lambda x.kx(\mathit{equal}\, x\, x))x \triangleright^* kx(\mathit{equal}\, x\, x)$. Find, however, another reduction of $(\lambda x.kx(\mathit{equal}\, x\, x))x$ that does not pass through $kx(\mathit{equal}\, x\, x)$.

This exercise shows that, even though $(\lambda x.B)x$ may be reduced to B, it is also possible (in cases where B is reducible) to reduce in a way that bypasses B by carrying out the reductions in a different order. The indeterminism in reduction order arises because, when reducing XY, one has a choice of whether to reduce X first, or reduce Y first, or interleave the reductions. Indeed, this example reveals why the indeterminism is necessary. If the order were deterministic then the reduction mechanism would have to be clairvoyant to decide the reduction order that ensures that $(\lambda x.B)x$ reduces to B in all cases. Only by allowing a free choice of order in reducing subterms can one ensure that at least *some* of the reductions of $(\lambda x.B)x$ pass through B.

EXAMPLE. Let x and y be two variables. Then $(\lambda xy.yx)(xy)\left[{\mathit{true},\mathit{false} \atop x,\ \ y}\right] \triangleright^* yx\left[{\mathit{true},\mathit{false} \atop x,\ \ y}\right]$, since

$$(\lambda xy.yx)(xy)\left[{\mathit{true},\mathit{false} \atop x,\ \ y}\right]$$

$$\mapsto s(s(k(\lambda x.(\lambda y.yx)))\mathit{former})\mathit{latter}(xy)\left[{\mathit{true},\mathit{false} \atop x,\ \ y}\right]$$

$$\overset{*}{\mapsto} s(s(k(s(s(ks)(k\ id))(s(kk)id)))\mathit{former})\mathit{latter}(xy)\left[{\mathit{true},\mathit{false} \atop x,\ \ y}\right]$$

$$\overset{*}{\mapsto} s(s(k(s(s(ks)(k\ id))(s(kk)id)))\mathit{former})\mathit{latter}(\mathit{true}\,\mathit{false})$$

$$\triangleright^* s(s(ks)(k\ id))(s(kk)id)\ \mathit{true}\,\mathit{false}$$

$$\triangleright^* \mathit{false}\ \mathit{true} \overset{*}{\mapsfrom} yx\left[{\mathit{true},\mathit{false} \atop x,\ \ y}\right].$$

THEOREM 17.

- $(\lambda x.T)X \;\triangleright^*\; T\begin{bmatrix}X\\x\end{bmatrix}$, if $X \not\triangleright$;
- $(\lambda x.(\lambda y.T))XY \;\triangleright^*\; T\begin{bmatrix}X\\x\end{bmatrix}\begin{bmatrix}Y\\y\end{bmatrix}$, if $X, Y \not\triangleright$ and $y \notin X, x$.

Proof. Theorem 16 gives $(\lambda x.T)x \;\triangleright^*\; T$ and hence $(\lambda x.(\lambda y.T))xy \;\triangleright^*\; T$, so the conclusions follow by instantiating x by X and y by Y (Theorem 14). ∎

EXAMPLE. Note that, if X is irreducible, $(\lambda x.(\lambda y.T))X$ reduces to $(\lambda y.T)\begin{bmatrix}X\\x\end{bmatrix}$, not to $(\lambda y.T\begin{bmatrix}X\\x\end{bmatrix})$. For example,

$$(\lambda x.(\lambda y.sxy))(true\,false) \;\triangleright^*\; (\lambda y.sxy)\begin{bmatrix}true\,false\\x\end{bmatrix} \stackrel{*}{\mapsto} s(s(ks)(kx))id\begin{bmatrix}true\,false\\x\end{bmatrix}$$
$$\stackrel{*}{\mapsto} s(s(ks)(k(true\,false)))id,$$

assuming x and y are two variables, which is different from $(\lambda y.sxy\begin{bmatrix}true\,false\\x\end{bmatrix})$ $\stackrel{*}{\mapsto} s(s(ks)(s(k\,true)(k\,false)))id$. The difference is that where the former has $k(true\,false)$ the latter has $s(k\,true)(k\,false)$. The former represents a function that takes an input Y and produces a term $s(true\,false)Y$ using the construction *true false*; the latter actually *builds* the construction *true false* out of its atomic components *true* and *false*, whereas in the former case *true false* was supplied prefabricated.

The distinction, then, between $(\lambda y.T)\begin{bmatrix}X\\x\end{bmatrix}$ and $(\lambda y.T\begin{bmatrix}X\\x\end{bmatrix})$ arises from the strictly intensional view of functions. Providing X as a prefabricated component and providing instructions for building X are two very different things.

EVALUABLE TERMS

An evaluable term is simply a term that reduces to an irreducible term. Evaluable terms have many useful properties similar to those of irreducible terms.

DEFINITION. A term T is *evaluable* iff *equal* $T\;T \;\triangleright^*\; true$, or, equivalently, iff $T \;\triangleright^*\; T' \not\triangleright$ for some term T'.

THEOREM 18. If $A \;\triangleright^*\!\triangleleft\; B$ then A is evaluable iff B is evaluable.

THEOREM 19. If AB is evaluable then so are A and B.

Proof. The only way AB can reduce to an irreducible term is by first reducing A and B independently to irreducible terms A' and B', and then reducing $A'B'$ to an irreducible term; this implies that A and B are evaluable. ∎

THEOREM 20. If $T\begin{bmatrix}X\\x\end{bmatrix}$ is evaluable and $x \in T$ then X is evaluable.

Proof. If $T\begin{bmatrix}X\\x\end{bmatrix}$ is evaluable then so is $T^*\begin{bmatrix}X\\x\end{bmatrix}$, since $T\begin{bmatrix}X\\x\end{bmatrix} \overset{*}{\mapsto} T^*\begin{bmatrix}X\\x\end{bmatrix}$. The conclusion then follows from the previous theorem by structural induction on the simple term T^*. ∎

THEOREM 21. If X is evaluable then $(\lambda x.T)X \vartriangleright^* \vartriangleleft T\begin{bmatrix}X\\x\end{bmatrix}$.

Proof. Let $X \vartriangleright^* X' \not\vartriangleright$. Then $(\lambda x.T)X \vartriangleright^* (\lambda x.T)X' \vartriangleright^* T\begin{bmatrix}X'\\x\end{bmatrix} \vartriangleleft^* T\begin{bmatrix}X\\x\end{bmatrix}$ by Theorems 17 and 14. ∎

THEOREM 22. If T and X are evaluable then so are $T\begin{bmatrix}X\\x\end{bmatrix}$ and $(\lambda x.T)X$.

Proof. Let $T \vartriangleright^* T' \not\vartriangleright$ and $X \vartriangleright^* X' \not\vartriangleright$. Then $T\begin{bmatrix}X\\x\end{bmatrix} \vartriangleright^* T\begin{bmatrix}X'\\x\end{bmatrix} \vartriangleright^* T'\begin{bmatrix}X'\\x\end{bmatrix} \not\vartriangleright$ by Theorem 14, so $T\begin{bmatrix}X\\x\end{bmatrix}$ is evaluable. $(\lambda x.T)X$ is also evaluable, since $(\lambda x.T)X \vartriangleright^* \vartriangleleft T\begin{bmatrix}X\\x\end{bmatrix}$. ∎

THEOREM 23. If X is evaluable and $A \vartriangleright^* \vartriangleleft B$ then $A\begin{bmatrix}X\\x\end{bmatrix} \vartriangleright^* \vartriangleleft B\begin{bmatrix}X\\x\end{bmatrix}$.

Proof. Let $X \vartriangleright^* X' \not\vartriangleright$ and $A \vartriangleright^* C \vartriangleleft^* B$. Then $A\begin{bmatrix}X\\x\end{bmatrix} \vartriangleright^* A\begin{bmatrix}X'\\x\end{bmatrix} \vartriangleright^* C\begin{bmatrix}X'\\x\end{bmatrix} \vartriangleleft^* B\begin{bmatrix}X'\\x\end{bmatrix} \vartriangleleft^* B\begin{bmatrix}X\\x\end{bmatrix}$ by Theorem 14. ∎

MORE ON λ-ABSTRACTIONS

EXAMPLE. Theorem 16 shows that $(\lambda X.T)$ is a function that expects its argument to be an instantiation of X. But what does $(\lambda X.T)$ do if provided with an argument Y that is not of this form? The answer is that it insists on interpreting Y as an instantiation of X, without noticing that it is not, and produces the corresponding instantiation of T. For example,

$$\begin{aligned}(\lambda kx.xy)(sss) &\overset{*}{\mapsto} s(s(k(\lambda k.(\lambda x.xy)))\mathit{former})\mathit{latter}(sss)\\ &\vartriangleright^* (\lambda k.(\lambda x.xy))(ss)s \vartriangleright sy,\end{aligned}$$

assuming x and y are two variables. In this case the function $(\lambda kx.xy)$ has tried to interpret sss as an instantiation of kx, which means matching s against x and producing the result sy; it doesn't notice that ss fails to match k. In general, the result of applying $(\lambda X.T)$ to a construction Y is *always* an instantiation of T, as the next two theorems establish.

THEOREM 24. If a is a constant and $X \not\triangleright$ then $(\lambda a.T)X \triangleright^* T$.

Proof. $(\lambda a.T)X \overset{*}{\mapsto} (\lambda a.T^*)X \triangleright^* T^* \overset{*}{\leftarrow\!\shortmid} T$, as required, where the second step is verified by structural induction on the simple term T^* as follows.

Case 1: T^* is an atom, b. Then $(\lambda a.T^*)X \mapsto kbX \triangleright b$, as required.

Case 2: T^* is UV. Then

$$(\lambda a.T^*)X \mapsto s(\lambda a.U)(\lambda a.V)X \triangleright ((\lambda a.U)X)((\lambda a.V)X) \triangleright^* UV,$$

by inductive hypothesis, as required. ▮

THEOREM 25. If $(\lambda X.T)Y$ is evaluable, where X has free variables $\underline{v}$ and $\underline{v} \notin Y$, then $(\lambda X.T)Y \triangleright^* T\left[\frac{\underline{V}}{\underline{v}}\right]$ for some irreducible terms $\underline{V}$ whose free variables are included in those of Y. In the special case where Y is a construction the condition that $(\lambda X.T)Y$ be evaluable may be dropped.

Proof. Since $(\lambda X.T)Y$ is evaluable, so is Y; thus $(\lambda X.T)Y \triangleright^* (\lambda X^*.T)Y'$ for some irreducible term Y'. I shall show that $(\lambda X^*.T)Y' \triangleright^* T\left[\frac{\underline{V}}{\underline{v}}\right]$ using structural induction on the simple term X^*. The inductive hypothesis is a slight generalisation of this: if $(\lambda X^*.T)\left[\frac{\underline{U}}{\underline{u}}\right] Y'$ is evaluable then $(\lambda X^*.T)\left[\frac{\underline{U}}{\underline{u}}\right] Y'$ $\triangleright^* T\left[\frac{\underline{V}}{\underline{v}}\right]\left[\frac{\underline{U}}{\underline{u}}\right]$, provided $\underline{U} \not\triangleright$ and $\underline{u} \notin Y'$.

Case 1: X^* is a constant, a. Then $(\lambda a.T)Y' \triangleright^* T$, by the previous theorem, so $(\lambda a.T)\left[\frac{\underline{U}}{\underline{u}}\right] Y' \triangleright^* T\left[\frac{\underline{U}}{\underline{u}}\right]$, by Theorem 14, as required.

Case 2: X^* is a variable, v. Then $(\lambda v.T)Y' \triangleright^* T\left[\frac{Y'}{v}\right]$, by Theorem 17, so $(\lambda v.T)\left[\frac{\underline{U}}{\underline{u}}\right] Y' \triangleright^* T\left[\frac{Y'}{v}\right]\left[\frac{\underline{U}}{\underline{u}}\right]$, by Theorem 14, as required.

Case 3: X^* is PQ. Then, from the compilation clause for $(\lambda PQ.T)$,

$$(\lambda PQ.T)\left[\frac{\underline{U}}{\underline{u}}\right] Y' \triangleright^* (\lambda P.(\lambda Q.T))\left[\frac{\underline{U}}{\underline{u}}\right] (\mathit{former}\ Y')(\mathit{latter}\ Y')$$

and since this is evaluable we have $\mathit{former}\ Y' \triangleright^* Y_1 \not\triangleright$ and $\mathit{latter}\ Y' \triangleright^*$ $Y_2 \not\triangleright$ for some Y_1, Y_2. The reduction continues

$$\begin{aligned}(\lambda P.(\lambda Q.T))\left[\frac{\underline{U}}{\underline{u}}\right] (\mathit{former}\ Y')(\mathit{latter}\ Y') &\triangleright^* (\lambda P.(\lambda Q.T))\left[\frac{\underline{U}}{\underline{u}}\right] Y_1 Y_2 \\ &\triangleright^* (\lambda Q.T)\left[\frac{\underline{M}}{\underline{m}}\right]\left[\frac{\underline{U}}{\underline{u}}\right] Y_2\end{aligned}$$

by inductive hypothesis, where $\underline{m}$ are the free variables of P and $\underline{M}$ are irreducible terms whose free variables are included in those of Y_1 (and hence in those of Y). Applying the inductive hypothesis again,

$$(\lambda Q.T)\left[\frac{\underline{M}}{\underline{m}}\right]\left[\frac{\underline{U}}{\underline{u}}\right] Y_2 \triangleright^* T\left[\frac{\underline{N}}{\underline{n}}\right]\left[\frac{\underline{M}}{\underline{m}}\right]\left[\frac{\underline{U}}{\underline{u}}\right] \overset{*}{\leftrightarrow} T\left[\frac{\underline{V}}{\underline{v}}\right]\left[\frac{\underline{U}}{\underline{u}}\right]$$

where $\underline{n}$ are the free variables of Q, $\underline{N}$ are irreducible terms whose free variables are included in those of Y_2 (and hence in those of Y), and $\underline{V}$ is a rearrangement of $\underline{N}, \underline{M}$, eliminating repetitions of instantiations of the same variable. This gives $(\lambda PQ.T)\left[\frac{\underline{U}}{\underline{u}}\right] Y' \ \rhd^* \ T\left[\frac{\underline{V}}{\underline{v}}\right]\left[\frac{\underline{U}}{\underline{u}}\right]$ as required.

This completes the proof of the general case. In the special case where Y is a construction we may drop the condition that $(\lambda X.T)Y$ be evaluable since the two places where it is required in the above argument are (i) to show that Y is evaluable (but we already know this if Y is a construction), and (ii) to show that *former* Y' and *latter* Y' are evaluable (but this is always true if Y is a construction). ∎

BRANCH METANOTATION

This section sets up a conditional construct which is useful in defining functions that branch depending on the value of their argument.

DEFINITION. For any terms C, F and G, the metanotation $(branch\ C\ F\ G)$ means

$$s(s(s(s(k\ if)C)(k\ F))(k\ G))id.$$

THEOREM 26. If $C, F, G \not\rhd$ then $(branch\ C\ F\ G) \not\rhd$.

THEOREM 27. If $C, F, G, X \not\rhd$ then $(branch\ C\ F\ G)X \ \rhd^* \ if\ (CX)\ F\ G\ X$.

PATTERN-MATCHING METANOTATIONS

The $(\lambda A.B)$ construct allows one to define functions by pattern matching. For example, a function *swap*, satisfying the relation

$$swap(u, v) \ \rhd^* \ (v, u),$$

could be defined simply as $(\lambda(u, v).(v, u))$. (Here I am using a pairing notation, '$(\ldots, \ldots)$', which will be formally introduced later.) However, this is a very limited form of pattern matching. In most pattern-matching definitions one needs several defining clauses, to be applied in order, possibly with recursion. For example, suppose one represents a list of constructions A, B, C, D by nested pairing: $(A, (B, (C, (D, nil))))$. Now suppose one wishes to define a function *member* that detects whether a given construction is an element of a given list. One would like to be able to define *member* by simply stipulating its reduction clauses

$$\begin{aligned} member(x, (x, rest)) \ &\rhd^* \ true \\ member(x, (first, rest)) \ &\rhd^* \ member(x, rest) \\ member(x, nil) \ &\rhd^* \ false, \end{aligned}$$

where these clauses are to be applied in order. Thus, to determine whether C is a member of $(A,(B,(C,(D,nil))))$, we reduce

$$member(C,(A,(B,(C,(D,nil))))) \rhd^* member(C,(B,(C,(D,nil)))) \rhd^* member(C,(C,(D,nil))) \rhd^* true.$$

Here, the argument pair $(C,(A,(B,(C,(D,nil)))))$ is first matched against the pattern $(x,(x,rest))$. It doesn't match, so it is next matched against the pattern $(x,(first,rest))$. This time it does match, with x matching C, *first* matching A, and *rest* matching $(B,(C,(D,nil)))$, so we apply this clause, giving $member(C,(B,(C,(D,nil))))$. Then we apply the definition recursively, and so on. As a second example,

$$member(D,(A,(B,(C,nil)))) \rhd^* member(D,(B,(C,nil))) \rhd^* member(D,(C,nil)) \rhd^* member(D,nil) \rhd^* false.$$

As the first step towards being able to set up such definitions, we need, for each pattern X in a defining clause, a function $check_X$ that determines whether a given argument is of the form X (for some instantiation of the free variables in X). (Recall that $(\lambda X.T)$ simply *assumes* its argument is of the form X without checking.)

DEFINITION. For any term X define a construction $check_X$ by

$check_a$ is *equal* a if a is a constant

$check_v$ is k *true* if v is a variable

$check_{PQ}$ is $(branch\ (s(k\ check_P)former)\ (s(k\ check_Q)latter)\ (k\ false))$

$check_X$ is $check_{X'}$ if $X \mapsto X'$.

THEOREM 28. The '$check_X$' notation is binding-independent; in fact, the choice of variables in X makes no difference at all to $check_X$.

THEOREM 29. Suppose $XY \not\rhd$. Then

- $check_{PQ}(XY) \rhd^* check_Q Y$, if $check_P X \rhd^* true$
- $check_{PQ}(XY) \rhd^* false$, if $check_P X \rhd^* false$.

Proof. $check_{PQ}(XY)$ is

$$\begin{aligned} & (branch\ (s(k\ check_P)former)\ (s(k\ check_Q)latter)\ (k\ false))(XY) \\ \rhd^*\ & if\ (s(k\ check_P)former(XY))\ (s(k\ check_Q)latter)\ (k\ false)\ (XY) \\ \rhd^*\ & if\ (check_P X)\ (s(k\ check_Q)latter)\ (k\ false)\ (XY) \\ \rhd^*\ & \begin{cases} s(k\ check_Q)latter\ (XY) \rhd^* check_Q Y & \text{if } check_P X \rhd^* true \\ k\ false\ (XY) \rhd false & \text{if } check_P X \rhd^* false \end{cases} \end{aligned}$$

as required. ∎

THEOREM 30. $check_X X\left[\frac{\underline{V}}{\underline{v}}\right] \rhd^* true$, provided $\underline{V} \not\rhd$ and $X\left[\frac{\underline{V}}{\underline{v}}\right] \not\rhd$.

Proof. $check_X X\left[\frac{\underline{V}}{\underline{v}}\right] \overset{*}{\mapsto} check_{X^*}(X^*\left[\frac{\underline{V}}{\underline{v}}\right]) \rhd^* true$, as required, where the second step is verified by structural induction on the simple term X^* as follows.

Case 1: X^* is a constant, a. Then $check_a(a\left[\frac{\underline{V}}{\underline{v}}\right]) \overset{*}{\mapsto} equal\ a\ a \rhd true.$

Case 2: X^* is a variable, x. Then $check_x(x\left[\frac{\underline{V}}{\underline{v}}\right])$ is $k\ true\ (x\left[\frac{\underline{V}}{\underline{v}}\right]) \rhd true.$

Case 3: X^* is PQ. Then $check_{PQ}(PQ\left[\frac{\underline{V}}{\underline{v}}\right]) \overset{*}{\mapsto} check_{PQ}((P\left[\frac{\underline{V}}{\underline{v}}\right])(Q\left[\frac{\underline{V}}{\underline{v}}\right]))$ $\rhd^* check_Q(Q\left[\frac{\underline{V}}{\underline{v}}\right])$ (by the previous theorem, since $check_P(P\left[\frac{\underline{V}}{\underline{v}}\right]) \rhd^* true$ by inductive hypothesis) $\rhd^* true$ by inductive hypothesis again. ∎

EXAMPLES. Let u, v, w, x, y be five variables.

(1) $check_{kx}(kx) \rhd^* true,$
(2) $check_{sxy}(s(kz)z) \rhd^* true,$
(3) $check_{(uv)(sw)}((true\,false)(nil\,(sx))) \rhd^* false$ since s mismatches nil,
(4) $check_{suu}(s\,true\,false) \rhd^* true,$
(5) $check_{nil\,nil}(id) \rhd^* true.$

Examples (4) and (5) show the defects of the '$check_X$' construct. In (4) *true* and *false* are both matched against u: there is no attempt to verify that constructions matching the same variable are the same. In (5), *former* and *latter* are applied to the argument *id* to extract its 'components', giving *nil* and *nil*: the $check_{nil\,nil}$ function doesn't notice that *id* is an atom.

These problems are solved by the '$match_X$' notation, which is a refinement of $check_X$.

DEFINITION. For any term X define a construction $match_X$ as

$$(branch\ check_X\ (s(s(k\ equal)(\lambda X.X))id)\ (k\,false)).$$

THEOREM 31. The '$match_X$' notation is binding-independent.

Proof. This follows since the $check_X$ notation and all ET constructs are binding-independent. ∎

THEOREM 32. If $check_X Y \rhd^* false$ then $match_X Y \rhd^* false.$

Proof. If $check_X Y \rhd^* false$ then $Y \rhd^* Y' \not\rhd$, for some Y', and $check_X Y'$ $\rhd^* false$, so $match_X Y \rhd^* match_X Y' \rhd^* if\ (check_X Y')\ (\ldots)\ (k\,false)\ Y' \rhd^*$ $k\,false\ Y' \rhd false.$ ∎

THEOREM 33. $match_X X\left[\frac{V}{\underline{v}}\right] \triangleright^* true$, provided $\underline{V} \not\triangleright$ and $X\left[\frac{V}{\underline{v}}\right] \not\triangleright$.

Proof.

$$\begin{aligned} match_X X\left[\tfrac{V}{\underline{v}}\right] &\triangleright^* if\ (check_X X\left[\tfrac{V}{\underline{v}}\right])\ (s(s(k\ equal)(\lambda X.X))id)\ (k\ false)\ (X\left[\tfrac{V}{\underline{v}}\right]) \\ &\triangleright^* (s(s(k\ equal)(\lambda X.X))id)\ (X\left[\tfrac{V}{\underline{v}}\right]) \\ &\triangleright^* equal\ ((\lambda X.X)X\left[\tfrac{V}{\underline{v}}\right])\ (X\left[\tfrac{V}{\underline{v}}\right]) \\ &\triangleright^* equal\ (X\left[\tfrac{V}{\underline{v}}\right])\ (X\left[\tfrac{V}{\underline{v}}\right]) \triangleright true. \end{aligned}$$

■

THEOREM 34. If $match_X Y \triangleright^* true$, where X has free variables $\underline{v}$ and $\underline{v} \notin Y$, then $Y \triangleright^* \triangleleft\ X\left[\frac{V}{\underline{v}}\right]$ for some irreducible terms $\underline{V}$ whose free variables are included in those of Y.

Proof. Since $match_X Y \triangleright^* true$ we must have $Y \triangleright^* Y' \not\triangleright$ for some Y'. Thus

$$\begin{aligned} match_X Y &\triangleright^* match_X Y' \\ &\triangleright^* if\ (check_X Y')\ (s(s(k\ equal)(\lambda X.X))id)\ (k\ false)\ Y'. \end{aligned}$$

If this term reduces to *true* then $check_X Y' \triangleright^* true$ and

$$s(s(k\ equal)(\lambda X.X))\ id\ Y' \triangleright^* equal\ ((\lambda X.X)Y')\ Y' \triangleright^* true.$$

From the latter it follows that $(\lambda X.X)Y' \triangleright^* Y'$. But by Theorem 25 $(\lambda X.X)Y' \triangleright^* X\left[\frac{V}{\underline{v}}\right]$ for some irreducible terms $\underline{V}$ whose free variables are included in those of Y, hence $Y \triangleright^* Y' \triangleleft^* X\left[\frac{V}{\underline{v}}\right]$, as required. ■

EXAMPLES. Repeating our $check_X$ examples,

(1) $match_{kx}(kx) \triangleright^* true$, by Theorem 33,
(2) $match_{sxy}(s(kz)z) \triangleright^* true$, by Theorem 33,
(3) $match_{(uv)(sw)}((true\ false)(nil\ (sx))) \triangleright^* false$, by Theorem 32,
(4) $match_{suu}(s\ true\ false) \triangleright^* equal\ ((\lambda(suu).suu)(s\ true\ false))\ (s\ true\ false)$ $\triangleright^* equal\ (s\ false\ false)\ (s\ true\ false) \triangleright false$,
(5) $match_{nil\ nil}(id) \triangleright^* equal\ ((\lambda(nil\ nil).nil\ nil)id)\ id \triangleright^* equal\ (nil\ nil)\ id$ $\triangleright false$.

Clearly $match_X$ does not suffer from the same problems as $check_X$. Why not just define $match_X$ as $s(s(k\ equal)(\lambda X.X))id$, since it is this subterm that seems to be doing all the work in the previous theorem? The answer is that although this would give $match_X Y \rhd^* true$ just when we want, it would fail to give a value when we want *false*. For example, $match_{(true\,(nil\,nil))}(false\ x)$ would fail to give a value with this definition, whereas with the definition actually used it successfully reduces to *false*.

The first application of $match_X$ is to introduce a refinement of the $(\lambda X.T)$ construct. Recall that $(\lambda X.T)$ construes its argument as an instantiation of X (even if it is not) and outputs the corresponding instantiation of T. We would prefer a function that produces no result when the argument is not an instantiation of X.

DEFINITION. For any terms X and T define the metanotation $(\lambda X{:}\,T)$ as

$$(branch\ match_X\ (\lambda X.T)\ (fxpt\ id)).$$

THEOREM 35. $(\lambda X{:}\,T) \not\rhd$; and $y \in (\lambda X{:}\,T)$ iff $y \in T$ and $y \notin X$.

THEOREM 36. The '$(\lambda X{:}\,T)$' notation is binding-independent.

Proof. This follows since the '$match_X$' notation and all ET constructs are binding-independent. ∎

THEOREM 37. $(\lambda X{:}\,T)X\left[\frac{\underline{V}}{\underline{v}}\right] \rhd^* T\left[\frac{\underline{V}}{\underline{v}}\right]$, if $\underline{V} \not\rhd$ and $X\left[\frac{\underline{V}}{\underline{v}}\right] \not\rhd$.

Proof. Using Theorems 33 and 16,

$$\begin{aligned}(\lambda X{:}\,T)X\left[\tfrac{\underline{V}}{\underline{v}}\right] &\rhd^* (if\ (match_X X\left[\tfrac{\underline{V}}{\underline{v}}\right])\ ((\lambda X.T)\left[\tfrac{\underline{V}}{\underline{v}}\right])\ (fxpt\ id))\ (X\left[\tfrac{\underline{V}}{\underline{v}}\right])\\ &\rhd^* ((\lambda X.T)\left[\tfrac{\underline{V}}{\underline{v}}\right])(X\left[\tfrac{\underline{V}}{\underline{v}}\right]) \rhd^* T\left[\tfrac{\underline{V}}{\underline{v}}\right]\end{aligned}$$

as required. ∎

THEOREM 38. If $(\lambda X{:}\,T)\left[\frac{\underline{U}}{\underline{u}}\right] Y$ is evaluable, where $\underline{U}$ are irreducible terms, X has free variables $\underline{v}$ and $\underline{v} \notin Y$, then $Y \rhd^*\lhd X\left[\frac{\underline{V}}{\underline{v}}\right]$ for some irreducible terms $\underline{V}$ whose free variables are included in those of Y.

Proof. Since $(\lambda X{:}\,T)\left[\frac{\underline{U}}{\underline{u}}\right] Y$ is evaluable we have, for some irreducible Y', $(\lambda X{:}\,T)\left[\frac{\underline{U}}{\underline{u}}\right] Y \rhd^* (\lambda X{:}\,T)\left[\frac{\underline{U}}{\underline{u}}\right] Y' \rhd^*\ if\ (match_X Y')\ ((\lambda X.T)\left[\frac{\underline{U}}{\underline{u}}\right])\ (fxpt\ id)\ Y'$, which is only evaluable if $match_X Y'$ reduces to *true*, in which case $Y \rhd^* Y' \rhd^*\lhd X\left[\frac{\underline{V}}{\underline{v}}\right]$ by Theorem 34, as required. ∎

DEFINING FUNCTIONS USING PATTERN MATCHING

This section introduces a more powerful mechanism for defining functions with several defining clauses and recursion. The special symbol '$\triangleq$' will be used to identify definitions of this sort.

DEFINITION. The expression

$$\begin{aligned} f\,X_1 &\triangleq T_1 \\ f\,X_2 &\triangleq T_2 \\ &\vdots \\ f\,X_k &\triangleq T_k, \end{aligned}$$

where $f \notin X_1, \ldots X_k$, means that a metaconstant is about to be introduced to denote the term *fxpt* Φ, where Φ is

$$(\lambda f.(\mathit{branch}\ \mathit{match}_{X_1}\ (\lambda X_1.T_1)\ (\mathit{branch}\ \mathit{match}_{X_2}\ (\lambda X_2.T_2)\ \ldots\ (\mathit{branch}\ \mathit{match}_{X_k}\ (\lambda X_k.T_k)f)\ \ldots))).$$

The following theorems will show that *fxpt* Φ is a function that, applied to a construction Y, matches Y successively against the patterns $X_1, X_2, \ldots$, until it finds the first pattern X_i that matches (that is, $Y \overset{*}{\leftrightarrow} X_i\left[\frac{\underline{V}}{\underline{v}}\right]$ for some constructions $\underline{V}$ and variables $\underline{v}$); then the result is the corresponding instantiation $T_i\left[\frac{\underline{V}}{\underline{v}}\right]$. If no pattern matches then there is no result.

THEOREM 39. $\mathit{fxpt}\,\Phi \not\triangleright$; $y \in \mathit{fxpt}\,\Phi$ iff y is not f and, for some i, $y \in T_i$ but $y \notin X_i$.

THEOREM 40. The definition of *fxpt* Φ is binding-independent (that is, independent of the choice of variables for the metavariables in $X_1, \ldots X_k$).

Proof. This follows since the 'match_X' notation and all ET constructs are binding-independent. ∎

THEOREM 41. Let $1 \leq i \leq k$, let $\underline{v} \in X_i$, and let $\underline{V}$ be any terms such that $f \notin \underline{V}$, $\underline{V} \not\triangleright$, $X_i\left[\frac{\underline{V}}{\underline{v}}\right] \not\triangleright$ and, for each $j < i$, $\mathit{match}_{X_j} X_i\left[\frac{\underline{V}}{\underline{v}}\right] \triangleright^* \mathit{false}$; then

$$\mathit{fxpt}\,\Phi\,(X_i\left[\tfrac{\underline{V}}{\underline{v}}\right]) \triangleright^* T_i\left[\tfrac{\underline{V}}{\underline{v}}\right]\left[\tfrac{\mathit{fxpt}\,\Phi}{f}\right].$$

Proof. By Theorems 33 and 16,

$$\Phi f\,(X_i\begin{bmatrix}\underline{V}\\ \underline{v}\end{bmatrix})$$

$$\rhd^*\ (\textit{branch match}_{X_1}\ (\lambda X_1.T_1)\ \ldots\ (\textit{branch match}_{X_i}\ (\lambda X_i.T_i)\ldots)\ldots)(X_i\begin{bmatrix}\underline{V}\\ \underline{v}\end{bmatrix})$$

$$\rhd^*\ (\lambda X_i.T_i)(X_i\begin{bmatrix}\underline{V}\\ \underline{v}\end{bmatrix})\ \overset{*}{\leftrightarrow}\ (\lambda X_i.T_i)X_i\begin{bmatrix}\underline{V}\\ \underline{v}\end{bmatrix}\ \rhd^*\ T_i\begin{bmatrix}\underline{V}\\ \underline{v}\end{bmatrix};$$

so

$$\textit{fxpt}\,\Phi\,(X_i\begin{bmatrix}\underline{V}\\ \underline{v}\end{bmatrix})\ \rhd\ \Phi(\textit{fxpt}\,\Phi)(X_i\begin{bmatrix}\underline{V}\\ \underline{v}\end{bmatrix})\ \overset{*}{\leftrightarrow}\ \Phi f\,(X_i\begin{bmatrix}\underline{V}\\ \underline{v}\end{bmatrix})\begin{bmatrix}\textit{fxpt}\,\Phi\\ f\end{bmatrix}\ \rhd^*\ T_i\begin{bmatrix}\underline{V}\\ \underline{v}\end{bmatrix}\begin{bmatrix}\textit{fxpt}\,\Phi\\ f\end{bmatrix}$$

by the instantiation theorem (14), as required. ∎

THEOREM 42. If $(\textit{fxpt}\,\Phi)\begin{bmatrix}\underline{U}\\ \underline{u}\end{bmatrix}Y$ is evaluable, where $\underline{U}$ are irreducible terms, X_i has free variables $\underline{v_i}$, and $\underline{v_i} \notin Y$ (for each $i = 1,\ldots k$), then $Y\ \rhd^*\!\lhd\ X\begin{bmatrix}\underline{V}\\ \underline{v_i}\end{bmatrix}$ for some i and some irreducible terms $\underline{V}$ whose free variables are included in those of Y.

Proof. Since $(\textit{fxpt}\,\Phi)\begin{bmatrix}\underline{U}\\ \underline{u}\end{bmatrix}Y$ is evaluable we must have, for some irreducible Y',

$$(\textit{fxpt}\,\Phi)\begin{bmatrix}\underline{U}\\ \underline{u}\end{bmatrix}Y\ \rhd^*\ (\textit{fxpt}\,\Phi)\begin{bmatrix}\underline{U}\\ \underline{u}\end{bmatrix}Y'\ \rhd\ \Phi\begin{bmatrix}\underline{U}\\ \underline{u}\end{bmatrix}(\textit{fxpt}\,\Phi\begin{bmatrix}\underline{U}\\ \underline{u}\end{bmatrix})Y'$$

$$\rhd^*\ (\textit{branch match}_{X_1}\ \ldots)Y',$$

which is evaluable only if $\textit{match}_{X_i}Y'\ \rhd^*\ \textit{true}$ for some i, in which case $Y\ \rhd^*\ Y'\ \rhd^*\!\lhd\ X_i\begin{bmatrix}\underline{V}\\ \underline{v_i}\end{bmatrix}$ by Theorem 34, as required. ∎

The usual way of using this definition mechanism will be to say 'Define f by $f\,X_1 \triangleq T_1,\ldots f\,X_k \triangleq T_k$'. Such a definition means, from now on, use f as a metaconstant to denote *fxpt* Φ (rather than the bound variable in Φ it denoted above). The reduction $\textit{fxpt}\,\Phi\,(X_i\begin{bmatrix}\underline{V}\\ \underline{v}\end{bmatrix})\ \rhd^*\ T_i\begin{bmatrix}\underline{V}\\ \underline{v}\end{bmatrix}\begin{bmatrix}\textit{fxpt}\,\Phi\\ f\end{bmatrix}$ may then usually be written simply as $f\,(X_i\begin{bmatrix}\underline{V}\\ \underline{v}\end{bmatrix})\ \rhd^*\ T_i\begin{bmatrix}\underline{V}\\ \underline{v}\end{bmatrix}$. The most commonly used instance of this will be $f\,X_i\ \rhd^*\ T_i$; but bear in mind that this is only applicable when X_i is irreducible and $\textit{match}_{X_j}X_i\ \rhd^*\ \textit{false}$ for each $j < i$.

EXAMPLE. The list membership function referred to above can be defined as follows (given a suitable pairing operation (…,…)). Let *x*, *first* and *rest* be three variables; define a construction *member* by

$$\textit{member}(x,(x,\textit{rest})) \triangleq \textit{true},$$

$$\textit{member}(x,(\textit{first},\textit{rest})) \triangleq \textit{member}(x,\textit{rest}),$$

$$\textit{member}(x,\textit{nil}) \triangleq \textit{false}.$$

This gives $member(x,(x,rest)) \rhd^* true$; and $member(X,(First,Rest)) \rhd^* member(X,Rest)$ for any constructions X, $First$ and $Rest$, provided X is not $First$; and $member(x,nil) \rhd^* false$, as required.

EXAMPLE. Suppose we define a coding of binary trees as constructions as follows: *false* represents an empty tree, and *true* X Y represents a tree consisting of a node and two subtrees coded as X and Y. (This is rather an inelegant coding scheme; better coding techniques will be introduced shortly.) We can define a function *tree* that checks whether a given construction is a tree code by

$$\begin{aligned} tree(false) &\triangleq true, \\ tree(true\ x\ y) &\triangleq if\ (tree\,x)\ (tree\,y)\ false, \\ tree(x) &\triangleq false, \end{aligned}$$

where x and y are two variables. Then, for example,

$$\begin{aligned} &tree(true\ (true\,false\,false)\,false) \\ &\rhd^*\ if\ (tree(true\,false\,false))\,(tree\,false)\,false \\ &\rhd^*\ if\ (if\ (tree\,false)\,(tree\,false)\,false)\ true\,false \\ &\rhd^*\ if\ true\ true\,false\ \rhd\ true. \end{aligned}$$

(Check that the conditions of Theorem 41 are satisfied.)

EXERCISES. Using the same binary tree coding,

(1) define a function *reflect* that swaps the left and right subtrees of each node, for example

$$\begin{aligned} &reflect(true\ (true\,false\,(true\,false\,false))\ false) \\ &\rhd^*\ true\ false\ (true\,(true\,false\,false)\,false); \end{aligned}$$

(2) define a function *flatten* that rearranges a binary tree into a right-branching tree with the same number of nodes, for example

$$\begin{aligned} &flatten(true\ (true\,false\,false)\ (true\,false\,false)) \\ &\rhd^*\ true\,false\,(true\,false\,(true\,false\,false)). \end{aligned}$$

REGULAR FUNCTION DEFINITIONS AND CONTINUITY

Theorem 41 shows that a function defined by pattern matching satisfies its defining 'equations'. But that is only half the story. We would also like to be able to argue by induction with respect to such recursively defined functions. For example, we would like to be able to argue that any binary tree has a certain property by induction on the structure of the tree. We shall see in Chapter 19 that such arguments can be formalised within protologic using a coding of binary trees (as above), the function *tree* defined by pattern matching, and the Fxpt Rules of protologic. But to apply the Fxpt Rules it is necessary to check that the function Φ used in the pattern-matching definition is continuous (in the sense defined in the Term Language). This section establishes a simple and useful sufficient condition for continuity in pattern-matching definitions.

The partial reduction relations $\rightharpoonup$, $\overset{*}{\rightharpoonup}$, $\leftharpoonup$, $\overset{*}{\leftharpoonup}$, $\overset{*}{\leftrightharpoons}$, $\not\leftrightharpoons$, $\rightharpoondown$, $\overset{*}{\rightharpoondown}$, $\not\rightharpoondown$ defined in the Term Language may be extended to the Expanded Term Language by

$$
\begin{array}{llll}
A \rightharpoonup B & \text{iff} \quad A^* \rightharpoonup B^* & A \overset{*}{\rightharpoonup} B & \text{iff} \quad A^* \overset{*}{\rightharpoonup} B^* \\
A \leftharpoonup B & \text{iff} \quad A^* \leftharpoonup B^* & A \overset{*}{\leftharpoonup} B & \text{iff} \quad A^* \overset{*}{\leftharpoonup} B^* \\
A \overset{*}{\leftrightharpoons} B & \text{iff} \quad A^* \overset{*}{\leftrightharpoons} B^* & A \not\leftrightharpoons & \text{iff} \quad A^* \not\leftrightharpoons \\
A \rightharpoondown B & \text{iff} \quad A^* \rightharpoondown B^* & A \overset{*}{\rightharpoondown} B & \text{iff} \quad A^* \overset{*}{\rightharpoondown} B^* \\
A \not\rightharpoondown & \text{iff} \quad A^* \not\rightharpoondown\,. & &
\end{array}
$$

DEFINITION. A function definition $f X_1 \triangleq T_1, \ldots f X_k \triangleq T_k$ is *regular* iff, for each $i = 1, \ldots k$, $X_i \not\triangleright$ and $vT_i \overset{*}{\rightharpoonup} v$, where v is a fresh variable and $\overset{*}{\rightharpoonup}$ is the partial reduction relation with respect to f and v.

(Note: when I speak of a 'fresh' variable, as in the above definition, I mean a variable that has not been used so far in the current definition or theorem.)

THEOREM 43. In a regular function definition of a term *fxpt* Φ, Φ is continuous.

Proof. Given the two variables f and v, call a term T *functional* iff $v(Tv) \overset{*}{\rightharpoonup} v$. The proof of the theorem is divided into the following steps.

(1) If $f \notin X$ and $X \not\triangleright$ then $vX \overset{*}{\rightharpoonup} v$.
   Proof: since $X \not\triangleright$, comparing the $\triangleright$ and $\rightharpoonup$ clauses we see that the only $\rightharpoonup$ rules that might apply to X are $a \rightharpoonup v$ (where a is *equal*, *true*, *false*, *former*, *latter*, *fxpt* or *nil*) or $x \rightharpoonup v$ (where x is a variable other than f and v). After applying these rules wherever possible we apply $v(AB) \rightharpoonup vAB$, $va \rightharpoonup v$ (where a is a constant), and $vv \rightharpoonup v$ repeatedly, to give $vX \overset{*}{\rightharpoonup} v$.

(2) $(\lambda f.T)f \overset{*}{\leftrightharpoons} T$.
Proof: by structural induction on T^*, using the cases

$$\begin{aligned}
(\lambda f.AB)f &\mapsto s(\lambda f.A)(\lambda f.B)f \overset{*}{\leftrightharpoons} ((\lambda f.A)f)((\lambda f.B)f) \overset{*}{\leftrightharpoons} AB \\
(\lambda f.f)f &\mapsto id\,f \rightharpoonup f \\
(\lambda f.a)f &\mapsto kaf \overset{*}{\leftrightharpoons} a \quad \text{if } a \text{ is a constant or a variable other than } f.
\end{aligned}$$

(3) $(branch\ C\ A\ B)$ is functional if C, B and A are.
Proof: suppose $v(Cv) \overset{*}{\rightharpoonup} v$, $v(Av) \overset{*}{\rightharpoonup} v$ and $v(Bv) \overset{*}{\rightharpoonup} v$. Then $C \overset{*}{\rightharpoonup} C' \not\rightharpoonup$, $C'v \overset{*}{\rightharpoonup} C'' \not\rightharpoonup$ and $vC'' \overset{*}{\rightharpoonup} v$, for some terms C' and C''. Similarly $A \overset{*}{\rightharpoonup} A' \not\rightharpoonup$, $A'v \overset{*}{\rightharpoonup} A'' \not\rightharpoonup$ and $vA'' \overset{*}{\rightharpoonup} v$, for some terms A' and A''; and $B \overset{*}{\rightharpoonup} B' \not\rightharpoonup$, $B'v \overset{*}{\rightharpoonup} B'' \not\rightharpoonup$ and $vB'' \overset{*}{\rightharpoonup} v$, for some terms B' and B''. Then

$$\begin{aligned}
v((branch\ C\ A\ B)v) &\overset{*}{\rightharpoonup} v((branch\ C'\ A'\ B')v) \overset{*}{\rightharpoonup} v(if\ (C'v)\ A'\ B'\ v) \\
&\overset{*}{\rightharpoonup} v(if\ C''\ A'\ B'\ v) \rightharpoonup v(if\ C''\ (A'v)\,(B'v)) \\
&\overset{*}{\rightharpoonup} v(if\ C''\ A''\ B'') \overset{*}{\rightharpoonup} v\,if\ C''\ A''\ B'' \overset{*}{\rightharpoonup} v.
\end{aligned}$$

(4) $(\lambda X.T)v \overset{*}{\leftrightharpoons} T$, if $f \notin X$.
Proof: by structural induction on T^* within structural induction on X^*, using the cases

$$\begin{aligned}
(\lambda PQ.T)v &\overset{*}{\rightharpoonup} (\lambda P.(\lambda Q.T))(vv)(vv) \overset{*}{\rightharpoonup} (\lambda P.(\lambda Q.T))vv \overset{*}{\leftrightharpoons} (\lambda Q.T)v \\
&\overset{*}{\leftrightharpoons} T \\
(\lambda a.AB)v &\mapsto s(\lambda a.A)(\lambda a.B)v \overset{*}{\leftrightharpoons} ((\lambda a.A)v)((\lambda a.B)v) \overset{*}{\leftrightharpoons} AB \\
&\quad \text{if } a \text{ is an atom} \\
(\lambda v.v)v &\mapsto id\,v \rightharpoonup v \\
(\lambda x.x)v &\mapsto id\,v \rightharpoonup v \leftharpoonup x \quad \text{if } x \text{ is a variable other than } v \text{ and } f \\
(\lambda a.b)v &\mapsto kbv \overset{*}{\leftrightharpoons} b \quad \text{if } a \text{ and } b \text{ are atoms but are not both} \\
&\qquad \text{the same variable.}
\end{aligned}$$

(5) $check_X$ is functional.
Proof: by structural induction on X^*. If X^* is an atom we have

$$\begin{aligned}
&v(check_a v) \text{ is } v(equal\ a\ v) \rightharpoonup v(vav) \overset{*}{\rightharpoonup} v \quad \text{if } a \text{ is a constant} \\
&v(check_x v) \text{ is } v(k\ true\ v) \rightharpoonup v(kvv) \overset{*}{\rightharpoonup} v \quad \text{if } x \text{ is a variable}
\end{aligned}$$

while if X^* is PQ then $check_{PQ}$ is functional by (3) since its components $s(k\ check_P)former$, $s(k\ check_Q)latter$ and $k\ false$ are functional, as is verified by

$$\begin{aligned} v(s(k\ check_P)former\ v) &\overset{*}{\leftharpoondown} v(check_P v) \overset{*}{\rightharpoonup} v \\ v(s(k\ check_Q)latter\ v) &\overset{*}{\leftharpoondown} v(check_Q v) \overset{*}{\rightharpoonup} v \\ v(k\ false\ v) &\rightharpoonup v(kvv) \overset{*}{\rightharpoonup} v \end{aligned}$$

assuming that $check_P$ and $check_Q$ are functional and using Theorem 6 of the Term Language.

(6) $match_X$ is functional if $f \notin X$ and $X \not\triangleright$.
Proof: $check_X$ and $k false$ are functional, as was shown in (5). Also, $s(s(k\ equal)(\lambda X.X))id$ is functional, since by (4) and (1)

$$v(s(s(k\ equal)(\lambda X.X))id\ v) \overset{*}{\rightharpoonup} v(v((\lambda X.X)v)v) \overset{*}{\leftharpoondown} v(vXv) \overset{*}{\rightharpoonup} v(vv) \overset{*}{\rightharpoonup} v,$$

which establishes that $v(s(s(k\ equal)(\lambda X.X))id\ v) \overset{*}{\rightharpoonup} v$. The result follows by (3).

(7) Φf is functional.
Proof: f is functional, since $v(fv) \rightharpoonup vv \rightharpoonup v$; so is $(\lambda X_i.T_i)$ for each i, since $v((\lambda X_i.T_i)v) \overset{*}{\leftharpoondown} vT_i \overset{*}{\rightharpoonup} v$ by (4). The result follows by (6), (3) and (2).

(8) Φ is continuous.
Proof: by (7), $v(\Phi fv) \overset{*}{\rightharpoonup} v$. ∎

EXERCISE. Check that the definitions of *tree*, *reflect* and *flatten* are regular.

METANOTATION

DEFINITION. Equality notation: $A = B$ is *equal* A B.

DEFINITION. The construction *boolean* is ($\lambda x.$*if x true true*).

THEOREM 44. *boolean(true)* $\triangleright^*$ *true* and *boolean(false)* $\triangleright^*$ *true*.

DEFINITION. Truth-functional conjunction: $A \,\&\, B$ is *if* A B *false*; and $A_1 \,\&\ldots\&\, A_k$ is $A_1 \,\&\, (A_2 \,\&\, \ldots (A_{k-1} \,\&\, A_k)\ldots)$.

THEOREM 45. If $X \not\triangleright$ then *true* & $X \triangleright X$ and *false* & $X \triangleright$ *false*.

DEFINITION. Pairs: (A, B) is sAB. The constructions *left* and *right* are $(\lambda(a, b).a)$ and $(\lambda(a, b).b)$, where a and b are two variables.

(The choice of the variables a and b makes no difference, up to compilation, due to binding-independence.)

DEFINITION. Tuples: $(A_1, A_2, A_3 \ldots A_{k-1}, A_k)$ is

$$(A_1, (A_2, (A_3, \ldots (A_{k-1}, A_k) \ldots))),$$

for $k \geq 2$. This notation is used for grouping a known number of terms in order.

DEFINITION. Lists: $[A_1, A_2, \ldots A_k]$ is $(A_1, [A_2, \ldots A_k])$ and $[\,]$ is *nil*. Thus $[A_1, A_2, \ldots A_k]$ is $(A_1, A_2, \ldots A_k, \mathit{nil})$ for $k \geq 1$. This notation is used instead of tuples when the number of terms varies.

We would like to use tuples and lists in pattern-matching definitions, so it is necessary to verify that the pattern-matching functions check_X and match_X are able to recognise them and decompose them correctly. This is established by the following two theorems.

THEOREM 46. Suppose $U, V \not\triangleright$. Then

$$\mathit{check}_{(X,Y)}(U, V) \triangleright^* \begin{cases} \mathit{check}_Y V & \text{if } \mathit{check}_X U \triangleright^* \mathit{true}, \\ \mathit{false} & \text{if } \mathit{check}_X U \triangleright^* \mathit{false}. \end{cases}$$

Proof. By Theorem 29, if $\mathit{check}_X U \triangleright^* \mathit{true}$ then $\mathit{check}_{sX}(sU) \triangleright^* \mathit{check}_X U \triangleright^* \mathit{true}$, so $\mathit{check}_{sXY}(sUV) \triangleright^* \mathit{check}_Y V$, as required.

If, on the other hand, $\mathit{check}_X U \triangleright^* \mathit{false}$ then $\mathit{check}_{sX}(sU) \triangleright^* \mathit{check}_X U \triangleright^* \mathit{false}$, so $\mathit{check}_{sXY}(sUV) \triangleright^* \mathit{false}$. ∎

THEOREM 47. If $U, V \not\triangleright$ then $\mathit{check}_{nil}(U, V) \triangleright^* \mathit{false}$ and $\mathit{check}_{(U,V)}\mathit{nil} \triangleright^* \mathit{false}$.

Proof. $\mathit{check}_{nil}(U, V)$ is $\mathit{equal}\ \mathit{nil}\ (sUV) \triangleright \mathit{false}$, as required.

For the second part,

$$s(k\,\mathit{check}_s)\mathit{former}\ \mathit{nil} \triangleright^* \mathit{check}_s(\mathit{former}\ \mathit{nil}) \triangleright \mathit{equal}\ s\ \mathit{nil} \triangleright \mathit{false}$$

and hence

$$\begin{aligned} s(k\,\mathit{check}_{sU})\mathit{former}\ \mathit{nil} &\triangleright^* \mathit{check}_{(sU)}(\mathit{former}\ \mathit{nil}) \triangleright \mathit{check}_{(sU)}\mathit{nil} \\ &\triangleright^* \mathit{if}\ (s(k\,\mathit{check}_s)\mathit{former}\ \mathit{nil})\,(\ldots)\,(k\,\mathit{false})\,\mathit{nil} \\ &\triangleright^* \mathit{false} \end{aligned}$$

and hence

$$\mathit{check}_{(sUV)}\mathit{nil} \triangleright^* \mathit{if}\ (s(k\,\mathit{check}_{sU})\mathit{former}\ \mathit{nil})\,(\ldots)\,(k\,\mathit{false})\,\mathit{nil} \triangleright^* \mathit{false}$$

as required. ∎

EXERCISES. Let u, v, w, x, y, z be six variables. Show that
(1) $check_{(x,false,z)}(true, false, w) \ \triangleright^* \ true$,
(2) $check_{[x,y,z]}[u, v] \ \triangleright^* \ false$,
(3) $check_{[x,y]}[u, v, nil] \ \triangleright^* \ false$.

EXAMPLE. We can define a function *concat* that concatenates two lists by

$$\begin{aligned} concat(nil, other) &\stackrel{\triangle}{=} other \\ concat((first, rest), other) &\stackrel{\triangle}{=} (first, concat(rest, other)) \end{aligned}$$

where *first*, *rest* and *other* are three variables. Let us show that

$$concat([a,b],[c,d,e]) \ \triangleright^* \ [a,b,c,d,e],$$

where a, b, c, d, e are five variables. First, $match_{(nil,other)}([a,b],[c,d,e]) \ \triangleright^*$ *false* (by Theorems 46, 47 and 32). Hence, by Theorem 41,

$$\begin{aligned} concat([a,b],[c,d,e]) \ \triangleright^* \ & (first, concat(rest, other))\left[\begin{smallmatrix} a, & [b], & [c,d,e] \\ first, & rest, & other \end{smallmatrix}\right] \\ \stackrel{*}{\mapsto} \ & (a, concat([b],[c,d,e])). \end{aligned}$$

Repeating the argument gives $(a,(b, concat(nil,[c,d,e])))$, which reduces to $(a,b,[c,d,e])$, which is the desired result $[a,b,c,d,e]$.

EXERCISES.
(1) Define a function *reverse* that reverses the order of elements in a list and verify that $reverse[a,b] \ \triangleright^* \ [b,a]$, where a and b are two variables.
(2) Define a function *insert* that inserts a given element in a given list, unless it is already present in which case it just returns the list unaltered. (Hint: don't use *member*.)
(3) Define a function *remove* that removes a given element from a given list, unless it is not present in which case it just returns the list unaltered. (Again, don't use *member*.)
(4) Define a function *union* that combines two lists into a single list, removing any repetition of elements.

CONSTRUCTORS

The purpose of this section is to introduce readable notations for representing algebraic data structures as constructions, following as closely as possible the standard notations of functional programming. Consider *stacks* for example. Given an *element* data type, it is usual to introduce two 'constructors', *empty* and *push*, which generate stacks of elements as follows:

- *empty* is a stack (informally regarded as the stack with no elements);
- if e is an element and s is a stack then $push(e, s)$ is a stack (informally regarded as a stack obtained by adding e at the top of s).

All stacks are generated by applying *push* repeatedly to *empty*; a typical stack is $push(e_1, push(e_2, push(e_3, empty)))$. Having defined stacks one then defines functions *pop*, *top* and *null*:

$$\begin{aligned} pop(empty) &= empty, & null(empty) &= true, \\ pop(push(e,s)) &= s, & null(push(e,s)) &= false, \\ top(push(e,s)) &= e, \end{aligned}$$

(note that $top(empty)$ is undefined). Now, I wish to follow this notation exactly, except with '$=$' replaced by '$\triangleq$'. Hence I need to define *constructors*. There are two kinds of constructor: 0-ary constructors, such as *empty*, which take no arguments; and 1-ary constructors, such as *push*, which take a tuple of arguments. I shall choose sets of constructions with suitable properties to act as 0-ary and 1-ary constructors.

DEFINITION. A *1-ary constructor* is a construction of the form $true\,[id, \ldots id]$. A *0-ary constructor* is a construction of the form $false\,false\,[id, \ldots id]$.

There is an infinite supply of constructors of each type. Whenever I say something like 'Let A, B, C be three fresh 1-ary constructors' I mean take the next three unused 1-ary constructors from the infinite supply. I shall use constructors for forming terms like

$$A(B(x, M), y, C(B(N, N))),$$

built up out of 0-ary constructors (M and N) and variables (x and y) by applying 1-ary constructors (A, B and C) to tuples. Terms of this sort are used for coding trees and other structured objects. I shall show that all such terms are irreducible and distinguishable using $check_X$ and hence using $match_X$, allowing them to be used freely as patterns in function definitions.

THEOREM 48. If F is a 1-ary constructor and $X \not\triangleright$ then $FX \not\triangleright$.

THEOREM 49. If P and Q are two 0-ary constructors or two 1-ary constructors then $check_P Q \triangleright^* false$.

Proof. Let X and Y be two lists of the form $[id, \ldots id]$ (with different numbers of elements). From Theorems 46 and 47,

$$check_{(id,X)}(id, Y) \triangleright^* check_X Y \qquad check_{nil}(id, X) \triangleright^* false$$
$$check_{(id,X)} nil \triangleright^* false$$

and hence by induction $check_X Y \triangleright^* false$.

Thus for two 0-ary constructors, *false false X* and *false false Y*, by Theorem 29,

$$check_{(false\, false\, X)}(false\, false\, Y) \triangleright^* check_X Y \triangleright^* false$$

and for two 1-ary constructors, *true X* and *true Y*,

$$check_{(true\, X)}(true\, Y) \triangleright^* check_X Y \triangleright^* false$$

as required. ∎

THEOREM 50. If F and G are two 1-ary constructors, and $X, Y \not\triangleright$, then $check_{FX}(GY) \triangleright^* false$, $FX = GY \triangleright false$, and $check_{FX}(FY) \triangleright^* check_X Y$.

Proof. The previous theorem says that $check_F G \triangleright^* false$, and hence by Theorems 29 and 48, $check_{FX}(GY) \triangleright^* false$. By the reduction clauses in the Term Language, $F = G \triangleright false$ and hence $FX = GY \triangleright false$. By Theorem 30, $check_F F \triangleright^* true$, and hence, by Theorem 29, $check_{FX}(FY) \triangleright^* check_X Y$. ∎

THEOREM 51. If B is a 0-ary constructor, F is a 1-ary constructor, and $X \not\triangleright$, then $check_B(FX) \triangleright^* false$, $check_{FX} B \triangleright^* false$, $B = FX \triangleright false$, and $FX = B \triangleright false$.

Proof. Using Theorem 29, $check_B(FX) \triangleright^* false$ since $check_{false} true \triangleright false$. Similarly, $check_{FX} B \triangleright^* false$ since $check_{true} false \triangleright false$. The other two reductions, $B = FX \triangleright false$, and $FX = B \triangleright false$, follow from the reduction clauses in the Term Language since $false = true \triangleright false$ and $true = false \triangleright false$. ∎

EXAMPLE. Define a new coding of trees, in which every node has 0, 1 or 2 subtrees, as follows. Let *none* be a fresh 0-ary constructor, and *one* and *two* be two fresh 1-ary constructors.

- *none* represents a leaf of the tree;
- *one*(X) represents a tree with one subtree, X;
- *two*(X, Y) represents a tree with two subtrees, X and Y.

EXERCISE. Redefine the functions *tree*, *reflect* and *flatten*, described earlier, for this new tree coding. Check that the conditions of Theorem 41 are satisfied. What does Theorem 42 tells us in this case?

The difference between this tree coding and the previous coding of binary trees, which did not use constructors, may appear slight. However, the use of constructors makes a big difference to the readability of complex arguments involving tree codes: see the proofs of Theorems 14 and 15 in Chapter 21, for example.

EXERCISE. Define a coding of predicate calculus formulae as constructions. You should assume that a supply of object variables $x_1, x_2, x_3, \ldots$ and a supply of predicate variables $P_1, P_2, P_3, \ldots$ are provided, that the atomic formulae are of the form $P_m(x_n)$, and that the first-order quantifiers $\forall x_n$ and $\exists x_n$ are the only quantifiers allowed. Define the following functions:

(1) *vbls*, which lists the index numbers of the free object variables in a formula; thus *vbls* applied to the code of $\forall x_5\,(P_1(x_1) \wedge P_2(x_5) \wedge P_1(x_3))$ would give $[1,3]$;
(2) *sub*, which, when applied to m, n and the code of a formula, gives the code of the formula with x_m substituted for all free occurrences of x_n (ignore the problem of variable clashes).

NUMBERS, PRIMITIVE RECURSIVE FUNCTIONS, AND NUMERIC TERMS

This section introduces (natural) numbers and arithmetic. Numbers are simply a certain recursively defined class of constructions, like stacks or binary trees. They have no privileged status; their properties are simply special cases of the properties of constructions in general.

DEFINITION. Let 0 be a fresh 0-ary constructor, let S be a fresh 1-ary constructor, and let *pre* be *latter*.

DEFINITION. The *numbers* are 0, $S0$, $S(S0)$, $S(S(S0))$, ... Let 1 be $S0$, 2 be $S1$, and so on up to 34. (These are all the numbers we shall need names for in this book.)

THEOREM 52. If $N \not\triangleright$ then $SN \not\triangleright$. Thus all numbers are constructions.

THEOREM 53. $pre(Sn) \triangleright n$ and $Sn = 0 \triangleright$ *false*.

DEFINITION. Define a number predicate *num* by

$$\begin{aligned} num\, 0 &\triangleq true \\ num\,(Sn) &\triangleq num\, n. \end{aligned}$$

THEOREM 54. If m and n are numbers then

$$check_m n \ \rhd^* \begin{cases} true & \text{if } m \text{ is } n, \\ false & \text{if } m \text{ is not } n \end{cases}$$

and the same is true for $match_m n$.

Proof. Note that, for any numbers m and n,

- $check_m\, m \ \rhd^*\ true$ and $match_m\, m \ \rhd^*\ true$ (Theorems 30 & 33);
- $check_{Sm}(Sn) \ \rhd^*\ check_m n$ (Theorem 50);
- $check_{Sm}\, 0 \ \rhd^*\ false$ (Theorem 51);
- $check_0\, Sn \ \rhd\ false$ (Theorem 51).

These facts imply that if m is not n then $check_m n \ \rhd^*\ false$ and hence (by Theorem 32) $match_m n \ \rhd^*\ false$. ∎

EXAMPLE. Addition and subtraction may be defined using pattern-matching by

$$\begin{aligned} plus(0,y) &\triangleq y, & minus(x,0) &\triangleq x, \\ plus(Sx,y) &\triangleq S(plus(x,y)), & minus(x,Sy) &\triangleq pre(minus(x,y)) \end{aligned}$$

where x and y are two variables.

EXERCISE. Define multiplication and integer division (with upward rounding) similarly.

EXAMPLE. We can define a function *list*, which checks whether a given construction is a list, and a function *length*, which determines the numbers of elements in a list, by

$$\begin{aligned} list(nil) &\triangleq true, & length(nil) &\triangleq 0, \\ list(first, rest) &\triangleq list(rest), & length(first, rest) &\triangleq S(length(rest)), \\ list(x) &\triangleq false, \end{aligned}$$

where *first* and *rest* are two variables.

EXERCISE. Modify the coding of trees defined earlier, in which each node had 0, 1 or 2 subtrees, so that each leaf holds a number, and define a function *sum* that adds up all the numbers in a given tree.

DEFINITION. A *k-ary primitive recursive (k-PR) function* is a term defined as follows.

0 is 0-PR

S is 1-PR

$proj_i^k$ is k-PR for any numbers i, k satisfying $1 \leq i \leq k$

$comp_l(H, F_1, \ldots F_l)$ is k-PR if H is l-PR
and $F_1, \ldots F_l$ are k-PR (where $k, l \geq 1$)

$rec_k(F, G)$ is $(k+1)$-PR if F is k-PR and G is $(k+2)$-PR (where $k \geq 0$).

Here,

$proj_i^k$ is $(\lambda(n_1, \ldots n_k).n_i)$ for each i, k satisfying $1 \leq i \leq k$

$comp_l$ is $(\lambda(h, f_1, \ldots f_l).(\lambda x.h(f_1 x, \ldots f_l x)))$

rec_k is $(\lambda(f, g).T_k)$ where T_k is defined by

$$\begin{cases} T_k(0, n_1, \ldots n_k) & \triangleq f(n_1, \ldots n_k), \\ T_k(Sm, n_1, \ldots n_k) & \triangleq g(m, n_1, \ldots n_k, T_k(m, n_1, \ldots n_k)) \end{cases}$$

where $n_1, \ldots n_k, h, f_1, \ldots f_l, x, f, g, m$ are all different variables. (Note that $f(n_1, \ldots n_k)$ is to be understood as f when $k = 0$. This convention will apply throughout; that is, $F()$ is a metanotation for F.)

EXERCISE. Redefine addition, subtraction and multiplication as primitive recursive functions. (The answer for multiplication is given in the section on finitary and constructive reasoning in Chapter 9.)

THEOREM 55. Primitive recursive functions have no free variables and are evaluable.

Proof. It is clear that they have no free variables. The proof that they are evaluable is by structural induction on the primitive recursive function.

0, S and $proj_i^k$ are constructions and so are evaluable.

If the primitive recursive functions $H, F_1, \ldots F_l$ are evaluable then

$$comp_l(H, F_1, \ldots F_l) \triangleright^* \triangleleft (\lambda x.h(f_1 x, \ldots f_l x))\begin{bmatrix} H, F_1, \ldots F_l \\ h, f_1, \ldots f_l \end{bmatrix},$$

which is also evaluable by Theorems 22 and 18.

If the primitive recursive functions F and G are evaluable then $rec_k(F, G)$ $\triangleright^* \triangleleft$ $T_k\begin{bmatrix} F, G \\ f, g \end{bmatrix}$ which is also evaluable. ∎

DEFINITION. A *numeric term* is 0, a variable, or $F(N_1, \ldots N_k)$, where $k \geq 1$, F is k-ary primitive recursive and $N_1, \ldots N_k$ are numeric terms.

Numeric terms will be used in the syntax of Heyting Arithmetic and Peano Arithmetic.

CHURCH'S THESIS

In future I want to be able to implement any 'mechanically effective' operation on constructions as a construction in the Expanded Term Language. Often the desired construction cannot be defined concisely using the notation introduced so far, so I simply want to say, by appeal to Church's Thesis (the weak form – see Chapter 5), that a construction implementing the operation exists and to introduce a metaconstant to denote it.

To justify this procedure it is necessary to introduce a notion of recursive function applicable to constructions. Recursive functions were defined originally on numbers; modern treatments of the subject often generalise them to free monoids (Eilenberg & Elgot, 1970); from this it is easy to see how to apply them to any recursively generated data structure. Simply define the recursive functions to include the atomic structures (analogous to 0 in the definition of primitive recursive functions above); functions building a composite structure out of its immediate components (analogous to S); projection functions $proj_i^n$; the composition of any recursive functions; and functions defined by primitive recursion and unbounded iteration from recursive functions.

To apply this procedure to constructions, recall a characterisation of constructions given in the Term Language: a (simple) construction is a constant or *sA*, *sAB*, *kA*, *equal A*, *if A*, *if A B*, *true A B* $\cdots$ *C*, *false A B* $\cdots$ *C*, *fxpt A* or *nil A B* $\cdots$ *C*, where $A, B, \ldots C$ are (simple) constructions. Hence we may regard the constructions as a recursively generated system in their own right, not merely as a subclass of the terms.

DEFINITION. A *recursive function* is a construction defined as follows. (In this definition, $x_1, \ldots x_n, u, v, y$ are $n+3$ variables and $\underline{x}$ is short for $x_1, \ldots x_n$.)

- Each constant is a recursive function.
- (Construction-building functions.) The following constructions are recursive functions, for any $n = 1, 2, \ldots$: $(\lambda(x, y).sxy)$, $(\lambda(x, y).\mathit{if}\, x\, y)$, $(\lambda(\underline{x}).\mathit{true}\ x_1 \cdots x_n)$, and $(\lambda(\underline{x}).\mathit{false}\ x_1 \cdots x_n)$, $(\lambda(\underline{x}).\mathit{nil}\ x_1 \cdots x_n)$.
- The projection functions $proj_i^n$ are recursive functions, for $1 \leq i \leq n$.
- (Composition.) If $A, B_1, \ldots B_l$, where $l \geq 1$, are recursive functions then so is $(\lambda y.A(B_1(y), \ldots B_l(y)))$.

- (Primitive recursion.) If $A_1, \ldots A_{11}, B$ are recursive functions then so is F, defined by

$$\begin{aligned} F(s, \underline{x}) &\triangleq A_1(\underline{x}) \\ F(id, \underline{x}) &\triangleq A_2(\underline{x}) \\ &\ldots \text{ and so on for the other constants } \ldots \\ F(nil, \underline{x}) &\triangleq A_{11}(\underline{x}) \\ F(uv, \underline{x}) &\triangleq B(u, v, \underline{x}, F(u, \underline{x}), F(v, \underline{x})). \end{aligned}$$

- (Unbounded iteration.) If A, B, C are recursive functions then so is F, defined as

$$fxpt\,(\lambda f.(branch\ A\ B\ (s(kf)(\lambda(y, \underline{x}).(Cy, \underline{x}))))).$$

EXERCISE. Show that, when F is defined by unbounded iteration from A, B, C,

$$F(Y, \underline{X}) \triangleright^* \begin{cases} B(Y, \underline{X}) & \text{if } A(Y, \underline{X}) \triangleright^* \ true \\ F(CY, \underline{X}) & \text{if } A(Y, \underline{X}) \triangleright^* \ false, \end{cases}$$

for any constructions $Y, \underline{X}$.

EXAMPLE. Unbounded iteration works by iterating C until A is satisfied, then applying B. As a special case, let B be $proj_1^{n+1}$ and C be S; then $F(0, \underline{X}) \triangleright^*$ the least number Y such that $A(Y, \underline{X}) \triangleright^*$ *true*, provided $A(N, \underline{X}) \triangleright^*$ *false* for $N = 1, \ldots Y - 1$. This represents μ-recursion for numeric functions.

THEOREM 56. (Weak Church's Thesis: first version.) For any mechanically effective partial operation mapping constructions to constructions, invariant under compilation of the argument, there is a recursive function F such that for any constructions A, B,

$FA \triangleright^* B$ iff the operation maps A to a term $\overset{*}{\leftrightarrow} B$.

Proof. This says that any effective operation can be expressed in terms of constants, construction-building functions, projection functions, composition, primitive recursion, and unbounded iteration, and so can be implemented as a recursive function. The justification for this is just as in the traditional numeric version of Church's Thesis. ∎

This theorem needs to be strengthened. Often I want not merely to map a construction to a construction but to map any irreducible term 'of the form A' to an irreducible term 'of the form B'. That is to say, for any choice of constructions $\underline{X}$ for the free variables $\underline{x}$ of A, I want to map $A\left[\frac{X}{x}\right]$ to $B\left[\frac{X}{x}\right]$. The above version of Church's Thesis allows me to implement this as a recursive function F such that, for any constructions $\underline{X}$, $F(A\left[\frac{X}{x}\right]) \triangleright^* B\left[\frac{X}{x}\right]$, but I would like the stronger reduction $FA \triangleright^* B$.

THEOREM 57. (Weak Church's Thesis: second version.) For any mechanically effective partial operation mapping constructions to constructions, invariant under compilation of the argument, there is a recursive function F such that for any irreducible terms A, B, with free variables $\underline{x}$,

$FA \rhd^* B$ iff, for any constructions $\underline{X}$, the operation maps $A\left[\frac{\underline{X}}{\underline{x}}\right]$ to a term $\overset{*}{\leftrightarrow} B\left[\frac{\underline{X}}{\underline{x}}\right]$.

Proof. The first version of Church's Thesis provides a recursive function F such that, for any constructions $\underline{X}$, $F(A\left[\frac{\underline{X}}{\underline{x}}\right]) \rhd^* B\left[\frac{\underline{X}}{\underline{x}}\right]$ iff the operation maps $A\left[\frac{\underline{X}}{\underline{x}}\right]$ to a term $\overset{*}{\leftrightarrow} B\left[\frac{\underline{X}}{\underline{x}}\right]$.

Now, first suppose that $A\left[\frac{\underline{X}}{\underline{x}}\right]$ is mapped to the corresponding instantiation $B\left[\frac{\underline{X}}{\underline{x}}\right]$ (up to $\overset{*}{\leftrightarrow}$), regardless of the constructions $\underline{X}$. It follows that F does not need to examine the internal composition of the constructions $\underline{X}$ occurring in $A\left[\frac{\underline{X}}{\underline{x}}\right]$: it may move them around, copy them, insert them in new terms, and compare different occurrences of the same construction to check that they are equal, but it does not need to take them apart or compare them with anything else. In other words, F may be defined in such a way that it only manipulates each of the constructions $\underline{X}$ as a whole. A precise way of saying this is that the reduction $F(A\left[\frac{\underline{X}}{\underline{x}}\right]) \rhd^* B\left[\frac{\underline{X}}{\underline{x}}\right]$ is of the form $F(A\left[\frac{\underline{X}}{\underline{x}}\right]) \rhd T_1\left[\frac{\underline{X}}{\underline{x}}\right] \rhd T_2\left[\frac{\underline{X}}{\underline{x}}\right] \rhd \cdots \rhd B\left[\frac{\underline{X}}{\underline{x}}\right]$, where each reduction step works regardless of $\underline{X}$, and hence $FA \rhd T_1 \rhd T_2 \rhd \cdots \rhd B$, as required.

The converse is easy. If $FA \rhd^* B$ then, for any constructions $\underline{X}$, $F(A\left[\frac{\underline{X}}{\underline{x}}\right]) \rhd^* B\left[\frac{\underline{X}}{\underline{x}}\right]$ by the instantiation theorem (14), and so, by the first version of Church's Thesis, the effective operation maps $A\left[\frac{\underline{X}}{\underline{x}}\right]$ to a term $\overset{*}{\leftrightarrow} B\left[\frac{\underline{X}}{\underline{x}}\right]$. ∎

In future, whenever I say something like 'define a recursive function F such that $FA \rhd^* B$' I shall be invoking the appropriate version of Church's Thesis. To paraphrase the second version of the Thesis, such a definition is legitimate provided there is an effective operation, invariant under compilation, mapping any instantiation of the free variables of A by constructions to the corresponding instantiation of B. This condition needs to be verified by inspection in each case.

CODING OF TERMS

Church's Thesis will often be used in conjunction with a coding of terms, that is, a representation of terms as irreducible terms. A term T may be represented by $(\lambda[\underline{x}].T)$ where $\underline{x}$ are some of its free variables. The point of this is that it inhibits the reduction of T, since $(\lambda[\underline{x}].T)$ is always irreducible, and it also binds the variables $\underline{x}$. As a special case, if we choose $\underline{x}$ to be *all* the free variables of T then $(\lambda[\underline{x}].T)$ is a construction representing T.

We may, however, wish to bind only some of the free variables. Suppose T has two free variables, x and y; we might wish y to be instantiated by its value Y before coding, and the resulting term $T\begin{bmatrix}Y\\y\end{bmatrix}$ (with free variable x) to be coded as a construction. We can achieve this effect by taking the code to be $(\lambda[x].T)$, which has free variable y.

I shall show that the $(\lambda[\underline{x}].T)$ representation is unique, up to compilation and replacement of the variables $\underline{x}$.

THEOREM 58. If a is an atom then $(\lambda a.A) \stackrel{*}{\leftrightarrow} (\lambda a.B)$ iff $A \stackrel{*}{\leftrightarrow} B$.

Proof. If $A \stackrel{*}{\leftrightarrow} B$ then it follows immediately that $(\lambda a.A) \stackrel{*}{\leftrightarrow} (\lambda a.B)$. Conversely, suppose $(\lambda a.A) \stackrel{*}{\leftrightarrow} (\lambda a.B)$; then $(\lambda a.A^*) \stackrel{*}{\leftrightarrow} (\lambda a.B^*)$. I shall show, by structural induction on the simple term A^*, that this implies A^* is B^*.

$(\lambda a.A^*) \mapsto id$, kb or $s(\lambda a.U)(\lambda a.V)$, depending on whether A^* is a (and a is a variable), whether A^* is an atom b (where a and b are not both the same variable), or whether A^* is UV, respectively. A similar remark applies to $(\lambda a.B^*)$. Hence there are three cases in which $(\lambda a.A^*) \stackrel{*}{\leftrightarrow} (\lambda a.B^*)$ holds.

Case 1: A^* and B^* are both a, and a is a variable.

Case 2: A^* and B^* are both an atom b, where a and b are not both the same variable.

Case 3: A^* is UV and B^* is PQ, for some simple terms U, V, P, Q, where $(\lambda a.U) \stackrel{*}{\leftrightarrow} (\lambda a.P)$ and $(\lambda a.V) \stackrel{*}{\leftrightarrow} (\lambda a.Q)$, so that by inductive hypothesis U is P and V is Q.

In each case A^* is B^*. This gives $A \stackrel{*}{\leftrightarrow} B$, as required. ∎

THEOREM 59. Let $x_1, \ldots x_m$ and $y_1, \ldots y_n$ be sequences of variables without repetitions. Then $(\lambda[x_1, \ldots x_m].A) \stackrel{*}{\leftrightarrow} (\lambda[y_1, \ldots y_n].B)$ iff m is n and $A' \stackrel{*}{\leftrightarrow} B'$, where A' and B' are the results of replacing $x_1, \ldots x_m$ and $y_1, \ldots y_n$ by $z_1, \ldots z_n$ in A and B, respectively, for any sequence of variables $z_1, \ldots z_n$ (without repetitions) that do not occur in $A, B, x_1, \ldots x_m, y_1, \ldots y_n$.

Proof. Recall the variable replacement notation $T\left\{{y \atop x}\right\}$ used in the section on binding-independence: A' may be written as $A\left\{{z_1 \atop x_1}\right\} \cdots \left\{{z_n \atop x_m}\right\}$ and B' as $B\left\{{z_1 \atop y_1}\right\} \cdots \left\{{z_n \atop y_n}\right\}$, provided m is n.

First suppose m is n and $A' \stackrel{*}{\leftrightarrow} B'$ for any such sequence $z_1, \ldots z_n$. Choosing one such sequence, $(\lambda[x_1, \ldots x_m].A) \stackrel{*}{\leftrightarrow} (\lambda[z_1, \ldots z_n].A') \stackrel{*}{\leftrightarrow} (\lambda[z_1, \ldots z_n].B') \stackrel{*}{\leftrightarrow} (\lambda[y_1, \ldots y_n].B)$, using binding-independence.

Conversely, suppose $(\lambda[x_1, \ldots x_m].A) \overset{*}{\leftrightarrow} (\lambda[y_1, \ldots y_n].B)$. Let $z_1, \ldots z_n$ be a sequence of variables, without repetitions, not occurring in $A, B, x_1, \ldots x_m$, $y_1, \ldots y_n$. The proof that m is n and $A' \overset{*}{\leftrightarrow} B'$ is by induction on m. If m is not 0 then

$$\begin{aligned}&(\lambda[x_1, \ldots x_m].A)\\ &\mapsto s(s(k(\lambda s x_1.(\lambda[x_2, \ldots x_m].A)))\mathit{former})\mathit{latter}\\ &\mapsto s(s(k(s(s(k(\lambda s.(\lambda x_1.(\lambda[x_2, \ldots x_m].A))))\mathit{former})\mathit{latter}))\mathit{former})\mathit{latter}\end{aligned}$$

whereas if m is 0 then $(\lambda[x_1, \ldots x_m].A)$ is $(\lambda \mathit{nil}.A)$ and hence

$$(\lambda[x_1, \ldots x_m].A) \mapsto kA \text{ or } \mapsto s(\lambda \mathit{nil}.P)(\lambda \mathit{nil}.Q)$$

for some P and Q. Since $(\lambda \mathit{nil}.Q) \overset{*}{\not\leftrightarrow} \mathit{latter}$, all these three cases are distinguishable by inspecting the structure of $(\lambda[x_1, \ldots x_m].A)^*$.

Similar remarks apply to $(\lambda[y_1, \ldots y_n].B)$. Hence m is 0 iff n is 0. In the case where m and n are not 0, we have

$$(\lambda s.(\lambda x_1.(\lambda[x_2, \ldots x_m].A))) \overset{*}{\leftrightarrow} (\lambda s.(\lambda y_1.(\lambda[y_2, \ldots y_n].B)))$$

and hence by the previous theorem

$$(\lambda x_1.(\lambda[x_2, \ldots x_m].A)) \overset{*}{\leftrightarrow} (\lambda y_1.(\lambda[y_2, \ldots y_n].B))$$

and so, by binding-independence,

$$(\lambda z_1.(\lambda[x_2, \ldots x_m].A\left\{{}^{z_1}_{x_1}\right\})) \overset{*}{\leftrightarrow} (\lambda z_1.(\lambda[y_2, \ldots y_n].B\left\{{}^{z_1}_{y_1}\right\}))$$

and hence, by the previous theorem again,

$$(\lambda[x_2, \ldots x_m].A\left\{{}^{z_1}_{x_1}\right\}) \overset{*}{\leftrightarrow} (\lambda[y_2, \ldots y_n].B\left\{{}^{z_1}_{y_1}\right\}).$$

Then, by inductive hypothesis, m is n and

$$A\left\{{}^{z_1}_{x_1}\right\}\left\{{}^{z_2}_{x_2}\right\}\cdots\left\{{}^{z_n}_{x_m}\right\} \overset{*}{\leftrightarrow} B\left\{{}^{z_1}_{y_1}\right\}\left\{{}^{z_2}_{y_2}\right\}\cdots\left\{{}^{z_n}_{y_n}\right\},$$

as required.

In the other case, where m and n are 0, we have $(\lambda \mathit{nil}.A) \overset{*}{\leftrightarrow} (\lambda \mathit{nil}.B)$, and so, by the previous theorem, $A \overset{*}{\leftrightarrow} B$, as required. ∎

CHAPTER 16

THE EXPANDED TERM LANGUAGE

This chapter summarises the Expanded Term Language developed in the previous chapter. As explained in Chapter 12, the present chapter is a 'theory' chapter, whereas the previous one was an 'intermediate chapter'. This means that the present chapter summarises everything about the Expanded Term Language that is needed for the rest of the book, omitting proofs, the details of some definitions, commentary, and some theorems of purely local significance. Thus this chapter presents the Expanded Term Language as a high-level programming language, independently of how it is 'implemented' in the low-level Term Language.

SYNTAX OF THE EXPANDED TERM LANGUAGE (ET)

The notions of constant, variable and atom are understood as in the Term Language.

DEFINITION. A *term* of ET is a sentence in the following language.

- The alphabet is that of T plus { 'λ', '.', '$\Big[$', '$\Big]$' }.
- The lexicon is that of T plus { 'λ', '.', '$\Big[$', '$\Big]$' }.
- Lexical analysis consists of recognising constants and replacing them by the token '*con*', recognising variables and replacing them by the token '*vbl*', and recognising every other token.
- The grammar of the language is as follows.
  - The terminals are the tokens.
  - The sole nonterminal is T, which is the start symbol.
  - The production rules are $T \rightarrow con \mid (\ vbl\ ) \mid (\ T\ T\ ) \mid (\ \lambda\ T\ .\ T\ ) \mid (\ T \left[{T \atop vbl} \right])$

From now on, the word 'term' will mean a term of ET; a term of T will be called a *simple* term. A term of the form $(\ \lambda\ T\ .\ T\ )$ is called a λ*-abstraction*; a term of the form $(\ T \left[{T \atop vbl} \right])$ is called an *instantiation*.

METATERMS

Metaterms are defined as expressions that are like terms except that they may contain metaconstants, metavariables and the metanotation introduced later in this chapter, and that some brackets may be omitted. Optional brackets and spaces may also be added. To state it more precisely, the alphabet is that of metaterms of T plus the characters needed for metanotation; the lexicon is that of metaterms of T plus { 'λ', '.', '[', ']' }; lexical analysis is the same as for metaterms of T; and the grammar is

$$T \to T\,L \mid T\left[{T \atop vbl}\right] \mid L$$

$$L \to con \mid vbl \mid (\,T\,) \mid (\,\lambda\,T\,.\,T\,) \mid termcon \mid vblvbl \mid termvbl \mid \ldots$$

where the start symbol is T and '...' represents production rules for all the metanotation introduced below. *Instances* of metaterms are defined and used as in T – to obtain an instance, replace each metaconstant or metavariable by a term or variable, add and remove brackets as required, remove spaces, and rewrite all metanotation.

The first piece of metanotation is that a multiple instantiation, $A\left[{X_1 \atop x_1}\right]\cdots\left[{X_k \atop x_k}\right]$, will often be abbreviated to $A\left[{X_1,\ldots X_k \atop x_1,\ldots x_k}\right]$ or $A\left[{\underline{X} \atop \underline{x}}\right]$, where $\underline{x}$ is the (possibly empty) sequence of variables $x_1,\ldots x_k$ and $\underline{X}$ is the sequence of terms $X_1,\ldots X_k$.

INTERPRETATION OF THE EXPANDED TERM LANGUAGE IN THE TERM LANGUAGE

The compilation relation $\mapsto$ is defined as follows, where the clauses may be applied in any order.

$$AB \mapsto A'B \quad \text{if } A \mapsto A'$$

$$AB \mapsto AB' \quad \text{if } B \mapsto B'$$

$$(\lambda A.B) \mapsto (\lambda A'.B) \quad \text{if } A \mapsto A'$$

$$(\lambda A.B) \mapsto (\lambda A.B') \quad \text{if } B \mapsto B'$$

$$(\lambda PQ.B) \mapsto s(s(k(\lambda P.(\lambda Q.B)))former)latter$$

$$(\lambda a.UV) \mapsto s(\lambda a.U)(\lambda a.V) \quad \text{if } a \text{ is an atom}$$

$$(\lambda x.x) \mapsto id \quad \text{if } x \text{ is a variable}$$

$$(\lambda a.b) \mapsto kb \quad \text{if } a \text{ and } b \text{ are atoms and not both the same variable}$$

$$A\left[{B \atop x}\right] \mapsto A'\left[{B \atop x}\right] \quad \text{if } A \mapsto A'$$

$$A\left[{B \atop x}\right] \mapsto A\left[{B' \atop x}\right] \quad \text{if } B \mapsto B'$$

$$PQ\left[{B \atop x}\right] \mapsto (P\left[{B \atop x}\right])(Q\left[{B \atop x}\right])$$

$$x\begin{bmatrix}B\\x\end{bmatrix} \mapsto B$$

$$a\begin{bmatrix}B\\x\end{bmatrix} \mapsto a \quad \text{if } a \text{ is an atom other than } x$$

Let $\overset{*}{\mapsto}$ (read as 'compiles to') be the reflexive and transitive closure of $\mapsto$. Let $\leftarrow\!\!\shortmid$ and $\overset{*}{\leftarrow\!\!\shortmid}$ be the converses of $\mapsto$ and $\overset{*}{\mapsto}$. Let $\overset{*}{\leftrightarrow}$ (read as 'is equivalent to') be the equivalence relation generated by $\mapsto$. Let $A \not\mapsto$ mean that there is no B such that $A \mapsto B$.

THEOREM 1. $A \not\mapsto$ iff A is simple.

THEOREM 2. Any compilation sequence $A \mapsto B \mapsto C \mapsto \cdots$ halts. For any term A there is a unique term A^* such that $A \overset{*}{\mapsto} A^* \not\mapsto$.

THEOREM 3. $A \overset{*}{\leftrightarrow} B$ iff A^* is B^*.

THEOREM 4. $A\begin{bmatrix}x\\x\end{bmatrix} \overset{*}{\leftrightarrow} A$.

FREE VARIABLES

DEFINITION. The relation $v \in T$ (read informally as 'the variable v occurs free in the term T') is defined as follows.

$v \in a$	iff v is a, for any atom a
$v \in AB$	iff $v \in A$ or $v \in B$
$v \in (\lambda A.B)$	iff $v \notin A$ and $v \in B$
$v \in A\begin{bmatrix}B\\x\end{bmatrix}$	iff ($v \in A$ and x is not v) or ($v \in B$ and $x \in A$).

Then $x, y, \ldots z \in A, B, \ldots C$ means that *each* of the variables $x, y, \ldots z$ occurs free in *each* of the terms $A, B, \ldots C$; while $x, y, \ldots z \notin A, B, \ldots C$ means that *none* of the variables occurs free in *any* of the terms. The *free variables* of T are the variables v such that $v \in T$.

THEOREM 5. Any term has finitely many free variables.

THEOREM 6. If $A \mapsto B$ then $x \in A$ iff $x \in B$. Hence $x \in A$ iff $x \in A^*$.

THEOREM 7. $A\begin{bmatrix}X\\x\end{bmatrix} \overset{*}{\leftrightarrow} A$, if $x \notin A$.

BINDING INDEPENDENCE

DEFINITION. A notation (whether metanotation or an official part of the language) is *binding-independent* iff the choice of variables for its bound metavariables makes no difference, up to compilation, provided different metavariables denote different variables not otherwise occurring in the notation. More precisely, a notation (denoting a term) is binding-independent iff replacing all occurrences of a variable that does not occur free in the term by a variable that does not occur in the notation leaves the term unchanged, up to compilation.

THEOREM 13. All constructs of the Expanded Term Language are binding-independent.

REDUCTION IN THE EXPANDED TERM LANGUAGE

DEFINITION. The reduction relations $\rhd$, $\rhd^*$, $\lhd$, $\lhd^*$, $\rhd^*\lhd$ and $\not\rhd$ in the Term Language are extended to the expanded language by

$$\begin{array}{llll} A \rhd B & \text{iff}\ A^* \rhd B^* & A \rhd^* B & \text{iff}\ A^* \rhd^* B^* \\ A \lhd B & \text{iff}\ A^* \lhd B^* & A \rhd^*\lhd B & \text{iff}\ A^* \rhd^*\lhd B^* \\ A \not\rhd & \text{iff}\ A^* \not\rhd\ . & & \end{array}$$

DEFINITION. A *construction* is an irreducible term with no free variables.

THEOREM 14. (The instantiation theorem.)

- If $X \not\rhd$ and $A \rhd^* B$ then $A\begin{bmatrix} X \\ x \end{bmatrix} \rhd^* B\begin{bmatrix} X \\ x \end{bmatrix}$.
- If $X \not\rhd$ and $A \not\rhd$ then $A\begin{bmatrix} X \\ x \end{bmatrix} \not\rhd$.
- If $X \rhd^* Y$ then $A\begin{bmatrix} X \\ x \end{bmatrix} \rhd^* A\begin{bmatrix} Y \\ x \end{bmatrix}$.

THEOREM 15. $(\lambda A.B) \not\rhd$.

THEOREM 16. $(\lambda A.B)A\begin{bmatrix} \underline{V} \\ \underline{v} \end{bmatrix} \rhd^* B\begin{bmatrix} \underline{V} \\ \underline{v} \end{bmatrix}$, provided $\underline{V} \not\rhd$ and $A\begin{bmatrix} \underline{V} \\ \underline{v} \end{bmatrix} \not\rhd$.

THEOREM 17.

- $(\lambda x.T)X \rhd^* T\begin{bmatrix} X \\ x \end{bmatrix}$, if $X \not\rhd$;
- $(\lambda x.(\lambda y.T))XY \rhd^* T\begin{bmatrix} X \\ x \end{bmatrix}\begin{bmatrix} Y \\ y \end{bmatrix}$, if $X, Y \not\rhd$ and $y \notin X, x$.

EVALUABLE TERMS

DEFINITION. A term T is *evaluable* iff *equal* $T\ T \triangleright^*$ *true*, or, equivalently, iff $T \triangleright^* T' \not\triangleright$ for some term T'.

THEOREM 18. If $A \triangleright^*\triangleleft B$ then A is evaluable iff B is evaluable.

THEOREM 20. If $T\begin{bmatrix}X\\x\end{bmatrix}$ is evaluable and $x \in T$ then X is evaluable.

THEOREM 21. If X is evaluable then $(\lambda x.T)X \triangleright^*\triangleleft T\begin{bmatrix}X\\x\end{bmatrix}$.

THEOREM 22. If T and X are evaluable then so are $T\begin{bmatrix}X\\x\end{bmatrix}$ and $(\lambda x.T)X$.

THEOREM 23. If X is evaluable and $A \triangleright^*\triangleleft B$ then $A\begin{bmatrix}X\\x\end{bmatrix} \triangleright^*\triangleleft B\begin{bmatrix}X\\x\end{bmatrix}$.

MORE ON λ-ABSTRACTIONS

THEOREM 24. If a is a constant and $X \not\triangleright$ then $(\lambda a.T)X \triangleright^* T$.

THEOREM 25. If $(\lambda X.T)Y$ is evaluable, where X has free variables $\underline{v}$ and $\underline{v} \notin Y$, then $(\lambda X.T)Y \triangleright^* T\begin{bmatrix}\underline{V}\\\underline{v}\end{bmatrix}$ for some irreducible terms $\underline{V}$ whose free variables are included in those of Y. In the special case where Y is a construction the condition that $(\lambda X.T)Y$ be evaluable may be dropped.

BRANCH METANOTATION

DEFINITION. For any terms C, F and G, the metanotation (*branch* $C\ F\ G$) means

$$s(s(s(s(k\ \mathit{if})C)(k\ F))(k\ G))\mathit{id}.$$

THEOREM 26. If $C, F, G \not\triangleright$ then (*branch* $C\ F\ G$) $\not\triangleright$.

THEOREM 27. If $C, F, G, X \not\triangleright$ then (*branch* $C\ F\ G$)$X \triangleright^*$ *if* $(CX)\ F\ G\ X$.

PATTERN-MATCHING METANOTATIONS

For any term X, a construction $check_X$ is defined. This notation is binding-independent.

THEOREM 29. Suppose $XY \not\triangleright$. Then

- $check_{PQ}(XY) \triangleright^* check_Q Y$, if $check_P X \triangleright^* true$
- $check_{PQ}(XY) \triangleright^* false$, if $check_P X \triangleright^* false$.

THEOREM 30. $check_X X\left[\frac{\underline{V}}{\underline{v}}\right] \triangleright^* true$, provided $\underline{V} \not\triangleright$ and $X\left[\frac{\underline{V}}{\underline{v}}\right] \not\triangleright$.

DEFINITION. For any term X define a construction $match_X$ as

$$(branch\ check_X\ (s(s(k\ equal)(\lambda X.X))id)\ (k\ false)).$$

THEOREM 31. The '$match_X$' notation is binding-independent.

THEOREM 32. If $check_X Y \triangleright^* false$ then $match_X Y \triangleright^* false$.

THEOREM 33. $match_X X\left[\frac{\underline{V}}{\underline{v}}\right] \triangleright^* true$, provided $\underline{V} \not\triangleright$ and $X\left[\frac{\underline{V}}{\underline{v}}\right] \not\triangleright$.

THEOREM 34. If $match_X Y \triangleright^* true$, where X has free variables $\underline{v}$ and $\underline{v} \notin Y$, then $Y \triangleright^*\triangleleft X\left[\frac{\underline{V}}{\underline{v}}\right]$ for some irreducible terms $\underline{V}$ whose free variables are included in those of Y.

DEFINITION. For any terms X and T define the metanotation $(\lambda X{:}\ T)$ as

$$(branch\ match_X\ (\lambda X.T)\ (fxpt\ id)).$$

THEOREM 35. $(\lambda X{:}\ T) \not\triangleright$; and $y \in (\lambda X{:}\ T)$ iff $y \in T$ and $y \notin X$.

THEOREM 36. The '$(\lambda X{:}\ T)$' notation is binding-independent.

THEOREM 37. $(\lambda X{:}\ T)X\left[\frac{\underline{V}}{\underline{v}}\right] \triangleright^* T\left[\frac{\underline{V}}{\underline{v}}\right]$, if $\underline{V} \not\triangleright$ and $X\left[\frac{\underline{V}}{\underline{v}}\right] \not\triangleright$.

THEOREM 38. If $(\lambda X{:}\ T)\left[\frac{\underline{U}}{\underline{u}}\right] Y$ is evaluable, where $\underline{U}$ are irreducible terms, X has free variables $\underline{v}$ and $\underline{v} \notin Y$, then $Y \triangleright^*\triangleleft X\left[\frac{\underline{V}}{\underline{v}}\right]$ for some irreducible terms $\underline{V}$ whose free variables are included in those of Y.

DEFINING FUNCTIONS USING PATTERN MATCHING

DEFINITION. The expression

$$\begin{aligned} f\,X_1 &\triangleq T_1 \\ f\,X_2 &\triangleq T_2 \\ &\vdots \\ f\,X_k &\triangleq T_k, \end{aligned}$$

where $f \notin X_1, \ldots X_k$, means that a metaconstant is about to be introduced to denote the term *fxpt* Φ, where Φ is

$$\begin{aligned} (\lambda f.(\mathit{branch}\ \mathit{match}_{X_1}\ (\lambda X_1.T_1)\ & \\ & (\mathit{branch}\ \mathit{match}_{X_2}\ (\lambda X_2.T_2)\ \ldots \\ & \qquad (\mathit{branch}\ \mathit{match}_{X_k}\ (\lambda X_k.T_k)\, f)\ \ldots))). \end{aligned}$$

THEOREM 39. $\mathit{fxpt}\,\Phi \not\triangleright$; $y \in \mathit{fxpt}\,\Phi$ iff y is not f and, for some i, $y \in T_i$ but $y \notin X_i$.

THEOREM 40. The definition of *fxpt* Φ is binding-independent (that is, independent of the choice of variables for the metavariables in $X_1, \ldots X_k$).

THEOREM 41. Let $1 \leq i \leq k$, let $\underline{v} \in X_i$, and let $\underline{V}$ be any terms such that $f \notin \underline{V}$, $\underline{V} \not\triangleright$, $X_i\left[\frac{\underline{V}}{\underline{v}}\right] \not\triangleright$ and, for each $j < i$, $\mathit{match}_{X_j} X_i\left[\frac{\underline{V}}{\underline{v}}\right] \triangleright^*$ *false*; then

$$\mathit{fxpt}\,\Phi\ (X_i\left[\frac{\underline{V}}{\underline{v}}\right])\ \triangleright^*\ T_i\left[\frac{\underline{V}}{\underline{v}}\right]\left[\frac{\mathit{fxpt}\,\Phi}{f}\right].$$

THEOREM 42. If $(\mathit{fxpt}\,\Phi)\left[\frac{\underline{U}}{\underline{u}}\right] Y$ is evaluable, where $\underline{U}$ are irreducible terms, X_i has free variables $\underline{v_i}$, and $\underline{v_i} \notin Y$ (for each $i = 1, \ldots k$), then $Y \triangleright^*\!\triangleleft\ X\left[\frac{\underline{V}}{\underline{v_i}}\right]$ for some i and some irreducible terms $\underline{V}$ whose free variables are included in those of Y.

The usual way of using this definition mechanism will be to say 'Define f by $f\,X_1 \triangleq T_1, \ldots f\,X_k \triangleq T_k$'. Such a definition means, from now on, use f as a metaconstant to denote *fxpt* Φ (rather than the bound variable in Φ it denoted above).

REGULAR FUNCTION DEFINITIONS AND CONTINUITY

The partial reduction relations $\rightharpoonup$, $\stackrel{*}{\rightharpoonup}$, $\leftharpoonup$, $\stackrel{*}{\leftharpoonup}$, $\stackrel{*}{\leftrightharpoons}$, $\not\rightharpoonup$, $\rightharpoondown$, $\stackrel{*}{\rightharpoondown}$, $\not\rightharpoondown$ defined in the Term Language may be extended to the Expanded Term Language by

$$
\begin{array}{llll@{\qquad}llll}
A \rightharpoonup B & \text{iff} & A^* \rightharpoonup B^* & & A \stackrel{*}{\rightharpoonup} B & \text{iff} & A^* \stackrel{*}{\rightharpoonup} B^* \\
A \leftharpoonup B & \text{iff} & A^* \leftharpoonup B^* & & A \stackrel{*}{\leftharpoonup} B & \text{iff} & A^* \stackrel{*}{\leftharpoonup} B^* \\
A \stackrel{*}{\leftrightharpoons} B & \text{iff} & A^* \stackrel{*}{\leftrightharpoons} B^* & & A \not\rightharpoonup & \text{iff} & A^* \not\rightharpoonup \\
A \rightharpoondown B & \text{iff} & A^* \rightharpoondown B^* & & A \stackrel{*}{\rightharpoondown} B & \text{iff} & A^* \stackrel{*}{\rightharpoondown} B^* \\
A \not\rightharpoondown & \text{iff} & A^* \not\rightharpoondown . & & & &
\end{array}
$$

DEFINITION. A function definition $f X_1 \triangleq T_1, \ldots f X_k \triangleq T_k$ is *regular* iff, for each $i = 1, \ldots k$, $X_i \not\triangleright$ and $vT_i \stackrel{*}{\rightharpoonup} v$, where v is a fresh variable and $\stackrel{*}{\rightharpoonup}$ is the partial reduction relation with respect to f and v.

THEOREM 43. In a regular function definition of a term *fxpt* Φ, Φ is continuous.

METANOTATION

DEFINITION. Equality notation: $A = B$ is *equal* A B.

DEFINITION. The construction *boolean* is ($\lambda x.$*if* x *true true*).

THEOREM 44. *boolean*(*true*) $\triangleright^*$ *true* and *boolean*(*false*) $\triangleright^*$ *true*.

DEFINITION. Truth-functional conjunction: $A \,\&\, B$ is *if* A B *false*; and $A_1 \,\&\ldots\&\, A_k$ is $A_1 \,\&\, (A_2 \,\&\, \ldots (A_{k-1} \,\&\, A_k)\ldots)$.

THEOREM 45. If $X \not\triangleright$ then *true* & $X \triangleright X$ and *false* & $X \triangleright$ *false*.

DEFINITION. Pairs: (A, B) is sAB. The constructions *left* and *right* are $(\lambda(a,b).a)$ and $(\lambda(a,b).b)$, where a and b are two variables.

DEFINITION. Tuples: $(A_1, A_2, A_3 \ldots A_{k-1}, A_k)$ is

$$(A_1, (A_2, (A_3, \ldots (A_{k-1}, A_k) \ldots))),$$

for $k \geq 2$.

DEFINITION. Lists: $[A_1, A_2, \ldots A_k]$ is $(A_1, [A_2, \ldots A_k])$ and $[\,]$ is *nil*.

THEOREM 46. Suppose $U, V \not\triangleright$. Then

$$check_{(X,Y)}(U,V) \triangleright^* \begin{cases} check_Y V & \text{if } check_X U \triangleright^* \ true, \\ false & \text{if } check_X U \triangleright^* \ false. \end{cases}$$

THEOREM 47. If $U, V \not\triangleright$ then $check_{nil}(U,V) \triangleright^*$ *false* and $check_{(U,V)}nil \triangleright^*$ *false*.

CONSTRUCTORS

Special types of constructions called *0-ary constructors* and *1-ary constructors* are defined; there is an infinite supply of each.

THEOREM 48. If F is a 1-ary constructor and $X \not\triangleright$ then $FX \not\triangleright$.

THEOREM 49. If P and Q are two 0-ary constructors or two 1-ary constructors then $check_P Q \;\triangleright^*\; false$.

THEOREM 50. If F and G are two 1-ary constructors, and $X, Y \not\triangleright$, then $check_{FX}(GY) \;\triangleright^*\; false$, $FX = GY \;\triangleright\; false$, and $check_{FX}(FY) \;\triangleright^*\; check_X Y$.

THEOREM 51. If B is a 0-ary constructor, F is a 1-ary constructor, and $X \not\triangleright$, then $check_B(FX) \;\triangleright^*\; false$, $check_{FX} B \;\triangleright^*\; false$, $B = FX \;\triangleright\; false$, and $FX = B \;\triangleright\; false$.

NUMBERS, PRIMITIVE RECURSIVE FUNCTIONS, AND NUMERIC TERMS

Three constructions, 0, S and *pre*, are defined.

DEFINITION. The *numbers* are 0, $S0$, $S(S0)$, $S(S(S0))$, ... Let 1 be $S0$, 2 be $S1$, and so on up to 34.

THEOREM 52. If $N \not\triangleright$ then $SN \not\triangleright$. Thus all numbers are constructions.

THEOREM 53. $pre(Sn) \;\triangleright\; n$ and $Sn = 0 \;\triangleright\; false$.

DEFINITION. Define a number predicate *num* by

$$num\, 0 \;\stackrel{\Delta}{=}\; true$$
$$num\,(Sn) \;\stackrel{\Delta}{=}\; num\, n.$$

THEOREM 54. If m and n are numbers then

$$check_m n \;\triangleright^*\; \begin{cases} true & \text{if } m \text{ is } n, \\ false & \text{if } m \text{ is not } n \end{cases}$$

and the same is true for $match_m n$.

DEFINITION. A *k-ary primitive recursive (k-PR) function* is a term defined as follows.

0 is 0-PR

S is 1-PR

$proj_i^k$ is k-PR for any numbers i, k satisfying $1 \le i \le k$

$comp_l(H, F_1, \ldots F_l)$ is k-PR if H is l-PR
and $F_1, \ldots F_l$ are k-PR (where $k, l \ge 1$)

$rec_k(F, G)$ is $(k+1)$-PR if F is k-PR and G is $(k+2)$-PR (where $k \ge 0$).

Here,

$$\begin{aligned}
proj_i^k &\text{ is } (\lambda(n_1,\dots n_k).n_i) \quad \text{for each } i,k \text{ satisfying } 1 \le i \le k\\
comp_l &\text{ is } (\lambda(h,f_1,\dots f_l).(\lambda x.h(f_1x,\dots f_lx)))\\
rec_k &\text{ is } (\lambda(f,g).T_k) \quad \text{where } T_k \text{ is defined by}
\end{aligned}$$

$$\begin{cases} T_k(0,n_1,\dots n_k) \stackrel{\triangle}{=} f(n_1,\dots n_k),\\ T_k(Sm,n_1,\dots n_k) \stackrel{\triangle}{=} g(m,n_1,\dots n_k,T_k(m,n_1,\dots n_k)) \end{cases}$$

where $n_1,\dots n_k,h,f_1,\dots f_l,x,f,g,m$ are all different variables. (Note that $f(n_1,\dots n_k)$ is to be understood as f when $k = 0$. This convention will apply throughout; that is, $F()$ is a metanotation for F.)

THEOREM 55. Primitive recursive functions have no free variables and are evaluable.

DEFINITION. A *numeric term* is 0, a variable, or $F(N_1,\dots N_k)$, where $k \ge 1$, F is k-ary primitive recursive and $N_1,\dots N_k$ are numeric terms.

CHURCH'S THESIS

Recursive functions are defined as a special kind of construction.

THEOREM 56. (Weak Church's Thesis: first version.) For any mechanically effective partial operation mapping constructions to constructions, invariant under compilation of the argument, there is a recursive function F such that for any constructions A, B,

$FA \rhd^* B$ iff the operation maps A to a term $\stackrel{*}{\leftrightarrow} B$.

THEOREM 57. (Weak Church's Thesis: second version.) For any mechanically effective partial operation mapping constructions to constructions, invariant under compilation of the argument, there is a recursive function F such that for any irreducible terms A, B, with free variables $\underline{x}$,

$FA \rhd^* B$ iff, for any constructions $\underline{X}$, the operation maps $A\left[\frac{\underline{X}}{\underline{x}}\right]$ to a term $\stackrel{*}{\leftrightarrow} B\left[\frac{\underline{X}}{\underline{x}}\right]$.

CODING OF TERMS

THEOREM 59. Let $x_1,\dots x_m$ and $y_1,\dots y_n$ be sequences of variables without repetitions. Then $(\lambda[x_1,\dots x_m].A) \stackrel{*}{\leftrightarrow} (\lambda[y_1,\dots y_n].B)$ iff m is n and $A' \stackrel{*}{\leftrightarrow} B'$, where A' and B' are the results of replacing $x_1,\dots x_m$ and $y_1,\dots y_n$ by $z_1,\dots z_n$ in A and B, respectively, for any sequence of variables $z_1,\dots z_n$ (without repetitions) that do not occur in $A, B, x_1,\dots x_m, y_1,\dots y_n$.

CHAPTER 17

THE PROTOLOGICAL SEQUENT CALCULUS

Protologic is a system for elementary reasoning with constructions, as sketched in Chapter 8. See the following chapter for a detailed justification of the axioms and rules.

SEQUENTS

A *protological sequent* is an irreducible term of the form

$$([F, \ldots G], H)$$

where $F, \ldots G$ is a (possibly empty) sequence of irreducible terms and H is an irreducible term. Most sequents I use will be of the form

$$([(\lambda[\underline{z}].A), \ldots (\lambda[\underline{z}].B)], (\lambda[\underline{z}].C))$$

where $\underline{z}$ is a (possibly empty) sequence of variables and $A, \ldots B, C$ are terms; such sequents will be expressed by the metanotation

$$(\underline{z})\, A,\ \ldots\ B \rightarrow C.$$

If a sequent has no free variables, its informal meaning is that, for any construction X, the evaluations of $FX, \ldots GX$ are related to the evaluation of HX in such a way as to guarantee that the value of HX is *true* if the values of $FX, \ldots GX$ are *true*. In the case of a sequent of the form $(\underline{z})\, A,\ \ldots\ B \rightarrow C$, this means that, after instantiating $\underline{z}$ by any constructions, the evaluation of C is related to the evaluations of $A, \ldots B$ in such a way as to guarantee that the value of C is *true* if the values of $A, \ldots B$ are *true*. A sequent *with* free variables has no meaning as it stands but becomes meaningful if its free variables are instantiated by constructions.

NAMING CONVENTIONS IN AXIOMATIC SYSTEMS

Protologic is defined by a list of axiom schemata and rules of inference. Before giving this list, I had better explain some naming conventions. Each axiom schema and rule of inference is given a descriptive name, followed usually by an abbreviated name in brackets. Occasionally two similar axiom schemata or rules share a name. In formal derivations, each axiom used will be labelled with the abbreviated name of the schema it comes from, and each inference step will be labelled with the abbreviated name of the rule of inference being applied; premises of the derivation will sometimes be labelled 'p0', 'p1', 'p2', ..., and sometimes left unlabelled. In informal arguments, full names will be used instead of the abbreviated names.

Theorems and derived rules of protologic will also be named and used in the same style.

These conventions will apply also to the other formal systems, CPF, LPT, HA and PA, introduced in Part III, and their second-order versions in Part IV.

AXIOM SCHEMATA OF PROTOLOGIC

Evaluation (eval): $\rightarrow T$ where $T \rhd^* \mathit{true}$

Truth (tr): $(\underline{z})\ T \rightarrow T = \mathit{true}$ $\quad (\underline{z})\ T = \mathit{true} \rightarrow T$

Reduction (red): $(\underline{z})\ A \rightarrow B$ where $A \rhd B$ or $B \rhd A$

Transitivity (trans): $(\underline{z})\ A = B,\ B = C \rightarrow A = C$

In addition, any instantiation of the above axioms by irreducible terms is an axiom. That is, if $([F, \ldots G], H)$ is an instance of an axiom schema and $X \not\rhd$ then $([F\begin{bmatrix}X\\x\end{bmatrix}, \ldots G\begin{bmatrix}X\\x\end{bmatrix}], H\begin{bmatrix}X\\x\end{bmatrix})$ counts as an instance of the same schema.

RULES OF INFERENCE

Rules with a double line are reversible; Γ and Δ are sequences of terms.

Exchange (exch): $\dfrac{(\underline{z})\ \Gamma,\ A,\ B,\ \Delta \rightarrow C}{(\underline{z})\ \Gamma,\ B,\ A,\ \Delta \rightarrow C}$ $\quad \dfrac{(\underline{w}, x, y, \underline{z})\ \Gamma \rightarrow T}{(\underline{w}, y, x, \underline{z})\ \Gamma \rightarrow T}$

Contraction (con): $\dfrac{(\underline{z})\ A,\ A,\ \Gamma \rightarrow C}{(\underline{z})\ A,\ \Gamma \rightarrow C}$ (double line) $\quad \dfrac{(x, x, \underline{z})\ \Gamma \rightarrow T}{(x, \underline{z})\ \Gamma \rightarrow T}$ (double line)

Thinning (thin): $\dfrac{(\underline{z})\ \Gamma \rightarrow T}{(\underline{z})\ A,\ \Gamma \rightarrow T}$ $\quad \dfrac{(\underline{z})\ \Gamma \rightarrow T}{(x, \underline{z})\ \Gamma \rightarrow T}$ where $x \notin \Gamma, T$

Cut: $\dfrac{(\underline{z})\ \Gamma \rightarrow A \qquad (\underline{z})\ A,\ \Delta \rightarrow B}{(\underline{z})\ \Gamma,\ \Delta \rightarrow B}$

Decomposition (de):

$$\frac{(\underline{z})\; AB = T,\ \Gamma \to C}{(a,\underline{z})\; aB = T,\ A = a,\ \Gamma \to C} \qquad \frac{(\underline{z})\; AB = T,\ \Gamma \to C}{(b,\underline{z})\; Ab = T,\ B = b,\ \Gamma \to C}$$

where $a, b \notin A, B, T, \Gamma, C$.

If: $$\frac{(\underline{z})\; C = \mathit{true},\ A = X,\ \Gamma \to T \qquad (\underline{z})\; C = \mathit{false},\ B = X,\ \Gamma \to T}{(\underline{z})\; (\mathit{if}\; C\, A\, B) = X,\ \Gamma \to T}$$

Fxpt:

$$\frac{(u, f, x, \underline{z})\; \Phi(Xf)x = u,\ \Gamma \to X(\Psi f)x = u}{(u, x, \underline{z})\; \mathit{fxpt}\, \Phi\, x = u,\ \Gamma \to X(\mathit{fxpt}\, \Psi)x = u}$$

$$\frac{(u, f, x, \underline{z})\; X(\Psi f)x = u,\ \Gamma \to \Phi(Xf)x = u}{(u, x, \underline{z})\; X(\mathit{fxpt}\, \Psi)x = u,\ \Gamma \to \mathit{fxpt}\, \Phi\, x = u}$$

where: u, f, x are three variables not occurring free in Φ, Ψ, X and Γ; Xf is evaluable; Φ, Ψ and X are continuous; and in the second rule X is non-constant.

In addition, any instantiation of the above inferences by irreducible terms is an inference. That is, if

$$\frac{([U_1, \ldots V_1], W_1) \quad \cdots \quad ([U_k, \ldots V_k], W_k)}{([F, \ldots G], H)}$$

is an instance of a rule of inference, and $X \not\succ$, then

$$\frac{([U_1\begin{bmatrix}X\\x\end{bmatrix}, \ldots V_1\begin{bmatrix}X\\x\end{bmatrix}], W_1\begin{bmatrix}X\\x\end{bmatrix}) \quad \cdots \quad ([U_k\begin{bmatrix}X\\x\end{bmatrix}, \ldots V_k\begin{bmatrix}X\\x\end{bmatrix}], W_k\begin{bmatrix}X\\x\end{bmatrix})}{([F\begin{bmatrix}X\\x\end{bmatrix}, \ldots G\begin{bmatrix}X\\x\end{bmatrix}], H\begin{bmatrix}X\\x\end{bmatrix})}$$

counts as an instance of the same rule.

Reflection principles are also allowed in Protologic (for the reasons given in Chapters 9 and 10). Consideration of reflection principles will be postponed to Chapter 21.

CHAPTER 18

COMMENTARY ON THE PROTOLOGICAL AXIOMS AND RULES

Protologic is defined by a list of axiom schemata and rules of inference. Since these are taken as primitive constituents of constructive reasoning there is no question of proving their correctness formally; however, it is still useful to motivate them informally by showing how they serve to pin down the meaning of a sequent and the constants of the Term Language, by relating them to the underlying philosophical principles, and by showing that when translated into conventional mathematical statements they are provable by conventional mathematical arguments.

First there are the *structural* axioms and rules (Evaluation, Truth, Reduction, Exchange, Contraction, Thinning and Cut), which make explicit the intended meaning of the sequent form $(\underline{z})\ A, \ldots B \rightarrow C$: they express, for example, the facts that the order of the variables $\underline{z}$ and of the terms $A, \ldots B$ doesn't matter, that repetition of variables or terms doesn't matter, and that all that matters about a term is whether it reduces to *true*.

The remaining axioms and rules are elimination rules for various features of the Term Language. In general, let $\mathcal{A}$ be a syntactic constructor in some formal language, so that given terms $T_1, \ldots T_k$ of the language one can form a larger term $\mathcal{A}(T_1, \ldots T_k)$. Then an *elimination rule* for $\mathcal{A}$ is anything that says, 'if $\mathcal{A}(T_1, \ldots T_k)$ has some property than $T_1, \ldots T_k$ have some other properties'. An *introduction rule* for $\mathcal{A}$ says the converse: 'if $T_1, \ldots T_k$ have some properties then $\mathcal{A}(T_1, \ldots T_k)$ has a property'. The semantics of $\mathcal{A}$ is characterised by its introduction and elimination rules.

The reduction clauses of the Term Language serve as introduction rules; for example, they tell us that (*if true A B*) has the value X if A has the value X and (*if false A B*) has the value X if B has the value X. We also need an elimination rule, saying that (*if C A B*) *only* has a value if C has the value *true* or *false*. The reduction clause *if C A B* $\triangleright$ *if C A B* achieves this effect, since if neither of the previous *if* clauses applies then this clause will cause reduction to loop. However, it does not *say* explicitly that (*if C A B*) is undefined, only that it has the same value as itself. This is inexpressible as a reduction clause, hence the elimination rule for *if* is provided by the If Rule of protologic, which says that if (*if C A B*) has a value X this can only be because C has the value *true* and A has the value X or C has the value *false* and B has the value X.

Likewise with *equal*. The reduction clauses state circumstances under which $A = B$ reduces to *true* or *false*: they constitute an introduction rule for equality. The corresponding elimination rule is that if $A = B$ reduces to *true* then A and B are interchangeable in all extensional contexts. This will be expressed in the Equality Theorem of Expanded Protologic (see below); this theorem is a consequence of the Transitivity Axiom Schema of Protologic, so the latter may be regarded as the elimination rule for *equal*.

The Decomposition Rules constitute an elimination rule for concatenation of terms. The reduction clauses

$$AB \rhd A'B \quad \text{if } A \rhd A', \qquad AB \rhd AB' \quad \text{if } B \rhd B'$$

tell us that AB may be evaluated by evaluating A and B to irreducible terms a and b, and then evaluating ab. The Decomposition Rules say that AB can *only* be evaluated by this route: if AB has the value T this can only be because A has a value a and aB has the value T, and likewise for the other rule.

The final case is the Fxpt Rules. The reduction clause $\mathit{fxpt}\, A\, B \rhd A(\mathit{fxpt}\, A)B$ is an introduction rule since it allows us to find the value of $\mathit{fxpt}\, A\, B$ assuming we can evaluate expressions involving A and B. The Fxpt Rules say that $\mathit{fxpt}\, A\, B$ *only* has a value when obtained by applying $\mathit{fxpt}\, A\, B \rhd A(\mathit{fxpt}\, A)B$ repeatedly.

I shall show that the Fxpt Rules, when translated into conventional mathematical notation, can be justified by conventional mathematical arguments. The inspiration for this is Loeckx & Sieber (1984); see this book for more details of the notation and standard results used below. Let Φ, Ψ and X be continuous, let $\bot$ be an empty function (undefined for all arguments) and let $f \sqsubseteq g$ mean that the partial function g is an extension of the partial function f.

THEOREM 1. Let X, Φ and Ψ be continuous. If $\Phi(Xf) \sqsubseteq X(\Psi f)$ for any partial function f then $\mathit{fxpt}\,\Phi \sqsubseteq X(\mathit{fxpt}\,\Psi)$.

Proof. The condition $\Phi(Xf) \sqsubseteq X(\Psi f)$ (for any f) implies by induction $\Phi^n(Xf) \sqsubseteq X(\Psi^n f)$ for any natural number n and any f (using the monotonicity of Φ, which follows from its continuity). Now, $\mathit{fxpt}\,\Phi$ is interpreted as the least fixed-point of Φ, which is $\bigcup_{n=0}^{\infty} \Phi^n(\bot)$, and likewise for $\mathit{fxpt}\,\Psi$, so

$$\forall n \quad \Phi^n(\bot) \sqsubseteq \Phi^n(X\bot) \sqsubseteq X(\Psi^n \bot) \sqsubseteq X(\mathit{fxpt}\,\Psi)$$

using the monotonicity of Φ and X, so

$$\mathit{fxpt}\,\Phi = \bigcup_{n=0}^{\infty} \Phi^n(\bot) \sqsubseteq X(\mathit{fxpt}\,\Psi)$$

as required. ∎

THEOREM 2. Let X, Φ and Ψ be continuous and X be non-constant (meaning $X\bot = \bot$). If $X(\Psi f) \sqsubseteq \Phi(Xf)$ for any partial function f then $X(\mathit{fxpt}\,\Psi) \sqsubseteq \mathit{fxpt}\,\Phi$.

Proof. Similar to Theorem 1. ∎

The Fxpt Rules follow from these theorems. Theorem 1 may be expressed in sequent notation as

$$\frac{(u,f,x)\ \Phi(Xf)x = u \to X(\Psi f)x = u}{(u,x)\ \mathit{fxpt}\,\Phi x = u \to X(\mathit{fxpt}\,\Psi)x = u}$$

Now, if we imagine X, Φ and Ψ as depending on some new variables $\underset{\sim}{z}$, which are constrained to satisfy some conditions Γ (not involving u,f,x) and are held constant throughout the argument, then we obtain a more general rule

$$\frac{(u,f,x,\underset{\sim}{z})\ \Phi(Xf)x = u,\ \Gamma \to X(\Psi f)x = u}{(u,x,\underset{\sim}{z})\ \mathit{fxpt}\,\Phi x = u,\ \Gamma \to X(\mathit{fxpt}\,\Psi)x = u}$$

which is the first Fxpt Rule, as required. The second Fxpt Rule is obtained from Theorem 2 in a similar way.

The above is a conventional mathematical justification of the Fxpt Rules. More illuminating, however, is an informal heuristic argument relating them to the philosophical discussion of induction in Chapter 8. For this, I shall continue to use conventional mathematical notation and I shall skate over many technical details to make the main points stand out. Suppose Φ, Ψ and X are of the form

$$\Phi fx = \begin{cases} Rx & \text{if } Cx = \mathit{true} \\ f(Hx) & \text{if } Cx = \mathit{false} \end{cases} \qquad \Psi fx = \begin{cases} R'x & \text{if } C'x = \mathit{true} \\ f(H'x) & \text{if } C'x = \mathit{false} \end{cases}$$

$$Xf = f \circ I$$

for some partial functions C, R, H, C', R', H', I. Then the first Fxpt Rule, omitting $\underset{\sim}{z}$ and Γ, boils down to

$$\frac{(u,f,x)\ \Phi(f \circ I)x = u \to \Psi f(Ix) = u}{(u,x)\ \mathit{fxpt}\,\Phi x = u \to \mathit{fxpt}\,\Psi(Ix) = u}$$

using the definition of X. The premise of this follows by the If Rule from the two cases

$$(u,f,x)\ Cx = \mathit{true},\ Rx = u \to C'(Ix) = \mathit{true}\ \&\ R'(Ix) = u$$

$$(u,f,x)\ Cx = \mathit{false},\ f(I(Hx)) = u \to C'(Ix) = \mathit{false}\ \&\ f(H'(Ix)) = u$$

using the definitions of Φ and Ψ, and these sequents follow from

$$C \sqsubseteq C' \circ I$$
$$R \sqsubseteq R' \circ I$$
$$I \circ H \sqsubseteq H' \circ I.$$

This is depicted as a 'one-way' commutative diagram in Figure 3.

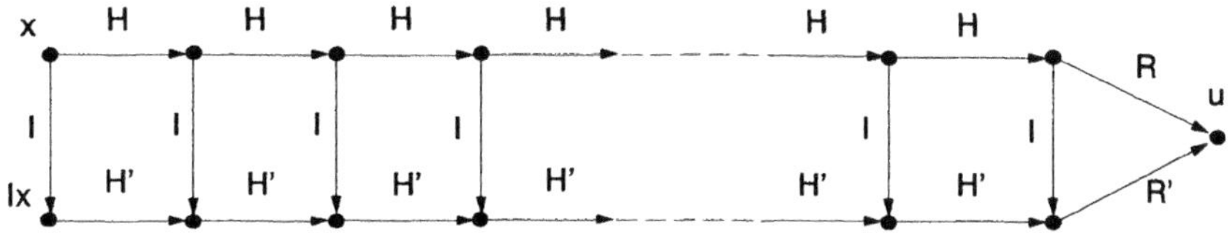

Figure 3: the relations between the computations in the Fxpt Rules.

As in the discussion of induction in Chapter 8, imagine *fxpt* Φ as an algorithm in English and *fxpt* Ψ as the 'same' algorithm in French; let *I* be English-to-French translation. Then the first Fxpt Rule says that if C', R' and H' are the French translations of *C*, *R* and *H*, then the whole computation *fxpt* Ψ is the French translation of *fxpt* Φ, and so produces the same outcome, *u*. The premises of the rule assert a *local* relation between any step of *fxpt* Φ and a step of *fxpt* Ψ, while the conclusion asserts a *global* relation concerning the whole computations. Thus the effect of the rule is to piece together the previously constructed local relations into a global relation. This local-to-global transition is just the function of induction, as explained in Chapter 8. Similar considerations apply to the second Fxpt Rule.

Protologic is a program correctness calculus. Of all the various systems of this sort in the literature, the one it most closely resembles is Goodman's (1972) combinatory logic. This is a sequent calculus with the usual structural rules and special axioms and rules that correspond, in my system, to instances of the Reduction axiom schema and the derived rules of Instantiation, Symmetry, Equality and Self-Equality (see the next chapter). Goodman's rule that equality is a decidable predicate corresponds to my If Rule. There is nothing in the system corresponding to my Fxpt Rules (though the version in Goodman (1970) includes a structural induction rule). The reason for the similarity between our systems is that my intended semantics of terms follows Goodman's in requiring that a term *AB* be evaluated by first evaluating *A* and *B* (see Chapter 13). The principal difference between our systems is that, in order to give an interpretation of the logical constants, Goodman (1970) needs to add a primitive evidence predicate π, constants concerning 'grasped domains' (*B*, *E* and *G*), and two reducibility operators (*F* and $\wp$). In my system π (or *DT*, as I call it) is defined explicitly (see Chapter 21) and the other constants are dispensed with.

CHAPTER 19

FROM PROTOLOGIC TO EXPANDED PROTOLOGIC

Expanded Protologic (EP) consists of Protologic plus the following derived sequents and rules of inference. Just as Protologic provides elimination rules for the constructs of the Term Language, so Expanded Protologic will provide elimination rules for the higher-level constructs and metanotation of the Expanded Term Language.

I shall give formal derivations in full, displaying every step and marking all axioms and inferences according to the naming conventions explained in Protologic, except that occasionally two simple steps will be combined into one or a use of the Exchange Rule will go without saying where this makes the derivation easier to read. In all derivations Γ is a sequence of terms.

PRELIMINARY DERIVATIONS

THEOREM 1. Tautology (taut): $(\underset{\sim}{z})\ \Gamma,\ T \to T$.

Proof.

$$\dfrac{\dfrac{(\underset{\sim}{z})\ T \to T = \mathit{true}\ \text{(tr)} \qquad (\underset{\sim}{z})\ T = \mathit{true} \to T\ \text{(tr)}}{(\underset{\sim}{z})\ T \to T}\ \text{cut}}{(\underset{\sim}{z})\ \Gamma,\ T \to T}\ \text{thin}$$

∎

THEOREM 2. Compilation (comp): $(\underset{\sim}{z})\ A \to B \qquad \dfrac{(\underset{\sim}{z})\ \Gamma \to A}{(\underset{\sim}{z})\ \Gamma \to B} \qquad \dfrac{(\underset{\sim}{z})\ A,\ \Gamma \to C}{(\underset{\sim}{z})\ B,\ \Gamma \to C}$

if $A \overset{*}{\leftrightarrow} B$.

Proof. Let C be a term such that $A \rhd C$ or $C \rhd A$ (if $A \not\rhd$ then take C as $id\,A$; otherwise take C such that $A \rhd C$). Then $B \rhd C$ or $C \rhd B$. Now, the sequent $(\underset{\sim}{z})\ A \to B$ is derived by

$$\dfrac{(\underset{\sim}{z})\ A \to C\ \text{(red)} \qquad (\underset{\sim}{z})\ C \to B\ \text{(red)}}{(\underset{\sim}{z})\ A \to B}\ \text{cut}$$

and the rules are derived from this by Cut. ∎

THEOREM 3. Reduction (red): $(\underset{\sim}{z})\ A \to B \quad \dfrac{(\underset{\sim}{z})\ \Gamma \to A}{(\underset{\sim}{z})\ \Gamma \to B} \quad \dfrac{(\underset{\sim}{z})\ A,\ \Gamma \to C}{(\underset{\sim}{z})\ B,\ \Gamma \to C}$

if $A \rhd^{*}\lhd B$.

Proof. From Tautology and Compilation by Cutting with instances of the Reduction Axiom Schema. ∎

THEOREM 4. Separation rule (sep):

$$\frac{(\underset{\sim}{z})\ A = B,\ \Gamma \to T}{(v,\underset{\sim}{z})\ A = v,\ B = v,\ \Gamma \to T}$$

if $v \notin A, B, \Gamma, T$.

Proof. Recall that $X = Y$ is *equal* $X\ Y$.

$$\cfrac{\cfrac{\cfrac{(\underset{\sim}{z})\ \textit{equal}\ A\ B,\ \Gamma \to T}{(\underset{\sim}{z})\ \textit{equal}\ A\ B = \textit{true},\ \Gamma \to T}\ \text{tr, cut}}{(v,\underset{\sim}{z})\ \textit{equal}\ A\ v = \textit{true},\ B = v,\ \Gamma \to T}\ \text{de}}{(v,\underset{\sim}{z})\ \textit{equal}\ A\ v,\ B = v,\ \Gamma \to T}\ \text{tr, cut}$$

∎

THEOREM 5. Symmetry (symm): $(\underset{\sim}{z})\ A = B \to B = A$.

Proof. Let v be a fresh variable, that is, a variable that does not occur in any term mentioned in the theorem so far.

$$\cfrac{\cfrac{\cfrac{(\underset{\sim}{z})\ B = A \to B = A\ \text{(taut)}}{(v,\underset{\sim}{z})\ B = v,\ A = v \to B = A}\ \text{sep}}{(v,\underset{\sim}{z})\ A = v,\ B = v \to B = A}\ \text{exch}}{(\underset{\sim}{z})\ A = B \to B = A}\ \text{sep}$$

∎

PROPERTIES OF THE INSTANTIATION NOTATION

The instantiation notation $T\begin{bmatrix}X\\x\end{bmatrix}$ was described in Chapter 13 as 'semantic substitution', in contrast with ordinary 'textual substitution', since it respects the principle that the value of $T\begin{bmatrix}X\\x\end{bmatrix}$ depends on the value of X, not on any other aspect of X. Thus $T\begin{bmatrix}X\\x\end{bmatrix}$ behaves exactly as T would in a context in which $x = X$. This is encapsulated in protologic in the following theorems and derived rules.

THEOREM 6. Equality theorem and rules (eq): $(\underline{z})\, A = B,\ T\left[{B \atop x}\right] \to T\left[{A \atop x}\right]$

$$\frac{(\underline{z})\, A = B,\ \Gamma \to T\left[{A \atop x}\right]}{(\underline{z})\, A = B,\ \Gamma \to T\left[{B \atop x}\right]} \qquad \frac{(\underline{z})\, A = B,\ T\left[{A \atop x}\right], \Gamma \to C}{(\underline{z})\, A = B,\ T\left[{B \atop x}\right], \Gamma \to C}$$

Proof. The Equality Theorem $(\underline{z})\, A = B,\ T\left[{B \atop x}\right] \to T\left[{A \atop x}\right]$ follows from

$$(\underline{z})\, A = B,\ T^*\left[{B \atop x}\right] = U \to T^*\left[{A \atop x}\right] = U$$

by Compilation and taking U as *true*; the latter sequent is derived by structural induction on the simple term T^* as follows.

Case 1: T^* is x. Then the desired sequent compiles to an instance of Transitivity: $(\underline{z})\, A = B,\ B = U \to A = U$.

Case 2: T^* is any other atom, a. Then the desired sequent compiles to an instance of Tautology: $(\underline{z})\, A = B,\ a = U \to a = U$.

Case 3: T^* is PQ. Then let p and q be two fresh variables; the derivation is as follows (with 'ih' standing for the inductive hypothesis).

$$\begin{array}{cl}
(\underline{z})\, A = B,\ P\left[{A \atop x}\right] Q\left[{A \atop x}\right] = U \to P\left[{A \atop x}\right] Q\left[{A \atop x}\right] = U \ \text{(taut)} & \\
\hline
(p, q, \underline{z})\, A = B,\ pq = U,\ P\left[{A \atop x}\right] = p,\ Q\left[{A \atop x}\right] = q \to P\left[{A \atop x}\right] Q\left[{A \atop x}\right] = U & \text{de} \\
\hline
(p, q, \underline{z})\, A = B,\ pq = U,\ P\left[{B \atop x}\right] = p,\ Q\left[{B \atop x}\right] = q \to P\left[{A \atop x}\right] Q\left[{A \atop x}\right] = U & \text{ih, cut} \\
\hline
(\underline{z})\, A = B,\ P\left[{B \atop x}\right] Q\left[{B \atop x}\right] = U \to P\left[{A \atop x}\right] Q\left[{A \atop x}\right] = U & \text{de} \\
\hline
(\underline{z})\, A = B,\ T^*\left[{B \atop x}\right] = U \to T^*\left[{A \atop x}\right] = U & \text{comp}
\end{array}$$

The Equality Rules are derived from the Equality Theorem using Cut and Symmetry. ∎

THEOREM 7. Self-equality (se): $(\underline{z})\ T\left[{X \atop x}\right] \to X = X$ if X is evaluable or $x \in T$.

Proof. If X is evaluable then $X = X \ \rhd^* \ \textit{true}$, so the sequent follows by Reduction and Thinning from $\to \textit{true}$. In the other case, where $x \in T$, the sequent is obtained from

$$(\underline{z})\ T^*\left[{X \atop x}\right] = U \to X = X$$

by Compilation and taking U as *true*. This latter sequent is derived by structural induction on the simple term T^* as follows.

Case 1: T^* is an atom. Then it must be x, so the derivation is

$$\dfrac{\dfrac{(\underline{z})\, X = U \to U = X \ \text{(symm)} \qquad (\underline{z})\, X = U,\ U = X \to X = X \ \text{(trans)}}{(\underline{z})\, X = U \to X = X}\ \text{cut, con}}{(\underline{z})\, T^*\left[{X \atop x}\right] = U \to X = X}\ \text{comp}$$

as required.

Case 2: T^* is PQ. Then $x \in P$ or $x \in Q$. Let p and q be two fresh variables; then the derivation is

$$\dfrac{\dfrac{\dfrac{(p, q, \underline{z})\, P\left[{X \atop x}\right] = p \to X = X \quad \text{or} \quad (p, q, \underline{z})\, Q\left[{X \atop x}\right] = q \to X = X \ \text{(ih)}}{(p, q, \underline{z})\, pq = U,\ P\left[{X \atop x}\right] = p,\ Q\left[{X \atop x}\right] = q \to X = X}\ \text{thin}}{(\underline{z})\, P\left[{X \atop x}\right] Q\left[{X \atop x}\right] = U \to X = X}\ \text{de}}{(\underline{z})\, T^*\left[{X \atop x}\right] = U \to X = X}\ \text{comp}$$

(where again 'ih' is the inductive hypothesis). ∎

THEOREM 8. Pre-instantiation rule (pre-inst): $\dfrac{(y, \underline{z})\, A\left[{y \atop x}\right], \ldots B\left[{y \atop x}\right] \to C\left[{y \atop x}\right]}{(\underline{z})\, A\left[{X \atop x}\right], \ldots B\left[{X \atop x}\right] \to C\left[{X \atop x}\right]}$

where: $y \notin X, A\left[{X \atop x}\right], \ldots B\left[{X \atop x}\right], C\left[{X \atop x}\right]$; and X is evaluable or $x \in A$ or $\ldots$ $x \in B$.

Proof.

$$\dfrac{\dfrac{\dfrac{\dfrac{(y, \underline{z})\, A\left[{y \atop x}\right], \ldots B\left[{y \atop x}\right] \to C\left[{y \atop x}\right]}{(y, \underline{z})\, X = y,\ X = y,\ A\left[{y \atop x}\right], \ldots B\left[{y \atop x}\right] \to C\left[{y \atop x}\right]}\ \text{thin}}{(y, \underline{z})\, X = y,\ X = y,\ A\left[{X \atop x}\right], \ldots B\left[{X \atop x}\right] \to C\left[{X \atop x}\right]}\ \text{eq}}{(\underline{z})\, X = X,\ A\left[{X \atop x}\right], \ldots B\left[{X \atop x}\right] \to C\left[{X \atop x}\right]}\ \text{sep}}{(\underline{z})\, A\left[{X \atop x}\right], \ldots B\left[{X \atop x}\right] \to C\left[{X \atop x}\right]}\ \text{se, cut}$$

∎

THEOREM 9. Instantiation rule (inst): $$\frac{(x,\underline{z})\,A,\ldots B \to C}{(\underline{z})\,A\begin{bmatrix}X\\x\end{bmatrix},\ldots B\begin{bmatrix}X\\x\end{bmatrix} \to C\begin{bmatrix}X\\x\end{bmatrix}}$$

where X is evaluable or $x \in A$ or ... $x \in B$.

Proof. Let u be a fresh variable.

$$\cfrac{\cfrac{\cfrac{\cfrac{(x,\underline{z})\,A,\ldots B \to C}{(x,u,\underline{z})\,A,\ldots B \to C}\ \text{thin}}{(x,u,\underline{z})\,A\begin{bmatrix}x\\x\end{bmatrix},\ldots B\begin{bmatrix}x\\x\end{bmatrix} \to C\begin{bmatrix}x\\x\end{bmatrix}}\ \text{comp}}{(u,\underline{z})\,A\begin{bmatrix}u\\x\end{bmatrix},\ldots B\begin{bmatrix}u\\x\end{bmatrix} \to C\begin{bmatrix}u\\x\end{bmatrix}}\ \text{pre-inst}}{(\underline{z})\,A\begin{bmatrix}X\\x\end{bmatrix},\ldots B\begin{bmatrix}X\\x\end{bmatrix} \to C\begin{bmatrix}X\\x\end{bmatrix}}\ \text{pre-inst}$$

∎

THEOREM 10. Extraction rule (ext): $$\frac{(x,\underline{z})\,x=X,\ A,\ldots B \to C}{(\underline{z})\,A\begin{bmatrix}X\\x\end{bmatrix},\ldots B\begin{bmatrix}X\\x\end{bmatrix} \to C\begin{bmatrix}X\\x\end{bmatrix}}$$ (double line)

where $x \notin X$, and X is evaluable or $x \in A$ or ... $x \in B$.

Proof. In one direction:

$$\cfrac{\cfrac{(x,\underline{z})\,x=X,\ A,\ldots B \to C}{(\underline{z})\,X=X,\ A\begin{bmatrix}X\\x\end{bmatrix},\ldots B\begin{bmatrix}X\\x\end{bmatrix} \to C\begin{bmatrix}X\\x\end{bmatrix}}\ \text{inst}}{(\underline{z})\,A\begin{bmatrix}X\\x\end{bmatrix},\ldots B\begin{bmatrix}X\\x\end{bmatrix} \to C\begin{bmatrix}X\\x\end{bmatrix}}\ \text{se, cut}$$

In the other direction:

$$\cfrac{\cfrac{(\underline{z})\,A\begin{bmatrix}X\\x\end{bmatrix},\ldots B\begin{bmatrix}X\\x\end{bmatrix} \to C\begin{bmatrix}X\\x\end{bmatrix}}{(x,\underline{z})\,x=X,\ A\begin{bmatrix}X\\x\end{bmatrix},\ldots B\begin{bmatrix}X\\x\end{bmatrix} \to C\begin{bmatrix}X\\x\end{bmatrix}}\ \text{thin}}{(x,\underline{z})\,x=X,\ A,\ldots B \to C}\ \text{eq}$$

∎

THEOREM 11. Conversion rules (conv): $$\frac{(\underline{z})\,\Gamma \to A\begin{bmatrix}\underline{X}\\\underline{x}\end{bmatrix}}{(\underline{z})\,\Gamma \to B\begin{bmatrix}\underline{X}\\\underline{x}\end{bmatrix}} \qquad \frac{(\underline{z})\,\Gamma,\ A\begin{bmatrix}\underline{X}\\\underline{x}\end{bmatrix} \to C}{(\underline{z})\,\Gamma,\ B\begin{bmatrix}\underline{X}\\\underline{x}\end{bmatrix} \to C}$$ (double lines)

where $A \ \triangleright^{*}\!\triangleleft\ B$ and the variables $\underline{x}$ are all different and occur free in A and B.

Proof. The rules follow by Cut with

$$(\underline{z})\, A\begin{bmatrix}X\\ \underline{x}\end{bmatrix} \to B\begin{bmatrix}X\\ \underline{x}\end{bmatrix} \quad \text{and} \quad (\underline{z})\, B\begin{bmatrix}X\\ \underline{x}\end{bmatrix} \to A\begin{bmatrix}X\\ \underline{x}\end{bmatrix},$$

which follow by Instantiation from

$$(\underline{x}, \underline{z})\, A \to B \quad \text{and} \quad (\underline{x}, \underline{z})\, B \to A,$$

which follow by Reduction. ∎

PROPERTIES OF TRUTH-FUNCTIONAL OPERATORS

THEOREM 12. Falsity elimination (*false*-el): $(\underline{z})\, \mathit{false} \to A$.

(This is where the principle of *ex falso quodlibet* makes its appearance in the system. As promised in Chapter 5, this principle arises naturally and inevitably from a general theory of constructions, without need for any special stipulations.)

Proof.

$$\dfrac{\dfrac{\dfrac{\dfrac{\to \mathit{true}\ \text{(eval)}}{(\underline{z})\, \mathit{false} = \mathit{true} \to \mathit{true}}\ \text{thin}}{(\underline{z})\, \mathit{false} = \mathit{true} \to (\mathit{if}\ \mathit{false}\ (\lambda \mathit{nil}.A)\ (k\ \mathit{true}))\mathit{nil}}\ \text{red}}{(\underline{z})\, \mathit{false} = \mathit{true} \to (\mathit{if}\ \mathit{true}\ (\lambda \mathit{nil}.A)\ (k\ \mathit{true}))\mathit{nil}}\ \text{eq}}{(\underline{z})\, \mathit{false} \to A}\ \text{red}$$

∎

THEOREM 13. Boolean rule (bool):

$$\dfrac{(\underline{z})\, x = \mathit{true},\ \Gamma \to C \qquad (\underline{z})\, x = \mathit{false},\ \Gamma \to C}{(\underline{z})\, \mathit{boolean}\, x,\ \Gamma \to C}$$

Proof.

$$\dfrac{\dfrac{\dfrac{\dfrac{(\underline{z})\, x = \mathit{true},\ \Gamma \to C}{(\underline{z})\, x = \mathit{true},\ \mathit{true} = \mathit{true},\ \Gamma \to C}\ \text{thin} \qquad \dfrac{(\underline{z})\, x = \mathit{false},\ \Gamma \to C}{(\underline{z})\, x = \mathit{false},\ \mathit{true} = \mathit{true},\ \Gamma \to C}\ \text{thin}}{(\underline{z})\, (\mathit{if}\ x\ \mathit{true}\ \mathit{true}) = \mathit{true},\ \Gamma \to C}\ \text{if}}{(\underline{z})\, \mathit{if}\ x\ \mathit{true}\ \mathit{true},\ \Gamma \to C}\ \text{tr, cut}}{(\underline{z})\, \mathit{boolean}\, x,\ \Gamma \to C}\ \text{red}$$

∎

THEOREM 14. Branch rule (branch):

$$\frac{(\underline{z})\ CX,\ FX = U,\ \Gamma \to T \qquad (\underline{z})\ CX = \mathit{false},\ GX = U,\ \Gamma \to T}{(\underline{z})\ (\mathit{branch}\ C\ F\ G)X = U,\ \Gamma \to T}$$

Proof. Let p be a fresh variable. Then we can derive

$$\cfrac{\cfrac{(\underline{z})\ CX,\ FX = U,\ \Gamma \to T}{(p,\underline{z})\ CX,\ F = p,\ pX = U,\ \Gamma \to T}\ \text{de}}{(p,\underline{z})\ CX = \mathit{true},\ F = p,\ pX = U,\ \Gamma \to T}\ \text{tr, cut}$$

and similarly

$$\frac{(\underline{z})\ CX = \mathit{false},\ GX = U,\ \Gamma \to T}{(p,\underline{z})\ CX = \mathit{false},\ G = p,\ pX = U,\ \Gamma \to T}\ \text{de}$$

and hence, applying the If Rule to the above two,

$$\cfrac{\cfrac{(p,\underline{z})\ (\mathit{if}\ (CX)\ F\ G) = p,\ pX = U,\ \Gamma \to T\ \ \text{(if)}}{(\underline{z})\ (\mathit{if}\ (CX)\ F\ G)X = U,\ \Gamma \to T}\ \text{de}}{(\underline{z})\ (\mathit{branch}\ C\ F\ G)X = U,\ \Gamma \to T}\ \text{conv}$$

as required. ∎

THEOREM 15. Conjunction (&):

$$(\underline{z})\ A_1,\ \ldots\ A_k \to A_1 \,\&\, \ldots \,\&\, A_k, \qquad (\underline{z})\ A_1 \,\&\, \ldots \,\&\, A_k \to A_i$$

for $i = 1, \ldots k$.

Proof. These sequents may be derived by Exchange and Cut from instances of the schemata $(\underline{z})\ X,\ Y \to X \,\&\, Y$, $(\underline{z})\ X \,\&\, Y \to X$, and $(\underline{z})\ X \,\&\, Y \to Y$, which are derived as follows.

$$\cfrac{\cfrac{\cfrac{\cfrac{(\underline{z})\ Y \to Y\ \ \text{(taut)}}{(\underline{z})\ Y \to \mathit{true} \,\&\, Y}\ \text{conv}}{(\underline{z})\ X = \mathit{true},\ Y \to \mathit{true} \,\&\, Y}\ \text{thin}}{(\underline{z})\ X = \mathit{true},\ Y \to X \,\&\, Y}\ \text{eq}}{(\underline{z})\ X,\ Y \to X \,\&\, Y}\ \text{tr, cut}$$

$$
\dfrac{
\dfrac{(\underline{z})\ X = \mathit{true} \to X\ \text{(tr)}}{(\underline{z})\ X = \mathit{true},\ Y = \mathit{true} \to X}\,\text{thin}
\qquad
\dfrac{\dfrac{(\underline{z})\ \mathit{false} \to X\ (\mathit{false}\text{-el})}{(\underline{z})\ \mathit{false} = \mathit{true} \to X}\,\text{red}}{(\underline{z})\ X = \mathit{false},\ \mathit{false} = \mathit{true} \to X}\,\text{thin}
}{
\dfrac{(\underline{z})\ (X \,\&\, Y) = \mathit{true} \to X}{(\underline{z})\ X \,\&\, Y \to X}\,\text{tr, cut}
}\,\text{if}
$$

$$
\dfrac{
\dfrac{(\underline{z})\ Y = \mathit{true} \to Y\ \text{(tr)}}{(\underline{z})\ X = \mathit{true},\ Y = \mathit{true} \to Y}\,\text{thin}
\qquad
\dfrac{\dfrac{(\underline{z})\ \mathit{false} \to Y\ (\mathit{false}\text{-el})}{(\underline{z})\ \mathit{false} = \mathit{true} \to Y}\,\text{red}}{(\underline{z})\ X = \mathit{false},\ \mathit{false} = \mathit{true} \to Y}\,\text{thin}
}{
\dfrac{(\underline{z})\ (X \,\&\, Y) = \mathit{true} \to Y}{(\underline{z})\ X \,\&\, Y \to Y}\,\text{tr, cut}
}\,\text{if}
$$

∎

RULES FOR THE PATTERN-MATCHING NOTATIONS

The rules in this section are essential for proving the correctness of functions defined by pattern matching. They formalise some of the theorems of the Expanded Term Language. For example, Theorems 33 and 34 of Chapter 16 show that $\mathit{match}_X Y \rhd^* \mathit{true}$ iff Y is an instantiation of X (for any construction Y). This is formalised in the Match Rule (Theorem 17, below), which says that the condition $\mathit{match}_X T$ is equivalent to $X = T$ for suitable values of the free variables of X.

THEOREM 16. $(\lambda X.T)$ rule $((\lambda X.T))$: $\dfrac{(\underline{v}, \underline{z})\ T = U,\ \Gamma \to C}{(\underline{z})\ (\lambda X.T)Y = U,\ \Gamma \to C}$

where $\underline{v}$ are the free variables of X and $\underline{v} \notin Y, U, \Gamma, C$.

Proof. This is derived by compiling X to X^* and applying structural induction on the simple term X^*, as follows.

Case 1: X^* is a constant, a. Then let y be a fresh variable; the derivation is

$$
\dfrac{\dfrac{\dfrac{(\underline{z})\ T = U,\ \Gamma \to C}{(y, \underline{z})\ T = U,\ \Gamma \to C}\,\text{thin}}{(y, \underline{z})\ (\lambda a.T)y = U,\ \Gamma \to C}\,\text{red}}{(\underline{z})\ (\lambda a.T)Y = U,\ \Gamma \to C}\,\text{inst}
$$

Case 2: X^* is a variable, v. The derivation is

$$\begin{array}{cl}
(v,\underline{z})\ T = U,\ \Gamma \to C & \\ \hline
(v,\underline{z})\ (\lambda v.T)v = U,\ \Gamma \to C & \text{red} \\ \hline
(\underline{z})\ (\lambda v.T)Y = U,\ \Gamma \to C & \text{inst}
\end{array}$$

Case 3: X^* is PQ. Then let v_p and v_q the free variables of P and Q respectively, and let m be a fresh variable; the derivation is

$$\begin{array}{cl}
(\underline{v},\underline{z})\ T = U,\ \Gamma \to C & \\ \hline
(\underline{v_q},\underline{v_p},\underline{z})\ T = U,\ \Gamma \to C & \text{con, exch} \\ \hline
(\underline{v_p},\underline{z})\ (\lambda Q.T)(\mathit{latter}\ Y) = U,\ \Gamma \to C & \text{ih} \\ \hline
(m,\underline{v_p},\underline{z})\ (\lambda Q.T) = m,\ m(\mathit{latter}\ Y) = U,\ \Gamma \to C & \text{de} \\ \hline
(m,\underline{z})\ (\lambda P.(\lambda Q.T))(\mathit{former}\ Y) = m,\ m(\mathit{latter}\ Y) = U,\ \Gamma \to C & \text{ih} \\ \hline
(\underline{z})\ (\lambda P.(\lambda Q.T))(\mathit{former}\ Y)(\mathit{latter}\ Y) = U,\ \Gamma \to C & \text{de} \\ \hline
(\underline{z})\ (\lambda PQ.T)Y = U,\ \Gamma \to C & \text{conv}
\end{array}$$

where 'ih' indicates a use of the inductive hypothesis. ∎

THEOREM 17. Match rule (match): $\dfrac{(\underline{z})\ \mathit{match}_X T,\ \Gamma \to C}{(\underline{v},\underline{z})\ X = T,\ \Gamma \to C}$ (double line)

where X is irreducible, with free variables $\underline{v}$, and $\underline{v} \notin T, \Gamma, C$.

Proof. Recall from the Expanded Term Language that match_X is

$$(\mathit{branch}\ \mathit{check}_X\ (s(s(k\ \mathit{equal})(\lambda X.X))\mathit{id})\ (k\ \mathit{false})).$$

The derivation in one direction is

$$\dfrac{\dfrac{\to \mathit{true}\ \text{(eval)}}{(\underline{v},\underline{z})\ \to \mathit{true}}\,\text{thin} \qquad \dfrac{\dfrac{\dfrac{(\underline{z})\ \mathit{match}_X T,\ \Gamma \to C}{(\underline{v},\underline{z})\ X = T,\ \mathit{match}_X T,\ \Gamma \to C}\,\text{thin}}{(\underline{v},\underline{z})\ X = T,\ \mathit{match}_X X,\ \Gamma \to C}\,\text{eq}}{(\underline{v},\underline{z})\ X = T,\ \mathit{true},\ \Gamma \to C}\,\text{red}}{(\underline{v},\underline{z})\ X = T,\ \Gamma \to C}\,\text{cut}$$

Conversely,

$$\begin{array}{cl}
(\underline{v},\underline{z})\ X = T,\ \Gamma \to C & \\ \hline
(\underline{z})\ (\lambda X.X)T = T,\ \Gamma \to C & (\lambda X.T)\ \text{Rule} \\ \hline
(\underline{z})\ \mathit{check}_X T,\ (\lambda X.X)T = T,\ \Gamma \to C & \text{thin} \\ \hline
(\underline{z})\ \mathit{check}_X T,\ s(s(k\ \mathit{equal})(\lambda X.X))\mathit{id}\ T,\ \Gamma \to C & \text{conv} \\ \hline
(\underline{z})\ \mathit{check}_X T,\ s(s(k\ \mathit{equal})(\lambda X.X))\mathit{id}\ T = \mathit{true},\ \Gamma \to C & \text{tr, cut}
\end{array}$$

and we also have, for a fresh variable t,

$$\cfrac{\cfrac{\cfrac{(t,\underline{z})\,\mathit{false} \to C \;(\mathit{false}\text{-el})}{(t,\underline{z})\,k\,\mathit{false}\;t = \mathit{true} \to C}\,\text{red}}{(\underline{z})\,k\,\mathit{false}\;T = \mathit{true} \to C}\,\text{inst}}{(\underline{z})\,\mathit{check}_X T = \mathit{false},\; k\,\mathit{false}\;T = \mathit{true},\; \Gamma \to C}\,\text{thin}$$

and hence, merging the previous two derivations using the Branch Rule,

$$\cfrac{(\underline{z})\,(\mathit{branch}\;\mathit{check}_X\;(s(s(k\;\mathit{equal})(\lambda X.X))\mathit{id})\;(k\,\mathit{false}))\;T = \mathit{true},\; \Gamma \to C}{(\underline{z})\,\mathit{match}_X T,\; \Gamma \to C}\,\text{tr, cut}$$

as required. ∎

THEOREM 18. Fxpt induction rule (fxpt-ind):

$$\cfrac{(u,x,\underline{z})\;\Phi A x = u,\; \Gamma \to A x = u}{(u,x,\underline{z})\,\mathit{fxpt}\,\Phi x = u,\; \Gamma \to A x = u}$$

where $A \not\succ$, Φ is continuous, and u, x are two variables not occurring free in Φ, A, Γ.

Proof. Let f, v, a be three fresh variables, let X be the term ka, and let Ψ be the term id. Now, Xf is evaluable, and Ψ and X are continuous since

$$\begin{array}{l} v(\Psi f v) \rightharpoonup v(fv) \rightharpoonup vv \rightharpoonup v \\ v(Xfv) \rightharpoonup v(kvfv) \rightharpoonup v(vv) \rightharpoonup vv \rightharpoonup v, \end{array}$$

where $\rightharpoonup$ is defined with respect to f and v, so the first Fxpt Rule may be applied, as follows.

$$\cfrac{\cfrac{\cfrac{\cfrac{\cfrac{\cfrac{(u,x,\underline{z})\;\Phi A x = u,\; \Gamma \to A x = u}{(u,x,a,\underline{z})\;\Phi a x = u,\; a = A,\; \Gamma \to a x = u}\,\text{ext}}{(u,f,x,a,\underline{z})\;\Phi a x = u,\; a = A,\; \Gamma \to a x = u}\,\text{thin}}{(u,f,x,a,\underline{z})\;\Phi(Xf)x = u,\; a = A,\; \Gamma \to X(\Psi f)x = u}\,\text{red}}{(u,x,a,\underline{z})\,\mathit{fxpt}\,\Phi x = u,\; a = A,\; \Gamma \to X(\mathit{fxpt}\,\Psi)x = u}\,\text{fxpt}}{(u,x,a,\underline{z})\,\mathit{fxpt}\,\Phi x = u,\; a = A,\; \Gamma \to a x = u}\,\text{red}}{(u,x,\underline{z})\,\mathit{fxpt}\,\Phi x = u,\; \Gamma \to A x = u}\,\text{ext}$$

∎

THEOREM 19. $(\lambda X\colon T)$ rule $((\lambda X\colon T))$:

$$\frac{(\underline{z})\ (\lambda X\colon T)Y = U,\ \Gamma \to C}{(\underline{v},\underline{z})\ X = Y,\ T = U,\ \Gamma \to C}$$

where X is irreducible, with free variables $\underline{v}$, and $\underline{v} \notin Y, U, \Gamma, C$.

Proof. Recall from the Expanded Term Language that $(\lambda X\colon T)$ is

$$(\mathit{branch}\ \mathit{match}_X\ (\lambda X.T)\ (\mathit{fxpt}\ \mathit{id})).$$

The derivation in one direction is

$$\begin{array}{cl}
(\underline{z})\ (\lambda X\colon T)Y = U,\ \Gamma \to C & \\ \hline
(\underline{v},\underline{z})\ X = Y,\ (\lambda X\colon T)Y = U,\ \Gamma \to C & \text{thin} \\ \hline
(\underline{v},\underline{z})\ X = Y,\ (\lambda X\colon T)X = U,\ \Gamma \to C & \text{eq} \\ \hline
(\underline{v},\underline{z})\ X = Y,\ T = U,\ \Gamma \to C & \text{red}
\end{array}$$

Conversely, let u, f, x, v be four fresh variables; then since id is continuous ($v(id\, f\, v) \rightharpoonup v(fv) \rightharpoonup vv \rightharpoonup v$) we can derive

$$\begin{array}{cl}
(u,x,\underline{z})\ \mathit{id}(ka)x = u \to kax = u\ \text{(red)} & \\ \hline
(u,x,\underline{z})\ \mathit{fxpt}\ \mathit{id}\ x = u \to kax = u & \text{fxpt-ind} \\ \hline
(u,x,\underline{z})\ \mathit{fxpt}\ \mathit{id}\ x = u \to a = u & \text{red}
\end{array}$$

for any constant a, and hence

$$\begin{array}{c}
\left\{\begin{array}{ll}
(u,x,\underline{z})\ \mathit{fxpt}\ \mathit{id}\ x = u \to \mathit{true} = u & \text{an instance of the above} \\
(u,x,\underline{z})\ \mathit{fxpt}\ \mathit{id}\ x = u \to \mathit{false} = u & \text{another instance} \\
(u,x,\underline{z})\ \mathit{false} = u \to u = \mathit{false} & \text{symm} \\
(u,x,\underline{z})\ \mathit{true} = u,\ u = \mathit{false} \to \mathit{true} = \mathit{false} & \text{trans} \\
(u,x,\underline{z})\ \mathit{true} = \mathit{false} \to \mathit{false} & \text{red} \\
(u,x,\underline{z})\ \mathit{false} \to C & \mathit{false}\text{-el}
\end{array}\right. \\ \hline
(u,x,\underline{z})\ \mathit{fxpt}\ \mathit{id}\ x = u \to C \quad \text{cut} \\ \hline
(\underline{z})\ \mathit{fxpt}\ \mathit{id}\ Y = U \to C \quad \text{inst} \\ \hline
(\underline{z})\ \mathit{match}_X Y = \mathit{false},\ \mathit{fxpt}\ \mathit{id}\ Y = U,\ \Gamma \to C \quad \text{thin}
\end{array}$$

Also, we can derive

$$\begin{array}{cl}
(\underline{v},\underline{z})\ X = Y,\ T = U,\ \Gamma \to C & \\ \hline
(\underline{v},\underline{z})\ X = Y,\ (\lambda X.T)X = U,\ \Gamma \to C & \text{red} \\ \hline
(\underline{v},\underline{z})\ X = Y,\ (\lambda X.T)Y = U,\ \Gamma \to C & \text{eq} \\ \hline
(\underline{z})\ \mathit{match}_X Y,\ (\lambda X.T)Y = U,\ \Gamma \to C & \text{match}
\end{array}$$

Merging the previous two derivations using the Branch Rule gives

$$(\underline{z})\ (\mathit{branch\ match}_X\ (\lambda X.T)\ (\mathit{fxpt\ id}))Y = U,\ \Gamma \to C$$

as required. ▮

ELIMINATION RULE FOR FUNCTIONS DEFINED USING PATTERN MATCHING

Consider a term *fxpt* Φ defined by

$$f X_1 \triangleq T_1$$
$$\vdots$$
$$f X_k \triangleq T_k$$

where $f \notin X_1, \ldots X_k$. Recall from the Expanded Term Language that

$$\Phi \text{ is } (\lambda f.(\mathit{branch\ match}_{X_1}\ (\lambda X_1.T_1) \cdots (\mathit{branch\ match}_{X_k}\ (\lambda X_k.T_k) f) \cdots)).$$

THEOREM 20. (Elimination rule for pattern matching.)

$$\frac{(u, \underline{v_1}, \underline{z})\ T_1\left[{}^A_f\right] = u,\ \Gamma \to AX_1 = u \quad \cdots \quad (u, \underline{v_k}, \underline{z})\ T_k\left[{}^A_f\right] = u,\ \Gamma \to AX_k = u}{(u, x, \underline{z})\ \mathit{fxpt}\ \Phi x = u,\ \Gamma \to Ax = u}$$

where: $A \not\succ$; $\underline{v_1}, \ldots \underline{v_k}$ are the variables occurring free in $X_1, \ldots X_k$, respectively; $f, u, x, \underline{v_1}, \ldots \underline{v_k} \notin \Gamma, A, \Phi$; $u \notin f, \underline{v_1}, \ldots \underline{v_k}$; and the function definition is regular.

Proof. Since the function definition is regular, Φ is continuous by Theorem 43 of the Expanded Term Language. Let y be a fresh variable; then, for i=1 ... k, we have the derivation

$$\cfrac{\cfrac{\cfrac{\cfrac{(u, \underline{v_i}, \underline{z})\ T_i\left[{}^A_f\right] = u,\ \Gamma \to AX_i = u}{(f, u, \underline{v_i}, \underline{z}) f = A,\ T_i = u,\ \Gamma \to AX_i = u}\ \text{ext}}{(f, u, \underline{v_i}, \underline{z}) f = A,\ (\lambda X_i.T_i)X_i = u,\ \Gamma \to AX_i = u}\ \text{red}}{(f, u, y, \underline{v_i}, \underline{z}) f = A,\ X_i = y,\ (\lambda X_i.T_i)y = u,\ \Gamma \to Ay = u}\ \text{ext}}{(f, u, y, \underline{z}) f = A,\ \mathit{match}_{X_i} y,\ (\lambda X_i.T_i)y = u,\ \Gamma \to Ay = u}\ \text{match}$$

Combining these derivations with $(f, u, y, \underline{z})\ f = A,\ fy = u,\ \Gamma \to Ay = u$ (derived by Equality) using the Branch Rule gives

$$\frac{\dfrac{\dfrac{\dfrac{(f,u,y,\underline{z})\,f=A,\ \Phi fy=u,\ \Gamma\to Ay=u\ \text{(branch, red)}}{(u,y,\underline{z})\ \Phi Ay=u,\ \Gamma\to Ay=u}\,\text{ext}}{(u,y,\underline{z})\,\mathit{fxpt}\,\Phi\, y=u,\ \Gamma\to Ay=u}\,\text{fxpt-ind}}{(u,x,y,\underline{z})\,\mathit{fxpt}\,\Phi\, y=u,\ \Gamma\to Ay=u}\,\text{thin or con}}{(u,x,\underline{z})\,\mathit{fxpt}\,\Phi\, x=u,\ \Gamma\to Ax=u}\,\text{inst}$$

as required. ∎

NUMERIC INDUCTION

THEOREM 21. *num*-elimination rule (*num*-el):

$$\frac{(u,\underline{z})\ \mathit{true}=u,\ \Gamma\to A0=u \qquad (u,n,\underline{z})\ An=u,\ \Gamma\to A(Sn)=u}{(u,n,\underline{z})\ \mathit{num}\,n=u,\ \Gamma\to An=u}$$

where $A \not\succ$, $u,n \notin \Gamma, A$, and u is not n.

Proof. This is simply the Elimination Rule for Pattern Matching, in the case of *num*. (The regularity condition is satisfied since $v\,\mathit{true} \rightharpoonup vv \rightharpoonup v$ and $v(\mathit{num}\,n) \rightharpoonup v(\mathit{num}\,v) \rightharpoonup vv \rightharpoonup v$, where v is a fresh variable and $\rightharpoonup$ is the partial reduction relation with respect to *num* and v.) ∎

THEOREM 22. Induction rule (ind): $$\frac{(\underline{z})\ \Gamma\to B\begin{bmatrix}0\\x\end{bmatrix} \qquad (n,\underline{z})\ B\begin{bmatrix}n\\x\end{bmatrix},\ \Gamma\to B\begin{bmatrix}Sn\\x\end{bmatrix}}{(n,\underline{z})\ \mathit{num}\,n,\ \Gamma\to B\begin{bmatrix}n\\x\end{bmatrix}}$$

where $n \notin \Gamma, B$.

Proof. Let u be a fresh variable and let F be the construction $(\lambda \mathit{true}\colon \mathit{true})$. The $(\lambda X\colon T)$-rule, in the case of F, gives

$$\frac{(\underline{w})\ \mathit{true}=Y,\ \mathit{true}=u\to C}{(\underline{w})\ FY=u\to C}$$

from which we can derive immediately the three sequents

$$(\underline{w})\ Y,\ \mathit{true}=u\to FY=u, \tag{1}$$

$$(\underline{w})\ FY=u\to Y, \tag{2}$$

$$(\underline{w})\ FY=u\to \mathit{true}=u \tag{3}$$

for any term Y. Now let A be the irreducible term $(\lambda x.FB)$. The Induction Rule is derived by beginning with

$$\begin{array}{cl}
(n,\underline{z})\; B\begin{bmatrix} n \\ x \end{bmatrix},\; \Gamma \to B\begin{bmatrix} Sn \\ x \end{bmatrix} & \\ \hline
(u,n,\underline{z})\; \mathit{true} = u,\; B\begin{bmatrix} n \\ x \end{bmatrix},\; \Gamma \to B\begin{bmatrix} Sn \\ x \end{bmatrix} & \text{thin} \\ \hline
(u,n,\underline{z})\; \mathit{true} = u,\; B\begin{bmatrix} n \\ x \end{bmatrix},\; \Gamma \to F(B\begin{bmatrix} Sn \\ x \end{bmatrix}) = u & \text{cut,(1)} \\ \hline
(u,n,\underline{z})\; F(B\begin{bmatrix} n \\ x \end{bmatrix}) = u,\; \Gamma \to F(B\begin{bmatrix} Sn \\ x \end{bmatrix}) = u & \text{cut,(2),(3)} \\ \hline
(u,n,\underline{z})\; An = u,\; \Gamma \to A(Sn) = u & \text{red}
\end{array}$$

and then using this in the following derivation

$$\dfrac{\begin{array}{cl}
(\underline{z})\; \Gamma \to B\begin{bmatrix} 0 \\ x \end{bmatrix} & \\ \hline
(u,\underline{z})\; \mathit{true} = u,\; \Gamma \to B\begin{bmatrix} 0 \\ x \end{bmatrix} & \text{thin} \\ \hline
(u,\underline{z})\; \mathit{true} = u,\; \Gamma \to F(B\begin{bmatrix} 0 \\ x \end{bmatrix}) = u & \text{cut,(1)} \\ \hline
(u,\underline{z})\; \mathit{true} = u,\; \Gamma \to A0 = u & \text{red}
\end{array} \qquad (u,n,\underline{z})\; An = u,\; \Gamma \to A(Sn) = u}{(u,n,\underline{z})\; \mathit{num}\, n = u,\; \Gamma \to An = u}\;\mathit{num}\text{-el}$$

$$\begin{array}{cl}
(u,n,\underline{z})\; \mathit{num}\, n = u,\; \Gamma \to An = u & \\ \hline
(u,n,\underline{z})\; \mathit{num}\, n = u,\; \Gamma \to F(B\begin{bmatrix} n \\ x \end{bmatrix}) = u & \text{red} \\ \hline
(u,n,\underline{z})\; \mathit{num}\, n = u,\; \Gamma \to B\begin{bmatrix} n \\ x \end{bmatrix} & \text{cut,(2)} \\ \hline
(n,\underline{z})\; \mathit{num}\, n = \mathit{true},\; \Gamma \to B\begin{bmatrix} n \\ x \end{bmatrix} & \text{inst} \\ \hline
(n,\underline{z})\; \mathit{num}\, n,\; \Gamma \to B\begin{bmatrix} n \\ x \end{bmatrix} & \text{tr, cut}
\end{array}$$

∎

EXERCISES. Derive the sequents

(1) $(\underline{z})\; X = Y \to SX = SY$
(2) $(n)\; \mathit{num}\, n \to n = \mathit{plus}(n, 0)$
(3) $(m,n)\; \mathit{num}\, n \to S(\mathit{plus}(n,m)) = \mathit{plus}(n, Sm)$
(4) $(m,n)\; \mathit{num}\, m,\; \mathit{num}\, n \to \mathit{plus}(m,n) = \mathit{plus}(n,m)$
(5) $(n_1, \ldots n_k)\; \mathit{num}\, n_1, \ldots\; \mathit{num}\, n_k \to \mathit{num}(f(n_1, \ldots n_k))$ where f is k-ary primitive recursive.

(Hint: follow the usual Peano Arithmetic proofs. Recall that *plus* was defined in an example in Chapter 15.)

EXERCISES. Derive the *List Rule*:

$$\frac{(\underline{z})\ \Gamma \to B\begin{bmatrix}nil\\x\end{bmatrix} \qquad (e,l,\underline{z})\ B\begin{bmatrix}l\\x\end{bmatrix},\ \Gamma \to B\begin{bmatrix}(e,l)\\x\end{bmatrix}}{(l,\underline{z})\ list(l),\ \Gamma \to B\begin{bmatrix}l\\x\end{bmatrix}}$$

where $e, l \notin \Gamma, B$. Use it to derive the following sequents.

(1) $(l_1, l_2)\ list(l_1),\ list(l_2) \to list(concat(l_1, l_2))$
(2) $(l)\ list(l) \to list(reverse(l))$
(3) $(l)\ list(l) \to num(length(l))$
(4) $(l_1, l_2)\ list(l_1),\ list(l_2) \to$
$length(concat(l_1, l_2)) = plus(length(l_1), length(l_2))$
(5) $(l)\ list(l) \to length(l) = length(reverse(l))$
(6) $(e, l)\ list(l),\ member(e, l) \to member(e, reverse(l))$
(7) $(e, l)\ list(l),\ member(e, reverse(l)) \to member(e, l)$

(Recall that the functions *list*, *concat*, *reverse*, *length* and *member* were defined in examples and exercises in Chapter 15.)

EXERCISE. State and derive a protological *Tree Rule*, analogous to the List Rule and Numeric Induction, for the coding of trees considered in an exercise in Chapter 15 in which each node has 0, 1 or 2 subtrees and each leaf holds a number. It should make use of a function *tree*, defined by pattern matching, such that $tree(t)\ \rhd^*\ true$ iff the construction t is the code of a tree. Use the Tree Rule to derive the sequent

$$(t)\ tree(t) \to num(sum(t)).$$

EXERCISE. State and derive a general structural induction rule for a construction f defined by

$$f(X_1) \triangleq C_1(\underline{v_1})\ \&\ f(A_1^1(\underline{v_1}))\ \&\ \cdots\ \&\ f(A_1^{n_1}(\underline{v_1}))$$
$$\vdots$$
$$f(X_k) \triangleq C_k(\underline{v_k})\ \&\ f(A_k^1(\underline{v_k}))\ \&\ \cdots\ \&\ f(A_k^{n_k}(\underline{v_k}))$$

where $\underline{v_i}$ are the free variables of the pattern X_i. Numeric Induction, the List Rule and the Tree Rule should be special cases of this rule.

CHAPTER 20

EXPANDED PROTOLOGIC

Expanded Protologic consists of Protologic plus the following theorems and derived rules of inference.

PRELIMINARY DERIVATIONS

THEOREM 1. Tautology (taut): $(\underline{z})\, \Gamma,\ T \to T.$

THEOREM 3. Reduction (red): $(\underline{z})\ A \to B \quad \dfrac{(\underline{z})\ \Gamma \to A}{(\underline{z})\ \Gamma \to B} \quad \dfrac{(\underline{z})\ A,\ \Gamma \to C}{(\underline{z})\ B,\ \Gamma \to C}$ if $A \ \triangleright^{*}\triangleleft\ B$.

THEOREM 5. Symmetry (symm): $(\underline{z})\ A = B \to B = A.$

PROPERTIES OF THE INSTANTIATION NOTATION

THEOREM 6. Equality theorem and rules (eq): $(\underline{z})\ A = B,\ T\begin{bmatrix} B \\ x \end{bmatrix} \to T\begin{bmatrix} A \\ x \end{bmatrix}$

$$\dfrac{(\underline{z})\ A = B,\ \Gamma \to T\begin{bmatrix} A \\ x \end{bmatrix}}{(\underline{z})\ A = B,\ \Gamma \to T\begin{bmatrix} B \\ x \end{bmatrix}} \qquad \dfrac{(\underline{z})\ A = B,\ T\begin{bmatrix} A \\ x \end{bmatrix}, \Gamma \to C}{(\underline{z})\ A = B,\ T\begin{bmatrix} B \\ x \end{bmatrix},\ \Gamma \to C}$$

THEOREM 7. Self-equality (se): $(\underline{z})\ T\begin{bmatrix} X \\ x \end{bmatrix} \to X = X$ if X is evaluable or $x \in T$.

THEOREM 9. Instantiation rule (inst): $\dfrac{(x, \underline{z})\ A, \ldots B \to C}{(\underline{z})\ A\begin{bmatrix} X \\ x \end{bmatrix}, \ldots B\begin{bmatrix} X \\ x \end{bmatrix} \to C\begin{bmatrix} X \\ x \end{bmatrix}}$

where X is evaluable or $x \in A$ or $\ldots$ $x \in B$.

THEOREM 10. Extraction rule (ext): $\dfrac{(x, \underline{z})\ x = X,\ A, \ldots B \to C}{(\underline{z})\ A\begin{bmatrix} X \\ x \end{bmatrix}, \ldots B\begin{bmatrix} X \\ x \end{bmatrix} \to C\begin{bmatrix} X \\ x \end{bmatrix}}$

where $x \notin X$, and X is evaluable or $x \in A$ or $\ldots$ $x \in B$.

THEOREM 11. Conversion rules (conv):

$$\frac{(\underline{z})\ \Gamma \to A\left[\frac{X}{\underline{x}}\right]}{(\underline{z})\ \Gamma \to B\left[\frac{X}{\underline{x}}\right]} \qquad \frac{(\underline{z})\ \Gamma,\ A\left[\frac{X}{\underline{x}}\right] \to C}{(\underline{z})\ \Gamma,\ B\left[\frac{X}{\underline{x}}\right] \to C}$$

where $A \ \triangleright^* \triangleleft \ B$ and the variables $\underline{x}$ are all different and occur free in A and B.

PROPERTIES OF TRUTH-FUNCTIONAL OPERATORS

THEOREM 12. Falsity elimination (*false*-el): $(\underline{z})\, \mathit{false} \to A$.

THEOREM 13. Boolean rule (bool):

$$\frac{(\underline{z})\ x = \mathit{true},\ \Gamma \to C \qquad (\underline{z})\ x = \mathit{false},\ \Gamma \to C}{(\underline{z})\ \mathit{boolean}\, x,\ \Gamma \to C}$$

THEOREM 14. Branch rule (branch):

$$\frac{(\underline{z})\ CX,\ FX = U,\ \Gamma \to T \qquad (\underline{z})\ CX = \mathit{false},\ GX = U,\ \Gamma \to T}{(\underline{z})\ (\mathit{branch}\ C\ F\ G)X = U,\ \Gamma \to T}$$

THEOREM 15. Conjunction (&):

$$(\underline{z})\, A_1,\ \ldots\ A_k \to A_1 \,\&\ldots\&\, A_k, \qquad (\underline{z})\, A_1 \,\&\ldots\&\, A_k \to A_i$$

for $i = 1, \ldots k$.

RULES FOR THE PATTERN-MATCHING NOTATIONS

THEOREM 16. $(\lambda X.T)$ rule $((\lambda X.T))$:

$$\frac{(\underline{v}, \underline{z})\ T = U,\ \Gamma \to C}{(\underline{z})\ (\lambda X.T)Y = U,\ \Gamma \to C}$$

where $\underline{v}$ are the free variables of X and $\underline{v} \notin Y, U, \Gamma, C$.

THEOREM 17. Match rule (match):

$$\frac{(\underline{z})\ \mathit{match}_X T,\ \Gamma \to C}{(\underline{v}, \underline{z})\ X = T,\ \Gamma \to C}$$

where X is irreducible, with free variables $\underline{v}$, and $\underline{v} \notin T, \Gamma, C$.

THEOREM 18. Fxpt induction rule (fxpt-ind):

$$\frac{(u, x, \underline{z})\ \Phi A x = u,\ \Gamma \to Ax = u}{(u, x, \underline{z})\ \mathit{fxpt}\ \Phi x = u,\ \Gamma \to Ax = u}$$

where $A \not\triangleright$, Φ is continuous, and u, x are two variables not occurring free in Φ, A, Γ.

THEOREM 19. $(\lambda X\colon T)$ rule $((\lambda X\colon T))$:

$$\frac{(\underline{z})\ (\lambda X\colon T)Y = U,\ \Gamma \to C}{(\underline{v}, \underline{z})\ X = Y,\ T = U,\ \Gamma \to C}$$

where X is irreducible, with free variables $\underline{v}$, and $\underline{v} \notin Y, U, \Gamma, C$.

ELIMINATION RULE FOR FUNCTIONS DEFINED USING PATTERN MATCHING

Consider a term *fxpt* Φ defined by

$$\begin{aligned} f X_1 &\triangleq T_1 \\ &\vdots \\ f X_k &\triangleq T_k \end{aligned}$$

where $f \notin X_1, \ldots X_k$.

THEOREM 20. (Elimination rule for pattern matching.)

$$\frac{(u, \underline{v_1}, \underline{z})\ T_1\begin{bmatrix}A\\f\end{bmatrix} = u,\ \Gamma \to AX_1 = u \quad \cdots \quad (u, \underline{v_k}, \underline{z})\ T_k\begin{bmatrix}A\\f\end{bmatrix} = u,\ \Gamma \to AX_k = u}{(u, x, \underline{z})\ \mathit{fxpt}\ \Phi x = u,\ \Gamma \to Ax = u}$$

where: $A \not\triangleright$; $\underline{v_1}, \ldots \underline{v_k}$ are the variables occurring free in $X_1, \ldots X_k$, respectively; $f, u, x, \underline{v_1}, \ldots \underline{v_k} \notin \Gamma, A, \Phi$; $u \notin f, \underline{v_1}, \ldots \underline{v_k}$; and the function definition is regular.

NUMERIC INDUCTION

THEOREM 22. Induction rule (ind): $$\frac{(\underline{z})\ \Gamma \to B\begin{bmatrix}0\\x\end{bmatrix} \quad (n, \underline{z})\ B\begin{bmatrix}n\\x\end{bmatrix},\ \Gamma \to B\begin{bmatrix}Sn\\x\end{bmatrix}}{(n, \underline{z})\ \mathit{num}\, n,\ \Gamma \to B\begin{bmatrix}n\\x\end{bmatrix}}$$

where $n \notin \Gamma, B$.

CHAPTER 21

FROM EXPANDED PROTOLOGIC TO THE CODING OF TREES

Protologic has been set up as a formal axiomatic system, and hence it is possible to code protological derivation trees as terms, to reason about them within protologic, and to formulate reflection principles. I shall define a construction *DT* such that $DT(D,X) \rhd^* true$ iff D is a coded protological derivation of the sequent X, allowing reflection principles. I shall also describe the encoding of trees in general and prove some general theorems about well-foundedness. These concepts and theorems constitute the next theory, the Coding of Trees (CT).

REFLECTION PRINCIPLES

A *reflection principle* is a sequent of the form

$$(\underline{z})\, DT(U, [\![\to C]\!]) \to C$$

for any terms U and C, where *DT* is the construction to be defined below. Here, $\to C$ is a sequent with no quantified variables and no terms on the left-hand side, in other words $(nil, (\lambda nil.C))$. I enclose it in '$[\![\ldots]\!]$' brackets to mark it as a sequent; whenever I use the $(\underline{z})\, A, \ldots B \to C$ metanotation for a sequent occurring as a subterm in a larger term I shall always enclose it in these special brackets.

In addition, any instantiation of the above reflection principle counts as a reflection principle: that is, $[\![(\underline{z})\ DT(U, [\![\to C]\!]) \to C]\!]\left[\frac{\underline{X}}{\underline{x}}\right]$ is the general form for a reflection principle, for any irreducible terms $\underline{X}$ and variables $\underline{x}$.

CODING OF PROTOLOGICAL DERIVATION TREES

A *derivation tree* (from a given list of premises $X_0, \ldots X_{k-1}$) is a tree of sequents whose leaves are protological axioms, reflection principles, and premises from the given premise list, and whose other nodes are related by the protological inference rules. A *derivation* (from a given premise list) is a derivation tree with its conclusion sequent (the root of the tree) removed.

Let *premise*, *none*, *one*, *two* and *rp* be five fresh 1-ary constructors. Number the axiom schemata and rules of protologic $1, \ldots 17$. A derivation tree is coded as an irreducible term (D, X), where D is the code of a derivation and X is the conclusion sequent; the derivation code D is

$premise(i)$	if X is the i'th premise, X_i (numbering them from 0)
$none(n)$	if X is an instance of axiom schema n
$one(n, P, Y)$	if X is inferred by rule n from a coded derivation tree, (P, Y)
$two(n, P, Y, Q, Z)$	if X is inferred by rule n from coded derivation trees, (P, Y) and (Q, Z)
$rp(\lambda[\underline{z}].U)\left[\begin{smallmatrix}X\\ \underline{x}\end{smallmatrix}\right]$	if X is a reflection principle, $\llbracket(\underline{z})\ DT(U, \llbracket\to C\rrbracket) \to C\rrbracket\left[\begin{smallmatrix}X\\ \underline{x}\end{smallmatrix}\right]$.

EXAMPLE. Recall from Chapter 17 that a sequent such as $(x, y, z)\ U,\ V,\ W \to X$ is really a term $([(\lambda[x, y, z].U), (\lambda[x, y, z].V), (\lambda[x, y, z].W)], (\lambda[x, y, z].X))$. The protological derivation

$$\dfrac{P\ \text{(premise 0)} \qquad \dfrac{\dfrac{A\ \text{(axiom 5)} \qquad B\ \text{(axiom 2)}}{C}\ \text{rule 12}}{Q}\ \text{rule 13}}{E}\ \text{rule 15}$$

is coded as

$$two(15, premise(0), P, one(13, two(12, none(5), A, none(2), B), C), Q).$$

The free variables of the code of a derivation or derivation tree are the free variables of its constituent sequents.

INTERNAL REPRESENTATION OF PROTOLOGICAL INFERENCE

By appeal to Church's Thesis (see the Expanded Term Language), let inf_0, inf_1, inf_2 and RP be four recursive functions such that for any irreducible terms X, Y, Z and F, and any construction D,

$$inf_0(n, X) \rhd^* true \quad \text{iff } X \text{ is an instance of axiom schema } n$$

$$inf_1(n, Y, X) \rhd^* true \quad \text{iff } \frac{Y}{X} \text{ is an instance of rule } n$$

$$inf_2(n, Y, Z, X) \rhd^* true \quad \text{iff } \frac{Y\ Z}{X} \text{ is an instance of rule } n$$

$$RP(F, D, Y) \rhd^* true \quad \text{iff } F \text{ is } (\lambda[\underline{z}].U)\left[\begin{smallmatrix}X\\ \underline{x}\end{smallmatrix}\right] \text{ and}$$

$$Y \text{ is } \llbracket(\underline{z})\ D(U, \llbracket\to C\rrbracket) \to C\rrbracket\left[\begin{smallmatrix}X\\ \underline{x}\end{smallmatrix}\right].$$

inf_0 should be defined in such a way that $\mathit{inf}_0(1, [\![\rightarrow T]\!]) \rhd^* T$, in accordance with the Evaluation Axiom Schema. The definitions are legitimate applications of Church's Thesis since the protological axioms, rules and reflection principles all have the property that a sequent, pair of sequents or triple of sequents represents an axiom, reflection principle or inference iff any instantiation of its free variables by constructions does.

THE DERIVATION TREE PREDICATE, *DT*

Define a construction *DT* by

$$\begin{aligned} DT(\mathit{none}(n), x) &\triangleq \mathit{inf}_0(n, x) \\ DT(\mathit{one}(n, p, y), x) &\triangleq \mathit{inf}_1(n, y, x) \;\&\; DT(p, y) \\ DT(\mathit{two}(n, p, y, q, z), x) &\triangleq \mathit{inf}_2(n, y, z, x) \;\&\; DT(p, y) \;\&\; DT(q, z) \\ DT(\mathit{rp}(f), x) &\triangleq RP(f, DT, x) \end{aligned}$$

where n, x, y, z, p, q, f are seven variables. Then $DT(D, X) \rhd^* \mathit{true}$ iff (D, X) is the code of a derivation tree with no premises, or in other words iff D is the code of a derivation with no premises for the sequent X. Note that *RP*, with *DT* supplied as its second argument, detects reflection principles.

EXAMPLE. Consider the derivation $\mathit{one}(13, \mathit{two}(12, \mathit{none}(5), A, \mathit{none}(2), B), C)$ of the sequent Q (part of the derivation in the previous example). Applying *DT* gives

$$\begin{aligned} & DT(\mathit{one}(13, \mathit{two}(12, \mathit{none}(5), A, \mathit{none}(2), B), C), Q) \\ & \rhd^* \mathit{inf}_1(13, C, Q) \;\&\; DT(\mathit{two}(12, \mathit{none}(5), A, \mathit{none}(2), B), C) \\ & \rhd^* \mathit{true} \;\&\; \mathit{inf}_2(12, A, B, C) \;\&\; DT(\mathit{none}(5), A) \;\&\; DT(\mathit{none}(2), B) \\ & \rhd^* \mathit{true} \;\&\; \mathit{true} \;\&\; \mathit{inf}_0(5, A) \;\&\; \mathit{inf}_0(2, B) \\ & \rhd^* \mathit{true} \;\&\; \mathit{true} \;\&\; \mathit{true} \;\&\; \mathit{true} \\ & \rhd^* \mathit{true} \end{aligned}$$

where $\mathit{inf}_1(13, C, Q)$, $\mathit{inf}_2(12, A, B, C)$, $\mathit{inf}_0(5, A)$ and $\mathit{inf}_0(2, B)$ all reduce to *true* since (we suppose) $\frac{C}{Q}$ is an instance of protological rule 13, $\frac{A \quad B}{C}$ is an instance of rule 12, A is an instance of axiom schema 5, and B is an instance of axiom schema 2.

This establishes that $\mathit{one}(13, \mathit{two}(12, \mathit{none}(5), A, \mathit{none}(2), B), C)$ is a derivation of Q, or in other words that the pair $(\mathit{one}(13, \mathit{two}(12, \mathit{none}(5), A, \mathit{none}(2), B), C), Q)$ is a derivation tree.

DEFINITION. Let $\mathcal{E}$ be the construction *none*(1).

THEOREM 1. $DT(\mathcal{E}, [\![\rightarrow T]\!]) \rhd^* T$.

THEOREM 2. $DT(rp(\lambda[z].U), [\![(\underline{z})\ DT(U, [\![\rightarrow C]\!]) \rightarrow C]\!]) \rhd^*$ *true*.

TREE CODES AND WELL-FOUNDEDNESS

Recall the discussion of the interpretation of constructions as trees in Chapter 10. This section provides a formal apparatus for defining tree codings.

DEFINITION. A *type symbol* is a construction that has been given an interpretation as representing a coding of a class of trees as constructions.

DEFINITION. A well-foundedness relation, $T : \Pi$, is defined in the following sequence of steps.

- Let T be a construction and Π be a type symbol; then $T : \Pi$ means that T is a well-founded tree, according to the coding Π.
- Let T be a term with no free variables and Π be a type symbol; then $T : \Pi$ means that if $T \rhd^* T' \not\rhd$ (T' thus being a construction) then $T' : \Pi$ (in the previous sense).

(Note that as a consequence of this, if the reduction of T fails to halt then $T : \Pi$ holds vacuously.)

- Let T be an arbitrary term, with free variables $\underline{x}$, and Π be a type symbol; then $T : \Pi$ means that, for any constructions $\underline{X}$, $T\left[\frac{\underline{X}}{\underline{x}}\right] : \Pi$ (in the previous sense).
- Let T be an arbitrary term and Π be a term that reduces to a type symbol Π'; then $T : \Pi$ means that $T : \Pi'$ (in the previous sense).

To summarise, $T : \Pi$ means that, for any constructions $\underline{X}$, if $T\left[\frac{\underline{X}}{\underline{x}}\right]$ reduces to a construction T' then T' is a well-founded tree of type Π', where Π' is the type symbol to which Π reduces.

The quantifier 'for any constructions $\underline{X}$' is to be understood in the protological sense as explained in Chapter 8. Note that the order of the instantiations $\left[\frac{\underline{X}}{\underline{x}}\right]$ makes no difference, up to $\overset{*}{\leftrightarrow}$, nor would it make a difference if we included extra instantiations $\left[\frac{Y}{y}\right]$, where $y \notin T$.

The relation $X : \Pi$ cannot be represented in the term language (that is, one cannot define a construction *WF* such that $WF(X, \Pi) \rhd^*$ *true* iff $X : \Pi$), for two reasons. First, well-foundedness is undecidable except for trivial classes of trees. Secondly, the relation depends on the interpretation of the construction Π' as a coding of trees, but the assignment of tree-codings to constructions is a piecemeal process and is never completed.

From now on the identifiers 'Π', 'Σ', 'Φ', 'Ψ', 'Ω' and 'Θ', sometimes with subscripts or primes, will always be used as metavariables denoting type symbols (or terms that reduce to type symbols). I shall say 'X is of type Π' iff the construction X is a tree according to the Π coding. Likewise if I refer to 'X (of type Π)' then I am regarding X as a tree according to the Π coding.

THEOREM 3. (Well-foundedness reduction rule.)

- If $A \ \triangleright^*\triangleleft\ B$ then $A\ :\ \Pi$ iff $B\ :\ \Pi$.
- If $\Pi\ \triangleright^*\triangleleft\ \Sigma$ then $A\ :\ \Pi$ iff $A\ :\ \Sigma$.

Proof. For the first part, let $\Pi\ \triangleright^*\ \Pi' \not\triangleright$, let $\underline{x}$ be the free variables of A and B, and let $\underline{X}$ be any constructions. Then $A\left[\frac{\underline{X}}{\underline{x}}\right]\ \triangleright^*\triangleleft\ B\left[\frac{\underline{X}}{\underline{x}}\right]$, so $A\left[\frac{\underline{X}}{\underline{x}}\right]$ and $B\left[\frac{\underline{X}}{\underline{x}}\right]$ reduce to the same construction, if any. Thus $A\left[\frac{\underline{X}}{\underline{x}}\right]\ :\ \Pi'$ iff $B\left[\frac{\underline{X}}{\underline{x}}\right]\ :\ \Pi'$. This establishes that $A\ :\ \Pi$ iff $B\ :\ \Pi$, as required.

The second part follows since Π and Σ reduce to the same type symbol. ∎

DEFINITION. Let *leaf* be a fresh 0-ary constructor. Make it a type symbol by letting it represent the following (trivial) coding of trees: any construction X represents a tree with no subtrees.

THEOREM 4. (*leaf* rule.) For any term T, $T\ :\ \mathit{leaf}$.

Proof. Let $\underline{x}$ be the free variables of T, and let $\underline{X}$ be constructions. If $T\left[\frac{\underline{X}}{\underline{x}}\right]$ reduces to a construction T' then $T'\ :\ \mathit{leaf}$ since T' codes a tree with no subtrees; so $T\ :\ \mathit{leaf}$. ∎

DEFINITION. Let *map* be a fresh 1-ary constructor. For any type symbols Π and Σ, define a type symbol $\mathit{map}(\Pi,\Sigma)$ representing trees F whose subtrees are the constructions T such that $FX\ \triangleright^*\ T$ for some construction $X\ :\ \Pi$. The subtrees T are coded according to Σ. (See Chapter 10.)

THEOREM 5. (*map* rule.)

- If $A\ :\ \mathit{map}(\Pi,\Sigma)$ and $B\ :\ \Pi$ then $AB\ :\ \Sigma$.
- If, for any construction $B\ :\ \Pi$, $AB\ :\ \Sigma$, then $A\ :\ \mathit{map}(\Pi,\Sigma)$.
- If $X \not\triangleright$, $\underline{u} \notin X$, $\underline{U}$ are constructions, and $T\left[\frac{\underline{U}}{\underline{u}}\right]\ :\ \Sigma$, then $(\lambda X.T)\left[\frac{\underline{U}}{\underline{u}}\right]\ :\ \mathit{map}(\mathit{leaf},\Sigma)$.

Proof. Let $\Pi \triangleright^* \Pi' \not\triangleright$ and $\Sigma \triangleright^* \Sigma' \not\triangleright$. For the first part, let $\underline{x}$ be the free variables of AB and let $\underline{X}$ be any constructions. Suppose $AB\left[\frac{\underline{X}}{\underline{x}}\right]$ reduces to a construction, C'. Then $A\left[\frac{\underline{X}}{\underline{x}}\right]$ and $B\left[\frac{\underline{X}}{\underline{x}}\right]$ must reduce to constructions, A' and B', such that $A'B' \triangleright^* C'$. The hypotheses that $A : map(\Pi, \Sigma)$ and $B : \Pi$ mean that $A\left[\frac{\underline{X}}{\underline{x}}\right] : map(\Pi', \Sigma')$ and $B\left[\frac{\underline{X}}{\underline{x}}\right] : \Pi'$, which in turn mean that $A' : map(\Pi', \Sigma')$ and $B' : \Pi'$. Hence, by the *map* coding, C' (of type Σ') is a subtree of A' (of type $map(\Pi', \Sigma')$). The latter tree is well-founded and hence so is the former: this is expressed as $C' : \Sigma'$. This establishes that $AB : \Sigma$, as required.

For the second part, let $\underline{x}$ be the free variables of A and let $\underline{X}$ be any constructions. Suppose $A\left[\frac{\underline{X}}{\underline{x}}\right]$ reduces to a construction, A'. Consider any construction $B : \Pi'$. If $A'B \triangleright^* T \not\triangleright$ then by the *map* coding T (of type Σ') is a subtree of A' (of type $map(\Pi', \Sigma')$). By hypothesis, $AB : \Sigma'$, which means, since $AB\left[\frac{\underline{X}}{\underline{x}}\right] \triangleright^* A'B \triangleright^* T \not\triangleright$, that $T : \Sigma'$. This shows that an arbitrary subtree T of A' is well-founded, and therefore A' is well-founded, which is written as $A' : map(\Pi', \Sigma')$. This establishes that $A : map(\Pi, \Sigma)$, as required.

For the third part, let $\underline{z}$ be the free variables of X, let $\underline{y}$ be the free variables of $(\lambda X.T)\left[\frac{\underline{U}}{\underline{u}}\right]$, and let $\underline{Y}$ be constructions. Consider any construction $B : leaf$. Then $(\lambda X.T)\left[\frac{\underline{U}}{\underline{u}}\right]\left[\frac{\underline{Y}}{\underline{y}}\right]B \overset{*}{\leftrightarrow} (\lambda X.T)B\left[\frac{\underline{U}}{\underline{u}}\right]\left[\frac{\underline{Y}}{\underline{y}}\right] \triangleright^* T\left[\frac{\underline{Z}}{\underline{z}}\right]\left[\frac{\underline{U}}{\underline{u}}\right]\left[\frac{\underline{Y}}{\underline{y}}\right] \overset{*}{\leftrightarrow} T\left[\frac{\underline{U}}{\underline{u}}\right]\left[\frac{\underline{Z}}{\underline{z}}\right]\left[\frac{\underline{Y}}{\underline{y}}\right]$ for some constructions $\underline{Z}$ by Theorem 25 of the Expanded Term Language. Now, by hypothesis, $T\left[\frac{\underline{U}}{\underline{u}}\right]\left[\frac{\underline{Z}}{\underline{z}}\right]\left[\frac{\underline{Y}}{\underline{y}}\right] : \Sigma'$. Hence, by the second part of this theorem, $(\lambda X.T)\left[\frac{\underline{U}}{\underline{u}}\right]\left[\frac{\underline{Y}}{\underline{y}}\right] : map(leaf, \Sigma')$. This establishes that $(\lambda X.T)\left[\frac{\underline{U}}{\underline{u}}\right] : map(leaf, \Sigma)$, as required. ∎

THEOREM 6. (Well-foundedness instantiation rule.) If $A : \Pi$ and $X \not\triangleright$ then $A\left[\frac{X}{x}\right] : \Pi$.

Proof. Using the *map* rule, if $A : \Pi$ then $(\lambda x.A) : map(leaf, \Pi)$ and $X : leaf$ so $A\left[\frac{X}{x}\right] \triangleleft^* (\lambda x.A)X : \Pi$. ∎

DEFINITION. Let *product* be a fresh 1-ary constructor. For any type symbols Π and Σ, define a type symbol $product(\Pi, \Sigma)$ representing trees of the form

(X, Y) with two subtrees, X of type Π and Y of type Σ,

for any constructions X and Y.

THEOREM 7. (*product* rule.)

- Any construction T such that $T\ :\ product(\Pi,\Sigma)$ is of the form (X,Y) for some constructions $X\ :\ \Pi$ and $Y\ :\ \Sigma$.
- $left\ :\ map(product(\Pi,\Sigma),\Pi)$.
- $right\ :\ map(product(\Pi,\Sigma),\Sigma)$.
- For any terms A and B, if $A\ :\ \Pi$ and $B\ :\ \Sigma$ then $(A,B)\ :\ product(\Pi,\Sigma)$.

Proof. Let $\Pi \triangleright^* \Pi' \not\triangleright$ and $\Sigma \triangleright^* \Sigma' \not\triangleright$.

The first part follows immediately from the definition of $product(\Pi',\Sigma')$.

For the second part, consider any construction $T\ :\ product(\Pi,\Sigma)$; then T is (X,Y), for some constructions $X\ :\ \Pi'$ and $Y\ :\ \Sigma'$, so $left\ T \triangleright^* X\ :\ \Pi' \triangleleft^* \Pi$. By the *map* rule this implies $left\ :\ map(product(\Pi,\Sigma),\Pi)$, as required. The third part is similar.

For the fourth part, let $\underline{x}$ be the free variables of A and B, and let $\underline{X}$ be any constructions. Suppose $(A,B)\left[\frac{\underline{X}}{\underline{x}}\right]$ reduces to a construction T. Then $A\left[\frac{\underline{X}}{\underline{x}}\right]$ and $B\left[\frac{\underline{X}}{\underline{x}}\right]$ must also reduce to constructions, A' and B', and T must be (A',B'). The hypotheses that $A\ :\ \Pi$ and $B\ :\ \Sigma$ mean that $A'\ :\ \Pi'$ and $B'\ :\ \Sigma'$. By the *product* coding this implies that $T\ :\ product(\Pi',\Sigma')$. This establishes that $(A,B)\ :\ product(\Pi,\Sigma)$, as required. ∎

EXAMPLES.

(1) $((nil,true),ss)\ :\ product(product(leaf,leaf),leaf)$. The tree represented by the construction $((nil,true),ss)$ according to this tree coding is

```
            ((nil, true), ss)
               /        \
        (nil, true)      ss
         /      \
       nil      true
```

In fact, any tree of the form $((X,Y),Z)$ is well-founded, for any constructions X,Y,Z, so we can say that $((x,y),z)\ :\ product(product(leaf,leaf), leaf)$.

(2) $(\lambda x.(x,(nil,x)))\ :\ map(\Pi,product(\Pi,product(leaf,\Pi)))$ for any type symbol Π. The tree is, in part, as follows

```
                    (λx.(x, (nil, x)))
              ...  /         |          \  ...
  (X1, (nil, X1))    (X2, (nil, X2))    (X3, (nil, X3))
    /     \            /     \            /     \
  X1    (nil, X1)    X2    (nil, X2)    X3    (nil, X3)
  /|\     / \        /|\     / \        /|\     / \
   :    nil  X1       :    nil  X2       :    nil  X3
             /|\                /|\                /|\
              :                  :                  :
```

where X_1, X_2, X_3 are three of the constructions that encode well-founded trees of type Π. This tree is clearly well-founded.

DEFINITION. Let *pi* be a fresh 1-ary constructor. For any type symbols $\Pi_1, \ldots \Pi_k$, define a type symbol $pi[\Pi_1, \ldots \Pi_k]$ representing trees of the form

$$[X_1, \ldots X_k] \quad \text{with subtrees } X_1, \ldots X_k \text{ of type } \Pi_1, \ldots \Pi_k \text{ respectively.}$$

(This includes the case of $k = 0$: *nil* represents a tree of type *pi*(*nil*) with no subtrees.)

THEOREM 8. (*pi* rule.)

- Any construction T such that $T : pi[\Pi, \ldots \Sigma]$ is of the form $[A, \ldots B]$, for some constructions $A : \Pi, \ldots B : \Sigma$.
- For any terms $A, \ldots B$, if $A : \Pi, \ldots B : \Sigma$ then $[A, \ldots B] : pi[\Pi, \ldots \Sigma]$.

Proof. The proof is similar to the proof of the *product* rule. ∎

EXAMPLE. $[(x, y), z, (u, v)] : pi[product(leaf, leaf), leaf, product(leaf, leaf)]$, since, for any constructions X, Y, Z, U, V the tree

```
        [(X, Y), Z, (U, V)]
         /       |       \
     (X, Y)      Z      (U, V)
      / \                / \
     X   Y              U   V
```

is well-founded.

THEOREM 9. (*if* well-foundedness rule.) $\mathit{if} : \mathit{map}(\mathit{leaf}, \mathit{map}(\Pi, \mathit{map}(\Pi, \Pi)))$.

Proof. Consider any constructions $X : \mathit{leaf}$, $A : \Pi$ and $B : \Pi$. Then there are three cases for the reduction of $\mathit{if}\, X\, A\, B$.

Case 1: X is *true*. Then $\mathit{if}\, X\, A\, B \rhd A : \Pi$.

Case 2: X is *false*. Then $\mathit{if}\, X\, A\, B \rhd B : \Pi$.

Case 3: X is any other construction. Then $\mathit{if}\, X\, A\, B$ reduces only to itself; thus $\mathit{if}\, X\, A\, B : \Pi$ by default.

In each case, $\mathit{if}\, X\, A\, B : \Pi$, so $\mathit{if} : \mathit{map}(\mathit{leaf}, \mathit{map}(\Pi, \mathit{map}(\Pi, \Pi)))$. ∎

THEOREM 10. (Recursion well-foundedness rule.)

$$\mathit{rec}_0 : \mathit{map}(\mathit{product}(\Pi, \mathit{map}(\mathit{product}(\mathit{leaf}, \Pi), \Pi)), \mathit{map}(\mathit{leaf}, \Pi)).$$

Proof. Recall from the Expanded Term Language that

$$\mathit{rec}_0 \text{ is } (\lambda(f,g).T_0) \quad \text{where} \begin{cases} T_0(0) & \stackrel{\Delta}{=} f, \\ T_0(Sm) & \stackrel{\Delta}{=} g(m, T_0(m)) \end{cases}$$

where f, g, m are three variables. This implies that $\mathit{rec}_0(f,g) \rhd^* T_0 \not\rhd$, $T_0(0) \rhd^* f$ and $T_0(Sm) \rhd^* g(m, T_0(m))$.

Consider any constructions $X : \mathit{product}(\Pi, \mathit{map}(\mathit{product}(\mathit{leaf}, \Pi), \Pi))$ and $N : \mathit{leaf}$. The task is to show that $\mathit{rec}_0\, X\, N : \Pi$.

X is (F, G), for some constructions $F : \Pi$ and $G : \mathit{map}(\mathit{product}(\mathit{leaf}, \Pi), \Pi)$. Thus $\mathit{rec}_0\, X\, N \rhd^* T_0\left[{F,G \atop f,g}\right] N$. Suppose the latter term reduces to a construction. By induction on the length of this reduction I shall show that $T_0\left[{F,G \atop f,g}\right] N : \Pi$.

By Theorem 42 of the Expanded Term Language, since $T_0\left[{F,G \atop f,g}\right] N$ reduces to a construction, $N \stackrel{*}{\leftrightarrow} 0$ or SM for some construction M.

In the case where $N \stackrel{*}{\leftrightarrow} 0$, we have $T_0\left[{F,G \atop f,g}\right] N \rhd^* F : \Pi$, as required.

In the case where $N \stackrel{*}{\leftrightarrow} SM$, we have $T_0\left[{F,G \atop f,g}\right] N \rhd^* G(M, T_0\left[{F,G \atop f,g}\right] M)$, in at least one reduction step; hence the reduction of $T_0\left[{F,G \atop f,g}\right] N$ to a construction involves, and so is longer than, a reduction of $T_0\left[{F,G \atop f,g}\right] M$ to a construction. Assuming, by inductive hypothesis, that $T_0\left[{F,G \atop f,g}\right] M : \Pi$, and recalling that $G : \mathit{map}(\mathit{product}(\mathit{leaf}, \Pi), \Pi)$ and $M : \mathit{leaf}$, it follows that $G(M, T_0\left[{F,G \atop f,g}\right] M) : \Pi$, and hence $T_0\left[{F,G \atop f,g}\right] N : \Pi$, as required.

This inductive argument establishes that $T\left[{F,G \atop f,g}\right] N : \Pi$ and hence $\mathit{rec}_0\, X\, N : \Pi$ and $\mathit{rec}_0 : \mathit{map}(\mathit{product}(\Pi, \mathit{map}(\mathit{product}(\mathit{leaf}, \Pi), \Pi)), \mathit{map}(\mathit{leaf}, \Pi))$, as required. ∎

THEOREM 11. (*fxpt id* well-foundedness rule.) $\mathit{fxpt\ id} : \mathit{map}(\Pi, \Sigma)$.

Proof. Consider any construction $X : \Pi$. Then $\mathit{fxpt\ id}\, X$ does not reduce to a construction; hence by default $\mathit{fxpt\ id}\, X : \Sigma$. By the *map* rule $\mathit{fxpt\ id} : \mathit{map}(\Pi, \Sigma)$, as required. ∎

It was explained in Chapter 10 that every protological derivation tree has an associated *reflection tree*, whose well-foundedness is essential for the protological derivation tree to be sound. I shall define the coding for reflection trees in such a way that the same construction can serve as the code for the derivation tree (via the coding defined at the start of this chapter) and as the code for the associated reflection tree (via the coding *rt*, defined below). The difference between the two codings is that, in the derivation tree coding, a reflection principle $rp(\lambda[\underline{z}].U)\left[\frac{\underline{X}}{\underline{x}}\right]$ is regarded as a leaf, whereas, in the reflection tree coding *rt*, the reflection principle is regarded as having all its instances $U\left[\frac{\underline{X}}{\underline{x}}\right]\left[\frac{\underline{Z}}{\underline{z}}\right]$ as subtrees.

DEFINITION. (Reflection trees.) Let *rt* be a fresh 0-ary constructor and let it represent the following coding of trees:

$none(N)$	has no subtrees
$one(N, P, Y)$	has one subtree, P, of type *rt*
$two(N, P, Y, Q, Z)$	has two subtrees, P and Q, of type *rt*
$rp(F)$	has subtrees, T, of type *rt*, where $FX \rhd^* T \not\rhd$ for some construction X

for any constructions N, P, Y, Q, Z, F. Informally, a tree of type *rt* is a protological derivation (with no premises) re-interpreted as a reflection tree. The notion of reflection tree here is essentially the same as the one in Chapter 10.

EXAMPLE. Here is (part of) a tree of type *rt*. X and Y are arbitrary constructions.

$$two(12, none(4), A, rp(\lambda[x,y].one(6, rp(\lambda[z].none(3)), B)), C)$$

/ \

$none(4)$ $\qquad$ $rp(\lambda[x,y].one(6, rp(\lambda[z].none(3)), B))$

… / | \ …

⋮ $\quad one(6, rp(\lambda[z].none(3)), B\left[\frac{X,Y}{x,y}\right])$ $\quad$ ⋮

|

$rp(\lambda[z].none(3))$

|

$none(3)$

THEOREM 12. (Properties of *rt*.)

- $none\ :\ map(leaf, rt)$,
- $\mathcal{E}\ :\ rt$,
- $one\ :\ map(product(leaf, product(rt, leaf)), rt)$,
- $two\ :$
  $map(product(leaf, product(rt, product(leaf, product(rt, leaf)))), rt)$,
- $rp\ :\ map(map(leaf, rt), rt)$.

Proof. From the definitions of *rt* and $\mathcal{E}$, and the *map* and *product* rules. ∎

THEOREM 13. (Soundness of protologic.) If $DT(D, ([F, \dots G], H)) \ \triangleright^* \ true$, $D\ :\ rt$, $F, \dots G, H, X$ are constructions, and $FX \ \triangleright^* \ true, \dots GX \ \triangleright^* \ true$, then $HX \ \triangleright^* \ true$.

This theorem affirms that protological derivations with well-founded reflection trees are sound with respect to the informal semantics in Chapter 17. This is the only theorem about protologic that assumes the philosophical soundness of protologic; hence I shall organise the arguments in Parts III and IV so that the dependence of the interpretation theorems on this theorem can be clearly isolated (see Chapter 35).

DEFINITION. Define a construction *join* by

$$
\begin{aligned}
join(premise(0)) &\triangleq left \\
join(premise(Sn)) &\triangleq s(k(join(premise(n))))right \\
join(none(n)) &\triangleq k(none(n)) \\
join(one(n, p, y)) &\triangleq (\lambda j.(\lambda l.one(n, jpl, y)))join \\
join(two(n, p, y, q, z)) &\triangleq (\lambda j.(\lambda l.two(n, jpl, y, jql, z)))join \\
join(rp(f)) &\triangleq k(rp(f))
\end{aligned}
$$

where n, p, y, q, z, j, l, f are eight variables.

The function *join* attaches a list of derivations to the premises of another derivation.

EXAMPLE. Let *D* be a coded derivation of the form

$$two(15, premise(0), X_0, one(13, two(12, none(5), A, premise(1), X_1), B), C).$$

for a sequent E. Note that the sequents X_0 and X_1 appear as premises in D. Now, suppose we have coded derivations D_0 and D_1 for X_0 and X_1 respectively, neither derivation having any premises; then combining the three derivations using *join* gives a derivation for E with no premises.

$$\begin{aligned} \mathit{join}(D)[D_0, D_1] \ &\rhd^* \ \mathit{two}(15, D_0, X_0, \\ &\qquad\qquad \mathit{one}(13, \mathit{two}(12, \mathit{none}(5), A, D_1, X_1), B), C) \\ \mathit{DT}(\mathit{join}(D)[D_0, D_1], E) \ &\rhd^* \ \mathit{true} \ \& \ \mathit{DT}(D_0, X_0) \\ &\qquad \& \ (\mathit{true} \ \& \ (\mathit{true} \ \& \ \mathit{true} \ \& \ \mathit{DT}(D_1, X_1))) \\ &\rhd^* \ \mathit{true} \end{aligned}$$

Moreover, $\mathit{join}(D)[D_0, D_1]$ is well-founded (in the *rt* coding) provided D_0 and D_1 are. The next theorem verifies that these properties hold in general.

THEOREM 14. (The *join*-lemma.) Suppose D is the code of a protological derivation without reflection principles of a sequent X from the premises $X_0, \ldots X_{k-1}$ ($k \geq 0$).

- For any terms $D_0, \ldots D_{k-1}$, if $\mathit{DT}(D_0, X_0) \rhd^* \mathit{true}, \ldots \mathit{DT}(D_{k-1}, X_{k-1}) \rhd^* \mathit{true}$ then $\mathit{DT}(\mathit{join}(D)[D_0, \ldots D_{k-1}], X) \rhd^* \mathit{true}$.
- For any variables $\underline{x}$ and terms $\underline{X}$, $\mathit{join}(D\left[\frac{\underline{X}}{\underline{x}}\right]) \ : \ \mathit{map}(\mathit{pi}[\underbrace{\mathit{rt}, \ldots \mathit{rt}}_{k \text{ times}}], \mathit{rt})$.

Proof. To prove the first part, suppose that $\mathit{DT}(D_i, X_i) \rhd^* \mathit{true}$, for each $i = 0, \ldots k-1$. It follows that, for each i, $D_i \rhd^* D'_i \not\rhd$, for some term D'_i. Let L be $[D'_0, \ldots D'_{k-1}]$.

The proof that $\mathit{DT}(\mathit{join}(D)[D_0, \ldots D_{k-1}], X) \rhd^* \mathit{true}$ is by structural induction on the derivation encoded by D; since the derivation has no reflection principles there are four cases.

Case 1: D is $\mathit{premise}(i)$ and X is X_i, for some number $i = 0, \ldots k-1$. Then

$$\mathit{join}(D)[D_0, \ldots D_{k-1}] \ \rhd^* \ \mathit{join}(\mathit{premise}(i))L \ \rhd^* \ D'_i,$$

using the first two clauses of the definition of *join*, and hence

$$\mathit{DT}(\mathit{join}(D)[D_0, \ldots D_{k-1}], X) \ \rhd^* \ \mathit{DT}(D'_i, X_i) \ \rhd^* \ \mathit{true},$$

since $\mathit{DT}(D_i, X_i) \rhd^* \mathit{true}$, as required.

Case 2: D is $\mathit{none}(n)$ and X is a instance of axiom schema number n. Then

$$\mathit{join}(D)[D_0, \ldots D_{k-1}] \ \rhd^* \ \mathit{join}(\mathit{none}(n))L \ \rhd^* \ \mathit{none}(n)$$

and hence

$$\mathit{DT}(\mathit{join}(D)[D_0, \ldots D_{k-1}], X) \ \rhd^* \ \mathit{DT}(\mathit{none}(n), X) \ \rhd^* \ \mathit{info}_0(n, X) \ \rhd^* \ \mathit{true}$$

as required.

Case 3: D is $one(n, P, Y)$, where $\frac{Y}{X}$ is an instance of rule n. Then by inductive hypothesis (applied to P, Y in place of D, X)

$$DT(join(P)[D_0, \ldots D_{k-1}], Y) \rhd^* DT(join(P)L, Y) \rhd^* DT(U, Y) \rhd^* true$$

for some irreducible U; consequently,

$$\begin{aligned} join(D)[D_0 \ldots D_{k-1}] &\rhd^* join(one(n, P, Y))L \rhd^* one(n, join(P)L, Y) \\ &\rhd^* one(n, U, Y) \not\rhd \end{aligned}$$

and hence

$$\begin{aligned} DT(join(D)[D_0 \ldots D_{k-1}], X) &\rhd^* DT(one(n, U, Y), X) \\ &\rhd^* inf_1(n, Y, X) \,\&\, DT(U, Y) \\ &\rhd^* true \,\&\, true \\ &\rhd^* true \end{aligned}$$

as required.

Case 4: D is $two(n, P, Y, Q, Z)$, where $\frac{Y\ Z}{X}$ is an instance of rule n. Then by inductive hypothesis (applied to P, Y in place of D, X)

$$DT(join(P)[D_0, \ldots D_{k-1}], Y) \rhd^* DT(join(P)L, Y) \rhd^* DT(U, Y) \rhd^* true$$

for some irreducible U, and likewise (applying the inductive hypothesis to Q, Z)

$$DT(join(Q)[D_0, \ldots D_{k-1}], Z) \rhd^* DT(join(Q)L, Z) \rhd^* DT(V, Z) \rhd^* true$$

for some irreducible V; consequently,

$$\begin{aligned} join(D)[D_0, \ldots D_{k-1}] &\rhd^* join(two(n, P, Y, Q, Z))L \\ &\rhd^* two(n, join(P)L, Y, join(Q)L, Z) \\ &\rhd^* two(n, U, Y, V, Z) \not\rhd \end{aligned}$$

and hence

$$\begin{aligned} DT(join(D)[D_0, \ldots D_{k-1}], X) &\rhd^* DT(two(n, U, Y, V, Z), X) \\ &\rhd^* inf_2(n, Y, Z, X) \,\&\, DT(U, Y) \,\&\, DT(V, Z) \\ &\rhd^* true \,\&\, true \,\&\, true \\ &\rhd^* true \end{aligned}$$

as required. This completes the proof of the first part of the theorem.

To prove the well-foundedness part, let $\underline{z}$ be the free variables of $D\left[\frac{\underline{X}}{\underline{x}}\right]$ and let $\underline{Z}$ be constructions. Suppose $join(D\left[\frac{\underline{X}}{\underline{x}}\right])\left[\frac{\underline{Z}}{\underline{z}}\right]$ reduces to a construction; then $D\left[\frac{\underline{X},\underline{Z}}{\underline{x},\underline{z}}\right]$ also reduces to a construction, D'. Consider any construction $L : pi[rt, \ldots rt]$. Then L is $[D_0, \ldots D_{k-1}]$, for some constructions $D_0, \ldots D_{k-1} : rt$. The task is to show that $join(D\left[\frac{\underline{X},\underline{Z}}{\underline{x},\underline{z}}\right])L : rt$. I shall show this by induction on the structure of the protological derivation encoded by D.

Case 1: D is *premise*(i), for some number $i = 0 \ldots k-1$. Then $join(premise(i)\left[\frac{\underline{X},\underline{Z}}{\underline{x},\underline{z}}\right])L \overset{*}{\mapsto} join(premise(i))L \triangleright^* D_i : rt$, as required.

Case 2: D is *none*(n). Then $join(D\left[\frac{\underline{X},\underline{Z}}{\underline{x},\underline{z}}\right])L \triangleright^* k(none(n))L \triangleright none(n) : rt$, as required.

Case 3: D is $one(n, P, Y)$. Then D' is $one(n, P', Y')$, where $P\left[\frac{\underline{X},\underline{Z}}{\underline{x},\underline{z}}\right] \triangleright^* P' \not\triangleright$ and $Y\left[\frac{\underline{X},\underline{Z}}{\underline{x},\underline{z}}\right] \triangleright^* Y' \not\triangleright$. Applying the inductive hypothesis to P,

$$join(P')L \triangleleft^* join(P\left[\frac{\underline{X},\underline{Z}}{\underline{x},\underline{z}}\right])L : rt$$

$$join(D\left[\frac{\underline{X},\underline{Z}}{\underline{x},\underline{z}}\right])L \triangleright^* join(one(n, P', Y'))L \triangleright^* one(n, join(P')L, Y') : rt,$$

as required.

Case 4: D is $two(n, P, Y, Q, Z)$. Then D' is $two(n, P', Y', Q', Z')$, where $P\left[\frac{\underline{X},\underline{Z}}{\underline{x},\underline{z}}\right] \triangleright^* P' \not\triangleright$, $Y\left[\frac{\underline{X},\underline{Z}}{\underline{x},\underline{z}}\right] \triangleright^* Y' \not\triangleright$, $Q\left[\frac{\underline{X},\underline{Z}}{\underline{x},\underline{z}}\right] \triangleright^* Q' \not\triangleright$ and $Z\left[\frac{\underline{X},\underline{Z}}{\underline{x},\underline{z}}\right] \triangleright^* Z' \not\triangleright$. Applying the inductive hypothesis to P and Q,

$$join(P')L \triangleleft^* join(P\left[\frac{\underline{X},\underline{Z}}{\underline{x},\underline{z}}\right])L : rt$$

$$join(Q')L \triangleleft^* join(Q\left[\frac{\underline{X},\underline{Z}}{\underline{x},\underline{z}}\right])L : rt$$

$$\begin{aligned} join(D\left[\frac{\underline{X},\underline{Z}}{\underline{x},\underline{z}}\right])L &\triangleright^* join(two(n, P', Y', Q', Z'))L \\ &\triangleright^* two(n, join(P')L, Y', join(Q')L, Z') : rt, \end{aligned}$$

as required. ▮

The argument in the *join*-lemma can be formalised within protologic. That is, not only can we show that if $D_0, \ldots D_{k-1}$ are derivations of $X_0, \ldots X_{k-1}$ respectively then $join(D)[D_0, \ldots D_{k-1}]$ reduces to a derivation of X, but we can also derive the sequent

$$\begin{aligned} (d_0, \ldots d_{k-1}, \underline{z})\ DT(d_0, X_0), \ \ldots \ DT(d_{k-1}, X_{k-1}) \\ \rightarrow DT(join(D)[d_0, \ldots d_{k-1}], X). \end{aligned}$$

The protological derivation required to do this can be generated automatically from the following information: $\underline{z}, X_0, \ldots X_{k-1}, D, X$. The function *Join* that generates the derivation is introduced in the next theorem.

THEOREM 15. (The *Join*-lemma.) If D is the code of a protological derivation without reflection principles of the sequent X from the premises $X_0, \ldots X_{k-1}$ ($k \geq 0$) then

- $DT(Join(Code), [\![(d_0, \ldots d_{k-1}, \underline{z})\ DT(d_0, X_0), \ \ldots\ DT(d_{k-1}, X_{k-1}) \rightarrow DT(join(D)[d_0, \ldots d_{k-1}], X)]\!]) \ \triangleright^*\ true$;
- $Join(Code)\ :\ rt$;

where the construction *Join* is defined below, *Code* is $((\lambda[\underline{z}].[X_0, \ldots X_{k-1}]), (\lambda[\underline{z}].(D, X)))$, and $d_0, \ldots d_{k-1} \notin D, X_0, \ldots X_{k-1}, X$.

Proof. As temporary metanotation, let $\underline{d}$ be the sequence of variables $d_0, \ldots d_{k-1}$, let $\mathbf{sq}\{T, W\}$ be the sequent

$$(\underline{d}, \underline{z})\ DT(d_0, X_0), \ldots DT(d_{k-1}, X_{k-1}) \rightarrow DT(join(T)[\underline{d}], W)$$

for any term T and sequent W, and abbreviate $DT(d_0, X_0), \ldots DT(d_{k-1}, X_{k-1})$ to *LHS*, and $(\lambda[\underline{z}].[X_0, \ldots X_{k-1}])$ to M.

I shall first show that $\mathbf{sq}\{D, X\}$ is derivable from the corresponding sequents for D's subtrees. Since the derivation encoded by D contains no reflection principles there are four cases.

Case 1: D is $premise(i)$ and X is X_i, for some number $i = 0, \ldots k-1$. Then $DT(join(D)[\underline{d}], X)\ \triangleright^*\ DT(d_i, X_i)$, so $\mathbf{sq}\{D, X\}$ is derivable by Reduction. Define a recursive function der_1 such that $der_1(i, (\lambda[\underline{z}].X), M)$ reduces to the code of this derivation of $\mathbf{sq}\{D, X\}$.

Case 2: D is $none(n)$ and X is an instance of axiom schema n. Then

$$DT(join(D)[\underline{d}], X)\ \triangleright^*\ DT(none(n), X)\ \triangleright^*\ info_0(n, X)\ \triangleright^*\ true,$$

so $\mathbf{sq}\{D, X\}$ is derivable by Reduction. Define a recursive function der_2 such that $der_2(n, (\lambda[\underline{z}].X), M)$ reduces to the code of this derivation of $\mathbf{sq}\{D, X\}$.

Case 3: D is $one(n, P, Y)$, where $\dfrac{Y}{X}$ is an instance of rule n. Then $\mathbf{sq}\{D, X\}$ is derivable from the premise $\mathbf{sq}\{P, Y\}$ (which is labelled 'p0', meaning premise number 0):

$$\dfrac{\dfrac{\dfrac{(\underline{d}, \underline{z})\ \rightarrow inf_1(n, Y, X)\ \text{(red)} \qquad \mathbf{sq}\{P, Y\}\ \text{(p0)}}{(\underline{d}, \underline{z})\ LHS \rightarrow inf_1(n, Y, X)\ \&\ DT(join(P)[\underline{d}], Y)}\ \&}{(\underline{d}, \underline{z})\ LHS \rightarrow DT(one(n, join(P)[\underline{d}], Y), X)}\ \text{conv}}{(\underline{d}, \underline{z})\ LHS \rightarrow DT(join(D)[\underline{d}], X)}\ \text{red}$$

Define a recursive function der_3 such that $der_3(n, (\lambda[\underline{z}].(P,Y)), (\lambda[\underline{z}].X), M)$ reduces to the code of this derivation.

Case 4: D is $two(n,P,Y,Q,Z)$, where $\frac{Y\ Z}{X}$ is an instance of rule n. Then $\mathbf{sq}\{D,X\}$ is derivable from the premises $\mathbf{sq}\{P,Y\}$ and $\mathbf{sq}\{Q,Z\}$ as in Case 3 (the premises are numbered 0 and 1 respectively):

$$\dfrac{\dfrac{\dfrac{(\underline{d},\underline{z}) \rightarrow inf_2(n,Y,Z,X)\ \text{(red)} \qquad \mathbf{sq}\{P,Y\}\ \text{(p0)} \qquad \mathbf{sq}\{Q,Z\}\ \text{(p1)}}{(\underline{d},\underline{z})\ LHS \rightarrow inf_2(n,Y,Z,X)\ \&\ DT(join(P)[\underline{d}],Y)\ \&\ DT(join(Q)[\underline{d}],Z)}\ \&}{(\underline{d},\underline{z})\ LHS \rightarrow DT(two(n,join(P)[\underline{d}],Y,join(Q)[\underline{d}],Z),X)}\ \text{conv}}{(\underline{d},\underline{z})\ LHS \rightarrow DT(join(D)[\underline{d}],X)}\ \text{red}$$

Define a recursive function der_4 such that $der_4(n, (\lambda[\underline{z}].(P,Y)), (\lambda[\underline{z}].(Q,Z)), (\lambda[\underline{z}].X), M)$ reduces to the code of this derivation.

Next define a term-rearranging recursive function F such that

$$\begin{aligned}
F(\lambda[\underline{z}].(premise(I),X)) &\rhd^* (1,I,(\lambda[\underline{z}].X))\\
F(\lambda[\underline{z}].(none(N),X)) &\rhd^* (2,N,(\lambda[\underline{z}].X))\\
F(\lambda[\underline{z}].(one(N,P,Y),X)) &\rhd^* (3,N,(\lambda[\underline{z}].(P,Y)),(\lambda[\underline{z}].X))\\
F(\lambda[\underline{z}].(two(N,P,Y,Q,Z),X)) &\rhd^* (4,N,(\lambda[\underline{z}].(P,Y)),(\lambda[\underline{z}].(Q,Z)),\\
&\qquad (\lambda[\underline{z}].X))
\end{aligned}$$

for any terms X,P,Y,Q,Z and any constructions I,N; and define a construction G by

$$\begin{aligned}
G((1,i,x),m) &\triangleq join(der_1(i,x,m))nil\\
G((2,n,x),m) &\triangleq join(der_2(n,x,m))nil\\
G((3,n,u,x),m) &\triangleq join(der_3(n,u,x,m))[G(Fu,m)]\\
G((4,n,u,v,x),m) &\triangleq join(der_4(n,u,v,x,m))[G(Fu,m),G(Fv,m)]
\end{aligned}$$

where i,n,x,u,v,m are six variables. Then I claim that

$$\begin{aligned}
&DT(G(F(\lambda[\underline{z}].(D,X)),M),\mathbf{sq}\{D,X\}) \rhd^* true\\
&\text{and } G(F(\lambda[\underline{z}].(D,X)),M) : rt.
\end{aligned} \qquad (*)$$

The proof is by structural induction on the derivation encoded by D. There are four cases.

Case 1: D is $premise(i)$ and X is X_i, for some number $i = 0,\ldots k-1$. Then

$$\begin{aligned}
G(F(\lambda[\underline{z}].(D,X)),M) &\rhd^* G((1,i,(\lambda[\underline{z}].X)),M)\\
&\rhd^* join(der_1(i,(\lambda[\underline{z}].X),M))nil
\end{aligned}$$

where, as stated above, $der_1(i, (\lambda[\underline{z}].X), M)$ reduces to the code of a derivation without reflection principles of $\mathbf{sq}\{D, X\}$ from no premises; so $(*)$ follows by the *join*-lemma.

Case 2: D is $none(n)$ and X is an instance of axiom schema n. Then

$$\begin{aligned} G(F(\lambda[\underline{z}].(D,X)),M) &\rhd^* G((2,n,(\lambda[\underline{z}].X)),M) \\ &\rhd^* join(der_2(n,(\lambda[\underline{z}].X),M))nil \end{aligned}$$

where, as stated above, $der_2(n, (\lambda[\underline{z}].X), M)$ reduces to the code of a derivation without reflection principles of $\mathbf{sq}\{D, X\}$ from no premises; so $(*)$ follows by the *join*-lemma.

Case 3: D is $one(n, P, Y)$, where $\frac{Y}{X}$ is an instance of rule n. Then

$$\begin{aligned} & G(F(\lambda[\underline{z}].(D,X)),M) \\ & \rhd^* G((3,n,(\lambda[\underline{z}].(P,Y)),(\lambda[\underline{z}].X)),M) \\ & \rhd^* join(der_3(n,(\lambda[\underline{z}].(P,Y)),(\lambda[\underline{z}].X),M))[G(F(\lambda[\underline{z}].(P,Y)),M)] \end{aligned}$$

where, as stated above, $der_3(n, (\lambda[\underline{z}].(P, Y)), (\lambda[\underline{z}].X), M)$ reduces to the code of a derivation without reflection principles of $\mathbf{sq}\{D, X\}$ from $\mathbf{sq}\{P, Y\}$. Also, by the inductive hypothesis $(*)$, applied to P, Y in place of D, X,

$$DT(G(F(\lambda[\underline{z}].(P,Y)),M),\mathbf{sq}\{P,Y\}) \rhd^* true$$
$$\text{and } G(F(\lambda[\underline{z}].(P,Y)),M) : rt.$$

So $(*)$ follows by the *join*-lemma.

Case 4: D is $two(n, P, Y, Q, Z)$, where $\frac{Y\ Z}{X}$ is an instance of rule n. Then

$$\begin{aligned} & G(F(\lambda[\underline{z}].(D,X)),M) \\ & \rhd^* G((4,n,(\lambda[\underline{z}].(P,Y)),(\lambda[\underline{z}].(Q,Z)),(\lambda[\underline{z}].X)),M) \\ & \rhd^* join(der_4(n,(\lambda[\underline{z}].(P,Y)),(\lambda[\underline{z}].(Q,Z)),(\lambda[\underline{z}].X),M)) \\ & \qquad [G(F(\lambda[\underline{z}].(P,Y)),M),G(F(\lambda[\underline{z}].(Q,Z)),M)] \end{aligned}$$

where, as stated above, $der_4(n, (\lambda[\underline{z}].(P, Y)), (\lambda[\underline{z}].(Q, Z)), (\lambda[\underline{z}].X), M)$ reduces to the code of a derivation without reflection principles of $\mathbf{sq}\{D, X\}$ from the premises $\mathbf{sq}\{P, Y\}$ and $\mathbf{sq}\{Q, Z\}$. So $(*)$ follows by the *join*-lemma and the inductive hypothesis.

This completes the proof of the claim. Finally, define *Join* as $(\lambda(m, d).\ G(Fd, m))$, where m, d are two variables. Then

$$Join(Code) \rhd^* G(F(\lambda[\underline{z}].(D,X)),M),$$

giving $DT(Join(Code), \mathbf{sq}\{D, X\}) \rhd^* true$ and $Join(Code) : rt$, as required. ∎

Theorem 13 says that protological derivations with well-founded reflection trees are sound. The following theorem shows what happens if the well-foundedness condition is dropped. The proof uses a protological version of the Zermelo-Russell paradox to construct an unsound derivation tree with ill-founded reflection tree.

THEOREM 16. There is a construction D^* such that $DT(D^*, \llbracket \to \mathit{false} \rrbracket) \rhd^*$ *true*.

Proof. Let x, d, q, d_0, d_1 be five variables, let X be the construction

$$(\lambda x.(\lambda d.DT(d, \llbracket (q)\ xxq \to \mathit{false} \rrbracket))),$$

and let D_0 be the code of the protological derivation

$$\dfrac{\to xxq\ \text{(p0)} \qquad \dfrac{(q)\ xxq \to \mathit{false}\ \text{(p1)}}{xxq \to \mathit{false}}\,\text{inst}}{\to \mathit{false}}\,\text{cut}$$

By the *Join*-lemma,

$$\begin{aligned} DT(\mathit{Join}(\mathit{Code}), \llbracket (d_0, d_1, q, x)\ & DT(d_0, \llbracket \to xxq \rrbracket), \\ & DT(d_1, \llbracket (q)\ xxq \to \mathit{false} \rrbracket) \to \\ & DT(\mathit{join}(D_0)[d_0, d_1], \llbracket \to \mathit{false} \rrbracket) \rrbracket) \rhd^*\ \mathit{true}, \end{aligned}$$

where *Code* is

$$((\lambda[q,x].[\llbracket \to xxq \rrbracket, \llbracket (q)\ xxq \to \mathit{false} \rrbracket]), (\lambda[q,x].(D_0, \llbracket \to \mathit{false} \rrbracket))).$$

Moreover, by Theorem 2,

$$\begin{aligned} & DT(rp(\lambda[q].\mathit{join}(D_0\begin{bmatrix} X \\ x \end{bmatrix})[\mathcal{E}, q]), \\ & \quad \llbracket (q)\ DT(\mathit{join}(D_0\begin{bmatrix} X \\ x \end{bmatrix})[\mathcal{E}, q], \llbracket \to \mathit{false} \rrbracket) \to \mathit{false} \rrbracket) \rhd^*\ \mathit{true}. \end{aligned}$$

Now let D_1 be the code of the protological derivation

$$\dfrac{\dfrac{\dfrac{\dfrac{\dfrac{\begin{array}{r}(d_0, d_1, q, x)\ DT(d_0, \llbracket \to xxq \rrbracket),\ DT(d_1, \llbracket (q)\ xxq \to \mathit{false} \rrbracket) \\ \to DT(\mathit{join}(D_0)[d_0, d_1], \llbracket \to \mathit{false} \rrbracket)\end{array}\ \text{(p0)}}{\begin{array}{r}(q,x)\ DT(\mathcal{E}, \llbracket \to xxq \rrbracket),\ DT(q, \llbracket (q)\ xxq \to \mathit{false} \rrbracket) \\ \to DT(\mathit{join}(D_0)[\mathcal{E}, q], \llbracket \to \mathit{false} \rrbracket)\end{array}}\,\text{inst}}{(q,x)\ xxq,\ Xxq \to DT(\mathit{join}(D_0)[\mathcal{E}, q], \llbracket \to \mathit{false} \rrbracket)}\,\text{red}}{(q)\ XXq,\ XXq \to DT(\mathit{join}(D_0\begin{bmatrix} X \\ x \end{bmatrix})[\mathcal{E}, q], \llbracket \to \mathit{false} \rrbracket)}\,\text{inst}}{(q)\ XXq \to DT(\mathit{join}(D_0\begin{bmatrix} X \\ x \end{bmatrix})[\mathcal{E}, q], \llbracket \to \mathit{false} \rrbracket)}\,\text{con}}{(q)\ XXq \to \mathit{false}}\,\text{cut with RP}$$

where the final step is a cut with the reflection principle

$$(q)\ DT(join(D_0\begin{bmatrix}X\\x\end{bmatrix})[\mathcal{E},q], [\![\ \to false]\!]) \to false,$$

which is premise 1. Now let D_2 be

$$join(D_1)[Join(Code), rp(\lambda[q].join(D_0\begin{bmatrix}X\\x\end{bmatrix})[\mathcal{E},q])].$$

Then by the *join*-lemma

$$DT(D_2, [\![(q)\ XXq \to false]\!]) \rhd^* \ true.$$

Next let D_3 be the code of the protological derivation

$$\cfrac{\cfrac{(q)\ XXq \to false\ \text{(p0)}}{(q)\ x = X,\ XXq \to false}\ \text{thin}}{(q)\ x = X,\ xxq \to false}\ \text{eq}$$

Then by the *join*-lemma

$$DT(join(D_3)[D_2], [\![(q)\ x = X,\ xxq \to false]\!]) \rhd^* \ true$$

and hence by instantiation

$$DT(join(D_3\begin{bmatrix}X\\x\end{bmatrix})[D_2], [\![(q)\ x = X,\ xxq \to false]\!]\begin{bmatrix}X\\x\end{bmatrix}) \rhd^* \ true.$$

Also let D_4 be the coded protological derivation

$$two(12, one(11, \mathcal{E}, [\![\ \to x = X]\!]\begin{bmatrix}X\\x\end{bmatrix}), [\![(q)\ \ \to x = X]\!]\begin{bmatrix}X\\x\end{bmatrix},$$
$$join(D_3\begin{bmatrix}X\\x\end{bmatrix})[D_2], [\![(q)\ x = X,\ xxq \to false]\!]\begin{bmatrix}X\\x\end{bmatrix})$$

(recall that rule 12 is Cut and rule 11 is a Thinning rule). By the definition of DT,

$$DT(D_4, [\![(q)\ xxq \to false]\!]\begin{bmatrix}X\\x\end{bmatrix}) \rhd^* \ true$$

and hence, by the definition of X,

$$XXD_4 \rhd^* \ true.$$

Now let D_5 be the code of the protological derivation

$$\cfrac{\to XXD_4\ \text{(eval)} \qquad \cfrac{(q)\ XXq \to false\ \text{(p0)}}{XXD_4 \to false}\ \text{inst}}{\to false}\ \text{cut}$$

giving, by the *join*-lemma,

$$DT(join(D_5)[D_2], [\![\ \to false]\!]) \rhd^* \ true.$$

This implies $join(D_5)[D_2] \rhd^* D^*$, for some construction D^*, and

$$DT(D^*, [\![\ \to false]\!]) \rhd^* \ true,$$

as required. ▮

CHAPTER 22

THE CODING OF TREES

REFLECTION PRINCIPLES

A *reflection principle* is a sequent of the form $[\![(\underline{z})\ DT(U, [\![\ \to C]\!]) \to C]\!]\left[\frac{\underline{X}}{\underline{x}}\right]$, where $\underline{X}$ are irreducible terms. (Anything enclosed within '$[\![\ldots]\!]$' brackets is a sequent.)

CODING OF PROTOLOGICAL DERIVATION TREES

A *derivation tree* (from a given list of premises $X_0, \ldots X_{k-1}$) is a tree of sequents whose leaves are protological axioms, reflection principles, and premises from the given premise list, and whose other nodes are related by the protological inference rules. A *derivation* (from a given premise list) is a derivation tree with its conclusion sequent (the root of the tree) removed.

Derivations and derivation trees may be coded as irreducible terms, using five 1-ary constructors *premise*, *none*, *one*, *two* and *rp*. The free variables of the code of a derivation or derivation tree are the free variables of its constituent sequents.

THE DERIVATION TREE PREDICATE, *DT*

Four constructions are defined: DT, $\mathcal{E}$, *join* and *Join*.

THEOREM 1. $DT(\mathcal{E}, [\![\ \to T]\!])\ \rhd^*\ T$.

THEOREM 2. $DT(rp(\lambda[z].U), [\![(\underline{z})\ DT(U, [\![\ \to C]\!]) \to C]\!])\ \rhd^*\ \mathit{true}$.

TREE CODES AND WELL-FOUNDEDNESS

DEFINITION. A *type symbol* is a construction that has been given an interpretation as representing a coding of a class of trees as constructions.

A well-foundedness relation $T : \Pi$ is defined, where T is any term and Π is a type symbol (or a term reducing to a type symbol).

From now on the identifiers 'Π', 'Σ', 'Φ', 'Ψ', 'Ω' and 'Θ', sometimes with subscripts or primes, will always be used as metavariables denoting type symbols (or terms that reduce to type symbols).

THEOREM 3. (Well-foundedness reduction rule.)

- If $A \;\triangleright^*\!\triangleleft\; B$ then $A : \Pi$ iff $B : \Pi$.
- If $\Pi \;\triangleright^*\!\triangleleft\; \Sigma$ then $A : \Pi$ iff $A : \Sigma$.

Two fresh 0-ary constructors, *leaf* and *rt*, are introduced, along with three fresh 1-ary constructors, *map*, *product* and *pi*. The following type symbols are defined: *leaf*, *map*(Π, Σ), *product*(Π, Σ), *pi*$[\Phi_1, \ldots \Phi_k]$, and *rt*, where $\Pi, \Sigma, \Phi_1, \ldots \Phi_k$ are type symbols.

THEOREM 4. (*leaf* rule.) For any term T, T : *leaf*.

THEOREM 5. (*map* rule.)

- If A : *map*(Π, Σ) and $B : \Pi$ then $AB : \Sigma$.
- If, for any construction $B : \Pi$, $AB : \Sigma$, then A : *map*(Π, Σ).
- If $X \not\triangleright$, $\underline{u} \notin X$, $\underline{U}$ are constructions, and $T\left[\frac{\underline{U}}{\underline{u}}\right] : \Sigma$, then $(\lambda X.T)\left[\frac{\underline{U}}{\underline{u}}\right]$: *map*(*leaf*, Σ).

THEOREM 6. (Well-foundedness instantiation rule.) If $A : \Pi$ and $X \not\triangleright$ then $A\left[\frac{X}{x}\right] : \Pi$.

THEOREM 7. (*product* rule.)

- Any construction T such that T : *product*(Π, Σ) is of the form (X, Y) for some constructions $X : \Pi$ and $Y : \Sigma$.
- *left* : *map*(*product*$(\Pi, \Sigma), \Pi)$.
- *right* : *map*(*product*$(\Pi, \Sigma), \Sigma)$.
- For any terms A and B, if $A : \Pi$ and $B : \Sigma$ then (A, B) : *product*(Π, Σ).

THEOREM 8. (*pi* rule.)

- Any construction T such that T : *pi*$[\Pi, \ldots \Sigma]$ is of the form $[A, \ldots B]$, for some constructions $A : \Pi, \ldots B : \Sigma$.
- For any terms $A, \ldots B$, if $A : \Pi, \ldots B : \Sigma$ then $[A, \ldots B]$: *pi*$[\Pi, \ldots \Sigma]$.

THEOREM 9. (*if* well-foundedness rule.) *if* : *map*(*leaf*, *map*$(\Pi$, *map*$(\Pi, \Pi)))$.

THEOREM 10. (Recursion well-foundedness rule.)

$$rec_0 : map(product(\Pi, map(product(leaf, \Pi), \Pi)), map(leaf, \Pi)).$$

THEOREM 11. (*fxpt id* well-foundedness rule.) *fxpt id* : *map*(Π, Σ).

THEOREM 12. (Properties of *rt*.)

- *none* : *map*(*leaf*, *rt*),
- $\mathcal{E}$: *rt*,
- *one* : *map*(*product*(*leaf*, *product*(*rt*, *leaf*)), *rt*),
- *two* :
  map(*product*(*leaf*, *product*(*rt*, *product*(*leaf*, *product*(*rt*, *leaf*)))), *rt*),
- *rp* : *map*(*map*(*leaf*, *rt*), *rt*).

THEOREM 13. (Soundness of protologic.) If $DT(D, ([F, \ldots G], H)) \rhd^* \mathit{true}$, $D : rt$, $F, \ldots G, H, X$ are constructions, and $FX \rhd^* \mathit{true}, \ldots GX \rhd^* \mathit{true}$, then $HX \rhd^* \mathit{true}$.

THEOREM 14. (The *join*-lemma.) Suppose D is the code of a protological derivation without reflection principles of a sequent X from the premises $X_0, \ldots X_{k-1}$ $(k \geq 0)$.

- For any terms $D_0, \ldots D_{k-1}$, if $DT(D_0, X_0) \rhd^* \mathit{true}, \ldots DT(D_{k-1}, X_{k-1}) \rhd^* \mathit{true}$ then $DT(join(D)[D_0, \ldots D_{k-1}], X) \rhd^* \mathit{true}$.
- For any variables $\underline{x}$ and terms $\underline{X}$, $join(D\left[\frac{\underline{X}}{\underline{x}}\right]) : map(pi[\underbrace{rt, \ldots rt}_{k \text{ times}}], rt)$.

THEOREM 15. (The *Join*-lemma.) If D is the code of a protological derivation without reflection principles of the sequent X from the premises $X_0, \ldots X_{k-1}$ $(k \geq 0)$ then

- $DT(Join(Code), [\![(d_0, \ldots d_{k-1}, \underline{z})\ DT(d_0, X_0), \ \ldots\ DT(d_{k-1}, X_{k-1}) \rightarrow DT(join(D)[d_0, \ldots d_{k-1}], X)]\!]) \rhd^* \mathit{true}$;
- $Join(Code) : rt$;

where $Code$ is $((\lambda[\underline{z}].[X_0, \ldots X_{k-1}]), (\lambda[\underline{z}].(D, X)))$, and $d_0, \ldots d_{k-1} \notin D, X_0, \ldots X_{k-1}, X$.

THEOREM 16. There is a construction D^* such that $DT(D^*, [\![\rightarrow \mathit{false}]\!]) \rhd^* \mathit{true}$.

LAYOUT OF WELL-FOUNDEDNESS ARGUMENTS

In all well-foundedness proofs using the theorems in this chapter, I shall adopt the following display conventions. Each step will be shown on a separate line, with indentation to indicate the scope of informal quantifiers, as in the following outline example.

$join(D_1)nil \,:\, rt$, by the *join*-lemma. (1)
For any construction $X \,:\, product(\Sigma, \Phi)$, (2)
  X is (A, B), for some constructions $A \,:\, \Sigma$ and $B \,:\, \Phi$. (3)
  For any construction $Y \,:\, \Psi$, (4)
    $FY \,:\, \Omega$. (5)
  $F \,:\, map(\Psi, \Omega)$ (6)
  $TX \triangleright^* C \,:\, \Pi$. (7)
$T \,:\, map(product(\Sigma, \Phi), \Pi)$. (8)

Here, each well-foundedness assertion follows from the preceding lines using the theorems in this chapter.

In line 1, I am supposing D_1 is the code of a protological derivation without reflection principles or premises. Then the assertion $join(D_1)nil \,:\, rt$ follows by the *join*-lemma with $k = 0$.

Line 2 is an informal quantifier ranging over well-founded trees X of type $product(\Sigma, \Phi)$. The scope of the quantifier is lines 3–7, as the indentation indicates.

Line 3 follows by the *product* rule from the hypothesis $X \,:\, product(\Sigma, \Phi)$.

Line 4 introduces a second quantifier, nested within the first. Its scope is line 5, indicated by the double indentation.

Line 5 deduces (in some way not indicated here) that $FY \,:\, \Omega$.

Line 6 (outside the quantification over Y) follows by the *map* rule from the fact if $Y \,:\, \Psi$ then $FY \,:\, \Omega$.

Line 7 deduces (somehow) from the preceding lines that $C \,:\, \Pi$, and since (we suppose) $TX \triangleright^* C$ the conclusion is drawn that $TX \,:\, \Pi$.

Line 8 (outside the quantification over X) follows by the *map* rule from the fact that if $X \,:\, product(\Sigma, \Phi)$ then $TX \,:\, \Pi$.

CHAPTER 23

THE EXPANDED TERM LANGUAGE AS A FUNCTIONAL PROGRAMMING LANGUAGE

The Expanded Term Language (ET) is a notation for expressing constructions, which are essentially data structures and algorithms. Thus ET may be regarded in computer science terms as a functional programming language, with the Term Language as its machine code and Protologic and Expanded Protologic as its correctness calculus. In this chapter I shall discuss the use of ET as a practical programming language, in comparison with other functional languages such as LISP (McCarthy, 1960), FP (Backus, 1978) and its successor FL (Backus, Williams & Wimmers, 1990), Standard ML (Milner, Tofte & Harper, 1990; Milner & Tofte, 1991), Miranda (Holyer, 1991), Haskell (ACM, 1992; Davie, 1992), and Hope (Burstall, MacQueen & Sannella, 1980). (See also Chapter 13 for a comparison of ET with λ-calculus.) The issues to be considered for each language are

- correctness proofs,
- functions as 'first-class' objects,
- data types, including polymorphism and overloading,
- pattern matching,
- lazy data structures, streams and interactive programming,
- modularity and programming in the large,
- efficiency of execution.

My main conclusion will be that ET, despite being a simple type-free language with strict evaluation, is able to fulfil most of the purposes that have led to the introduction of complex type systems and lazy evaluation in other languages.

CORRECTNESS PROOFS

Functional programming languages lend themselves readily to mathematical analysis; in the case of ET the properties of programs can be established either by informal argument (as in Chapter 15) or formally using Expanded Protologic. ET has a number of distinctive features that make such arguments more tractable and powerful than in the case of other functional languages.

First, ET's syntax is very much simpler than other languages': compare for example the grammar of ET with the eleven pages of syntax charts for

Standard ML in Paulson (1991). The main reason for this is that features such as numbers, tuples, lists, algebraic types and pattern-matching function definitions are treated as metanotation in ET, whereas in other languages they are taken as part of the core syntax. Treating these features as metanotation means that they are available when one wants to use them, yet they do not need to be taken into account when proving meta-theorems. In ET all metanotation is rewritten in terms of constants, variables, function applications, λ-abstractions and instantiations; moreover the λ-abstractions and instantiations are removed by compilation. Hence if one wants to prove a meta-theorem by structural induction on terms there are only three cases to consider: constants, variables and function applications (see Chapter 15 for many examples of such proofs). Structural induction would be very much more difficult in the other languages, with their more elaborate syntax. Similarly there is no need to include special axioms and rules for numbers, tuples, *etc.*, in Protologic; rather, theorems and rules for them can be *derived* as part of Expanded Protologic.

A second advantage of ET is that it is accompanied by a correctness calculus, Expanded Protologic, which encompasses the whole of constructive reasoning (if you believe the arguments in Chapters 8–10) and hence incorporates program transformation techniques such as those of Burstall & Darlington (1977), while prohibiting invalid transformations where a total function is transformed into a partial function (Darlington, 1987), and it also incorporates FP's functional algebra (Harrison & Khoshnevisan, 1987).

The third advantage of ET is its clear informal semantics. All functional languages are inspired by the principle of *referential transparency*, which states that the behaviour of any expression should be determined by its value, which is in turn determined by the values of its syntactic components (and not by the execution history, as in imperative languages). However several functional languages (LISP, Hope, Standard ML and FL) also include imperative features and hence do not live up to this principle. Recall from Chapter 13 the three principles on which ET is founded:

- that functions are intensional,
- compositionality – a strengthened form of referential transparency,
- that a (free) variable stands for an arbitrary construction.

These simplify the semantics considerably. In Miranda for example there are eight kinds of value: a value may be either data or a function, either finite or infinite, and either total or partial. In ET these is only one kind, since there is no distinction between data and functions, all values are finite, and all evaluation is strict.

The consequences of these principles will be expanded on below.

FUNCTIONS AS OBJECTS

An essential principle of the general-purpose computer is the unity of programs and data. Both programs and data are represented as bit strings in the machine's memory, and the same sequence of bits can be interpreted as either. Thus a bit string may be generated by a compiler (during which process it is viewed as a data structure), then treated as a program and executed; alternatively, instead of being executed it could be taken as an input to a program that analyses its storage requirements or optimises it. Thus programs can be constructed, manipulated and taken apart by other programs in arbitrary ways.

One would wish the same principle to be embodied in high-level language (albeit expressed in terms of high-level data structures rather than bit strings). Yet very few high-level languages allow this. Even functional languages, which claim to treat functions as 'first-class' objects (Milner & Tofte, 1991, Preface; Davie, 1992, p. 38), in fact treat functions as a special kind of object that can only be created individually by the user, using a special function definition notation, or produced by λ-abstraction or via higher-order functions (these all being direct or indirect ways of forming lambda-abstractions or partially evaluated lambda-abstractions such as $(\lambda x.(\lambda y.T))X$), and whose internal composition cannot be examined. The only exception is the earliest functional language, LISP, where a λ-expression is a list and can be created or decomposed like any other list. Unfortunately this feature was abandoned in later languages. It is reinstated in ET, where the only kind of object is a construction, any construction may act as a function, and any constructive function on constructions can be implemented as a construction.

In other languages the special treatment of functions is sometimes justified by taking an extensional interpretation: if a function is given only in extension then it makes no sense to decompose it into syntactic components (Davie, 1992, p. 19; Holyer, 1991, p. 183). My reply to this is that computing deals only with intensional functions (algorithms); one may wish to think of an algorithm as representing an extensional function, but the extensional function is in the eye of the beholder. The only circumstances in which it might be more natural to think of functions extensionally would be if it were possible to test functions for extensional equality, but of course this is not possible in any programming language.

DATA TYPES

Functional languages usually have elaborate type systems, including numbers, characters, booleans, tuples, lists, algebraic types, and polymorphic types. ET is somewhat unusual in being type-free, although of course all the aforementioned data structures can be defined. The view I take of types is that type-checking is a limited form of correctness proof and therefore should be considered as part of protologic. The conventional type apparatus can be translated into protologic as follows.

The first step is to identify a type with the function that tests whether a construction is an instance of the type. For example, the type of natural numbers is identified with the function *num* defined by

$$\begin{aligned} num\,0 &\triangleq true, \\ num(Sn) &\triangleq num(n). \end{aligned}$$

Hence I would interpret an algebraic type definition for binary trees of numbers

data tree == empty ++ tip(num) ++ node(tree # tree)

(in Hope syntax) as a function definition

$$\begin{aligned} tree(empty) &\triangleq true, \\ tree(tip(n)) &\triangleq num(n), \\ tree(node(t_1, t_2)) &\triangleq tree(t_1)\ \&\ tree(t_2), \end{aligned}$$

where *empty* is a 0-ary constructor, *tip* and *node* are two 1-ary constructors, and n, t_1, t_2 are three variables. Polymorphic type definitions are handled in a similar way. For example, the Hope definition

data treeof(alpha) == empty ++ tip(alpha) ++ node(tree # tree)

(where *alpha* is a type variable representing an arbitrary type) is interpreted as saying that *treeof* is $(\lambda\alpha.F)$, where F is defined by

$$\begin{aligned} F(empty) &\triangleq true, \\ F(tip(n)) &\triangleq \alpha(n), \\ F(node(t_1, t_2)) &\triangleq F(t_1)\ \&\ F(t_2). \end{aligned}$$

As a second polymorphic example, lists of any given type α make up the type $listof(\alpha)$, where *listof* is $(\lambda\alpha.F)$ and F is defined by

$$\begin{aligned} F(nil) &\triangleq true, \\ F(first, rest) &\triangleq \alpha(first)\ \&\ F(rest). \end{aligned}$$

In addition, *deep* polymorphism is possible, that is, polymorphic types parametrised by a polymorphic type rather than a monomorphic type as above. Deep polymorphism is usually not implemented in functional languages.

The second step is to identify a function type signature

$$f : A_1 \to A_2 \to \cdots \to A_k \to B$$

with a protological sequent

$$(\alpha, \ldots \beta, x_1, \ldots x_k)\; A_1 x_1, \; \ldots A_k x_k \to B(f x_1 \cdots x_k),$$

where $\alpha, \ldots \beta$ are the free type variables in $A_1, \ldots A_k, B$. This works with any kind of types $A_1, \ldots A_k, B$, except for function types. (Higher-order function types, therefore, cannot be represented using protologic, although of course higher-order functions can be defined. If one really wanted higher-order function types one would have to use the protological interpretation of predicate calculus in Part III.) For example, suppose we define a polymorphic function $extract : treeof(\alpha) \to listof(\alpha)$ that selects the items of type α from a binary tree and makes a list of them:

$$\begin{aligned} extract(empty) &\triangleq nil, \\ extract(tip(n)) &\triangleq [n], \\ extract(node(t_1, t_2)) &\triangleq concat(extract(t_1), extract(t_2)), \end{aligned}$$

where *concat* concatenates two lists. In this case the type signature means

$$(\alpha, t)\; treeof(\alpha)(t) \to listof(\alpha)(extract(t)).$$

The third step is to consider type checking as a protological derivation process. For example, to check the type signature $extract : treeof(\alpha) \to listof(\alpha)$ we apply the general structural induction rule for pattern matching, introduced in the final exercise of Chapter 19, to the type $treeof(\alpha)$, giving

$$\frac{\left\{\begin{array}{l} (\alpha) \to B\begin{bmatrix} empty \\ x \end{bmatrix} \\ (\alpha, n)\; \alpha(n) \to B\begin{bmatrix} tip(n) \\ x \end{bmatrix} \\ (\alpha, t_1, t_2)\; B\begin{bmatrix} t_1 \\ x \end{bmatrix}, B\begin{bmatrix} t_2 \\ x \end{bmatrix} \to B\begin{bmatrix} node(t_1, t_2) \\ x \end{bmatrix} \end{array}\right.}{(\alpha, t)\; treeof(\alpha)(t) \to B\begin{bmatrix} t \\ x \end{bmatrix}}$$

If we take B as $listof(\alpha)(extract(x))$ in this rule then the conclusion is the desired sequent and the premises are

$$(\alpha) \rightarrow listof(\alpha)(extract(empty))$$
$$(\alpha, n)\ \alpha(n) \rightarrow listof(\alpha)(extract(tip(n)))$$
$$(\alpha, t_1, t_2)\ listof(\alpha)(extract(t_1)),\ listof(\alpha)(extract(t_2)) \rightarrow listof(\alpha)(extract(node(t_1, t_2)))$$

of which the first two are derived by Reduction and the third is derived by Instantiation and Reduction from

$$(\alpha, l_1, l_2)\ listof(\alpha)(l_1),\ listof(\alpha)(l_2) \rightarrow listof(\alpha)(concat(l_1, l_2))$$

which is derivable by the general structural induction rule, applied to $listof(\alpha)$ (cf the List Rule in Chapter 19). Clearly this derivation procedure is determined by the syntactic structure of the definitions of $treeof(\alpha)$ and $listof(\alpha)$, and hence type checking is decidable. It would be possible to define general type checking and type inference algorithms for ET, resembling the ones used for Standard ML, Miranda and Haskell; however, I prefer to take a more general view and define type checking as any decidable subsystem of protologic, to be chosen by the language implementor or programmer.

Overloaded functions are easily implemented in ET. For example, suppose we define an *integer* type by

$$integer(int(m, n)) \triangleq num(m)\ \&\ num(n)$$

where *int* is a fresh 1-ary constructor. (In other words, an integer is essentially a pair of natural numbers.) Then we may wish to define functions *equal* and *plus* that apply both to natural numbers and integers (*equal* and *plus* are said to be 'overloaded'). We define *plus* first and then *equal*.

$$\begin{aligned}
plus(0, n) &\triangleq n, \\
plus(Sm, n) &\triangleq S(plus(m, n)), \\
plus(int(m, n), int(p, q)) &\triangleq int(plus(m, p), plus(n, q)) \\
equal(0, 0) &\triangleq true, \\
equal(0, Sn) &\triangleq false, \\
equal(Sm, 0) &\triangleq false, \\
equal(Sm, Sn) &\triangleq equal(m, n), \\
equal(int(m, n), int(p, q)) &\triangleq equal(plus(m, q), plus(n, p)).
\end{aligned}$$

The example can be extended to include rational numbers, the *rational* type being defined by

$$\begin{aligned} rational(rat(p, int(n,n))) &\triangleq false, \\ rational(rat(p,q)) &\triangleq integer(p) \;\&\; integer(q) \end{aligned}$$

where *rat* is a fresh 1-ary constructor. It is left as an exercise for the reader to define *equal*, *plus*, *minus*, *times* and *divide* as overloaded functions applicable to natural numbers, integers and rationals. (The use of 'laws', as in the first version of Miranda, to reduce integers and rationals to a unique representation (Davie, 1992, §4.9), seems unnecessary.)

A feature of complex type systems that is not usually implemented in functional languages (because it makes type checking undecidable) is *dependent* types. For example, consider a type *vector*(n) of n-dimensional vectors (expressed as tuples): in ordinary type notation, $vector(n) = \underbrace{real \times \cdots \times real}_{n \text{ times}}$.

This can be defined in ET by

$$\begin{aligned} vector(n)v &\triangleq vector'(n,v), \\ vector'(1,x) &\triangleq real(x), \\ vector'(S(Sn),(x,l)) &\triangleq real(x) \;\&\; vector(Sn,l), \end{aligned}$$

assuming the type *real* is already defined.

To conclude, the whole conventional type apparatus of functional programming, with the exception of higher-order function types, can be easily interpreted as metanotation in ET. The advantages of handling types in this way are as follows:

- there is no need to complicate the syntax with type variables, type declarations, or special notations for polymorphism or overloading;
- extensions to the type system may be made without fear of disrupting the existing type theory (for example, one might worry that deep polymorphism introduces self-referential paradoxes), since the semantics is taken care of by the translation into ET;
- types are constructions, so they can be manipulated in arbitrary ways (in the usual jargon, types are 'first-class citizens');
- a type discipline, requiring explicit type declarations and adherence to a limited set of types, can be enforced through a type-checking preprocessor, when desired;
- alternatively, the type system can be extended when desired to a more powerful (and undecidable) system to support correctness proofs, as in Martin-Löf's (1975) type theory.

The type system can be as restricted or as flexible as one likes, since the full power of protologic is available. Any type system involves a trade-off between flexibility and security, and the optimal trade-off will be different for different software development projects. Hence the application programmer should be free to load a suitable type-checking module for a particular application, or even different ones for different parts of a large system, just as one loads different library modules for different applications.

PATTERN MATCHING

One feature that ET shares with most other functional languages is the ability to define functions by pattern matching. The purpose of pattern matching is to allow function definitions to be written in a declarative equational style, avoiding the use of conditional expressions. Languages vary in the versatility of their pattern-matching mechanisms. In Hope, for example, one cannot rely on the patterns being applied in order, and hence one needs to avoid 'overlapping' patterns (that is, two patterns in one definition that can match the same argument), while in Standard ML no variable may occur more than once in a pattern. Consider for example the definition of the *member* function, which tests whether a given x is a member of a given list. In ET the definition is

$$\begin{aligned} member(x, nil) &\triangleq false, \\ member(x, (x, rest)) &\triangleq true, \\ member(x, (first, rest)) &\triangleq member(x, rest) \end{aligned}$$

the clauses being applied sequentially (see Chapter 15 for discussion of this and other examples). In Hope or Standard ML the definition would have to be rewritten as

$$\begin{aligned} member(x, nil) &\triangleq false, \\ member(x, (first, rest)) &\triangleq \textbf{if } x = first \textbf{ then } true \textbf{ else } member(x, rest) \end{aligned}$$

(using improvised syntax). Since the purpose of pattern matching is to replace conditional expressions with equations covering the separate cases, the rewritten version is surely undesirable. Overlapping patterns are controversial (Moor, 1987). The most important objection to them is that they make the semantics of the function definition dependent on the order in which the clauses are applied and not just on the meaning of the individual clauses as equations; thus they appear to undermine the declarative nature of the definition.

In ET this objection can be countered using the formal theory of pattern matching. The semantics of a pattern-matching function definition

$$
\begin{aligned}
f\,X_1 &\triangleq T_1,\\
&\vdots\\
f\,X_k &\triangleq T_k
\end{aligned}
$$

is given by Theorems 41 and 42 of the Expanded Term Language and by Theorem 20 of Expanded Protologic, in which

$$(u, x, \underline{z})\, \mathit{fxpt}\, \Phi x = u,\ \Gamma \to Ax = u$$

is derived from the premises

$$(u, \underline{v_i}, \underline{z})\ T_i\left[\begin{smallmatrix}A\\f\end{smallmatrix}\right] = u,\ \Gamma \to AX_i = u$$

for $i = 1, \ldots k$. Now, admittedly there is nothing in this rule that takes account of the fact that the clauses $f\,X_i \triangleq T_i$ are applied in order. However, the rule could be strengthened by writing the ith premise as

$$(u, \underline{v_i}, \underline{z})\, \mathit{match}_{X_1}(X_i){=}\mathit{false},\ \ldots\ \mathit{match}_{X_{i-1}}(X_i){=}\mathit{false},\ T_i\left[\begin{smallmatrix}A\\f\end{smallmatrix}\right]{=}u,\ \Gamma \to AX_i{=}u$$

(the proof in Chapter 19 is easily modified). This expresses in a purely declarative way the fact that the ith clause $f\,X_i \triangleq T_i$ only applies to arguments matching X_i but not matching $X_1, \ldots X_{i-1}$. I did not write the rule in this form in Expanded Protologic because I thought it was complicated enough as it was and the stronger form was not needed at the time.

A further variation on pattern-matching is that Haskell allows *guards* in the defining clauses. For example, the factorial function could be defined in Haskell by

```
fac n | n > 0 = n * fac(n-1)
fac 0 = 1
```

where the guard $n > 0$ must be satisfied if the first clause is to apply. A similar facility could easily be added to the pattern-matching notation of ET: one simply replaces match_n by ($\mathit{branch}\ \mathit{match}_n\ (\lambda n.n > 0)\ (k\ \mathit{false})$), or, in general, match_{X_i} by ($\mathit{branch}\ \mathit{match}_{X_i}\ (\lambda X_i.G_i)\ (k\ \mathit{false})$), where G_i is the guard condition on pattern X_i.

LAZY DATA STRUCTURES, STREAMS AND INTERACTIVE PROGRAMMING

Some functional languages (Miranda and Haskell, for example) use *lazy evaluation*, a method of evaluating expressions that is similar to normal order reduction in λ-calculus except that repeated sub-expressions are only evaluated once. Other languages (Hope, FP, FL and Standard ML) use *strict evaluation*, in which arguments are fully evaluated before functions are applied to them, although Hope does provide lazy lists and an intensional conditional expression. LISP is normally strict but allows evaluation to be suppressed for some function arguments.

Lazy evaluation is considered by some (Hughes, 1990) to be an essential feature of functional programming: its most important application is lazy lists, potentially infinite data structures that can be used to represent input or output streams in interactive programs, and that can allow complex functions to be decomposed cleanly into several pipelined processes. In intuitionistic terminology, infinite lazy lists are simply choice sequences and stream functions (such as *Add* and *Dup* in the examples below) are continuous functions, so it is highly desirable that this style of programming should be implementable in a language that is intended to serve as the foundation for intuitionistic mathematics.

In Chapter 13 I argued in favour of strict evaluation, on semantic grounds, so it might appear that lazy programming is not possible in ET. However, it is easy to implement the useful lazy facilities in a strict language. For example, an intensional conditional expression **if** *C* **then** *A* **else** *B*, in which *C* is evaluated, then either *A* or *B* depending on whether the value of *C* is *true* or *false*, can be defined in strict languages using 'functions of no arguments': the ET version is

$$if\ C\ (\lambda nil.A)\ (\lambda nil.B)\ nil.$$

I did not define such a metanotation in ET because it does not appear to be needed: the strict (*branch C F G*) construct (which is identical to the FP construct $C \rightarrow F\ ;\ G$) is better suited to the contexts in which conditionals are used (see Chapter 15 for examples).

Lazy lists can be implemented in ET in the same way as in other strict languages such as Standard ML (Myers, Clack & Poon, 1993, Appendix F). Define the metanotation

$$(A : L) \quad \text{is} \quad (\lambda nil.(A, L))$$

and two functions

$$head\ x \triangleq left(x\ nil),$$

$$tail\ x \triangleq right(x\ nil).$$

EXERCISE. Verify that for any terms A and L,

- $(A : L) \not\triangleright$, $(A : L)nil \;\triangleright^*\; (A, L)$,
- $head(A : L) \;\triangleright^*\!\triangleleft\; A$ and $tail(A : L) \;\triangleright^*\!\triangleleft\; L$, if A and L are evaluable.

As a simple example, define the lazy list *from n* by

$$from\ n \stackrel{\Delta}{=} (n : from(Sn)).$$

EXERCISE. Verify that

- $from \not\triangleright$, and $from\ n$ and $from(Sn)$ are evaluable,
- $head(from\ n) \;\triangleright^*\; n$ and $tail(from\ n) \;\triangleright^*\!\triangleleft\; from(Sn)$,

and hence that $head(tail^K(from\ N))$ reduces to $N + K$, for any numbers N, K (where $tail^K\ T$ is a metanotation for $\underbrace{tail(tail(\cdots(tail}_{K \text{ times}}\ T)\cdots)))$.

Stream functions, that is functions that act incrementally on lazy lists, may be defined in the usual way. For example, consider a function *Add* that receives as input a stream of numbers $n_0, n_1, n_2, n_3, \ldots$ and has to output the sums of consecutive pairs $n_0 + n_1, n_2 + n_3, \ldots$. This is implemented in ET by

$$Add\ l \stackrel{\Delta}{=} (plus(head\ l, head(tail\ l)) : Add(tail(tail\ l))).$$

EXERCISE. Verify that

- $Add \not\triangleright$,
- $Add\ l$ is evaluable, and hence $Add\ X$ is evaluable if X is,
- $head(Add\ X) \;\triangleright^*\!\triangleleft\; plus(head\ X, head(tail\ X))$ and $tail(Add\ X) \;\triangleright^*\!\triangleleft\; Add(tail(tail\ X))$, if $head\ X$ and $head(tail\ X)$ reduce to numbers and $tail(tail\ X)$ is evaluable.

Given as input a lazy list L, and assuming that L is irreducible and that $head(tail^K\ L)$ reduces to a number A_K for each number K, it then follows that $head(tail^K(Add\ L))$ reduces to $A_{2K} + A_{2K+1}$, for any number K. As a second example of a stream function, consider a function *Dup* that duplicates its inputs: given $n_0, n_1, n_2, \ldots$ it outputs $n_0, n_0, n_1, n_1, n_2, n_2, \ldots$. This is defined in ET by

$$Dup\ l \stackrel{\Delta}{=} (head\ l : (head\ l : Dup(tail\ l))).$$

EXERCISE. Verify that

- *Dup* $\not\triangleright$,
- *Dup l* is evaluable, and hence *Dup X* is evaluable if X is,
- $head(Dup\, X) \; \triangleright^* \triangleleft \; head\, X$ if $head\, X$ is evaluable,
- $head(tail(Dup\, X)) \; \triangleright^* \triangleleft \; head\, X$ and $tail(tail(Dup\, X)) \; \triangleright^* \triangleleft \; Dup(tail\, X)$, if $head\, X$ and $tail\, X$ are evaluable.

These techniques can be applied to interactive programming. Suppose we connect the output stream of *Add* to the input stream of *Dup* and connect the output stream of *Dup* to the input stream of *Add*. To start the interaction, let us put 1 on the input stream of *Dup*. This can be specified informally by

$$\begin{aligned} stream_1 &= (1 : Add(stream_2)), \\ stream_2 &= Dup(stream_1). \end{aligned}$$

We can eliminate the mutual recursion and express this formally in ET as

$$stream_1\; nil \;\triangleq\; (1, Add(Dup(stream_1))).$$

EXERCISE. Verify that

- $head(tail^K\, stream_1)$ reduces to 2^K,
- $tail^{K+1}\, stream_1$ is evaluable,
- $tail^{K+1}\, stream_1 \; \triangleright^* \triangleleft \; Add(Dup(tail^K\, stream_1))$,

by induction on K (take the conjunction of the three statements as inductive hypothesis). This shows that $stream_1$ consists of powers of two, $[1, 2, 4, 8, \ldots]$.

If one of the processes *Add* or *Dup* were replaced by a human user this would represent a model of interactive input-output, with the input and output appropriately synchronised. The above calculations show how from assumptions about the user's behaviour we can deduce the course of the interaction.

This examples indicate how the advantages of lazy programming can be achieved in a strict language without sacrificing the semantic clarity of strict evaluation.

MODULARITY AND PROGRAMMING IN THE LARGE

A practical programming language needs a mechanism for dividing a complex program into self-contained pieces, called *modules*, that interact with each other through narrowly defined interfaces. Within each module data structures and algorithms are encapsulated and may not be accessed from outside the module except through the interface; hence the interface of a module with the rest of the program serves as a specification for its behaviour. Modules should be stored in separate files, with the module specification and implementation in different files to allow separate compilation when a module implementation is changed. A common kind of module is an *abstract data type*, which defines a single data type together with functions for creating and manipulating instances of the type.

Several functional languages (Miranda, Hope and Haskell) provide a crude kind of modularity. However, the module specification and implementation are not separated into different files, and sometimes an unnecessary distinction is drawn between abstract data types and other modules.

Standard ML has a more developed module system, in which *signatures* play the role of module specifications and *structures* play the role of module implementations.

A modular organisation is used throughout most of this book, with the 'theory' chapters being module specifications and the 'intermediate' chapters being module implementations (see Chapter 12). However, I have not as yet described any formal module notation for ET. Here is an example of how modules in ET would look. Consider the problem of implementing a stack as an abstract data type, where the elements of the stack may be of any one type specified by the user. The specification would look as follows.

module specification *stacks*;
  export *stackof*, *empty*, *push*, *pop*, *top*, *null*;
  $(\alpha) \rightarrow stackof(\alpha)(empty)$
  $(e,s,\alpha)\ \alpha(e),\ stackof(\alpha)(s) \rightarrow stackof(\alpha)(push(e,s))$
  $(s,\alpha)\ stackof(\alpha)(s) \rightarrow stackof(\alpha)(pop(s))$
  $(s,\alpha)\ stackof(\alpha)(s),\ null(s) = false \rightarrow \alpha(top(s))$
  $(s,\alpha)\ stackof(\alpha)(s) \rightarrow boolean(null(s))$
  $(e,s,\alpha)\ \alpha(e),\ stackof(\alpha)(s) \rightarrow pop(push(e,s)) = s$
  $(e,s,\alpha)\ \alpha(e),\ stackof(\alpha)(s) \rightarrow top(push(e,s)) = e$
  $\rightarrow null(empty)$
  $(e,s,\alpha)\ \alpha(e),\ stackof(\alpha)(s) \rightarrow null(push(e,s)) = false$
end *stacks*.

The module implementation would contain definitions of *stackof*, *empty*, *push*, *pop*, *top* and *null*, and protological derivations of the sequents in the module specification. Any other module that wished to use, say, a stack of natural numbers would import the *stacks* module and use the type *stackof*(*num*) with instances such as *empty*, *push*(5, *empty*) and *push*(3, *push*(7, *empty*)).

Note that the module specification contains both sequents expressing type signatures (such as (s,α) $stackof(\alpha)(s) \rightarrow stackof(\alpha)(pop(s)))$ and sequents expressing correctness (such as (e,s,α) $\alpha(e)$, $stackof(\alpha)(s) \rightarrow pop(push(e,s)) = s$) and thus forms a complete specification of the abstract data type – unlike module interfaces in most other programming languages, where only type signatures are provided.

This shows that ET lends itself naturally to modular programming. It may also be possible to implement the *functors* of Standard ML or the *parameterised modules* of OBJ (Goguen, 1990) in ET. An object-oriented style could certainly be supported.

EFFICIENCY OF EXECUTION

Functional languages tend to be inefficient in terms of execution time, due to the very high level of the languages, their reliance on heap storage and hence the frequent need to pause for garbage collection, the lack of updatable data structures, and the complexities of implementing lazy evaluation. These difficulties (except for the last) also apply to ET. The simplest way of implementing ET is to compile ET terms into terms of T (the simple Term Language) and then to implement reduction of the T terms using an interpreter. This method works, but produces extremely large T terms, arising from the compilation of nested λ-abstractions, which are extremely slow to run. The size of the T terms can be reduced by introducing new combinators. Some common choices (Diller, 1988) are B, C, W, B', C', S', where

$$
\begin{array}{rlrlrl}
Bxyz & \rhd\ x(yz) & Cxyz & \rhd\ xzy & Wxy & \rhd\ xyy \\
B'wxyz & \rhd\ wx(yz) & C'wxyz & \rhd\ w(xz)y & S'wxyz & \rhd\ w(xz)(yz).
\end{array}
$$

For example, this allows $(\lambda x.w(xy))$ to compile to $Bw(C\,id\,y)$ rather than $s(kw)(s\,id\,(ky))$. A more powerful set of combinators s_n, k_n, id_n^i, u_n may be introduced as metanotation of T, as follows. For $n{=}0,1,2,\dots$ and $i{=}1,\dots n$,

$$
\begin{array}{rll@{\qquad}rll}
s_0XY & \text{is} & XY & s_{n+1}XY & \text{is} & s_n(s_n(k_ns)X)Y \\
k_0X & \text{is} & X & k_{n+1}X & \text{is} & s_n(k_nk)(k_nX) \\
id_{n+1}^{n+1} & \text{is} & k_nid & id_{n+1}^i & \text{is} & s_n(k_nk)id_n^i \\
u_nX & \text{is} & \multicolumn{4}{l}{s_{n+1}(s_{n+1}(s_n(k_nk)X)(k_nformer))(k_nlatter).}
\end{array}
$$

This gives, for any atoms $a_1, \dots a_n$, writing $(\lambda a_1.(\lambda a_2.\cdots(\lambda a_n.T)\cdots))$ as

$(\lambda \underline{a}.T)$,

$$\begin{aligned}
(\lambda \underline{a}.XY) &\overset{*}{\mapsto} s_n(\lambda \underline{a}.X)(\lambda \underline{a}.Y), \\
(\lambda \underline{a}.b) &\overset{*}{\mapsto} k_n b \quad \text{if } b \text{ is a constant, or a variable but not } a_1, \ldots a_n, \\
(\lambda \underline{a}.a_i) &\overset{*}{\mapsto} id_n^i \quad \text{if } a_i \text{ is a variable and not } a_{i+1}, \ldots a_n, \\
& \qquad\quad \text{for } i = 1, \ldots n, \\
(\lambda \underline{a}.(\lambda XY.T)) &\overset{*}{\mapsto} u_n(\lambda \underline{a}.(\lambda X.(\lambda Y.T))).
\end{aligned}$$

These combinators allow nested λ-abstractions to be expressed concisely.

EXAMPLE.

$$\begin{aligned}
(\lambda xy(uv).vuxk) &\overset{*}{\mapsto} u_0(\lambda xy.(\lambda uv.vuxk)) \overset{*}{\mapsto} u_0(u_0(\lambda x.(\lambda y.(\lambda uv.vuxk)))) \\
&\overset{*}{\mapsto} u_0(u_0(u_2(\lambda x.(\lambda y.(\lambda u.(\lambda v.vuxk)))))) \\
&\overset{*}{\mapsto} u_0(u_0(u_2(s_4(s_4(s_4(id_4^4\ id_4^3)id_4^1)(k_4 k))))).
\end{aligned}$$

EXERCISE. Derive formulae for the number of atoms in s_nXY, k_nX, id_n^i and u_nX in terms of the number of atoms in X and Y, and hence show that the term in the previous example has 302 atoms when compiled.

EXERCISE. Derive simple reduction rules for s_nXY, k_nX, id_n^i and u_nX.

Another way of mitigating the problem of enormous T terms is what might be called *lazy compilation*, that is, only compiling an ET term as far as is necessary for a particular computation. An example is

$$(\lambda x.(\lambda z.xxx)y) = (\lambda w.www) \overset{*}{\mapsto} s(\lambda x.(\lambda z.xxx))(ky) = s(\lambda w.ww)id \ \triangleright \ \mathit{false}.$$

In this case the two λ-abstractions have only been compiled as far as is necessary to determine that their final components (ky and id) are different and hence the equality can reduce to *false* without compiling further.

In a real implementation of ET one would also wish to implement certain metanotational constructs in a special way, particularly pattern matching and lazy lists, rather than simply rewriting them into ET syntax. The definitions given in this book for ET metanotation, the compilation mapping from ET to T, and the T reduction mechanism, should be regarded as a formal semantics rather than a practicable implementation method.

PART III: THE INTERPRETATION OF ARITHMETIC

CHAPTER 24

INTRODUCTION TO PART III

The Theory of Constructions can be used to give an interpretation of first-order number theory. First-order number theory consists of formulae built up from atomic formulae using logical connectives and quantifiers, where the atomic formulae express the results of primitive recursive computations. The standard formalisation of this theory is Peano Arithmetic.

The key idea in the interpretation is that the meaning of a number-theoretic formula is given by what constructions count as proofs of it. In contrast with classical semantics, where the notion of proof is defined in terms of a prior notion of truth, in intuitionism the semantics of a formula is given entirely by the proof relation, $\vdash$. As I argue in Chapter 6, the notion of truth does not enter into the semantics at all.

Formulae, then, are interpreted as *tests for constructions* or, as I shall say, as *proof functions* (functions testing whether a given construction is to count as a proof).

As argued in Chapter 10, the proof relation is undecidable; however this does not rule out a formal theory shedding some light on it. Every proof is a tree, under some encoding of trees as constructions. Every proof function has two components: a decidable test A to be applied to constructions, and a coding Π of a class of trees as constructions. For a construction to succeed as a proof it must pass the A test and it must represent a *well-founded* tree, under the coding Π. The division of labour between these two components of the proof function, A and Π, is such that if all the atomic formulae of the given formula were replaced by *true* then the A test would be trivially satisfiable, while the Π component would be unchanged. The tree coding Π depends on the logical structure of the formula in terms of connectives and quantifiers but is independent of the atomic formulae; whereas the A component looks after the atomic formulae but ignores questions of well-foundedness.

The following mathematical chapters are organised into theories and intermediate chapters (in the manner explained in Chapter 12), carrying straight on from the Coding of Trees theory in Part II. There are five theories, as follows.

- *Logic* (L) contains definitions of the basic logical constants, which are constructions acting on proof functions. Previous accounts of intuitionistic logic have aped classical logic in taking $\wedge$, $\vee$, $\supset$, $\exists$ and $\forall$ as logical primitives (though it is sometimes pointed out that $\vee$ can be defined in terms of $\exists$ (Dummett, 1977, §2.1), and Kreisel (1962) has speculated that there may be other intuitionistic propositional connectives than the usual ones). I depart from this tradition by starting with a different and more expressive supply of logical constants, *val*, *con*, *predify*, $\wedge$, $\exists$, $\Box$ and Δ. The intuitionistic proof relation $\vdash$ is defined in this theory as a relation between terms and proof functions.

- The *Calculus of Proof Functions* (CPF) formalises the relations between proof functions. It is an axiomatic theory dealing with sequent expressions of the form $I, \ldots J \Rightarrow K$; such a sequent means roughly that a proof of the proof function K may be obtained from proofs of the proof functions $I, \ldots J$. In particular, if $\Rightarrow K$ is derivable in CPF then K has an intuitionistic proof.

- *Logic of Partial Terms* (LPT) is an axiomatic theory essentially consisting of first-order intuitionistic predicate calculus with equality and induction, adapted to dealing with terms that may not have values. LPT is expressed wholly in terms of logical formulae rather than proof functions. The usual logical constants, $\vee$, $\supset$ and $\forall$, are defined in terms of the primitive logical constants (via another logical constant, denoted '$\bullet$'). An unexpected feature of these definitions is that two notions of implication emerge: there is one logical constant, $\supset$, which obeys the usual introduction and elimination rules and therefore truly deserves the name 'implication', and there is a stronger logical constant, $>$ ('super-implication'), which corresponds to the sequent arrow $\Rightarrow$ of CPF and which is closer to the traditional intuitionists' definition of implication. Of the two, $>$ is more primitive: $\supset$ is defined in terms of $>$. Every formula of LPT may be interpreted as a proof function; every theorem of LPT corresponds to a theorem of CPF and therefore has an intuitionistic proof.

- *Heyting Arithmetic* (HA) emerges as a restriction of LPT, in which the variables range over numbers and all terms are built out of primitive recursive functions. This gives the intended interpretation of intuitionistic arithmetic; the axioms all have proofs, the rules of inference all preserve provability, and intuitionistically unsound modes of reasoning, such as Markov's Principle, are unprovable.

- *Peano Arithmetic* (PA) is HA plus the Principle of Excluded Middle, and is interpreted in HA using a Gödel-Gentzen double-negation interpretation. This is obviously not the intended interpretation of PA, but it is the best interpretation an intuitionist can give to it.

Thus the outcome is an interpretation of Peano Arithmetic showing how every theorem has an intuitionistic proof. It is traditional in intuitionism to take HA rather than PA as the formalisation of number theory, but the proof-theoretic difference between them is so slight that I prefer to go all the way to PA. This may help to dispel the widespread impression that intuitionism is 'classical mathematics minus the Excluded Middle'.

CHAPTER 25

FROM THE CODING OF TREES TO LOGIC

In intuitionism a formula of predicate calculus is interpreted as a *proof function* (that is, a function that tests whether a given construction is to count as a proof), and will be represented here as a term, I. If $IX \triangleright^* \mathit{true}$ then I shall say that X passes the test, or is a *proof* of I. If this were the whole story then proofhood would be decidable. So to take account of the undecidable well-foundedness aspect of proof (Chapter 10) I is required to consist of a decidable part A and a tree coding Π. The definition of proof is amended to read: X is a proof of I iff $AX \triangleright^* \mathit{true}$ and $X : \Pi$. (Thus every proof is a tree.)

As a special case, suppose A is of the form $(\lambda \mathit{nil}.T)$ and Π is *leaf*; then *any* construction will count as a proof of I if $T \triangleright^* \mathit{true}$, and *no* construction will count otherwise. Such a proof function I may be regarded as representing the *atomic formula* T. This corresponds to the intuitionistic principle that atomic formulae are decidable and so have a truth value without need of proof.

A *predicate* is something that associates a proof function with any construction x. For simplicity I shall always assume that the Π part of the proof function does not depend on x.

The *logical constants* are functions that build a new proof function or predicate out of given proof functions or predicates. In this chapter I shall define formally the notions of proof function and predicate, introduce atomic formulae and the basic logical constants (*val*, *con*, *predify*, $\wedge$, $\exists$, $\Box$ and $\triangle$), and prove their properties.

PROOF FUNCTIONS, PROOFS AND PREDICATES

DEFINITION. Let *pfn* be a fresh 1-ary constructor. A *proof function* is a term I such that $I \triangleright^* \mathit{pfn}(A,\Pi) \not\triangleright$, for some term A and type symbol Π.

DEFINITION. Let *pred* be a fresh 1-ary constructor. A *predicate* is a term P such that $P \triangleright^* \mathit{pred}(F,\Pi) \not\triangleright$, where Fx is evaluable for any variable x and Π is a type symbol.

I shall use the letters 'I' and 'J' to denote proof functions and the letter 'P' to denote a predicate.

THEOREM 1. Let X be an irreducible term. If I is a proof function then so is $I\begin{bmatrix}X\\x\end{bmatrix}$, and if P is a predicate then so is $P\begin{bmatrix}X\\x\end{bmatrix}$.

Proof. First, suppose I is a proof function and let $I \rhd^* \mathit{pfn}(A,\Pi) \not\rhd$. Then $I\begin{bmatrix}X\\x\end{bmatrix} \rhd^* \mathit{pfn}(A\begin{bmatrix}X\\x\end{bmatrix},\Pi) \not\rhd$, which is also a proof function.

Secondly, suppose P is a predicate, let $P \rhd^* \mathit{pred}(F,\Sigma) \not\rhd$ and let u be a fresh variable; thus Fu is evaluable. Now, $P\begin{bmatrix}X\\x\end{bmatrix} \rhd^* \mathit{pred}(F\begin{bmatrix}X\\x\end{bmatrix},\Sigma) \not\rhd$, where, for any variable v, $F\begin{bmatrix}X\\x\end{bmatrix} v \overset{*}{\leftrightarrow} (Fu)\begin{bmatrix}X\\x\end{bmatrix}\begin{bmatrix}v\\u\end{bmatrix}$, which is evaluable since Fu is. This shows that $P\begin{bmatrix}X\\x\end{bmatrix}$ is a predicate. ∎

DEFINITION. The relation $\vdash$, between terms and proof functions, is defined by:

$$X \vdash I \quad \text{iff} \quad I \rhd^* \mathit{pfn}(A,\Pi) \not\rhd \,,\; AX \rhd^* \mathit{true} \quad \text{and} \quad X : \Pi.$$

If $X \vdash I$ holds then X is said to be an *intuitionistic proof* (or just a *proof*) of I.

The following theorem verifies the familiar intuitionistic principle that possessing a method for obtaining a proof is tantamount to possessing a proof.

THEOREM 2. (Extensionality of $\vdash$.) If $X \rhd^*\lhd Y$ and I is a proof function then $X \vdash I$ iff $Y \vdash I$. Moreover if $I \rhd^*\lhd J$ then $X \vdash I$ iff $X \vdash J$.

Proof. Immediate from the definition and the well-foundedness reduction rule. ∎

THEOREM 3. If $Q \vdash I$ and $X \not\rhd$ then $Q\begin{bmatrix}X\\x\end{bmatrix} \vdash I\begin{bmatrix}X\\x\end{bmatrix}$.

Proof. Let $I \rhd^* \mathit{pfn}(A,\Pi) \not\rhd$. The hypothesis that $Q \vdash I$ means that $AQ \rhd^* \mathit{true}$ and $Q : \Pi$. Hence $A\begin{bmatrix}X\\x\end{bmatrix} Q\begin{bmatrix}X\\x\end{bmatrix} \rhd^* \mathit{true}$ and $Q\begin{bmatrix}X\\x\end{bmatrix} : \Pi$. Thus $Q\begin{bmatrix}X\\x\end{bmatrix} \vdash \mathit{pfn}(A\begin{bmatrix}X\\x\end{bmatrix},\Pi) \lhd^* I\begin{bmatrix}X\\x\end{bmatrix}$. ∎

THE BASIC LOGICAL CONSTANTS

DEFINITION. (Atomic formulae.) For any term T let $\ulcorner T \urcorner$ be the proof function $\mathit{pfn}((\lambda \mathit{nil}.T), \mathit{leaf})$.

THEOREM 4. (Conservativeness of $\vdash$ over reduction.) Let T be any term, let $\underline{x}$ be any variables, and let $\underline{X}$ be any irreducible terms.

- If Q is a construction then $Q \vdash \ulcorner T \urcorner \left[\frac{X}{\underline{x}}\right]$ iff $T\left[\frac{X}{\underline{x}}\right] \rhd^*$ *true*.
- If Q is any term such that $Q \vdash \ulcorner T \urcorner \left[\frac{X}{\underline{x}}\right]$ then $T\left[\frac{X}{\underline{x}}\right] \rhd^*$ *true*.

Proof. Note that $(\lambda nil.T)\left[\frac{X}{\underline{x}}\right] Y \rhd^* T\left[\frac{X}{\underline{x}}\right]$, for any irreducible term Y, since $(\lambda nil.T)\left[\frac{X}{\underline{x}}\right] Y \overset{*}{\leftrightarrow} ((\lambda nil.T)y)\left[\frac{X}{\underline{x}}\right]\left[\frac{Y}{y}\right] \rhd^* T\left[\frac{X}{\underline{x}}\right]\left[\frac{Y}{y}\right] \overset{*}{\leftrightarrow} T\left[\frac{X}{\underline{x}}\right]$, where y is a fresh variable.

To prove the first part of the theorem, $Q \vdash \ulcorner T \urcorner \left[\frac{X}{\underline{x}}\right]$ iff $(\lambda nil.T)\left[\frac{X}{\underline{x}}\right] Q \rhd^*$ *true* and $Q : leaf$, and this is the case iff $T\left[\frac{X}{\underline{x}}\right] \rhd^*$ *true*.

For the second part, if $Q \vdash \ulcorner T \urcorner \left[\frac{X}{\underline{x}}\right]$ then $(\lambda nil.T)\left[\frac{X}{\underline{x}}\right] Q \rhd^*$ *true*, so $Q \rhd^* Q' \not\rhd$ for some Q' and $(\lambda nil.T)\left[\frac{X}{\underline{x}}\right] Q' \rhd^*$ *true*, so $T\left[\frac{X}{\underline{x}}\right] \rhd^*$ *true*. ∎

Next we need some functions relating proof functions to predicates. The 'value' of a predicate $pred(F,\Pi)$ at an 'argument' x is the proof function $pfn(Fx,\Pi)$; note that the decidable part Fx depends on x but the tree coding Π never does. Conversely, given a proof function $pfn(A,\Pi)$ and a variable x we can obtain a predicate $pred((\lambda x.A),\Pi)$ by λ-abstraction. In the special case where the proof function does not depend on x it is easier to form the predicate $pred(kA,\Pi)$ instead. These three operations will be accomplished by the functions *val*, *predify* and *con* respectively.

DEFINITION. Define the constructions *val* and *con* by

$$val \text{ is } (\lambda pred(f,u).(\lambda x.pfn(fx,u)))$$
$$con \text{ is } (\lambda pfn(a,u).pred(ka,u))$$

where f, u, x, a are four variables.

THEOREM 5. If P is a predicate and $X \not\rhd$ then $val\,P\,X$ is a proof function. If I is a proof function and $X \not\rhd$ then $con\,I$ is a predicate and $val(con\,I)X \rhd^* \lhd I$.

Proof. Let $P \rhd^* pred(F,\Pi) \not\rhd$, and let x be a fresh variable. Then Fx is evaluable so $Fx \rhd^* B \not\rhd$ for some term B. Hence $val\,P\,X \rhd^* pfn(FX,\Pi) \rhd^* pfn(B\left[\frac{X}{x}\right],\Pi) \not\rhd$. This establishes that $val\,P\,X$ is a proof function.

Let $I \rhd^* pfn(A,\Sigma) \not\rhd$. Then $con\,I \rhd^* pred(kA,\Sigma) \not\rhd$, and $kAx \rhd^* A \not\rhd$, for any variable x, so $con\,I$ is a predicate. Moreover, $val(con\,I)X \rhd^* pfn(A,\Sigma) \lhd^* I$, as required. ∎

DEFINITION. Define a construction *predify* as

$$(\lambda t.pred(s(k(\lambda pfn(a,u).a))t, (\lambda pfn(a,u).u)(t\ nil)))$$

where t, a, u are three variables.

THEOREM 6. If I is a proof function and x is a variable then $predify(\lambda x.I)$ is a predicate and $val(predify(\lambda x.I))x \rhd^* \lhd\ I$.

Proof. Let $I \rhd^*\ pfn(A,\Pi) \not\rhd$; let a, u be two fresh variables. Then

$$\begin{aligned}
&predify(\lambda x.I)\\
&\rhd^*\ pred(s(k(\lambda pfn(a,u).a))(\lambda x.I), (\lambda pfn(a,u).u)((\lambda x.I)\ nil))\\
&\rhd^*\ pred(s(k(\lambda pfn(a,u).a))(\lambda x.I), (\lambda pfn(a,u).u)(pfn(A\left[{}^{nil}_{x}\right], \Pi)))\\
&\rhd^*\ pred(s(k(\lambda pfn(a,u).a))(\lambda x.I), \Pi) \not\rhd\ .
\end{aligned}$$

Also, for any variable y,

$$\begin{aligned}
s(k(\lambda pfn(a,u).a))(\lambda x.I)y\ &\rhd^*\ (\lambda pfn(a,u).a)((\lambda x.I)y)\\
&\rhd^*\ (\lambda pfn(a,u).a)(pfn(A\left[{}^{y}_{x}\right], \Pi))\ \rhd^*\ A\left[{}^{y}_{x}\right] \not\rhd\ ,
\end{aligned}$$

so $s(k(\lambda pfn(a,u).a))(\lambda x.I)y$ is evaluable. This confirms that $predify(\lambda x.I)$ is a predicate. In addition,

$$\begin{aligned}
val(predify(\lambda x.I))x\ &\rhd^*\ pfn(s(k(\lambda pfn(a,u).a))(\lambda x.I)x, \Pi)\\
&\rhd^*\ pfn((\lambda pfn(a,u).a)((\lambda x.I)x), \Pi)\\
&\rhd^*\ pfn((\lambda pfn(a,u).a)(pfn(A,\Pi)), \Pi)\\
&\rhd^*\ pfn(A,\Pi)\ \lhd^*\ I
\end{aligned}$$

as required. ∎

Next I shall define two familiar intuitionistic logical constants, conjunction ($\wedge$) and existential quantification ($\exists$).

DEFINITION. Define constructions $\wedge_d$, $\wedge_w$ and $\wedge$ by

$$\begin{aligned}
\wedge_d &\text{ is } (\lambda(a,b).(\lambda(q,r){:}\ aq\ \&\ br))\\
\wedge_w &\text{ is } product\\
\wedge &\text{ is } (\lambda(pfn(a,u), pfn(b,v)).pfn(\wedge_d(a,b), \wedge_w(u,v)))
\end{aligned}$$

where a, b, u, v, q, r are six variables.

THEOREM 7. If I and J are proof functions then so is $\wedge(I,J)$.

Proof. Let $I \ \rhd^* \ pfn(A,\Pi) \not\rhd$ and $J \ \rhd^* \ pfn(B,\Sigma) \not\rhd$; let q,r,a,b be four fresh variables. Then

$$\begin{aligned}\wedge(I,J) \ &\rhd^* \ pfn(\wedge_d(A,B), \wedge_w(\Pi,\Sigma)) \\ &\rhd^* \ pfn((\lambda(q,r){:}\, aq \,\&\, br)\begin{bmatrix}A,B\\a,b\end{bmatrix}, product(\Pi,\Sigma)) \not\rhd\end{aligned}$$

which is clearly a proof function. ∎

THEOREM 8. If I and J are proof functions and Q is a term, then $Q \vdash \wedge(I,J)$ iff $Q \ \rhd^* \ (R_1,R_2) \not\rhd$, where $R_1 \vdash I$ and $R_2 \vdash J$.

Proof. As in the previous theorem, let $I \ \rhd^* \ pfn(A,\Pi) \not\rhd$, let $J \ \rhd^* \ pfn(B,\Sigma) \not\rhd$, and let q,r,a,b be four fresh variables. Then

$$\wedge(I,J) \ \rhd^* \ pfn((\lambda(q,r){:}\, aq \,\&\, br)\begin{bmatrix}A,B\\a,b\end{bmatrix}, product(\Pi,\Sigma)) \not\rhd \ .$$

Now, if $Q \ \vdash \ \wedge(I,J)$ then $(\lambda(q,r){:}\, aq \,\&\, br)\begin{bmatrix}A,B\\a,b\end{bmatrix} Q \ \ \rhd^* \ \ true$ and $Q \ : \ product(\Pi,\Sigma)$. Hence, by Theorem 38 of the Expanded Term Language, $Q \ \rhd^*\lhd \ (R_1,R_2)$ for some irreducible terms R_1,R_2, which means that $Q \ \rhd^* \ (R_1,R_2) \not\rhd$. It follows that $AR_1 \,\&\, BR_2 \ \rhd^* \ true$, and also $R_1 : \Pi$ and $R_2 : \Sigma$. Thus $R_1 \vdash I$ and $R_2 \vdash J$.

The converse is similar. ∎

DEFINITION. Define constructions $\exists_d$, $\exists_w$ and $\exists$ by

$$\begin{aligned}\exists_d \ &\text{is} \ (\lambda f.(\lambda(x,q){:}\, fxq))\\ \exists_w \ &\text{is} \ (\lambda u.product(leaf,u))\\ \exists \ &\text{is} \ (\lambda pred(f,u).pfn(\exists_d f, \exists_w u))\end{aligned}$$

where f,x,q,u are four variables.

THEOREM 9. If P is a predicate then $\exists P$ is a proof function.

Proof. Let $P \ \rhd^* \ pred(F,\Pi) \not\rhd$; let x,q,f be three fresh variables. Then

$$\exists P \ \rhd^* \ pfn(\exists_d F, \exists_w \Pi) \ \rhd^* \ pfn((\lambda(x,q){:}\, fxq)\begin{bmatrix}F\\f\end{bmatrix}, product(leaf,\Pi)) \not\rhd$$

which is clearly a proof function. ∎

THEOREM 10. If P is a predicate and Q is a term, then $Q \vdash \exists P$ iff $Q \ \rhd^* \ (X,R) \not\rhd$, where $R \vdash \mathit{val}\, P\, X$.

Proof. As in the previous theorem, let $P \ \rhd^* \ \mathit{pred}(F,\Pi) \not\rhd$ and let x,q,f be three fresh variables. Then $\exists P \ \rhd^* \ \mathit{pfn}((\lambda(x,q){:}\mathit{fxq})\left[{F \atop f}\right], \mathit{product}(\mathit{leaf},\Pi)) \not\rhd$. Now, if $Q \vdash \exists P$ then $(\lambda(x,q){:}\mathit{fxq})\left[{F \atop f}\right] Q \ \rhd^* \ \mathit{true}$ and $Q \ : \ \mathit{product}(\mathit{leaf},\Pi)$. Hence, by Theorem 38 of the Expanded Term Language, $Q \ \rhd^*\lhd \ (X,R)$ for some irreducible terms X,R, which means that $Q \ \rhd^* \ (X,R) \not\rhd$. It follows that $FXR \ \rhd^* \ \mathit{true}$ and $R \ : \ \Pi$. But $\mathit{val}\, P\, X \ \rhd^* \ \mathit{pfn}(FX,\Pi) \ \rhd^* \ \mathit{pfn}(A,\Pi) \not\rhd$ for some term A where $FX \ \rhd^* \ A \not\rhd$, so $AR \ \rhd^* \ \mathit{true}$ and hence $R \vdash \mathit{val}\, P\, X$.
The converse is similar. ▮

Next comes an unfamiliar logical constant, $\Box$. Loosely speaking, a proof of $\Box(I,P)$ is a function mapping any x and any proof of I to a proof of $\mathit{val}\, P\, x$, together with a protological derivation showing that this property holds. The logical constant $\Box$ combines implication (or, strictly speaking, super-implication – see Chapter 29) with universal quantification. The reason for combining them is simply that it saves duplication of arguments in later chapters.

DEFINITION. Define constructions $\Box_d$, $\Box_w$, $\Box$ and $\mathit{el}_\Box$ by

$$\begin{aligned}
\Box_d &\text{ is } (\lambda(a,f).(\lambda(d,t){:}\, DT(d, [\![(q,x)\ aq \to fx(tqx)]\!])))\\
\Box_w &\text{ is } (\lambda(u,v).\mathit{product}(rt, \mathit{map}(u, \mathit{map}(\mathit{leaf}, v))))\\
\Box &\text{ is } (\lambda(\mathit{pfn}(a,u), \mathit{pred}(f,v)).\mathit{pfn}(\Box_d(a,f), \Box_w(u,v)))\\
\mathit{el}_\Box &\text{ is } \mathit{right}
\end{aligned}$$

where a,f,d,t,q,x,u,v are eight variables.

THEOREM 11. If I is a proof function and P is a predicate then $\Box(I,P)$ is a proof function.

Proof. Let $I \ \rhd^* \ \mathit{pfn}(A,\Pi) \not\rhd$ and $P \ \rhd^* \ \mathit{pred}(F,\Sigma) \not\rhd$. Let a,f,d,t,q,x be six fresh variables and let $Fx \ \rhd^* \ B \not\rhd$. Then

$$\begin{aligned}
\Box(I,P) \ &\rhd^* \ \mathit{pfn}(\Box_d(A,F), \Box_w(\Pi,\Sigma))\\
&\rhd^* \ \mathit{pfn}((\lambda(d,t){:}\, DT(d, [\![(q,x)\ aq \to fx(tqx)]\!]))\left[{A,F \atop a,f}\right],\\
&\qquad \mathit{product}(rt, \mathit{map}(\Pi, \mathit{map}(\mathit{leaf}, \Sigma)))) \not\rhd
\end{aligned}$$

which is clearly a proof function. ▮

THEOREM 12. For any proof function I, predicate P, and terms Q, R, where I, P, Q, R have no free variables, and any construction X, if $Q \vdash I$ and $R \vdash \Box(I, P)$ then $el_{\Box}\, R\, Q\, X \vdash val\, P\, X$.

Proof. Let $I \;\triangleright^*\; pfn(A, \Pi) \not\triangleright$, let $P \;\triangleright^*\; pred(F, \Sigma) \not\triangleright$, and let a, f, t, q, x be five fresh variables. Then $val\, P\, X \;\triangleright^*\; pfn(FX, \Sigma) \;\triangleright^*\; pfn(B, \Sigma) \not\triangleright$ for some term B. Now, our hypotheses that $Q \vdash I$ and $R \vdash \Box(I, P)$ mean that $AQ \;\triangleright^*\; true$, $Q : \Pi$, $\Box_d(A, F)R \;\triangleright^*\; true$, and $R : \Box_w(\Pi, \Sigma)$. It follows that $Q \;\triangleright^*\; Q'$ and $AQ' \;\triangleright^*\; true$ for some construction Q'; and similarly $R \;\triangleright^*\; R'$ and $\Box_d(A, F)R' \;\triangleright^*\; true$ for some construction R'.

By Theorem 38 of the Expanded Term Language, $R' \;\triangleright^*\!\triangleleft\; (D, T)$, for some irreducible terms D and T; hence in fact $R' \;\triangleright^*\; (D, T) \not\triangleright$, and thus $DT(D, [\![(q, x)\ aq \rightarrow fx(tqx)]\!]\left[\begin{smallmatrix}A,F,T\\ a,f,t\end{smallmatrix}\right]) \;\triangleright^*\; true$, $D : rt$, and $T : map(\Pi, map(leaf, \Sigma))$.

Now, the sequent $[\![(q, x)\ aq \rightarrow fx(tqx)]\!]\left[\begin{smallmatrix}A,F,T\\ a,f,t\end{smallmatrix}\right]$ is really

$$([(\lambda[q, x].aq)\left[\begin{smallmatrix}A\\ a\end{smallmatrix}\right]], (\lambda[q, x].fx(tqx))\left[\begin{smallmatrix}F,T\\ f,t\end{smallmatrix}\right]),$$

and we have $(\lambda[q, x].aq)\left[\begin{smallmatrix}A\\ a\end{smallmatrix}\right][Q', X] \;\triangleright^*\; AQ' \;\triangleright^*\; true$, so by the soundness of protologic we may infer that $true \;\triangleleft^*\; (\lambda[q, x].fx(tqx))\left[\begin{smallmatrix}F,T\\ f,t\end{smallmatrix}\right][Q', X] \;\triangleright^*\; FX(TQ'X) \;\triangleright^*\!\triangleleft\; B(el_{\Box}\, R\, Q\, X)$.

Also, $el_{\Box}\, R\, Q\, X \;\triangleright^*\!\triangleleft\; TQX : \Sigma$, so we have verified that $el_{\Box}\, R\, Q\, X \vdash val\, P\, X$.
∎

EXERCISE. Why do we need to assume that I, P, Q, R, X have no free variables in the previous theorem? What is the most that may be concluded if I, P, Q, R, X have free variables?

A variation of $\Box$ that we shall also need is Δ, defined next. Loosely speaking, $\Delta(I, P)$ is a predicate with the property that a proof of $val(\Delta(I, P))x$ is a function transforming any proof of I to a proof of $val\, P\, x$, together with a protological derivation verifying that the function works as claimed.

DEFINITION. Define constructions Δ_d, Δ_w and Δ by

$$\Delta_d \text{ is } (\lambda(a, f).(\lambda x.\Box_d(a, k(fx))))$$
$$\Delta_w \text{ is } \Box_w$$
$$\Delta \text{ is } (\lambda(pfn(a, u), pred(f, v)).pred(\Delta_d(a, f), \Delta_w(u, v)))$$

where a, f, x, u, v are five variables.

THEOREM 13. If I is a proof function, P is a predicate, and x is a variable, then $\Delta(I,P)$ is a predicate and $val(\Delta(I,P))x \rhd^*\lhd \Box(I, con(val\,P\,x))$.

Proof. Let $I \rhd^* pfn(A,\Pi) \not\rhd$ and $P \rhd^* pred(F,\Sigma) \not\rhd$. Let a, f, u be three fresh variables. Then

$$\begin{aligned} &\Delta(I,P) \\ &\rhd^* pred(\Delta_d(A,F), \Delta_w(\Pi,\Sigma)) \\ &\rhd^* pred((\lambda x.\Box_d(a,k(fx)))\left[{A,F \atop a,f}\right], product(rt, map(\Pi, map(leaf,\Sigma)))) \not\rhd . \end{aligned}$$

Moreover, using the properties of evaluable terms in the Expanded Term Language, Fu is evaluable and hence so is $k(Fu)$; $\Box_d(a,f)$ is evaluable and hence so is $\Box_d(A,k(Fu)) \lhd^* (\lambda x.\Box_d(a,k(fx)))\left[{A,F \atop a,f}\right] u$. Thus, for any variable y, $(\lambda x.\Box_d(a,k(fx)))\left[{A,F \atop a,f}\right] y$ is evaluable. This establishes that $\Delta(I,P)$ is a predicate. Moreover,

$$\begin{aligned} val(\Delta(I,P))x &\rhd^*\lhd pfn(\Box_d(A,k(Fx)), \Box_w(\Pi,\Sigma)) \\ &\rhd^*\lhd \Box(pfn(A,\Pi), pred(k(Fx),\Sigma)) \\ &\rhd^*\lhd \Box(I, con(val\,P\,x)) \end{aligned}$$

as required. ∎

EXERCISE. Show that, if A, B, C, D are terms and x is a variable then

$$val(\Delta(\wedge(\ulcorner A\urcorner, \Box(\ulcorner B\urcorner, con\ulcorner C\urcorner)), predify(\lambda x.\ulcorner D\urcorner)))x$$

is a proof function.

The logical constants introduced above generate all the proof functions needed for the interpretation of arithmetic, though they certainly do not generate all proof functions possible.

The protological properties of the logical constants $\wedge$ and $\Box$ can be summed up in the following theorems.

THEOREM 14. Protological $\wedge_d$-rule: $$\frac{(q,r,\underline{z})\,Aq,\ Br,\ \Gamma \to C\left[{(q,r) \atop p}\right]}{(p,\underline{z})\,\wedge_d(A,B)p,\ \Gamma \to C}$$

where p, q, r are three variables, $p \notin A, B, \Gamma$, and $q, r \notin A, B, \Gamma, C$.

Proof. Let a and b be two fresh variables.

$$\dfrac{\dfrac{\dfrac{\dfrac{\dfrac{(q,r,\underline{z})\ Aq,\ Br,\ \Gamma \to C\begin{bmatrix}(q,r)\\ p\end{bmatrix}}{(a,b,p,q,r,\underline{z})\ a=A,\ b=B,\ (q,r)=p,\ aq,\ br,\ \Gamma \to C}\ \text{ext}}{(a,b,p,q,r,\underline{z})\ a=A,\ b=B,\ (q,r)=p,\ aq\,\&\,br,\ \Gamma \to C}\ \&,\text{cut}}{(a,b,p,\underline{z})\ a=A,\ b=B,\ (\lambda(q,r){:}\ aq\,\&\,br)p,\ \Gamma \to C}\ (\lambda X{:}\ T)}{(a,b,p,\underline{z})\ a=A,\ b=B,\ \wedge_d(a,b)p,\ \Gamma \to C}\ \text{red}}{(p,\underline{z})\ \wedge_d(A,B)p,\ \Gamma \to C}\ \text{ext}$$

∎

THEOREM 15. Protological $\Box_d$-rule:

$$\dfrac{(d,t,\underline{z})\ DT(d,[\![(q,x)\ aq \to fx(tqx)]\!]),\ \Gamma \to C\begin{bmatrix}(d,t)\\ p\end{bmatrix}}{(p,\underline{z})\ \Box_d(a,f)p,\ \Gamma \to C}$$

where q,x,a,f,d,t,p are seven variables, $p \notin \Gamma$, and $d,t \notin \Gamma, C$.

Proof.

$$\dfrac{\dfrac{\dfrac{(d,t,\underline{z})\ DT(d,[\![(q,x)\ aq \to fx(tqx)]\!]),\ \Gamma \to C\begin{bmatrix}(d,t)\\ p\end{bmatrix}}{(p,d,t,\underline{z})\ (d,t)=p,\ DT(d,[\![(q,x)\ aq \to fx(tqx)]\!]),\ \Gamma \to C}\ \text{ext}}{(p,\underline{z})\ (\lambda(d,t){:}\ DT(d,[\![(q,x)\ aq \to fx(tqx)]\!]))p,\ \Gamma \to C}\ (\lambda X{:}\ T)}{(p,\underline{z})\ \Box_d(a,f)p,\ \Gamma \to C}\ \text{red}$$

∎

THEOREM 16. $\Box$-lemma:

- $DT(El_{\Box}(Y)u,[\![(a,f,q,x,r)\ \ aq,\ \ \Box_d(a,f)(Yur)\ \ \to\ \ fx(el_{\Box}(Yur)qx)]\!])\ \ \rhd^*$ *true*
- $El_{\Box}\ :\ map(map(\Phi, map(leaf, \Box_w(\Pi,\Sigma))), map(\Phi, rt))$
- $el_{\Box}\ :\ map(\Box_w(\Pi,\Sigma), map(\Pi, map(leaf,\Sigma)))$

where the construction $El_{\Box}$ is defined in the proof, Y is any construction, Φ, Π and Σ are type symbols, and u,a,f,q,x,r are six variables.

Proof. Let d,d',t be three fresh variables. Let D_1 be the code of the protological derivation

$$\dfrac{\to aq\ \text{(p0)} \qquad \dfrac{(q,x)\ aq \to fx(tqx)\ \text{(p1)}}{aq \to fx(tqx)}\ \text{inst}}{\to fx(tqx)}\ \text{cut}$$

where the premises are labelled 'p0' and 'p1' to indicate their order in the premise list. By the *Join*-lemma

$$DT(Join(Code), [\![(d', d, t, a, f, q, x, r)\ DT(d', [\![\rightarrow aq]\!]), DT(d, [\![(q, x)\ aq \rightarrow fx(tqx)]\!]) \rightarrow DT(join(D_1)[d', d], [\![\rightarrow fx(tqx)]\!])]\!]) \rhd^* true$$

where *Code* is

$$((\lambda[t, a, f, q, x, r].[[\![\rightarrow aq]\!], [\![(q, x)\ aq \rightarrow fx(tqx)]\!]]), (\lambda[t, a, f, x, q, r].(D_1, [\![\rightarrow fx(tqx)]\!]))).$$

Let Jn be $join(D_1)[\mathcal{E}, left(Yur)]$ and let $\underline{z}$ be d, t, a, f, q, x, r. By Theorem 2 in the Coding of Trees,

$$DT(rp(\lambda[\underline{z}].Jn), [\![(\underline{z})\ DT(Jn, [\![\rightarrow fx(tqx)]\!]) \rightarrow fx(tqx)]\!]) \rhd^* true.$$

Now let T be the term $DT(d, [\![(q, x)\ aq \rightarrow fx(tqx)]\!])$, and let D_2 be the code of the protological derivation

$$\dfrac{\dfrac{\dfrac{\dfrac{\dfrac{\dfrac{\dfrac{\dfrac{\dfrac{(d', \underline{z})\ DT(d', [\![\rightarrow aq]\!]),\ T \rightarrow DT(join(D_1)[d', d], [\![\rightarrow fx(tqx)]\!])\ \text{(p0)}}{(\underline{z})\ DT(\mathcal{E}, [\![\rightarrow aq]\!]),\ T \rightarrow DT(join(D_1)[\mathcal{E}, d], [\![\rightarrow fx(tqx)]\!])}\text{inst}}{(\underline{z})\ aq,\ T \rightarrow DT(join(D_1)[\mathcal{E}, left(d, t)], [\![\rightarrow fx(tqx)]\!])}\text{red}}{(\underline{z})\ aq,\ (d, t) = Yur,\ T \rightarrow DT(join(D_1)[\mathcal{E}, left(d, t)], [\![\rightarrow fx(tqx)]\!])}\text{thin}}{(\underline{z})\ aq,\ (d, t) = Yur,\ T \rightarrow DT(Jn, [\![\rightarrow fx(tqx)]\!])}\text{eq}}{(\underline{z})\ aq,\ (d, t) = Yur,\ T \rightarrow fx(tqx)}\text{cut with p1}}{(\underline{z})\ aq,\ (d, t) = Yur,\ T \rightarrow fx(el_{\Box}(d, t)qx)}\text{red}}{(\underline{z})\ aq,\ (d, t) = Yur,\ T \rightarrow fx(el_{\Box}(Yur)qx)}\text{eq}}{(a, f, q, x, r)\ aq,\ (\lambda(d, t){:}\ T)(Yur) \rightarrow fx(el_{\Box}(Yur)qx)}(\lambda X{:}\ T)}{(a, f, q, x, r)\ aq,\ \Box_d(a, f)(Yur) \rightarrow fx(el_{\Box}(Yur)qx)}\text{red}$$

where the step marked 'cut with p1' is a Cut with the sequent

$$(\underline{z})\ DT(Jn, [\![\rightarrow fx(tqx)]\!]) \rightarrow fx(tqx),$$

which is premise 1. Define a recursive function $El_{\Box}$ such that

$$El_{\Box}(Y) \rhd^* (\lambda u.join(D_2)[Join(Code), rp(\lambda[\underline{z}].Jn)]).$$

Then by the *join*-lemma

$$DT(El_{\Box}(Y)u, [\![(a,f,q,x,r)\ aq,\ \Box_d(a,f)(Yur) \rightarrow fx(el_{\Box}(Yur)qx)]\!]) \rhd^* \ true$$

as required.

To check the well-foundedness conditions,
for any constructions $Y : map(\Phi, map(leaf, \Box_w(\Pi,\Sigma)))$ and $U : \Phi$,
$\mathcal{E} : rt$
$YUr : \Box_w(\Pi,\Sigma) \rhd^* product(rt, map(\Pi, map(leaf,\Sigma)))$
$left(YUr) : rt$
$Jn\begin{bmatrix} U \\ u \end{bmatrix} \overset{*}{\mapsto} join(D_1)[\mathcal{E}, left(YUr)] : rt$, by the *join*-lemma
$(\lambda[\underline{z}].Jn)\begin{bmatrix} U \\ u \end{bmatrix} : map(leaf, rt)$, by the *map* rule
$rp(\lambda[\underline{z}].Jn)\begin{bmatrix} U \\ u \end{bmatrix} : rt$
$Join(Code) : rt$, by the *Join*-lemma
$El_{\Box}(Y)U \rhd^* join(D_2\begin{bmatrix} U \\ u \end{bmatrix})[Join(Code), rp(\lambda[\underline{z}].Jn)\begin{bmatrix} U \\ u \end{bmatrix}] : rt$, by the *join*-lemma.
$El_{\Box} : map(map(\Phi, map(leaf,\Box_w(\Pi,\Sigma))), map(\Phi, rt))$, as required.
Also, for any construction $V : \Box_w(\Pi,\Sigma)$,
V is (D,T), for some constructions $D : rt$ and $T : map(\Pi, map(leaf,\Sigma))$
$el_{\Box}\, V \rhd^* T : map(\Pi, map(leaf,\Sigma))$.
$el_{\Box} : map(\Box_w(\Pi,\Sigma), map(\Pi, map(leaf,\Sigma)))$, as required.
∎

CHAPTER 26

LOGIC

PROOF FUNCTIONS, PROOFS AND PREDICATES

DEFINITION. Let *pfn* be a fresh 1-ary constructor. A *proof function* is a term I such that $I \rhd^* \mathit{pfn}(A, \Pi) \not\rhd$, for some term A and type symbol Π.

DEFINITION. Let *pred* be a fresh 1-ary constructor. A *predicate* is a term P such that $P \rhd^* \mathit{pred}(F, \Pi) \not\rhd$, where Fx is evaluable for any variable x and Π is a type symbol.

I shall use the letters 'I' and 'J' to denote proof functions and the letter 'P' to denote a predicate.

THEOREM 1. Let X be an irreducible term. If I is a proof function then so is $I\begin{bmatrix} X \\ x \end{bmatrix}$, and if P is a predicate then so is $P\begin{bmatrix} X \\ x \end{bmatrix}$.

DEFINITION. The relation $\vdash$, between terms and proof functions, is defined by:

$$X \vdash I \quad \text{iff} \quad I \rhd^* \mathit{pfn}(A, \Pi) \not\rhd \text{ , } AX \rhd^* \mathit{true} \quad \text{and} \quad X : \Pi.$$

If $X \vdash I$ holds then X is said to be an *intuitionistic proof* (or just a *proof*) of I.

THEOREM 2. (Extensionality of $\vdash$.) If $X \rhd^* \lhd Y$ and I is a proof function then $X \vdash I$ iff $Y \vdash I$. Moreover if $I \rhd^* \lhd J$ then $X \vdash I$ iff $X \vdash J$.

THEOREM 3. If $Q \vdash I$ and $X \not\rhd$ then $Q\begin{bmatrix} X \\ x \end{bmatrix} \vdash I\begin{bmatrix} X \\ x \end{bmatrix}$.

THE BASIC LOGICAL CONSTANTS

DEFINITION. (Atomic formulae.) For any term T let $\ulcorner T \urcorner$ be the proof function $pfn((\lambda nil.T), leaf)$.

THEOREM 4. (Conservativeness of $\vdash$ over reduction.) Let T be any term, let $\underline{x}$ be any variables, and let $\underline{X}$ be any irreducible terms.

- If Q is a construction then $Q \vdash \ulcorner T \urcorner \left[\frac{\underline{X}}{\underline{x}}\right]$ iff $T\left[\frac{\underline{X}}{\underline{x}}\right] \rhd^* \; true.$
- If Q is any term such that $Q \vdash \ulcorner T \urcorner \left[\frac{\underline{X}}{\underline{x}}\right]$ then $T\left[\frac{\underline{X}}{\underline{x}}\right] \rhd^* \; true.$

DEFINITION. Define the constructions *val* and *con* by

$$val \text{ is } (\lambda pred(f, u).(\lambda x.pfn(fx, u)))$$
$$con \text{ is } (\lambda pfn(a, u).pred(ka, u))$$

where f, u, x, a are four variables.

THEOREM 5. If P is a predicate and $X \not\rhd$ then $val\,P\,X$ is a proof function. If I is a proof function and $X \not\rhd$ then $con\,I$ is a predicate and $val(con\,I)X \rhd^*\lhd I$.

DEFINITION. A construction *predify* is defined.

THEOREM 6. If I is a proof function and x is a variable then $predify(\lambda x.I)$ is a predicate and $val(predify(\lambda x.I))x \rhd^*\lhd I$.

DEFINITION. Define constructions $\wedge_d$, $\wedge_w$ and $\wedge$ by

$$\wedge_d \text{ is } (\lambda(a, b).(\lambda(q, r){:}\, aq \,\&\, br))$$
$$\wedge_w \text{ is } product$$
$$\wedge \text{ is } (\lambda(pfn(a, u), pfn(b, v)).pfn(\wedge_d(a, b), \wedge_w(u, v)))$$

where a, b, u, v, q, r are six variables.

THEOREM 7. If I and J are proof functions then so is $\wedge(I, J)$.

THEOREM 8. If I and J are proof functions and Q is a term, then $Q \vdash \wedge(I, J)$ iff $Q \rhd^* (R_1, R_2) \not\rhd$, where $R_1 \vdash I$ and $R_2 \vdash J$.

DEFINITION. Define constructions $\exists_d$, $\exists_w$ and $\exists$ by

$$\exists_d \text{ is } (\lambda f.(\lambda(x, q){:}\, fxq))$$
$$\exists_w \text{ is } (\lambda u.product(leaf, u))$$
$$\exists \text{ is } (\lambda pred(f, u).pfn(\exists_d f, \exists_w u))$$

where f, x, q, u are four variables.

THEOREM 9. If P is a predicate then $\exists P$ is a proof function.

THEOREM 10. If P is a predicate and Q is a term, then $Q \vdash \exists P$ iff $Q \rhd^* (X, R) \not\rhd$, where $R \vdash val\, P X$.

DEFINITION. Define constructions $\Box_d$, $\Box_w$, $\Box$ and $el_\Box$ by

$$\begin{aligned}
\Box_d &\text{ is } (\lambda(a,f).(\lambda(d,t){:}\, DT(d, [\![(q,x)\ aq \to fx(tqx)]\!])))\\
\Box_w &\text{ is } (\lambda(u,v).product(rt, map(u, map(leaf, v))))\\
\Box &\text{ is } (\lambda(pfn(a,u), pred(f,v)).pfn(\Box_d(a,f), \Box_w(u,v)))\\
el_\Box &\text{ is } right
\end{aligned}$$

where a, f, d, t, q, x, u, v are eight variables.

THEOREM 11. If I is a proof function and P is a predicate then $\Box(I,P)$ is a proof function.

THEOREM 12. For any proof function I, predicate P, and terms Q, R, where I, P, Q, R have no free variables, and any construction X, if $Q \vdash I$ and $R \vdash \Box(I,P)$ then $el_\Box R Q X \vdash val\, P X$.

DEFINITION. Define constructions $\triangle_d$, $\triangle_w$ and $\triangle$ by

$$\begin{aligned}
\triangle_d &\text{ is } (\lambda(a,f).(\lambda x.\Box_d(a, k(fx))))\\
\triangle_w &\text{ is } \Box_w\\
\triangle &\text{ is } (\lambda(pfn(a,u), pred(f,v)).pred(\triangle_d(a,f), \triangle_w(u,v)))
\end{aligned}$$

where a, f, x, u, v are five variables.

THEOREM 13. If I is a proof function, P is a predicate, and x is a variable, then $\triangle(I,P)$ is a predicate and $val(\triangle(I,P))x \rhd^* \lhd \Box(I, con(val\, P\, x))$.

THEOREM 14. Protological $\wedge_d$-rule:
$$\frac{(q,r,\underset{\sim}{z})\ Aq,\ Br,\ \Gamma \to C\left[{}^{(q,r)}_{p}\right]}{(p,\underset{\sim}{z})\ \wedge_d(A,B)p,\ \Gamma \to C}$$

where p, q, r are three variables, $p \notin A, B, \Gamma$, and $q, r \notin A, B, \Gamma, C$.

THEOREM 15. Protological $\Box_d$-rule:
$$\frac{(d,t,\underset{\sim}{z})\ DT(d, [\![(q,x)\ aq \to fx(tqx)]\!]),\ \Gamma \to C\left[{}^{(d,t)}_{p}\right]}{(p,\underset{\sim}{z})\ \Box_d(a,f)p,\ \Gamma \to C}$$

where q, x, a, f, d, t, p are seven variables, $p \notin \Gamma$, and $d, t \notin \Gamma, C$.

THEOREM 16. $\Box$-lemma:

- $DT(El_\Box(Y)u, [\![(a,f,q,x,r)\ \ aq,\ \Box_d(a,f)(Yur)\ \to\ fx(el_\Box(Yur)qx)]\!])\ \rhd^*$ *true*
- $El_\Box\ :\ map(map(\Phi, map(leaf, \Box_w(\Pi,\Sigma))), map(\Phi, rt))$
- $el_\Box\ :\ map(\Box_w(\Pi,\Sigma), map(\Pi, map(leaf,\Sigma)))$

where the construction $El_\Box$ is defined in the proof, Y is any construction, Φ, Π and Σ are type symbols, and u, a, f, q, x, r are six variables.

CHAPTER 27

FROM LOGIC TO THE CALCULUS OF PROOF FUNCTIONS

The previous theory, Logic, defined the notions of proof function, proof and predicate, and the primitive logical constants that operate on them. The next theory, Calculus of Proof Functions (CPF), is an axiomatic system consisting of logical sequents of the form $I, \ldots J \Rightarrow K$; the informal meaning of such a sequent is roughly that if the proof functions $I, \ldots J$ have proofs then K has a proof. The present chapter will show how to interpret CPF in protologic, using the properties of the logical constants stated in Logic.

LOGICAL SEQUENTS

A *logical sequent* is a sentence in the following language.

- The alphabet is that of the Expanded Term Language plus ',' and '$\Rightarrow$'.
- The tokens are those of the Expanded Term Language plus ',' and '$\Rightarrow$'.
- Lexical analysis consists of recognising constants and replacing them by the token '*con*', recognising variables and replacing them by the token '*vbl*', and recognising every other token.
- The grammar of the language is as follows.
  - The terminals are the tokens.
  - The non-terminals are *S*, *LHS*, *Ps*, *P*, *T*; the start symbol is *S*.
  - The production rules are

$$S \rightarrow LHS \Rightarrow P$$
$$LHS \rightarrow Ps \mid \varepsilon$$
$$Ps \rightarrow P \, , \, Ps \mid P$$
$$P \rightarrow T$$
$$T \rightarrow con \mid (\, vbl \,) \mid (\, T\, T \,) \mid (\, \lambda\, T \,.\, T \,) \mid (\, T \left[{T \atop vbl} \right])$$

In addition there is a context-sensitive constraint, that all the substrings matching P must be proof functions.

METANOTATION

All the metanotation of the Expanded Term Language will be allowed in the terms in logical sequents. The letters 'I', 'J' and 'K' will be used as metavariables denoting proof functions, the letters 'Γ' and 'Δ' as metavariables denoting (possibly empty) sequences of proof functions, and the letters 'P', 'P_1' and 'P_2' as metavariables denoting predicates. *Metasequents* and their *instances* are defined by analogy with metaterms and their instances in the Expanded Term Language. The typical metanotation for a sequent will be '$\Gamma \Rightarrow I$'.

CODING OF LOGICAL SEQUENTS AS CONSTRUCTIONS

DEFINITION. If the proof functions in the sequent $\Gamma \Rightarrow J$ have free variables $\underline{z}$ then the *code* of the sequent, $\{\Gamma \Rightarrow J\}$, is the construction $(\lambda[\underline{z}].([\Gamma], J))$.

(Note that this definition requires that the variables $\underline{z}$ be listed in some particular order. I shall use standard order (defined in the Term Language), and indeed throughout this chapter whenever I refer to the free variables of some terms I shall assume the free variables are listed in standard order.)

THE PROOF RELATION, $\Vdash$, FOR LOGICAL SEQUENTS

DEFINITION. The relation $Q \Vdash I_1, \ldots I_k \Rightarrow J$ (read as 'Q proves $I_1, \ldots I_k \Rightarrow J$') where $I_1, \ldots I_k \Rightarrow J$ is a logical sequent, Q is a term with no free variables, and the proof functions $I_1, \ldots I_k, J$ have free variables $\underline{x}$, means that

- $Q \;\triangleright^*\; (D, (\lambda[\underline{x}].T)) \not\triangleright$, for some D and T,
- $I_1 \;\triangleright^*\; pfn(A_1, \Pi_1) \not\triangleright , \ldots\; I_k \;\triangleright^*\; pfn(A_k, \Pi_k) \not\triangleright , J \;\triangleright^*\; pfn(B, \Sigma) \not\triangleright$,
- $DT(D, [\![(q_1, \ldots q_k, \underline{x})\, A_1 q_1, \;\ldots\; A_k q_k \rightarrow B(T[q_1, \ldots q_k])]\!]) \;\triangleright^*\; true$,
- $D \;:\; rt$ and $T \;:\; map(pi[\Pi_1, \ldots \Pi_k], \Sigma)$,

where $q_1, \ldots q_k$ are k variables different from $\underline{x}$.

Note that the intended interpretation of the logical sequent arrow $\Rightarrow$ is not exactly intuitionistic implication. The meaning of $\Rightarrow$ is given precisely by the $\Vdash$ relation. To prove a sequent $I \Rightarrow J$ one has to find a function T such that (i) if Q satisfies the decidable part of the requirements for a proof of I then $T[Q]$ satisfies the decidable part of the requirements for a proof of J, and (ii) if Q is a well-founded tree (under I's tree coding) then

$T[Q]$ is also well-founded (under J's tree coding). Because the decidable and undecidable parts of proof are treated separately here this is a stronger condition than implication, in which one simply has to map any proof of I to a proof of J. This means that the logical constant corresponding to $\Rightarrow$ (which I shall introduce in Chapter 29 and call *super-implication*, $>$) is stronger than intuitionistic implication ($\supset$). In fact, $\supset$ will be defined in terms of $>$: $A \supset B$ is essentially $A > (\mathit{true} > B)$.

THEOREM 1. (Extensionality of $\VDash$.) If $Q \;\rhd^*\lhd\; Q'$ then $Q \VDash \Gamma \to I$ iff $Q' \VDash \Gamma \to I$.

DEFINITION. Define a construction *pr* as $(\lambda(d,t).t\,\mathit{nil}\,\mathit{nil})$, where d,t are two variables.

THEOREM 2. (Soundness of $\VDash$ with respect to $\vdash$.) If $Q \VDash \;\Rightarrow I$, where I has no free variables, then $\mathit{pr}(Q) \vdash I$.

Proof. Let $I \;\rhd^*\; \mathit{pfn}(A,\Pi) \not\rhd$. The hypothesis that $Q \VDash \;\Rightarrow I$ means that

- $Q \;\rhd^*\; (D, (\lambda \mathit{nil}.T)) \not\rhd$, for some D and T,
- $DT(D, [\![\to A(T\,\mathit{nil})]\!]) \;\rhd^*\; \mathit{true}$,
- $D \,:\, \mathit{rt}$ and $T \,:\, \mathit{map}(\mathit{pi}(\mathit{nil}),\Pi)$.

The sequent $[\![\to A(T\,\mathit{nil})]\!]$ is, more explicitly, $(\mathit{nil}, (\lambda \mathit{nil}.A(T\,\mathit{nil})))$. Hence, by the soundness theorem for protologic, $(\lambda \mathit{nil}.A(T\,\mathit{nil}))\mathit{nil} \;\rhd^*\; \mathit{true}$ and so $A(T\,\mathit{nil}) \;\rhd^*\; \mathit{true}$. In addition, $\mathit{nil} \,:\, \mathit{pi}(\mathit{nil})$ by the *pi* rule, so $T\,\mathit{nil} \,:\, \Pi$. Therefore $\mathit{pr}(Q) \;\rhd^*\; T\,\mathit{nil} \vdash I$. ▮

THEOREM 3. (Completeness of $\VDash$ with respect to $\vdash$.) If $Q \vdash I$ then $\mathit{in}(\lambda[\underline{z}].(Q,I)) \VDash I$, where $\underline{z}$ are the free variables of I and Q, for a certain construction *in*.

Proof. Let $I \;\rhd^*\; \mathit{pfn}(A,\Pi) \not\rhd$. The hypothesis that $Q \vdash I$ means that $AQ \;\rhd^*\; \mathit{true}$ and $Q \,:\, \Pi$. Now let $\underline{x}$ be the free variables of I, let $\underline{y}$ be the variables of $\underline{z}$ that are not in $\underline{x}$, let T be the term $(\lambda \mathit{nil}.Q\left[\frac{\underline{0}}{\underline{y}}\right])$, where $\underline{0}$ is a sequence of 0s with as many terms as $\underline{y}$, and let D_0 be the code of the protological derivation

$$\dfrac{\dfrac{\dfrac{\dfrac{\to \mathit{true}\ \text{(eval)}}{(\underline{z}) \to \mathit{true}}\,\text{thin}}{(\underline{z}) \to AQ}\,\text{red}}{(\underline{x}) \to AQ\left[\frac{\underline{0}}{\underline{y}}\right]}\,\text{inst}}{(\underline{x}) \to A(T\,\mathit{nil})}\,\text{red}$$

Then by the *join*-lemma

$$DT(join(D_0)nil, [\![(\underline{x}) \rightarrow A(T\,nil)]\!]) \triangleright^* true$$

and $join(D_0) : map(pi(nil), rt)$, giving $join(D_0)nil : rt$. Since $TX \triangleright^* Q\left[\frac{0}{\underline{y}}\right] : \Pi$ for any construction $X : pi(nil)$ we also have $T : map(pi(nil), \Pi)$. Now define the construction *in* such that

$$in(\lambda[\underline{z}].(Q, I)) \triangleright^* (join(D_0)nil, (\lambda[\underline{x}].T))$$

giving

$$in(\lambda[\underline{z}].(Q, I)) \Vdash\!\!= \Rightarrow I$$

as required. ∎

THEOREM 4. For any proof function I and predicate P, there is no term Q such that $Q \Vdash\!\!= \Box(I, P) \Rightarrow \ulcorner false \urcorner$.

Proof. Let $I \triangleright^* pfn(A, \Pi) \not\triangleright$ and $P \triangleright^* pred(F, \Sigma) \not\triangleright$. Then $\ulcorner false \urcorner \mapsto pfn(k\,false, leaf)$ and

$$\Box(I, P) \triangleright^* pfn(\Box_d(A, F), \Box_w(\Pi, \Sigma)) \triangleright^* pfn(B, \Phi) \not\triangleright$$

for some term B and type symbol Φ. Let $\underline{z}$ be the free variables of I and P, let q, r, x, a, f, t be six fresh variables, and suppose that $Q \Vdash\!\!= \Box(I, P) \Rightarrow \ulcorner false \urcorner$. This means that $Q \triangleright^* (D, (\lambda[\underline{z}].T)) \not\triangleright$ where

- $DT(D, [\![(r, \underline{z})\; Br \rightarrow k\,false\,(T[r])]\!]) \triangleright^* true$,
- $D : rt$ and $T : map(pi[\Phi], leaf)$.

Now let $\underline{0}$ be a sequence of 0s with as many terms as $\underline{z}$. We can find a construction R such that $B\left[\frac{\underline{0}}{\underline{z}}\right] R \triangleright^* true$, as follows. Recall the construction D^* in the Coding of Trees with the property that $DT(D^*, [\![\; \rightarrow false]\!]) \triangleright^* true$. Let D_0 be the code of the protological derivation

$$\dfrac{\dfrac{\rightarrow false \;(\text{p0})}{(q, x)\; aq \rightarrow false}\,\text{thin} \qquad (q, x)\, false \rightarrow fx(tqx) \;(\textit{false}\text{-el})}{(q, x)\; aq \rightarrow fx(tqx)}\,\text{cut}$$

Then by the *join*-lemma,

$$DT(join(D_0)[D^*], [\![(q, x)\; aq \rightarrow fx(tqx)]\!]) \triangleright^* true,$$

and hence, by the definition of $\Box_d$,

$$\Box_d(a,f)(join(D_0)[D^*],t) \ \triangleright^* \ true$$

and hence, by instantiation,

$$\Box_d(A,F)\begin{bmatrix}0\\\underline{z}\end{bmatrix}(join(D_0\begin{bmatrix}A,F,id\\a,f,t\end{bmatrix})[D^*],id) \ \triangleright^* \ true$$

so $(join(D_0\begin{bmatrix}A,F,id\\a,f,t\end{bmatrix})[D^*],id) \ \triangleright^* \ R$, for some construction R, and $B\begin{bmatrix}0\\\underline{z}\end{bmatrix}R \ \triangleright^* \ true$. On the other hand, $kfalse(T\begin{bmatrix}0\\\underline{z}\end{bmatrix}[R])$ certainly does not reduce to *true*: the only irreducible term it could reduce to is *false*. Hence the sequent $(r,\underline{z})\ Br \to kfalse(T[r])$ is unsound. (In detail, the sequent is really $([(\lambda[r,\underline{z}].Br)],(\lambda[r,\underline{z}].kfalse(T[r])))$, and we have $(\lambda[r,\underline{z}].Br)[R,\underline{0}] \ \triangleright^* \ B\begin{bmatrix}0\\\underline{z}\end{bmatrix}R \ \triangleright^* \ true$ but $(\lambda[r,\underline{z}].kfalse(T[r]))[R,\underline{0}] \ \triangleright^* \ kfalse(T\begin{bmatrix}0\\\underline{z}\end{bmatrix}[R])$, which does not reduce to *true*). This contradicts the soundness of protologic. Hence there is no such proof Q. ∎

CALCULUS OF PROOF FUNCTIONS (CPF)

The axioms and rules of CPF are all instances of the following schemata.

Truth introduction (*true*-in): $\Rightarrow \ulcorner true \urcorner$
Falsity elimination (*false*-el): $\ulcorner false \urcorner \Rightarrow I$
Reduction (red): $I \Rightarrow J$ where $I \triangleright J$ or $J \triangleright I$;
$\ulcorner X \urcorner \Rightarrow \ulcorner Y \urcorner$ where $X \triangleright Y$ or $Y \triangleright X$
$\wedge$-introduction ($\wedge$-in): $I,\ J \Rightarrow \wedge(I,J)$
$\wedge$-elimination ($\wedge$-el): $\wedge(I,J) \Rightarrow I \qquad \wedge(I,J) \Rightarrow J$
$\exists$-introduction ($\exists$-in): $val\,P\,x \Rightarrow \exists P$
$\Box$-expansion ($\Box$-exp): $I \Rightarrow \Box(J, con\,I)$
$\Box$-compression ($\Box$-comp): $\Box(\ulcorner true \urcorner, \Delta(I,P)) \Rightarrow \Box(I,P)$
Term existence (te): $\ulcorner T\begin{bmatrix}X\\x\end{bmatrix} \urcorner \Rightarrow \exists(predify(\lambda x.\ulcorner x = X \urcorner))$
where $x \in T$ but $x \notin X$

Equality (eq): $\ulcorner X = Y \urcorner,\ val\,P\,X \Rightarrow val\,P\,Y$ where $X, Y \not\triangleright$;
$\ulcorner X = Y \urcorner,\ \ulcorner T\begin{bmatrix}X\\u\end{bmatrix} \urcorner \Rightarrow \ulcorner T\begin{bmatrix}Y\\u\end{bmatrix} \urcorner$

Thinning (thin): $$\frac{\Gamma \Rightarrow J}{I,\ \Gamma \Rightarrow J}$$

Exchange (exch): $$\frac{\Gamma,\ I,\ J,\ \Delta \Rightarrow K}{\Gamma,\ J,\ I,\ \Delta \Rightarrow K}$$

Contraction (con): $\dfrac{I,\ I,\ \Gamma \Rightarrow J}{I,\ \Gamma \Rightarrow J}$

$\exists$-elimination ($\exists$-el): $\dfrac{val\,P\,x,\ \Gamma \Rightarrow I}{\exists P,\ \Gamma \Rightarrow I}$ where $x \notin P, \Gamma, I$

$\Box$-introduction & elimination ($\Box$-in & $\Box$-el): $\dfrac{I \Rightarrow val\,P\,x}{\Rightarrow \Box(I,P)}$

where $x \notin I, P$

$\Box$-transitivity ($\Box$-tr): $\dfrac{I,\ val\,P_1\,X \Rightarrow J}{\Box(I,P_1),\ \Box(J,P_2) \Rightarrow \Box(I,P_2)}$ where $X \not\in$

Cut: $\dfrac{\Gamma \Rightarrow I \qquad I,\ \Delta \Rightarrow J}{\Gamma,\ \Delta \Rightarrow J}$

Boolean rule (bool): $\dfrac{\ulcorner x = true \urcorner,\ \Gamma \Rightarrow I \qquad \ulcorner x = false \urcorner,\ \Gamma \Rightarrow I}{\ulcorner boolean\,x \urcorner,\ \Gamma \Rightarrow I}$

Induction (ind): $\dfrac{\Rightarrow val\,P\,0 \qquad val\,P\,n \Rightarrow val\,P\,(Sn)}{\ulcorner num\,n \urcorner \Rightarrow val\,P\,n}$ where $n \notin P$

INTERPRETATIONS OF THE CPF AXIOMS AND RULES

For each CPF axiom I shall produce a proof of it, and for each CPF rule of inference I shall show how to obtain a proof of the conclusion sequent from proofs of the premise sequents. It will then follow that any CPF derivation can be mechanically transformed into a proof (in the $\Vdash$ sense); I shall construct explicitly a function *spr* that carries out this transformation.

THEOREM 5. (CPF truth introduction axiom.)

$Pr_1(\{\Rightarrow \ulcorner true \urcorner\}) \Vdash\ \Rightarrow \ulcorner true \urcorner,$

where the construction Pr_1 is defined below.

Proof. Note that $\ulcorner true \urcorner \mapsto pfn(k\,true, leaf)$. Let T be id; then $k\,true\,(T\,nil)$ $\rhd^*$ $true$, so

$$DT(\mathcal{E}, [\![\to k\,true\,(T\,nil)]\!])\ \rhd^*\ true.$$

Now let Pr_1 be $k(\mathcal{E},(\lambda nil.T))$. Then $Pr_1(\{\Rightarrow \ulcorner true \urcorner\})$ $\rhd^*$ $(\mathcal{E},(\lambda nil.T))$, which satisfies the decidable part of the proof relation $Pr_1(\{\Rightarrow \ulcorner true \urcorner\}) \Vdash$ $\Rightarrow \ulcorner true \urcorner$.

To check the well-foundedness conditions,

$\mathcal{E}\ :\ rt.$

For any construction $X\ :\ pi(nil)$,

$TX\ :\ leaf$, by the *leaf* rule.

$T\ :\ map(pi(nil), leaf)$, as required.

∎

THEOREM 6. (CPF falsity elimination axiom.)

$$Pr_2(\{\ulcorner false \urcorner \Rightarrow I\}) \Vdash \ulcorner false \urcorner \Rightarrow I,$$

where the construction Pr_2 is defined below.

Proof. Let $I \;\triangleright^*\; pfn(A,\Pi) \not\triangleright$, and note that $\ulcorner false \urcorner \;\mapsto\; pfn(kfalse, leaf)$. Let $\underline{x}$ be the free variables of I, let q be a fresh variable, let T be *fxpt id*, and let D_1 be the code of the protological derivation

$$\frac{(q,\underline{x})\, false \rightarrow A(T[q])\ \ (false\text{-el})}{(q,\underline{x})\, kfalse\, q \rightarrow A(T[q])}\ \text{red}$$

Then by the *join*-lemma

$$DT(join(D_1)nil, [\![(q,\underline{x})\, kfalse\, q \rightarrow A(T[q])]\!]) \;\triangleright^*\; true$$

which implies that $join(D_1)nil \;\triangleright^*\; D$ for some construction D. So define a recursive function Pr_2 such that $Pr_2(\{\ulcorner false \urcorner \Rightarrow I\}) \;\triangleright^*\; (D,(\lambda[\underline{x}].T))$, thus satisfying the decidable part of the proof relation.

To check the well-foundedness conditions,

$D \;\triangleleft^*\; join(D_1)nil \;:\; rt$, by the *join*-lemma, as required.

T is *fxpt id* $:$ $map(pi[leaf],\Pi)$, by the *fxpt id* well-foundedness rule, as required.

∎

THEOREM 7. (First CPF reduction axiom.) $Pr_3(\{I \Rightarrow J\}) \Vdash I \Rightarrow J$, where $I \;\triangleright\; J$ or $J \;\triangleright\; I$, and the construction Pr_3 is defined below.

Proof. Let $I, J \;\triangleright^*\; pfn(A,\Pi) \not\triangleright$. Let $\underline{x}$ be the free variables of I, J, let q be a fresh variable, and let T be $(\lambda[q].q)$. Then $A(T[q]) \;\triangleright^*\; Aq$, so let D_1 be the code of the protological derivation of the sequent $(q,\underline{x})\, Aq \rightarrow A(T[q])$ by Reduction. Then by the *join*-lemma

$$DT(join(D_1)nil, [\![(q,\underline{x})\, Aq \rightarrow A(T[q])]\!]) \;\triangleright^*\; true$$

which implies that $join(D_1)nil \;\triangleright^*\; D$ for some construction D. So define a recursive function Pr_3 such that $Pr_3(\{I \Rightarrow J\}) \;\triangleright^*\; (D,(\lambda[\underline{x}].T))$, thus satisfying the decidable part of the proof relation.

To check the well-foundedness conditions,

$D \;\triangleleft^*\; join(D_1)nil \;:\; rt$, by the *join*-lemma, as required.

For any construction $X \;:\; pi[\Pi]$,

X is $[Q]$, for some construction $Q \;:\; \Pi$, by the *pi* rule

$TX \;\triangleright^*\; Q \;:\; \Pi$.

$T \;:\; map(pi[\Pi],\Pi)$, as required.

∎

THEOREM 8. (Second CPF reduction axiom.)

$Pr_4(\{\ulcorner X \urcorner \Rightarrow \ulcorner Y \urcorner\}) \Vdash\!\!= \ulcorner X \urcorner \Rightarrow \ulcorner Y \urcorner$,

where $X \rhd Y$ or $Y \rhd X$, and the construction Pr_4 is defined below.

Proof. Let $\underline{x}$ be the free variables of X, Y, let q be a fresh variable, let T be *id*, and let D_1 be the code of the protological derivation

$$\frac{(q, \underline{x})\ X \to Y \text{ (red)}}{(q, \underline{x})\ (\lambda nil.X)q \to (\lambda nil.Y)(T[q])}\ \text{red}$$

Then by the *join*-lemma

$$DT(join(D_1)nil, [\![(q, \underline{x})\ (\lambda nil.X)q \to (\lambda nil.Y)(T[q])]\!]) \rhd^* true$$

which implies that $join(D_1)nil \rhd^* D$ for some construction D. So define a recursive function Pr_4 such that $Pr_4(\{\ulcorner X \urcorner \Rightarrow \ulcorner Y \urcorner\}) \rhd^* (D, (\lambda[\underline{x}].T))$, thus satisfying the decidable part of the proof relation.

To check the well-foundedness conditions,

$D \lhd^* join(D_1)nil : rt$, by the *join*-lemma, as required.

For any construction $X : pi[leaf]$,

$TX : leaf$, by the *leaf* rule.

$T : map(pi[leaf], leaf)$, as required.

∎

THEOREM 9. (CPF $\wedge$-introduction axiom.)

$Pr_5(\{I, J \Rightarrow \wedge(I, J)\}) \Vdash\!\!= I, J \Rightarrow \wedge(I, J)$,

where the construction Pr_5 is defined below.

Proof. Let $I \rhd^* pfn(A, \Pi) \not\rhd$ and $J \rhd^* pfn(B, \Sigma) \not\rhd$. Then

$$\wedge(I, J) \rhd^* pfn(\wedge_d(A, B), \wedge_w(\Pi, \Sigma)) \rhd^* pfn(C, product(\Pi, \Sigma)) \not\rhd$$

for some term C. Let $\underline{x}$ be the free variables of I, J, let p, q, r be three fresh variables, let T be $(\lambda[q, r].(q, r))$, and let D_1 be the code of the protological derivation

$$\cfrac{\cfrac{(p, \underline{x})\ \wedge_d(A, B)p \to \wedge_d(A, B)p \text{ (taut)}}{(q, r, \underline{x})\ Aq,\ Br \to \wedge_d(A, B)(q, r)}\ \wedge_d\text{-rule}}{(q, r, \underline{x})\ Aq,\ Br \to C(T[q, r])}\ \text{red}$$

Then by the *join*-lemma

$$DT(join(D_1)nil, [\![(q,r,\underline{x})\, Aq,\ Br \to C(T[q,r])]\!]) \ \triangleright^*\ true$$

which implies that $join(D_1)nil \ \triangleright^*\ D$ for some construction D. So define a recursive function Pr_5 such that $Pr_5(\{I,\ J \Rightarrow \wedge(I,J)\}) \ \triangleright^*\ (D,(\lambda[\underline{x}].T))$, thus satisfying the decidable part of the proof relation.

To check the well-foundedness conditions,
$D \ \triangleleft^*\ join(D_1)nil\ :\ rt$, by the *join*-lemma, as required.
For any construction $X\ :\ pi[\Pi,\Sigma]$,
X is $[Q,R]$, for some constructions $Q\ :\ \Pi$ and $R\ :\ \Sigma$
$TX \ \triangleright^*\ (Q,R)\ :\ product(\Pi,\Sigma)$.
$T\ :\ map(pi[\Pi,\Sigma], product(\Pi,\Sigma))$, as required.
∎

THEOREM 10. (First CPF $\wedge$-elimination axiom.)
$Pr_6(\{\wedge(I,J) \Rightarrow I\}) \Vdash \wedge(I,J) \Rightarrow I$,
where the construction Pr_6 is defined below.

Proof. Let $I \ \triangleright^*\ pfn(A,\Pi) \not\triangleright$ and $J \ \triangleright^*\ pfn(B,\Sigma) \not\triangleright$. Then

$$\wedge(I,J) \ \triangleright^*\ pfn(\wedge_d(A,B), \wedge_w(\Pi,\Sigma)) \ \triangleright^*\ pfn(C, product(\Pi,\Sigma)) \not\triangleright$$

for some term C. Let $\underline{x}$ be the free variables of I, J, let p,q,r be three fresh variables, let T be $(\lambda[(q,r)].q)$, and let D_1 be the code of the protological derivation

$$\dfrac{\dfrac{\dfrac{(q,r,\underline{x})\, Aq,\ Br \to Aq \ \text{(taut)}}{(q,r,\underline{x})\, Aq,\ Br \to A(T[(q,r)])}\,\text{red}}{(p,\underline{x})\, \wedge_d(A,B)p \to A(T[p])}\,\wedge_d\text{-rule}}{(p,\underline{x})\, Cp \to A(T[p])}\,\text{red}$$

Then by the *join*-lemma

$$DT(join(D_1)nil, [\![(p,\underline{x})\, Cp \to A(T[p])]\!]) \ \triangleright^*\ true$$

which implies that $join(D_1)nil \ \triangleright^*\ D$ for some construction D. So define a recursive function Pr_6 such that $Pr_6(\{\wedge(I,J) \Rightarrow I\}) \ \triangleright^*\ (D,(\lambda[\underline{x}].T))$, thus satisfying the decidable part of the proof relation.

To check the well-foundedness conditions,
$D \ \triangleleft^*\ join(D_1)nil\ :\ rt$, by the *join*-lemma, as required.
For any construction $X\ :\ pi[product(\Pi,\Sigma)]$,
X is $[(Q,R)]$, for some constructions $Q\ :\ \Pi$ and $R\ :\ \Sigma$
$TX \ \triangleright^*\ Q\ :\ \Pi$.
$T\ :\ map(pi[product(\Pi,\Sigma)],\Pi)$, as required.
∎

THEOREM 11. (Second CPF $\wedge$-elimination axiom.)

$Pr_7(\{\wedge(I,J) \Rightarrow J\}) \models \wedge(I,J) \Rightarrow J,$

where the construction Pr_7 is defined below.

Proof. As in the previous theorem. ∎

THEOREM 12. (CPF $\exists$-introduction axiom.)

$Pr_8(\{val\,P\,x \Rightarrow \exists P\}) \models val\,P\,x \Rightarrow \exists P,$

where the construction Pr_8 is defined below.

Proof. Let $P \triangleright^* pred(F,\Pi) \not\triangleright$. Then $val\,P\,x \triangleright^* pfn(Fx,\Pi) \triangleright^* pfn(A,\Pi) \not\triangleright$ and $\exists P \triangleright^* pfn(\exists_d F, \exists_w \Pi) \triangleright^* pfn(B, product(leaf,\Pi)) \not\triangleright$, for some terms A and B. Let $\underline{z}$ be the free variables of P,x, let q be a fresh variable, and let T be $(\lambda[q].(x,q))$. Then $B(T[q]) \triangleright^*\triangleleft \exists_d F(x,q) \triangleright^* Fxq \triangleright^* Aq$, so let D_1 be the code of the protological derivation of the sequent $(q,\underline{z})\,Aq \rightarrow B(T[q])$ by Reduction. Then by the *join*-lemma

$$DT(join(D_1)nil, [\![(q,\underline{z})\,Aq \rightarrow B(T[q])]\!]) \triangleright^* true$$

which implies that $join(D_1)nil \triangleright^* D$ for some construction D. So define a recursive function Pr_8 such that $Pr_8(\{val\,P\,x \Rightarrow \exists P\}) \triangleright^* (D,(\lambda[\underline{z}].T))$, thus satisfying the decidable part of the proof relation.

To check the well-foundedness conditions,

$D \triangleleft^* join(D_1)nil : rt$, by the *join*-lemma, as required.

For any construction $Z : pi[\Pi]$,

Z is $[Q]$, for some construction $Q : \Pi$

$x : leaf$

$TZ \triangleright^* (x,Q) : product(leaf,\Pi)$.

T is $map(pi[\Pi], product(leaf,\Pi))$, as required.

∎

THEOREM 13. (CPF $\Box$-expansion axiom.)

$Pr_9(\{I \Rightarrow \Box(J, con\,I)\}) \models I \Rightarrow \Box(J, con\,I),$

where the construction Pr_9 is defined below.

Proof. Let $I \triangleright^* pfn(A,\Pi) \not\triangleright$ and $J \triangleright^* pfn(B,\Sigma) \not\triangleright$. Then

$$\Box(J, con\,I) \triangleright^* pfn(\Box_d(B,kA), \Box_w(\Sigma,\Pi)) \triangleright^* pfn(C,\Phi) \not\triangleright$$

for some term C and type symbol Φ.

Let $\underline{z}$ be the free variables of I,J, let d,t,f,a,b,q,r,x be eight fresh variables, let X be the term $(t = k(kq))\,\&\,(f = ka)\,\&\,aq$, and let D_1 be the code of the protological derivation

$$
\dfrac{\dfrac{\to X \ \text{(p0)}}{(r,x) \ \to X}\,\text{thin}
\qquad
\dfrac{\dfrac{\dfrac{\dfrac{(r,x)\ aq \to kax(k(kq)rx)\ \text{(red)}}{(r,x)\ t=k(kq),\ f=ka,\ aq,\ br \to kax(k(kq)rx)}\,\text{thin}}{(r,x)\ t=k(kq),\ f=ka,\ aq,\ br \to fx(trx)}\,\text{eq}}{(r,x)\ X,\ br \to fx(trx)}\,\&}
{(r,x)\ br \to fx(trx)}\,\text{cut}
$$

Now let T be $(\lambda[q].(join(D_1\begin{bmatrix}k(kq)\\t\end{bmatrix}\begin{bmatrix}ka\\f\end{bmatrix}\begin{bmatrix}A\\a\end{bmatrix}\begin{bmatrix}B\\b\end{bmatrix})[\mathcal{E}],k(kq))))$. Then by the *Join*-lemma

$$
\begin{aligned}
&DT(Join(Code),\llbracket(d,t,f,a,b,q,\underline{z})\ DT(d,\llbracket\ \to X\rrbracket)\to\\
&\qquad\qquad DT(join(D_1)[d],\llbracket(r,x)\ br\to fx(trx)\rrbracket)\rrbracket)\ \triangleright^*\ true
\end{aligned}
$$

where *Code* is

$$((\lambda[t,f,a,b,q,\underline{z}].[\llbracket\ \to X\rrbracket]),(\lambda[t,f,a,b,q,\underline{z}].(D_1,\llbracket(r,x)\ br\to fx(trx)\rrbracket))).$$

Now let D_2 be the code of the protological derivation

$$
\dfrac{\dfrac{\dfrac{\dfrac{\dfrac{\dfrac{\dfrac{(d,t,f,a,b,q,\underline{z})\ DT(d,\llbracket\to X\rrbracket)\to DT(join(D_1)[d],\llbracket(r,x)\,br\to fx(trx)\rrbracket)\ \text{(p0)}}{(t,f,a,b,q,\underline{z})\ DT(\mathcal{E},\llbracket\to X\rrbracket)\to DT(join(D_1)[\mathcal{E}],\llbracket(r,x)\,br\to fx(trx)\rrbracket)}\,\text{inst}}{(t,f,a,b,q,\underline{z})\ X\to DT(join(D_1)[\mathcal{E}],\llbracket(r,x)\ br\to fx(trx)\rrbracket)}\,\text{red}}{(t,f,a,b,q,\underline{z})\ X\to\Box_d(b,f)(join(D_1)[\mathcal{E}],t)}\,\text{conv}}{(t,f,a,b,q,\underline{z})\ t=k(kq),\ f=ka,\ aq\to\Box_d(b,f)(join(D_1)[\mathcal{E}],t)}\,\&}{(a,b,q,\underline{z})\ aq\to\Box_d(b,ka)(join(D_1\begin{bmatrix}k(kq)\\t\end{bmatrix}\begin{bmatrix}ka\\f\end{bmatrix})[\mathcal{E}],k(kq))}\,\text{ext}}{(q,\underline{z})\ Aq\to\Box_d(B,kA)(join(D_1\begin{bmatrix}k(kq)\\t\end{bmatrix}\begin{bmatrix}ka\\f\end{bmatrix}\begin{bmatrix}A\\a\end{bmatrix}\begin{bmatrix}B\\b\end{bmatrix})[\mathcal{E}],k(kq))}\,\text{inst}}{(q,\underline{z})\ Aq\to C(T[q])}\,\text{red}
$$

Then by the *join*-lemma

$$DT(join(D_2)[Join(Code)],\llbracket(q,\underline{z})\ Aq\to C(T[q])\rrbracket)\ \triangleright^*\ true$$

which implies that $join(D_2)[Join(Code)]\ \triangleright^*\ D$ for some construction D. So define a recursive function Pr_9 such that $Pr_9(\{I\Rightarrow\Box(J,con\,I)\})\ \triangleright^*\ (D,(\lambda[\underline{z}].T))$, thus satisfying the decidable part of the proof relation.

To check the well-foundedness conditions,
$Join(Code)\ :\ rt$, by the *Join*-lemma
$D\ \triangleleft^*\ join(D_2)[Join(Code)]\ :\ rt$, by the *join*-lemma, as required.

For any construction $V : pi[\Pi]$,

V is $[Q]$, for some construction $Q : \Pi$

$\mathcal{E} : rt$

$join(D_1\begin{bmatrix}k(kq)\\t\end{bmatrix}\begin{bmatrix}ka\\f\end{bmatrix}\begin{bmatrix}A\\a\end{bmatrix}\begin{bmatrix}B\\b\end{bmatrix}\begin{bmatrix}Q\\q\end{bmatrix})[\mathcal{E}] : rt$, by the *join*-lemma.

For any constructions $R : \Sigma$ and $X : leaf$,

$k(kQ)RX \rhd^* Q : \Pi.$

$k(kQ) : map(\Sigma, map(leaf, \Pi))$

$TV \rhd^* (join(D_1\begin{bmatrix}k(kq)\\t\end{bmatrix}\begin{bmatrix}ka\\f\end{bmatrix}\begin{bmatrix}A\\a\end{bmatrix}\begin{bmatrix}B\\b\end{bmatrix}\begin{bmatrix}Q\\q\end{bmatrix})[\mathcal{E}], k(kQ))$

$: product(rt, map(\Sigma, map(leaf, \Pi))) \lhd^* \Box_w(\Sigma, \Pi) \rhd^* \Phi.$

$T : map(pi[\Pi], \Phi)$, as required.

$\blacksquare$

THEOREM 14. (CPF $\Box$-compression axiom.)

$Pr_{10}(\{\Box(\ulcorner true \urcorner, \Delta(I,P)) \Rightarrow \Box(I,P)\}) \Vdash\!\!= \Box(\ulcorner true \urcorner, \Delta(I,P)) \Rightarrow \Box(I,P)$, where the construction Pr_{10} is defined below.

Proof. Let $I \rhd^* pfn(A,\Pi) \not\rhd$ and $P \rhd^* pred(F,\Sigma) \not\rhd$. Then

$$\Box(\ulcorner true \urcorner, \Delta(I,P)) \rhd^*\!\lhd\ pfn(\Box_d(k\,true, \Delta_d(A,F)), \Box_w(leaf, \Delta_w(\Pi,\Sigma)))$$
$$\rhd^*\ pfn(B,\Phi) \not\rhd$$
$$\Box(I,P) \rhd^*\ pfn(\Box_d(A,F), \Box_w(\Pi,\Sigma)) \rhd^*\ pfn(C,\Omega) \not\rhd$$

for some terms B, C and type symbols Φ, Ω.

Let $\underline{z}$ be the free variables of I, P, let $a,b,d,d',d'',f,g,h,m,q,r,t,v,w,x$ be fifteen fresh variables, let

Y be $(\lambda t.(\lambda x.txx))$,

M be $(\lambda w.(\lambda x.el_\Box(Ytx)wx))$,

X be $(m = M)\ \&\ (g = \Delta_d(a,f))\ \&\ (b = k\,true)$,

and let D_1 be the code of the protological derivation

$$\dfrac{\dfrac{\dfrac{(v,x)\,bv \to gx(tvx)\ (\mathrm{p2})}{(x)\,bx \to gx(txx)}\,\text{inst}}{(w,x)\,bx \to gx(txx)}\,\text{thin} \qquad \dfrac{\dfrac{\dfrac{\dfrac{\dfrac{(a,h,w,x,r)\,aw,\ \Box_d(a,h)(Ytr) \to hx(el_\Box(Ytr)wx)\ (\mathrm{p1})}{(w,x)\,aw,\ \Box_d(a,k(fx))(Ytx) \to k(fx)x(el_\Box(Ytx)wx)}\,\text{inst}}{(w,x)\,aw,\ \Delta_d(a,f)x(txx) \to fx(Mwx)}\,\text{conv}}{(w,x)\,X,\ aw,\ \Delta_d(a,f)x(txx) \to fx(Mwx)}\,\text{thin}}{(w,x)\,X,\ aw,\ gx(txx) \to fx(mwx)}\,\&,\text{eq}}{(w,x)\,bx,\ X,\ aw \to fx(mwx)}\,\text{cut}$$

$$\dfrac{\dfrac{\to X\ (\mathrm{p0})}{(w,x)\ \to X}\,\text{thin} \qquad \dfrac{\dfrac{(w,x)\,bx,\ X,\ aw \to fx(mwx)}{(w,x)\,k\,true\,x.\ X,\ aw \to fx(mwx)}\,\&,\text{eq}}{(w,x)\,X,\ aw \to fx(mwx)}\,\text{red}}{(w,x)\,aw \to fx(mwx)}\,\text{cut}$$

Then by the *Join*-lemma

$$DT(Join(Code), \llbracket (d'', d', d, m, t, g, b, a, f, \underline{z})\ DT(d'', \llbracket \ \rightarrow X \rrbracket),$$
$$DT(d', \llbracket (a, h, w, x, r)\ aw,\ \Box_d(a, h)(Ytr) \rightarrow hx(el_\Box(Ytr)wx) \rrbracket),$$
$$DT(d, \llbracket (v, x)\ bv \rightarrow gx(tvx) \rrbracket)$$
$$\rightarrow DT(join(D_1)[d'', d', d], \llbracket (w, x)\ aw \rightarrow fx(mwx) \rrbracket) \rrbracket) \ \rhd^* \ true$$

where *Code* is defined in the required way as $((\lambda[m, t, g, b, a, f, \underline{z}].[\llbracket \ \rightarrow X \rrbracket, \llbracket (a, h, w, x, r)\ aw,\ \Box_d(a, h)(Ytr) \rightarrow hx(el_\Box(Ytr)wx) \rrbracket, \llbracket (v, x)\ bv \rightarrow gx(tvx) \rrbracket]), (\lambda[m, t, g, b, a, f, \underline{z}].(D_1, \llbracket (w, x)\ aw \rightarrow fx(mwx) \rrbracket)))$. Also, by the $\Box$-lemma,

$$DT(El_\Box(Y)t, \llbracket (a, h, w, x, r)\ aw,\ \Box_d(a, h)(Ytr) \rightarrow hx(el_\Box(Ytr)wx) \rrbracket) \ \rhd^* \ true.$$

Let T be $(\lambda[(d, t)].(join(D_1 \begin{bmatrix} M \\ m \end{bmatrix} \begin{bmatrix} \Delta_d(a,f) \\ g \end{bmatrix} \begin{bmatrix} k\,true \\ b \end{bmatrix} \begin{bmatrix} A,F \\ a,f \end{bmatrix})[\mathcal{E}, El_\Box(Y)t, d], M))$, let Γ be the four terms $g = \Delta_d(a,f)$, $b = k\,true$, $a = A$, $f = F$, let E be the term $DT(d, \llbracket (v, x)\ bv \rightarrow gx(tvx) \rrbracket)$, and let D_2 be the code of the protological derivation

$$(d'', d', d, m, t, g, b, a, f, \underline{z})\ DT(d'', \llbracket \ \rightarrow X \rrbracket),$$
$$DT(d', \llbracket (a, h, w, x, r)\ aw,\ \Box_d(a, h)(Ytr) \rightarrow hx(el_\Box(Ytr)wx) \rrbracket), \quad \text{(p0)}$$
$$E \rightarrow DT(join(D_1)[d'', d', d], \llbracket (w, x)\ aw \rightarrow fx(mwx) \rrbracket)$$

——————————————————————— inst

$$(d, m, t, g, b, a, f, \underline{z})\ DT(\mathcal{E}, \llbracket \ \rightarrow X \rrbracket),$$
$$DT(El_\Box(Y)t, \llbracket (a, h, w, x, r)\ aw,\ \Box_d(a, h)(Ytr) \rightarrow hx(el_\Box(Ytr)wx) \rrbracket),$$
$$E \rightarrow DT(join(D_1)[\mathcal{E}, El_\Box(Y)t, d], \llbracket (w, x)\ aw \rightarrow fx(mwx) \rrbracket)$$

——————————————————————— red

$$(d, m, t, g, b, a, f, \underline{z})\ X,\ E \rightarrow$$
$$DT(join(D_1)[\mathcal{E}, El_\Box(Y)t, d], \llbracket (w, x)\ aw \rightarrow fx(mwx) \rrbracket)$$

——————————————————————— conv

$$(d, m, t, g, b, a, f, \underline{z})\ X,\ E \rightarrow \Box_d(a,f)(join(D_1)[\mathcal{E}, El_\Box(Y)t, d], m)$$

——————————————————————— &, thin

$$(d, m, t, g, b, a, f, \underline{z})\, m = M,\ \Gamma,\ E \rightarrow \Box_d(a,f)(join(D_1)[\mathcal{E}, El_\Box(Y)t, d], m)$$

——————————————————————— ext

$$(d, t, g, b, a, f, \underline{z})\ \Gamma,\ E \rightarrow \Box_d(a,f)(join(D_1 \begin{bmatrix} M \\ m \end{bmatrix})[\mathcal{E}, El_\Box(Y)t, d], M)$$

——————————————————————— eq

$$(d, t, g, b, a, f, \underline{z})\ \Gamma,\ E \rightarrow$$
$$\Box_d(a,f)(join(D_1 \begin{bmatrix} M \\ m \end{bmatrix} \begin{bmatrix} \Delta_d(a,f) \\ g \end{bmatrix} \begin{bmatrix} k\,true \\ b \end{bmatrix} \begin{bmatrix} A,F \\ a,f \end{bmatrix})[\mathcal{E}, El_\Box(Y)t, d], M)$$

——————————————————————— red

$$(d, t, g, b, a, f, \underline{z})\ \Gamma,\ E \rightarrow \Box_d(a,f)(T[(d, t)])$$

——————————————————————— $\Box_d$-rule

$$(q, g, b, a, f, \underline{z})\ \Gamma,\ \Box_d(b, g)q \rightarrow \Box_d(a,f)(T[q])$$

——————————————————————— ext

$$(q, a, f, \underline{z})\ a = A,\ f = F,\ \Box_d(k\,true, \Delta_d(a,f))q \rightarrow \Box_d(a,f)(T[q])$$

——————————————————————— ext

$$(q, \underline{z})\ \Box_d(k\,true, \Delta_d(A, F))q \rightarrow \Box_d(A, F)(T[q])$$

——————————————————————— red

$$(q, \underline{z})\ Bq \rightarrow C(T[q])$$

Then by the *join*-lemma

$$DT(join(D_2)[Join(Code)], \llbracket(q, \underline{z})\ Bq \to C(T[q])\rrbracket)\ \rhd^*\ true$$

which implies that $join(D_2)[Join(Code)]\ \rhd^*\ D$ for some construction D. So define a recursive function Pr_{10} such that

$$Pr_{10}(\{\Box(\ulcorner true\urcorner, \triangle(I,P)) \Rightarrow \Box(I,P)\})\ \rhd^*\ (D, (\lambda[\underline{z}].T)),$$

thus satisfying the decidable part of the proof relation.

To check the well-foundedness conditions,

$Join(Code)\ :\ rt$, by the *Join*-lemma

$D\ \lhd^*\ join(D_2)[Join(Code)]\ :\ rt$, by the *join*-lemma, as required.

For any construction $Z\ :\ pi[\Phi]$,

Z is $[Q]$, for some construction $Q\ :\ \Phi\ \lhd^*\ \Box_w(leaf, \triangle_w(\Pi,\Sigma))$

Q is (D_0, T_0), for some constructions $D_0\ :\ rt$ and $T_0\ :\ map(leaf, map(leaf, \triangle_w(\Pi,\Sigma)))$

$\mathcal{E}\ :\ rt$

For any constructions $W\ :\ map(leaf, map(leaf, \triangle_w(\Pi,\Sigma)))$ and $R\ :\ leaf$,

$YWR\ \rhd^*\ WRR\ :\ \triangle_w(\Pi,\Sigma)$, which is $\Box_w(\Pi,\Sigma)$.

$Y\ :\ map(map(leaf, map(leaf, \triangle_w(\Pi,\Sigma))), map(leaf, \Box_w(\Pi,\Sigma)))$

$El_\Box(Y)\ :\ map(map(leaf, map(leaf, \triangle_w(\Pi,\Sigma))), rt)$, by the $\Box$-lemma

$El_\Box(Y)T_0\ :\ rt$

$join(D_1\begin{bmatrix}M\\m\end{bmatrix}\begin{bmatrix}\triangle_d(a,f)\\g\end{bmatrix}\begin{bmatrix}k\ true\\b\end{bmatrix}\begin{bmatrix}A,F\\a,f\end{bmatrix}\begin{bmatrix}D_0,T_0\\d_0,t_0\end{bmatrix})[\mathcal{E}, El_\Box(Y)T_0, D_0]\ :\ rt$ by the *join*-lemma.

For any constructions $W\ :\ \Pi$ and $R\ :\ leaf$,

$M\begin{bmatrix}T_0\\t\end{bmatrix}WR\ \rhd^*\ el_\Box(YT_0R)WR\ \rhd^*\ el_\Box(T_0RR)WR\ :\ \Sigma$, by the $\Box$-lemma.

$M\begin{bmatrix}T_0\\t\end{bmatrix}\ :\ map(\Pi, map(leaf,\Sigma))$

$TZ\ \rhd^*\ (join(D_1\begin{bmatrix}M\\m\end{bmatrix}\begin{bmatrix}\triangle_d(a,f)\\g\end{bmatrix}\begin{bmatrix}k\ true\\b\end{bmatrix}\begin{bmatrix}A,F\\a,f\end{bmatrix}\begin{bmatrix}D_0,T_0\\d_0,t_0\end{bmatrix})[\mathcal{E}, El_\Box(Y)T_0, D_0], M\begin{bmatrix}T_0\\t\end{bmatrix})$
$:\ product(rt, map(\Pi, map(leaf,\Sigma)))\ \lhd^*\ \Box_w(\Pi,\Sigma)\ \rhd^*\ \Omega.$

$T\ :\ map(pi[\Phi], \Omega)$, as required.

∎

THEOREM 15. (CPF term existence axiom.)

$Pr_{11}(\{\ulcorner T\begin{bmatrix}X\\x\end{bmatrix}\urcorner \Rightarrow \exists(predify(\lambda x.\ulcorner x = X\urcorner))\})\ \Vdash$
$\ulcorner T\begin{bmatrix}X\\x\end{bmatrix}\urcorner \Rightarrow \exists(predify(\lambda x.\ulcorner x = X\urcorner)),$

where $x \in T$ but $x \notin X$, and the construction Pr_{11} is defined below.

Proof. Let $predify(\lambda x.\ulcorner x = X\urcorner) \rhd^* pred(F,\Pi) \not\triangleright$. Then

$$pfn(Fx,\Pi) \lhd^* val(pred(F,\Pi))x \lhd^* val(predify(\lambda x.\ulcorner x = X\urcorner))x \rhd^*\lhd \ulcorner x = X\urcorner,$$

using the properties of *val* and *predify* from Logic. This implies that $Fx \rhd^* (\lambda nil.x = X)$ and Π is *leaf*. Hence

$$\begin{aligned} \exists(predify(\lambda x.\ulcorner x = X\urcorner)) &\rhd^* pfn(\exists_d F, \exists_w leaf) \\ &\rhd^* pfn(B, product(leaf, leaf)) \not\triangleright \end{aligned}$$

for some term B.

Let $\underline{z}$ be the free variables of $T\begin{bmatrix}X\\x\end{bmatrix}$, let q be a fresh variable, let T' be the term $(\lambda nil.(X, nil))$, and let D_1 be the code of the protological derivation

$$\cfrac{\cfrac{\cfrac{\cfrac{(x,q,\underline{z})\ x = X \to F\,x\,nil \ \text{(red)}}{(x,q,\underline{z})\ x = X \to \exists_d F(x,nil)}\,\text{red}}{(x,q,\underline{z})\ x = X,\ T \to \exists_d F(x,nil)}\,\text{thin}}{(q,\underline{z})\ T\begin{bmatrix}X\\x\end{bmatrix} \to \exists_d F(X,nil)}\,\text{ext}}{(q,\underline{z})\ (\lambda nil.T\begin{bmatrix}X\\x\end{bmatrix})q \to B(T'[q])}\,\text{red}$$

Then by the *join*-lemma

$$DT(join(D_1)nil, [\![(q,\underline{z})\ (\lambda nil.T\begin{bmatrix}X\\x\end{bmatrix})q \to B(T'[q])]\!]) \rhd^* true$$

which implies that $join(D_1)nil \rhd^* D$ for some construction D. So define a recursive function Pr_{11} such that $Pr_{11}(\{\ulcorner T\begin{bmatrix}X\\x\end{bmatrix}\urcorner \Rightarrow \exists(predify(\lambda x.\ulcorner x = X\urcorner))\}) \rhd^* (D, (\lambda[\underline{z}].T'))$, thus satisfying the decidable part of the proof relation.

To check the well-foundedness conditions,
$D \lhd^* join(D_1)nil : rt$, by the *join*-lemma, as required.
For any construction $Z : pi[leaf]$,
$T'Z \rhd^* (X, nil) : product(leaf, leaf)$.
$T' : map(pi[leaf], product(leaf, leaf))$, as required.
∎

THEOREM 16. (First CPF equality axiom.)
$Pr_{12}(\{\ulcorner X = Y\urcorner, val\,P\,X \Rightarrow val\,P\,Y\}) \models\!\!\!\models \ulcorner X = Y\urcorner, val\,P\,X \Rightarrow val\,P\,Y$, where $X, Y \not\triangleright$ and the construction Pr_{12} is defined below.

Proof. Let $\underline{x}$ be the free variables of P, X, Y, and let q, r, u be three fresh variables. Let $P \rhd^* pred(F,\Pi) \not\triangleright$, where $Fu \rhd^* A \not\triangleright$, for some term A. Thus $val\,P\,X \rhd^* pfn(FX,\Pi) \rhd^* pfn(A\begin{bmatrix}X\\u\end{bmatrix},\Pi) \not\triangleright$, and likewise for $val\,P\,Y$. Let T be $(\lambda[q,r].r)$ and let D_1 be the code of the protological derivation

$$\frac{(q,r,\underline{x})\ X=Y,\ A\begin{bmatrix}X\\u\end{bmatrix} r \rightarrow A\begin{bmatrix}Y\\u\end{bmatrix} r \text{ (eq)}}{(q,r,\underline{x})\ (\lambda nil.X=Y)q,\ A\begin{bmatrix}X\\u\end{bmatrix} r \rightarrow A\begin{bmatrix}Y\\u\end{bmatrix} (T[q,r])}\text{ red}$$

Then by the *join*-lemma

$$DT(join(D_1)nil, [\![(q,r,\underline{x})\ (\lambda nil.X=Y)q,\ A\begin{bmatrix}X\\u\end{bmatrix} r \rightarrow A\begin{bmatrix}Y\\u\end{bmatrix} (T[q,r])]\!]) \triangleright^* true$$

which implies that $join(D_1)nil \triangleright^* D$ for some construction D. So define a recursive function Pr_{12} such that $Pr_{12}(\{\ulcorner X=Y\urcorner,\ val\,P\,X \Rightarrow val\,P\,Y\}) \triangleright^*$ $(D,(\lambda[\underline{x}].T))$, thus satisfying the decidable part of the proof relation.

To check the well-foundedness conditions,
$D \triangleleft^* join(D_1)nil : rt$, by the *join*-lemma, as required.
For any construction $Z : pi[leaf,\Pi]$,
Z is $[Q,R]$, for some construction $Q : leaf$ and $R : \Pi$
$TZ \triangleright^* R : \Pi$.
$T : map(pi[leaf,\Pi],\Pi)$, as required.
∎

THEOREM 17. (Second CPF equality axiom.)

$$Pr_{13}(\{\ulcorner X=Y\urcorner,\ \ulcorner T\begin{bmatrix}X\\u\end{bmatrix}\urcorner \Rightarrow \ulcorner T\begin{bmatrix}Y\\u\end{bmatrix}\urcorner\}) \Vdash \ulcorner X=Y\urcorner,\ \ulcorner T\begin{bmatrix}X\\u\end{bmatrix}\urcorner \Rightarrow \ulcorner T\begin{bmatrix}Y\\u\end{bmatrix}\urcorner,$$

where the construction Pr_{13} is defined below.

Proof. Let $\underline{x}$ be the free variables of $X, Y, T\begin{bmatrix}X\\u\end{bmatrix}, T\begin{bmatrix}Y\\u\end{bmatrix}$, let q, r be two fresh variables, let T' be *id*, and let D_1 be the code of the protological derivation

$$\frac{(q,r,\underline{x})\ X=Y,\ T\begin{bmatrix}X\\u\end{bmatrix} \rightarrow T\begin{bmatrix}Y\\u\end{bmatrix} \text{ (eq)}}{(q,r,\underline{x})\ (\lambda nil.X=Y)q,\ (\lambda nil.T\begin{bmatrix}X\\u\end{bmatrix})r \rightarrow (\lambda nil.T\begin{bmatrix}Y\\u\end{bmatrix})(T'[q,r])}\text{ red}$$

Then by the *join*-lemma

$$\begin{aligned}&DT(join(D_1)nil,\\ &\quad [\![(q,r,\underline{x})\ (\lambda nil.X=Y)q,\ (\lambda nil.T\begin{bmatrix}X\\u\end{bmatrix})r \rightarrow (\lambda nil.T\begin{bmatrix}Y\\u\end{bmatrix})(T'[q,r])]\!])\\ &\triangleright^*\ true\end{aligned}$$

which implies that $join(D_1)nil \triangleright^* D$ for some construction D. So define a recursive function Pr_{13} such that $Pr_{13}(\{\ulcorner X=Y\urcorner,\ \ulcorner T\begin{bmatrix}X\\u\end{bmatrix}\urcorner \Rightarrow \ulcorner T\begin{bmatrix}Y\\u\end{bmatrix}\urcorner\}) \triangleright^*$ $(D,(\lambda[\underline{x}].T'))$, thus satisfying the decidable part of the proof relation.

To check the well-foundedness conditions,

$D \mathrel{\lhd^*} join(D_1)nil \; : \; rt$, by the *join*-lemma, as required.
For any construction $X \; : \; pi[leaf, leaf]$,
$T'X \; : \; leaf$, by the *leaf* rule.
$T' \; : \; map(pi[leaf, leaf], leaf)$, as required.
▮

THEOREM 18. (CPF thinning rule.)
If $Q \mathrel{|\!\!\models} \Gamma \Rightarrow J$ then $Pr_{14}(Q, \{\Gamma \Rightarrow J\}, \{I, \; \Gamma \Rightarrow J\}) \mathrel{|\!\!\models} I, \Gamma \Rightarrow J$, where the construction Pr_{14} is defined below.

Proof. Let $I \mathrel{\rhd^*} pfn(A, \Pi) \not\rhd$, let $J \mathrel{\rhd^*} pfn(B, \Sigma) \not\rhd$, and let the proof functions of Γ reduce to $pfn(C_1, \Phi_1), \ldots pfn(C_n, \Phi_n) \not\rhd$. Let $\underline{x}$ be the free variables of Γ, J, let $\underline{y}$ be the free variables of I, Γ, J, let $q_1, \ldots q_n, r$ be $n+1$ fresh variables, and abbreviate $q_1, \ldots q_n$ to $\underline{q}$. The hypothesis that $Q \mathrel{|\!\!\models} \Gamma \Rightarrow J$ means that $Q \mathrel{\rhd^*} (D, (\lambda[\underline{x}].T)) \not\rhd$ where

- $DT(D, [\![(\underline{q}, \underline{x}) \; C_1 q_1, \; \ldots \; C_n q_n \rightarrow B(T[\underline{q}])]\!]) \mathrel{\rhd^*} true$,
- $D \; : \; rt$ and $T \; : \; map(pi[\Phi_1, \ldots \Phi_n], \Sigma)$.

Let T' be $(\lambda[r, \underline{q}].T[\underline{q}])$ and let D_1 be the code of the protological derivation

$$\cfrac{\cfrac{(\underline{q}, \underline{x}) \; C_1 q_1, \; \ldots \; C_n q_n \rightarrow B(T[\underline{q}]) \; (\text{p0})}{(r, \underline{q}, \underline{y}) \; Ar, \; C_1 q_1, \; \ldots \; C_n q_n \rightarrow B(T[\underline{q}])}\,\text{thin}}{(r, \underline{q}, \underline{y}) \; Ar, \; C_1 q_1, \; \ldots \; C_n q_n \rightarrow B(T'[r, \underline{q}])}\,\text{red}$$

Then by the *join*-lemma

$$DT(join(D_1)[D], [\![(r, \underline{q}, \underline{y}) \; Ar, \; C_1 q_1, \; \ldots \; C_n q_n \rightarrow B(T'[r, \underline{q}])]\!]) \mathrel{\rhd^*} true$$

which implies that $join(D_1)[D] \mathrel{\rhd^*} D'$ for some construction D'. So define a recursive function Pr_{14} such that $Pr_{14}((D, (\lambda[\underline{x}].T)), \{\Gamma \Rightarrow J\}, \{I, \; \Gamma \Rightarrow J\}) \mathrel{\rhd^*} (D', (\lambda[\underline{y}].T'))$, thus satisfying the decidable part of the proof relation.

To check the well-foundedness conditions,
$D' \mathrel{\lhd^*} join(D_1)[D] \; : \; rt$, by the *join*-lemma, as required.
For any construction $X \; : \; pi[\Pi, \Phi_1, \ldots \Phi_n]$,
X is $[R, Q_1, \ldots Q_n]$ for some constructions $R \; : \; \Pi, \; Q_1 \; : \; \Phi_1, \ldots Q_n \; : \; \Phi_n$
$T'X \mathrel{\rhd^*} T[Q_1, \ldots Q_n] \; : \; \Sigma$.
$T' \; : \; map(pi[\Pi, \Phi_1, \ldots \Phi_n], \Sigma)$, as required.
▮

THEOREM 19. (CPF exchange rule.) If $Q \mathrel{|\!\!\models} \Gamma, \; I, \; J, \; \Delta \Rightarrow K$ then $Pr_{15}(Q, \{\Gamma, \; I, \; J, \; \Delta \Rightarrow K\}, \{\Gamma, \; J, \; I, \; \Delta \Rightarrow K\}) \mathrel{|\!\!\models} \Gamma, \; J, \; I, \; \Delta \Rightarrow K$, where the construction Pr_{15} is defined below.

Proof. Let $I \rhd^* pfn(A,\Pi) \not\rhd$, let $J \rhd^* pfn(B,\Sigma) \not\rhd$, let $K \rhd^* pfn(C,\Phi) \not\rhd$, let the proof functions of Γ reduce to $pfn(E_1,\Psi_1),\dots pfn(E_m,\Psi_m) \not\rhd$, and let the proof functions of Δ reduce to $pfn(F_1,\Omega_1),\dots pfn(F_n,\Omega_n) \not\rhd$. Let $\underline{x}$ be the free variables of Γ, I, J, Δ, K, let $u_1,\dots u_m, q, r, v_1,\dots v_n$ be $m+n+2$ fresh variables, and abbreviate $u_1,\dots u_m$ to $\underline{u}$ and $v_1,\dots v_n$ to $\underline{v}$. The hypothesis that $Q \Vdash \Gamma,\ I,\ J,\ \Delta \Rightarrow J$ means that $Q \rhd^* (D,(\lambda[\underline{x}].T)) \not\rhd$ where

- $DT(D, [\![(\underline{u}, q, r, \underline{v}, \underline{x})\ E_1u_1,\dots E_mu_m,\ Aq,\ Br,\ F_1v_1,\dots F_nv_n \rightarrow C(T[\underline{u}, q, r, \underline{v}])]\!]) \rhd^*\ true,$
- $D\ :\ rt$ and $T\ :\ map(pi[\Psi_1,\dots \Psi_m, \Pi, \Sigma, \Omega_1,\dots \Omega_n], \Phi)$.

Let T' be $(\lambda[\underline{u}, r, q, \underline{v}].T[\underline{u}, q, r, \underline{v}])$, and let D_1 be the code of the protological derivation

$$\cfrac{\cfrac{(\underline{u}, q, r, \underline{v}, \underline{x})\ E_1u_1,\dots E_mu_m,\ Aq,\ Br,\ F_1v_1,\dots F_nv_n \rightarrow C(T[\underline{u}, q, r, \underline{v}])\ \text{(p0)}}{(\underline{u}, r, q, \underline{v}, \underline{x})\ E_1u_1,\dots E_mu_m,\ Br,\ Aq,\ F_1v_1,\dots F_nv_n \rightarrow C(T[\underline{u}, q, r, \underline{v}])}\ \text{exch}}{(\underline{u}, r, q, \underline{v}, \underline{x})\ E_1u_1,\dots E_mu_m,\ Br,\ Aq,\ F_1v_1,\dots F_nv_n \rightarrow C(T'[\underline{u}, r, q, \underline{v}])}\ \text{red}$$

Then by the *join*-lemma

$$DT(join(D_1)[D], [\![(\underline{u}, r, q, \underline{v}, \underline{x})\ E_1u_1,\dots E_mu_m,\ Br,\ Aq,\ F_1v_1,\dots F_nv_n \rightarrow C(T'[\underline{u}, r, q, \underline{v}])]\!]) \rhd^*\ true$$

which implies that $join(D_1)[D] \rhd^* D'$ for some construction D'. So define a recursive function Pr_{15} such that $Pr_{15}((D,(\lambda[\underline{x}].T)), \{\Gamma,\ I,\ J,\ \Delta \Rightarrow K\}, \{\Gamma,\ J,\ I,\ \Delta \Rightarrow K\}) \rhd^* (D',(\lambda[\underline{x}].T'))$, thus satisfying the decidable part of the proof relation.

To check the well-foundedness conditions,
$D' \lhd^* join(D_1)[D]\ :\ rt$, by the *join*-lemma, as required.
For any construction $X\ :\ pi[\Psi_1,\dots \Psi_m, \Sigma, \Pi, \Omega_1,\dots \Omega_n]$,
X is $[U_1,\dots U_m, R, Q, V_1,\dots V_n]$, for some constructions $U_1\ :\ \Psi_1,\dots U_m\ :\ \Psi_m, R\ :\ \Sigma, Q\ :\ \Pi, V_1\ :\ \Omega_1,\dots V_n\ :\ \Omega_n$
$T'X \rhd^* T[U_1,\dots U_m, Q, R, V_1,\dots V_n]\ :\ \Phi$.
$T'\ :\ map(pi[\Psi_1,\dots \Psi_m, \Sigma, \Pi, \Omega_1,\dots \Omega_n], \Phi)$, as required.
∎

THEOREM 20. (CPF contraction rule.) If $Q \Vdash I,\ I,\ \Gamma \Rightarrow J$
then $Pr_{16}(Q, \{I,\ I,\ \Gamma \Rightarrow J\}, \{I,\ \Gamma \Rightarrow J\}) \Vdash I,\ \Gamma \Rightarrow J$,
where the construction Pr_{16} is defined below.

Proof. Let $I \rhd^* pfn(A,\Pi) \not\rhd$, let $J \rhd^* pfn(B,\Sigma) \not\rhd$, and let the proof functions of Γ reduce to $pfn(C_1,\Phi_1),\dots pfn(C_n,\Phi_n) \not\rhd$. Let $\underline{x}$ be the free variables of I, Γ, J, let $q, r, u_1,\dots u_n$ be $n+2$ fresh variables, and abbreviate $u_1,\dots u_n$ to $\underline{u}$. The hypothesis that $Q \Vdash I,\ I,\ \Gamma \Rightarrow J$ means that $Q \rhd^* (D,(\lambda[\underline{x}].T))$, where

- $DT(D, [\![(q, r, \underline{u}, \underline{x})\ Aq,\ Ar,\ C_1u_1, \ldots C_nu_n \to B(T[q, r, \underline{u}])]\!]) \ \triangleright^*\ true$,
- $D\ :\ rt$ and $T\ :\ map(pi[\Pi, \Pi, \Phi_1, \ldots \Phi_n], \Sigma)$.

Let T' be $(\lambda[q, \underline{u}].T[q, q, \underline{u}])$ and let D_1 be the code of the protological derivation

$$\cfrac{\cfrac{\cfrac{(q, r, \underline{u}, \underline{x})\ Aq,\ Ar,\ C_1u_1, \ldots C_nu_n \to B(T[q, r, \underline{u}])\ \text{(p0)}}{(q, \underline{u}, \underline{x})\ Aq,\ Aq,\ C_1u_1, \ldots C_nu_n \to B(T[q, q, \underline{u}])}\ \text{inst}}{(q, \underline{u}, \underline{x})\ Aq,\ C_1u_1, \ldots C_nu_n \to B(T[q, q, \underline{u}])}\ \text{con}}{(q, \underline{u}, \underline{x})\ Aq,\ C_1u_1, \ldots C_nu_n \to B(T'[q, \underline{u}])}\ \text{red}$$

Then by the *join*-lemma

$$DT(join(D_1)[D], [\![(q, \underline{u}, \underline{x})\ Aq,\ C_1u_1, \ldots C_nu_n \to B(T'[q, \underline{u}])]\!]) \ \triangleright^*\ true$$

which implies that $join(D_1)[D]\ \triangleright^*\ D'$ for some construction D'. So define a recursive function Pr_{16} such that

$$Pr_{16}((D, (\lambda[\underline{x}].T)), \{I,\ I,\ \Gamma \Rightarrow J\}, \{I,\ \Gamma \Rightarrow J\}) \ \triangleright^*\ (D', (\lambda[\underline{x}].T')),$$

thus satisfying the decidable part of the proof relation.

To check the well-foundedness conditions,
$D'\ \triangleleft^*\ join(D_1)[D]\ :\ rt$, by the *join*-lemma, as required.
For any construction $X\ :\ pi[\Pi, \Phi_1, \ldots \Phi_n]$,
X is $[Q, U_1, \ldots U_n]$, for some constructions $Q\ :\ \Pi$, $U_1\ :\ \Phi_1, \ldots U_n\ :\ \Phi_n$
$T'X\ \triangleright^*\ T[Q, Q, U_1, \ldots U_n]\ :\ \Sigma$.
$T'\ :\ map(pi[\Pi, \Phi_1, \ldots \Phi_n], \Sigma)$, as required.
∎

THEOREM 21. (CPF $\exists$-elimination.) If $Q\ \models\ val\,P\,x,\ \Gamma \Rightarrow I$
then $Pr_{17}(Q, \{val\,P\,x,\ \Gamma \Rightarrow I\}, \{\exists P,\ \Gamma \Rightarrow I\})\ \models\ \exists P,\ \Gamma \Rightarrow I$,
where $x \notin P, \Gamma, I$, and the construction Pr_{17} is defined below.

Proof. Let $P\ \triangleright^*\ pred(F, \Pi)\ \not\triangleright$, let $I\ \triangleright^*\ pfn(A, \Sigma)\ \not\triangleright$, and let the proof functions of Γ reduce to $pfn(E_1, \Phi_1), \ldots pfn(E_n, \Phi_n)\ \not\triangleright$. Then

$$val\,P\,x\ \triangleright^*\ pfn(Fx, \Pi)\ \triangleright^*\ pfn(B, \Pi)\ \not\triangleright$$

$$\exists P\ \triangleright^*\ pfn(\exists_d F, \exists_w \Pi)\ \triangleright^*\ pfn(C, product(leaf, \Pi))\ \not\triangleright$$

for some terms B and C.

Let $\underline{y}$ be the free variables of P, x, Γ, I, let $\underline{z}$ be the free variables of P, Γ, I, let $p, f, q, r_1, \ldots r_n$ be $n+3$ fresh variables, and abbreviate $r_1, \ldots r_n$ to $\underline{r}$. The hypothesis that $Q\ \models\ val\,P\,x,\ \Gamma \Rightarrow I$ means that $Q\ \triangleright^*\ (D, (\lambda[\underline{y}].T))\ \not\triangleright$, where

- $DT(D, [\![(q, \underline{r}, \underline{y})\ Bq,\ E_1r_1,\ \ldots E_nr_n \to A(T[q, \underline{r}])]\!])\ \triangleright^*\ true$,
- $D\ :\ rt$ and $T\ :\ map(pi[\Pi, \Phi_1, \ldots \Phi_n], \Sigma)$.

Let T' be $(\lambda[(x, q), \underline{r}].T[q, \underline{r}])$ and let D_1 be the code of the protological derivation

$$\begin{array}{cl}
(q,\underline{r},\underline{y})\,Bq,\ E_1r_1,\ \dots E_nr_n \to A(T[q,\underline{r}])\ (\text{p0}) & \\ \hline
(q,\underline{r},\underline{y})\,Fxq,\ E_1r_1,\ \dots E_nr_n \to A(T'[(x,q),\underline{r}]) & \text{red} \\ \hline
(p,f,q,\underline{r},\underline{y})\,f=F,\ p=(x,q),\ fxq,\ E_1r_1,\ \dots E_nr_n \to A(T'[p,\underline{r}]) & \text{ext} \\ \hline
(p,f,\underline{r},\underline{z})\,f=F,\ (\lambda(x,q){:}fxq)p,\ E_1r_1,\ \dots E_nr_n \to A(T'[p,\underline{r}]) & (\lambda X{:}\,T) \\ \hline
(p,f,\underline{r},\underline{z})\,f=F,\ \exists_d fp,\ E_1r_1,\ \dots E_nr_n \to A(T'[p,\underline{r}]) & \text{red} \\ \hline
(p,\underline{r},\underline{z})\ \exists_d Fp,\ E_1r_1,\ \dots E_nr_n \to A(T'[p,\underline{r}]) & \text{ext} \\ \hline
(p,\underline{r},\underline{z})\ Cp,\ E_1r_1,\ \dots E_nr_n \to A(T'[p,\underline{r}]) & \text{red}
\end{array}$$

Then by the *join*-lemma

$$DT(join(D_1)[D], [\![(p,\underline{r},\underline{z})\ Cp,\ E_1r_1,\ \dots E_nr_n \to A(T'[p,\underline{r}])]\!]) \ \triangleright^*\ true$$

which implies that $join(D_1)[D]\ \triangleright^*\ D'$ for some construction D'. So define a recursive function Pr_{17} such that

$$Pr_{17}((D,(\lambda[\underline{y}].T)), \{val\,P\,x,\ \Gamma \Rightarrow I\}, \{\exists P,\ \Gamma \Rightarrow I\})\ \triangleright^*\ (D',(\lambda[\underline{z}].T')),$$

thus satisfying the decidable part of the proof relation.

To check the well-foundedness conditions,
$D'\ \triangleleft^*\ join(D_1)[D]\ :\ rt$, by the *join*-lemma, as required.
For any construction $Z\ :\ pi[product(leaf,\Pi),\Phi_1,\dots\Phi_n]$,
Z is $[(X,Q),R_1,\dots R_n]$, for some constructions $X\ :\ leaf$, $Q\ :\ \Pi$, $R_1\ :\ \Phi_1,\dots R_n\ :\ \Phi_n$
$T\begin{bmatrix}X\\x\end{bmatrix}\ :\ map(pi[\Pi,\Phi_1,\dots\Phi_n],\Sigma)$, by the well-foundedness instantiation rule
$T'Z\ \triangleright^*\ T\begin{bmatrix}X\\x\end{bmatrix}[Q,R_1,\dots R_n]\ :\ \Sigma$
$T'\ :\ map(pi[product(leaf,\Pi),\Phi_1,\dots\Phi_n],\Sigma)$, as required.
∎

THEOREM 22. (CPF $\Box$-introduction rule.) If $Q \models I \Rightarrow val\,P\,x$ then $Pr_{18}(Q,\{I \Rightarrow val\,P\,x\},\{\Rightarrow \Box(I,P)\}) \models\ \Rightarrow \Box(I,P)$, where $x \notin I,P$, and the construction Pr_{18} is defined below.

Proof. Let $I\ \triangleright^*\ pfn(A,\Pi)\ \not\triangleright$ and $P\ \triangleright^*\ pred(F,\Sigma)\ \not\triangleright$; then

$$val\,P\,x\ \triangleright^*\ pfn(Fx,\Sigma)\ \triangleright^*\ pfn(B,\Sigma)\ \not\triangleright$$
$$\Box(I,P)\ \triangleright^*\ pfn(\Box_d(A,F),\Box_w(\Pi,\Sigma))\ \triangleright^*\ pfn(C,\Phi)\ \not\triangleright$$

for some terms B, C and type symbol Φ. Let $\underline{y}$ be the free variables of I,P,x, let $\underline{z}$ be the free variables of I,P, and let q,d_1,d_2,a,f,u be six fresh variables. The hypothesis that $Q \models I \Rightarrow val\,P\,x$ means that $Q\ \triangleright^*\ (D,(\lambda[\underline{y}].T))\ \not\triangleright$, where

- $DT(D,[\![(q,\underline{y})\,Aq \to B(T[q])]\!])\ \triangleright^*\ true,$
- $D\ :\ rt$ and $T\ :\ map(pi[\Pi],\Sigma)$.

Let U be $(\lambda q.(\lambda x.T[q]))$, let X be $(a = A) \mathbin{\&} (f = F) \mathbin{\&} (u = U)$, and let D_1 be the code of the protological derivation

$$\dfrac{\dfrac{\to X\ (\mathrm{p0})}{(q,x)\ \to X}\,\mathrm{thin} \qquad \dfrac{\dfrac{\dfrac{\dfrac{\dfrac{(q,\underline{y})\ Aq \to B(T[q])\ (\mathrm{p1})}{(q,x)\ Aq \to B(T[q])}\,\mathrm{inst}}{(q,x)\ Aq \to Fx(Uqx)}\,\mathrm{red}}{(q,x)\ a = A,\ f = F,\ u = U,\ Aq \to Fx(Uqx)}\,\mathrm{thin}}{(q,x)\ a = A,\ f = F,\ u = U,\ aq \to fx(uqx)}\,\mathrm{eq}}{(q,x)\ X,\ aq \to fx(uqx)}\,\&}{(q,x)\ aq \to fx(uqx)}\,\mathrm{cut}$$

Then by the *Join*-lemma

$$\begin{aligned} &DT(Join(Code), \llbracket (d_1, d_2, a, f, u, \underline{z})\ DT(d_1, \llbracket \to X \rrbracket), \\ &\qquad DT(d_2, \llbracket (q,\underline{y})\ Aq \to B(T[q]) \rrbracket) \\ &\qquad \to DT(join(D_1)[d_1, d_2], \llbracket (q,x)\ aq \to fx(uqx) \rrbracket) \rrbracket) \ \rhd^* \ true \end{aligned}$$

where *Code* is

$$\begin{aligned} &((\lambda[a,f,u,\underline{z}].[\llbracket \to X \rrbracket, \llbracket (q,\underline{y})\ Aq \to B(T[q]) \rrbracket]), \\ &(\lambda[a,f,u,\underline{z}].(D_1, \llbracket (q,x)\ aq \to fx(uqx) \rrbracket))). \end{aligned}$$

Let T' be $(\lambda nil.(join(D_1\left[\begin{smallmatrix} A,F,U \\ a,f,u \end{smallmatrix}\right])[\mathcal{E}, D], U))$ and let D_2 be the code of the protological derivation

$$\dfrac{\dfrac{\dfrac{\dfrac{\dfrac{\dfrac{\begin{array}{l}(d_1, d_2, a, f, u, \underline{z})\ DT(d_1, \llbracket \to X \rrbracket),\ DT(d_2, \llbracket (q,\underline{y})\ Aq \to B(T[q]) \rrbracket) \\ \qquad \to DT(join(D_1)[d_1, d_2], \llbracket (q,x)\ aq \to fx(uqx) \rrbracket)\end{array}\ (\mathrm{p0})}{\begin{array}{l}(a,f,u,\underline{z})\ DT(\mathcal{E}, \llbracket \to X \rrbracket),\ DT(D, \llbracket (q,\underline{y})\ Aq \to B(T[q]) \rrbracket) \\ \qquad \to DT(join(D_1)[\mathcal{E}, D], \llbracket (q,x)\ aq \to fx(uqx) \rrbracket)\end{array}}\,\mathrm{inst}}{(a,f,u,\underline{z})\ X,\ true \to DT(join(D_1)[\mathcal{E}, D], \llbracket (q,x)\ aq \to fx(uqx) \rrbracket)}\,\mathrm{red}}{(a,f,u,\underline{z})\ X \to \Box_d(a,f)(join(D_1)[\mathcal{E}, D], u)}\,\mathrm{conv}}{(a,f,u,\underline{z})\ a = A,\ f = F,\ u = U \to \Box_d(a,f)(join(D_1)[\mathcal{E}, D], u)}\,\&}{(\underline{z})\ \to \Box_d(A,F)(join(D_1\left[\begin{smallmatrix} A,F,U \\ a,f,u \end{smallmatrix}\right])[\mathcal{E}, D], U)}\,\mathrm{ext}}{(\underline{z})\ \to C(T'\,nil)}\,\mathrm{red}$$

Then by the *join*-lemma

$$DT(join(D_2)[Join(Code)], \llbracket(\underline{z}) \rightarrow C(T'\, nil)\rrbracket) \rhd^* true$$

which implies that $join(D_2)[Join(Code)] \rhd^* D'$ for some construction D'. So define a recursive function Pr_{18} such that

$$Pr_{18}((D, (\lambda[\underline{y}].T)), \{I \Rightarrow val\, P\, x\}, \{\Rightarrow \Box(I, P)\}) \rhd^* (D', (\lambda[\underline{z}].T')),$$

thus satisfying the decidable part of the proof relation.

To check the well-foundedness conditions,
$Join(Code) : rt$, by the *Join*-lemma
$D' \lhd^* join(D_2)[Join(Code)] : rt$, by the *join*-lemma, as required.
$\mathcal{E} : rt$
$join(D_1\begin{bmatrix} A,F,U \\ a,f,u \end{bmatrix})[\mathcal{E}, D] : rt$, by the *join*-lemma.
For any constructions $Q : \Pi$ and $X : leaf$,
$T\begin{bmatrix} X \\ x \end{bmatrix} : map(pi[\Pi], \Sigma)$, by the well-foundedness instantiation rule
$UQX \rhd^* T\begin{bmatrix} X \\ x \end{bmatrix}[Q] : \Sigma$.
$U : map(\Pi, map(leaf, \Sigma))$.
For any construction $Z : pi(nil)$,
$T'Z \rhd^* (join(D_1\begin{bmatrix} A,F,U \\ a,f,u \end{bmatrix})[\mathcal{E}, D], U) : product(rt, map(\Pi, map(leaf, \Sigma)))$
$\lhd^* \Box_w(\Pi, \Sigma) \rhd^* \Phi$.
$T' : map(pi(nil), \Phi)$, as required.
∎

THEOREM 23. (CPF $\Box$-elimination rule.) If $Q \Vdash\!\!= \Rightarrow \Box(I, P)$ then $Pr_{19}(Q, \{\Rightarrow \Box(I, P)\}, \{I \Rightarrow val\, P\, x\}) \Vdash\!\!= I \Rightarrow val\, P\, x$, where $x \notin I, P$, and the construction Pr_{19} is defined below.

Proof. Let $I \rhd^* pfn(A, \Pi) \not\rhd$ and $P \rhd^* pred(F, \Sigma) \not\rhd$. Then

$$\Box(I, P) \rhd^* pfn(\Box_d(A, F), \Box_w(\Pi, \Sigma)) \rhd^* pfn(B, \Phi) \not\rhd$$
$$val\, P\, x \rhd^* pfn(Fx, \Sigma) \rhd^* pfn(C, \Sigma) \not\rhd$$

for some terms B, C and type symbol Φ. Let $\underline{y}$ be the free variables of I, P, let $\underline{z}$ be the free variables of I, P, x, and let a, f, q, r, u be five fresh variables. The hypothesis that $Q \Vdash\!\!= \Rightarrow \Box(I, P)$ means that $Q \rhd^* (D, (\lambda[\underline{y}].T)) \not\rhd$ where

- $DT(D, \llbracket(\underline{y}) \rightarrow B(T\, nil)\rrbracket) \rhd^* true$,
- $D : rt$ and $T : map(pi(nil), \Phi)$.

Now let T' be $(\lambda[q].el_{\Box}(T\,nil)qx)$ and Y be $(\lambda nil{:}\,(\lambda[\underline{y}]{:}\,T\,nil))$. By the $\Box$-lemma

$$DT(El_{\Box}(Y)u, [\![(a,f,q,x,r)\ aq,\ \Box_d(a,f)(Yur) \to fx(el_{\Box}(Yur)qx)]\!]) \rhd^*\ true.$$

Let D_1 be the code of the protological derivation

$$\frac{\dfrac{(\underline{y}) \to B(T\,nil)\ \text{(p0)}}{(q,\underline{z}) \to B(T\,nil)}\,\text{thin} \qquad \dfrac{\dfrac{\dfrac{(a,f,q,x,r)\ aq,\ \Box_d(a,f)(Yur) \to fx(el_{\Box}(Yur)qx)\ \text{(p1)}}{(a,f,q,\underline{z},r)\ aq,\ \Box_d(a,f)(Yur) \to fx(el_{\Box}(Yur)qx)}\,\text{thin}}{(q,\underline{z})\,Aq,\ \Box_d(A,F)(Yu[\underline{y}]) \to Fx(el_{\Box}(Yu[\underline{y}])qx)}\,\text{inst}}{(q,\underline{z})\,Aq,\ B(T\,nil) \to C(T'[q])}\,\text{red}}{(q,\underline{z})\,Aq \to C(T'[q])}\,\text{cut}$$

Then by the *join*-lemma

$$DT(join(D_1)[D, El_{\Box}(Y)u], [\![(q,\underline{z})\,Aq \to C(T'[q])]\!]) \rhd^*\ true$$

which implies, by the instantiation theorem (14) of the Expanded Term Language,

$$DT(join(D_1\left[\begin{smallmatrix}0\\u\end{smallmatrix}\right])[D, El_{\Box}(Y)0], [\![(q,\underline{z})\,Aq \to C(T'[q])]\!]) \rhd^*\ true$$

which implies that $join(D_1\left[\begin{smallmatrix}0\\u\end{smallmatrix}\right])[D, El_{\Box}(Y)0] \rhd^* D'$ for some construction D'. So define a recursive function Pr_{19} such that

$$Pr_{19}((D,(\lambda[\underline{y}].T)), \{\Rightarrow \Box(I,P)\}, \{I \Rightarrow val\,P\,x\}) \rhd^* (D',(\lambda[\underline{z}].T')),$$

thus satisfying the decidable part of the proof relation.

To check the well-foundedness conditions,
$nil\ :\ pi(nil)$, by the *pi* rule
$T\,nil\ :\ \Phi \lhd^* \Box_w(\Pi,\Sigma)$
$Y\ :\ map(leaf, map(leaf, \Box_w(\Pi,\Sigma)))$, applying the *map* rule twice
$El_{\Box}(Y)\ :\ map(leaf, rt)$, by the $\Box$-lemma
$El_{\Box}(Y)0\ :\ rt$
$D' \lhd^* join(D_1\left[\begin{smallmatrix}0\\u\end{smallmatrix}\right])[D, El_{\Box}(Y)0]\ :\ rt$, by the *join*-lemma, as required.
For any construction $Z\ :\ pi[\Pi]$,
Z is $[Q]$, for some construction $Q\ :\ \Pi$
$T'Z \rhd^* el_{\Box}(T\,nil)Qx\ :\ \Sigma$, by the $\Box$-lemma.
$T'\ :\ map(pi[\Pi],\Sigma)$, as required.
∎

THEOREM 24. (CPF $\Box$-transitivity rule.) If $Q \VDash I,\ val\,P_1\,X \Rightarrow J$ then $Pr_{20}(Q, \{I,\ val\,P_1\,X \Rightarrow J\}, \{\Box(I,P_1),\ \Box(J,P_2) \Rightarrow \Box(I,P_2)\}) \VDash$
$\Box(I,P_1),\ \Box(J,P_2) \Rightarrow \Box(I,P_2)$,
where $X \not\triangleright$ and the construction Pr_{20} is defined below.

Proof. Let $I \triangleright^* pfn(A,\Pi) \not\triangleright$, $J \triangleright^* pfn(B,\Sigma) \not\triangleright$, $P_1 \triangleright^* pred(F,\Phi) \not\triangleright$, and $P_2 \triangleright^* pred(G,\Psi) \not\triangleright$. Then

$$
\begin{aligned}
&val\,P_1\,X \ \triangleright^*\ pfn(FX,\Phi)\ \triangleright^*\ pfn(F',\Phi) \not\triangleright\\
&\Box(I,P_1)\ \triangleright^*\ pfn(C_1,\Omega_1) \not\triangleright\\
&\Box(J,P_2)\ \triangleright^*\ pfn(C_2,\Omega_2) \not\triangleright\\
&\Box(I,P_2)\ \triangleright^*\ pfn(C_3,\Omega_3) \not\triangleright
\end{aligned}
$$

for some terms F', C_1, C_2, C_3 and type symbols $\Omega_1, \Omega_2, \Omega_3$. Let $\underline{y}$ be the free variables of I, P_1, X, J, let $\underline{z}$ be the free variables of I, P_1, J, P_2, and let $\underline{u}$ be the free variables of X that are not in $\underline{z}$.

Now let $q, r, t_1, t_2, t_4, d_0, d_1, d_2, d_3, a, b, f, g, v, w$ be fifteen fresh variables. The hypothesis that $Q \VDash I,\ val\,P_1\,X \Rightarrow J$ means that $Q \triangleright^* (D, (\lambda[\underline{y}].T)) \not\triangleright$, where

- $DT(D, [\![(q,r,\underline{y})\,Aq,\ F'r \to B(T[q,r])]\!]) \triangleright^*$ *true*,
- $D\ :\ rt$ and $T\ :\ map(pi[\Pi,\Phi],\Sigma)$.

Let $\overline{X}$ be $X\begin{bmatrix}\underline{0}\\ \underline{u}\end{bmatrix}$, where $\underline{0}$ is a sequence of 0s, let $\overline{T}$ be $T\begin{bmatrix}\underline{0}\\ \underline{u}\end{bmatrix}$, let T_4 be $(\lambda q.t_2(\overline{T}[q, t_1 q\overline{X}]))$, let Z be $(a = A)\,\&\,(b = B)\,\&\,(f = F)\,\&\,(t_4 = T_4)$, and let D_4 be the code of the protological derivation that begins with

$$
\dfrac{
\dfrac{\dfrac{\dfrac{\dfrac{(q,r,\underline{y})\,Aq,\ F'r \to B(T[q,r])\ \text{(p3)}}{(q,r,\underline{y})\,Aq,\ FXr \to B(T[q,r])}\,\text{red}}{(q,r)\,Aq,\ F\overline{X}r \to B(\overline{T}[q,r])}\,\text{inst}}{(q,r,w)Z, Aq,\ F\overline{X}r \to B(\overline{T}[q,r])}\,\text{thin}}{(q,r,w)Z, aq, f\overline{X}r \to b(\overline{T}[q,r])}\,\text{\&,eq}
\qquad
\dfrac{\dfrac{(v,w)\,bv \to gw(t_2vw)\ \text{(p2)}}{(q,r,v,w)\,bv \to gw(t_2vw)}\,\text{thin}}{(q,r,w)b(\overline{T}[q,r]) \to gw(t_2(\overline{T}[q,r])w)}\,\text{inst}
}{
\dfrac{\dfrac{\dfrac{(q,r,w)\,Z,\ aq,\ f\overline{X}r \to gw(t_2(\overline{T}[q,r])w)}{(q,w)\,Z,\ aq,\ f\overline{X}(t_1q\overline{X}) \to gw(t_2(\overline{T}[q,t_1q\overline{X}])w)}\,\text{inst}}{(q,w)\,Z,\ aq,\ f\overline{X}(t_1q\overline{X}) \to gw(T_4qw)}\,\text{red}}{(q,w)\,Z,\ aq,\ f\overline{X}(t_1q\overline{X}) \to gw(t_4qw)}\,\text{\&,eq}
}\,\text{cut}
$$

and then continues as follows

$$\dfrac{\dfrac{\dfrac{(q,x)\;aq \to fx(t_1qx)\ (\mathrm{p1})}{(q)\;aq \to f\overline{X}(t_1q\overline{X})}\,\mathrm{inst}}{(q,w)aq\to f\overline{X}(t_1q\overline{X})}\,\mathrm{thin} \qquad (q,w)Z,\, aq,\, f\overline{X}(t_1q\overline{X}) \to gw(t_4qw) \qquad \dfrac{\to Z\ (\mathrm{p0})}{(q,w) \to Z}\,\mathrm{thin}}{(q,w)\;aq \to gw(t_4qw)}\,\mathrm{cut}$$

Now let $\underline{p}$ be the sequence of variables $a,b,f,g,\underline{z}$. Then by the *Join*-lemma we have

$$\begin{aligned}
&DT(Join(Code), \llbracket(d_0,d_1,d_2,d_3,t_1,t_2,t_4,\underline{p})\\
&\qquad\qquad DT(d_0, \llbracket \to Z\rrbracket),\\
&\qquad\qquad DT(d_1, \llbracket(q,x)\;aq \to fx(t_1qx)\rrbracket),\\
&\qquad\qquad DT(d_2, \llbracket(v,w)\;bv \to gw(t_2vw)\rrbracket),\\
&\qquad\qquad DT(d_3, \llbracket(q,r,\underline{y})\;Aq,\; F'r \to B(T[q,r])\rrbracket)\\
&\qquad\qquad \to DT(join(D_4)[d_0,d_1,d_2,d_3], \llbracket(q,w)\;aq \to gw(t_4qw)\rrbracket)\rrbracket)\\
&\rhd^*\ true
\end{aligned}$$

where *Code* is

$$\begin{aligned}
&((\lambda[t_1,t_2,t_4,\underline{p}].[\llbracket \to Z\rrbracket,\\
&\qquad\qquad \llbracket(q,x)\;aq \to fx(t_1qx)\rrbracket,\\
&\qquad\qquad \llbracket(v,w)\;bv \to gw(t_2vw)\rrbracket,\\
&\qquad\qquad \llbracket(q,r,\underline{y})\;Aq,\; F'r \to B(T[q,r])\rrbracket]),\\
&(\lambda[t_1,t_2,t_4,\underline{p}].(D_4, \llbracket(q,w)\;aq \to gw(t_4qw)\rrbracket))).
\end{aligned}$$

Let T' be $(\lambda[(d_1,t_1),(d_2,t_2)].(join(D_4\left[\begin{smallmatrix}T_4,A,B,F,G\\ t_4,a,b,f,g\end{smallmatrix}\right])[\mathcal{E},d_1,d_2,D],T_4))$, let Γ be the two terms

$$DT(d_1, \llbracket(q,x)\;aq \to fx(t_1qx)\rrbracket),\; DT(d_2, \llbracket(v,w)\;bv \to gw(t_2vw)\rrbracket),$$

let Δ be the four terms

$$a = A,\; b = B,\; f = F,\; g = G,$$

and D_5 be the code of the protological derivation

$$(d_0, d_1, d_2, d_3, t_1, t_2, t_4, \underline{p})\, DT(d_0, [\![\ \to Z]\!]),\ \Gamma,$$
$$DT(d_3, [\![(q, r, \underline{y})\, Aq,\ F'r \to B(T[q,r])]\!]) \qquad \text{(p0)}$$
$$\to DT(join(D_4)[d_0, d_1, d_2, d_3], [\![(q, w)\ aq \to gw(t_4qw)]\!])$$

inst

$$(d_1, d_2, t_1, t_2, t_4, \underline{p})\, DT(\mathcal{E}, [\![\ \to Z]\!]),\ \Gamma,$$
$$DT(D, [\![(q, r, \underline{y})\, Aq,\ F'r \to B(T[q,r])]\!])$$
$$\to DT(join(D_4)[\mathcal{E}, d_1, d_2, D], [\![(q, w)\ aq \to gw(t_4qw)]\!])$$

red

$$(d_1, d_2, t_1, t_2, t_4, \underline{p})\, Z,\ \Gamma,\ true$$
$$\to DT(join(D_4)[\mathcal{E}, d_1, d_2, D], [\![(q, w)\ aq \to gw(t_4qw)]\!])$$

conv

$$(d_1, d_2, t_1, t_2, t_4, \underline{p})\, Z,\ \Gamma \to \Box_d(a, g)(join(D_4)[\mathcal{E}, d_1, d_2, D], t_4)$$

thin, &

$$(d_1, d_2, t_1, t_2, t_4, \underline{p})\ t_4 = T_4,\ \Delta,\ \Gamma \to \Box_d(a, g)(join(D_4)[\mathcal{E}, d_1, d_2, D], t_4)$$

ext

$$(d_1, d_2, t_1, t_2, \underline{p})\ \Delta,\ \Gamma \to \Box_d(a, g)(join(D_4\begin{bmatrix} T_4 \\ t_4 \end{bmatrix})[\mathcal{E}, d_1, d_2, D], T_4)$$

eq

$$(d_1, d_2, t_1, t_2, \underline{p})\ \Delta,\ \Gamma \to \Box_d(a, g)(join(D_4\begin{bmatrix} T_4, A, B, F, G \\ t_4, a, b, f, g \end{bmatrix})[\mathcal{E}, d_1, d_2, D], T_4)$$

red

$$(d_1, d_2, t_1, t_2, \underline{p})\ \Delta,\ \Gamma \to \Box_d(a, g)(T'[(d_1, t_1), (d_2, t_2)])$$

$\Box_d$-rule

$$(q, r, \underline{p})\ \Delta,\ \Box_d(a, f)q,\ \Box_d(b, g)r \to \Box_d(a, g)(T'[q, r])$$

ext

$$(q, r, \underline{z})\ \Box_d(A, F)q,\ \Box_d(B, G)r \to \Box_d(A, G)(T'[q, r])$$

red

$$(q, r, \underline{z})\ C_1q,\ C_2r \to C_3(T'[q, r])$$

Then by the *join*-lemma

$$DT(join(D_5)[Join(Code)], [\![(q, r, \underline{z})\ C_1q,\ C_2r \to C_3(T'[q, r])]\!]) \ \rhd^*\ true$$

which implies that $join(D_5)[Join(Code)] \ \rhd^*\ D'$ for some construction D'. So define a recursive function Pr_{20} such that

$$Pr_{20}((D, (\lambda[\underline{y}].T)), \{I,\ val\, P_1\, X \Rightarrow J\}, \{\Box(I, P_1),\ \Box(J, P_2) \Rightarrow \Box(I, P_2)\})$$
$$\rhd^*\ (D', (\lambda[\underline{z}].T'))$$

thus satisfying the decidable part of the proof relation.

To check the well-foundedness conditions,
$Join(Code)\ :\ rt$, by the *Join*-lemma
$D'\ \lhd^*\ join(D_5)[Join(Code)]\ :\ rt$, by the *join*-lemma, as required.
For any construction $M\ :\ pi[\Omega_1, \Omega_2]\ \lhd^*\ pi[\Box_w(\Pi, \Phi), \Box_w(\Sigma, \Psi)]$,
M is $[Q, R]$, for some constructions $Q\ :\ \Box_w(\Pi, \Phi)$ and $R\ :\ \Box_w(\Sigma, \Psi)$
Q is (D_1, T_1) for some constructions $D_1\ :\ rt$ and $T_1\ :\ map(\Pi, map(leaf, \Phi))$
R is (D_2, T_2) for some constructions $D_2\ :\ rt$ and $T_2\ :\ map(\Sigma, map(leaf, \Psi))$

$\mathcal{E} : rt$
$join(D_4\left[{T_4,A,B,F,G \atop t_4,a,b,f,g}\right]\left[{T_1,T_2 \atop t_1,t_2}\right])[\mathcal{E},D_1,D_2,D] : rt$, by the *join*-lemma
$\overline{T} : map(pi[\Pi,\Phi],\Sigma)$, by the well-foundedness instantiation rule
For any constructions $U : \Pi$ and $V : leaf$,
$T_1 U\overline{X} : \Phi$
$\overline{T}[U, T_1 U\overline{X}] : \Sigma$
$T_4\left[{T_1,T_2 \atop t_1,t_2}\right] UV \rhd^* T_2(\overline{T}[U,T_1 U\overline{X}])V : \Psi$.
$T_4\left[{T_1,T_2 \atop t_1,t_2}\right] : map(\Pi, map(leaf,\Psi))$
$T'M \rhd^* (join(D_4\left[{T_4,A,B,F,G \atop t_4,a,b,f,g}\right]\left[{T_1,T_2 \atop t_1,t_2}\right])[\mathcal{E},D_1,D_2,D], T_4\left[{T_1,T_2 \atop t_1,t_2}\right])$
$: product(rt, map(\Pi, map(leaf,\Psi))) \lhd^* \Box_w(\Pi,\Psi) \rhd^* \Omega_3$.
$T' : map(pi[\Omega_1,\Omega_2],\Omega_3)$, as required.
∎

THEOREM 25. (CPF cut rule.)
If $Q_1 \models \Gamma \Rightarrow I$ and $Q_2 \models I,\ \Delta \Rightarrow J$
then $Pr_{21}(Q_1,Q_2,\{\Gamma \Rightarrow I\},\{I,\ \Delta \Rightarrow J\},\{\Gamma,\ \Delta \Rightarrow J\}) \models \Gamma,\ \Delta \Rightarrow J$,
where the construction Pr_{21} is defined below.

Proof. Let $I \rhd^* pfn(A,\Pi) \rhd^*$, let $J \rhd^* pfn(B,\Sigma) \not\rhd$, let the proof functions of Γ reduce to $pfn(C_1,\Phi_1),\dots pfn(C_m,\Phi_m) \not\rhd$, and let the proof functions of Δ reduce to $pfn(E_1,\Psi_1),\dots pfn(E_n,\Psi_n) \not\rhd$.

Let $\underline{x}$ be the free variables of Γ, I, let $\underline{y}$ be the free variables of I,Δ,J, let $\underline{z}$ be the free variables of Γ,Δ,J, let $r_1,\dots r_m, q, u_1,\dots u_n$ be $m+n+1$ fresh variables, and abbreviate $r_1,\dots r_m$ to $\underline{r}$ and $u_1,\dots u_n$ to $\underline{u}$. Let Γ' be the sequence of terms $C_1 r_1,\dots C_m r_m$ and Δ' be the sequence of terms $E_1 u_1,\dots E_n u_n$. The hypotheses that $Q_1 \models \Gamma \Rightarrow I$ and $Q_2 \models I,\ \Delta \Rightarrow J$ mean that $Q_1 \rhd^* (D_1,(\lambda[\underline{x}].T_1)) \not\rhd$ and $Q_2 \rhd^* (D_2,(\lambda[\underline{y}].T_2)) \not\rhd$, where

- $DT(D_1, [\![(\underline{r},\underline{x})\ \Gamma' \to A(T_1[\underline{r}])]\!]) \rhd^* true$
- $D_1 : rt$ and $T_1 : map(pi[\Phi_1,\dots\Phi_m],\Pi)$
- $DT(D_2, [\![(q,\underline{u},\underline{y})\ Aq,\ \Delta' \to B(T_2[q,\underline{u}])]\!]) \rhd^* true$
- $D_2 : rt$ and $T_2 : map(pi[\Pi,\Psi_1,\dots\Psi_n],\Sigma)$.

Let $\underline{w}$ be the variables of $\underline{x}$ and $\underline{y}$ that are not in $\underline{z}$, let T' be the term $(\lambda[\underline{r},\underline{u}].T_2[T_1[\underline{r}],\underline{u}]\left[{\underline{0} \atop \underline{w}}\right])$, where $\underline{0}$ is a sequence of 0s, and let D_0 be the code of the protological derivation

$$\dfrac{\dfrac{(\underline{r},\underline{x})\ \Gamma' \to A(T_1[\underline{r}])\ \text{(p0)}}{(\underline{r},\underline{u},\underline{x},\underline{y})\ \Gamma' \to A(T_1[\underline{r}])}\,\text{thin} \qquad \dfrac{\dfrac{(q,\underline{u},\underline{y})\ Aq,\ \Delta' \to B(T_2[q,\underline{u}])\ \text{(p1)}}{(\underline{r},q,\underline{u},\underline{x},\underline{y})\ Aq,\ \Delta' \to B(T_2[q,\underline{u}])}\,\text{thin}}{(\underline{r},\underline{u},\underline{x},\underline{y})\ A(T_1[\underline{r}]),\ \Delta' \to B(T_2[T_1[\underline{r}],\underline{u}])}\,\text{inst}}{\dfrac{\dfrac{(\underline{r},\underline{u},\underline{x},\underline{y})\ \Gamma',\ \Delta' \to B(T_2[T_1[\underline{r}],\underline{u}])}{(\underline{r},\underline{u},\underline{z})\ \Gamma',\ \Delta' \to B(T_2[T_1[\underline{r}],\underline{u}]\left[\frac{0}{\underline{w}}\right])}\,\text{inst}}{(\underline{r},\underline{u},\underline{z})\ \Gamma',\ \Delta' \to B(T'[\underline{r},\underline{u}])}\,\text{red}}\,\text{cut}$$

Then by the *join*-lemma

$$DT(join(D_0)[D_1,D_2],[\![(\underline{r},\underline{u},\underline{z})\ \Gamma',\ \Delta' \to B(T'[\underline{r},\underline{u}])]\!])\ \rhd^*\ true$$

which implies that $join(D_0)[D_1,D_2]\ \rhd^*\ D'$ for some construction D'. So define a recursive function Pr_{21} such that

$$Pr_{21}((D_1,(\lambda[\underline{x}].T_1)),(D_2,(\lambda[\underline{y}].T_2)),\{\Gamma \Rightarrow I\},\{I,\ \Delta \Rightarrow J\},\{\Gamma,\ \Delta \Rightarrow J\})$$
$$\rhd^*\ (D',(\lambda[\underline{z}].T')),$$

thus satisfying the decidable part of the proof relation.

To check the well-foundedness conditions,
$D'\ \lhd^*\ join(D_0)[D_1,D_2]\ :\ rt$, by the *join*-lemma, as required.
For any construction $X\ :\ pi[\Phi_1,\dots\Phi_m,\Psi_1,\dots\Psi_n]$,
X is $[R_1,\dots R_m,U_1,\dots U_n]$, for some constructions $R_1\ :\ \Phi_1,\dots R_m\ :\ \Phi_m$, $U_1\ :\ \Psi_1,\dots U_n\ :\ \Psi_n$
$T_2[T_1[R_1,\dots R_m],U_1,\dots U_n]\ :\ \Sigma$
$T'X\ \rhd^*\ T_2[T_1[R_1,\dots R_m],U_1,\dots U_n]\left[\frac{0}{\underline{w}}\right]\ :\ \Sigma$, by the well-foundedness instantiation rule.
$T'\ :\ map(pi[\Phi_1,\dots\Phi_m,\Psi_1,\dots\Psi_n],\Sigma)$, as required.
▮

THEOREM 26. (CPF *boolean*-elimination rule.)
If $Q_1 \Vdash \ulcorner x = true \urcorner,\ \Gamma \Rightarrow I$ and $Q_2 \Vdash \ulcorner x = false \urcorner,\ \Gamma \Rightarrow I$ then
$Pr_{22}(Q_1,Q_2,\{\ulcorner x = true \urcorner,\ \Gamma \Rightarrow I\},\{\ulcorner x = false \urcorner,\ \Gamma \Rightarrow I\},$
$\{\ulcorner boolean\,x \urcorner,\ \Gamma \Rightarrow I\}) \Vdash \ulcorner boolean\,x \urcorner,\ \Gamma \Rightarrow I,$
where the construction Pr_{22} is defined below.

Proof. Let $I\ \rhd^*\ pfn(A,\Pi)\ \not\rhd$ and let the proof functions Γ reduce to $pfn(B_1,\Sigma_1),\dots pfn(B_k,\Sigma_k)\ \not\rhd$. Note that

$$\ulcorner x = true \urcorner \text{ is } pfn((\lambda nil.x = true), leaf),$$
$$\ulcorner x = false \urcorner \text{ is } pfn((\lambda nil.x = false), leaf),$$
$$\ulcorner boolean\,x \urcorner \text{ is } pfn((\lambda nil.boolean\,x), leaf).$$

Let $\underline{z}$ be the free variables of x, Γ, I, let $q, r_1, \ldots r_k, y, v$ be $k+3$ fresh variables, and abbreviate $r_1, \ldots r_k$ to $\underline{r}$ and $B_1 r_1, \ldots B_k r_k$ to Γ'. The hypotheses that $Q_1 \Vdash \ulcorner x = true \urcorner, \Gamma \Rightarrow I$ and $Q_2 \Vdash \ulcorner x = false \urcorner, \Gamma \Rightarrow I$ mean that $Q_1 \rhd^* (D_1, (\lambda[\underline{z}].T_1)) \not\rhd$ and $Q_2 \rhd^* (D_2, (\lambda[\underline{z}].T_2)) \not\rhd$, where

- $DT(D_1, [\![(q, \underline{r}, \underline{z})\ (\lambda nil.x = true)q,\ \Gamma' \rightarrow A(T_1[q, \underline{r}])]\!]) \rhd^*\ true$ and $DT(D_2, [\![(q, \underline{r}, \underline{z})\ (\lambda nil.x = false)q,\ \Gamma' \rightarrow A(T_2[q, \underline{r}])]\!]) \rhd^*\ true$,
- $D_1, D_2\ :\ rt$ and $T_1, T_2\ :\ map(pi[leaf, \Sigma_1, \ldots \Sigma_k], \Pi)$.

Now, let T_1' be $(\lambda y.T_1 y)$, let T_2' be $(\lambda y.T_2 y)$, let F be $(\lambda v.if\ v\ T_1'\ T_2'\ [q, \underline{r}])$, let T be $(\lambda y.if\ x\ T_1'\ T_2'\ y)$ and let D_0 be the code of the protological derivation that starts with

$$\cfrac{\cfrac{(q, \underline{r}, \underline{z})\ (\lambda nil.x = true)q,\ \Gamma' \rightarrow A(T_1[q, \underline{r}])\ \text{(p0)}}{(q, \underline{r}, \underline{z})\ x = true,\ \Gamma' \rightarrow A(F\ true)}\ \text{red}}{(q, \underline{r}, \underline{z})\ x = true,\ \Gamma' \rightarrow A(Fx)}\ \text{eq}$$

and similarly derives

$$\cfrac{\cfrac{(q, \underline{r}, \underline{z})\ (\lambda nil.x = false)q,\ \Gamma' \rightarrow A(T_2[q, \underline{r}])\ \text{(p1)}}{(q, \underline{r}, \underline{z})\ x = false,\ \Gamma' \rightarrow A(F\ false)}\ \text{red}}{(q, \underline{r}, \underline{z})\ x = false,\ \Gamma' \rightarrow A(Fx)}\ \text{eq}$$

and then combines the two branches using the Boolean Rule to give

$$\cfrac{(q, \underline{r}, \underline{z})\ boolean\, x,\ \Gamma' \rightarrow A(Fx)\ \text{(bool)}}{(q, \underline{r}, \underline{z})\ (\lambda nil.boolean\, x)q,\ \Gamma' \rightarrow A(T[q, \underline{r}])}\ \text{red}$$

Then by the *join*-lemma

$$DT(join(D_0)[D_1, D_2], [\![(q, \underline{r}, \underline{z})\ (\lambda nil.boolean\, x)q,\ \Gamma' \rightarrow A(T[q, \underline{r}])]\!]) \rhd^*\ true$$

which implies that $join(D_0)[D_1, D_2] \rhd^* D$ for some construction D. So define a recursive function Pr_{22} such that

$$Pr_{22}((D_1, (\lambda[\underline{z}].T_1)), (D_2, (\lambda[\underline{z}].T_2)), \{\ulcorner x = true \urcorner, \Gamma \Rightarrow I\},$$
$$\{\ulcorner x = false \urcorner, \Gamma \Rightarrow I\}, \{\ulcorner boolean\, x \urcorner, \Gamma \Rightarrow I\}) \rhd^* (D, (\lambda[\underline{z}].T)),$$

thus satisfying the decidable part of the proof relation.

To check the well-foundedness conditions,
$D \lhd^*\ join(D_0)[D_1, D_2]\ :\ rt$, by the *join*-lemma, as required.
For any construction $Y\ :\ pi[leaf, \Sigma_1, \ldots \Sigma_k]$,
$T_1' Y \rhd^* T_1 Y\ :\ \Pi$ and $T_2' Y \rhd^* T_2 Y\ :\ \Pi$.

$T_1' : map(pi[leaf, \Sigma_1, \ldots \Sigma_k], \Pi)$ and $T_2' : map(pi[leaf, \Sigma_1, \ldots \Sigma_k], \Pi)$
$if\, x\, T_1'\, T_2' : map(pi[leaf, \Sigma_1, \ldots \Sigma_k], \Pi)$, by the *if* well-foundedness rule.
For any construction $Y : pi[leaf, \Sigma_1, \ldots \Sigma_k]$,
$\quad TY \triangleright^* if\, x\, T_1'\, T_2'\, Y : \Pi$.
$T : map(pi[leaf, \Sigma_1, \ldots \Sigma_k], \Pi)$, as required.
∎

THEOREM 27. (CPF induction rule.)
If $Q_1 \Vdash\!\!= \Rightarrow val\, P\, 0$ and $Q_2 \Vdash\!\!= val\, P\, n \Rightarrow val\, P\,(Sn)$ then
$Pr_{23}(Q_1, Q_2, \{\Rightarrow val\, P\, 0\}, \{val\, P\, n \Rightarrow val\, P\,(Sn)\}, \{\ulcorner num\, n \urcorner \Rightarrow val\, P\, n\})$
$\quad \Vdash\!\!= \ulcorner num\, n \urcorner \Rightarrow val\, P\, n$,
where $n \notin P$ and the construction Pr_{23} is defined below.

Proof. Let $P \triangleright^* pred(F, \Pi) \not\triangleright$, let $\underline{y}$ be the free variables of P, let $\underline{z}$ be the free variables of P, n, and let q, a, g, x be four fresh variables. Thus $val\, P\, x \triangleright^* pfn(Fx, \Pi) \triangleright^* pfn(B, \Pi) \not\triangleright$, for some term B. Hence $val\, P\, 0 \triangleright^* pfn(B\left[\begin{smallmatrix}0\\x\end{smallmatrix}\right], \Pi) \not\triangleright$, and similarly for $val\, P\, n$ and $val\, P\,(Sn)$. The hypotheses that $Q_1 \Vdash\!\!= \Rightarrow val\, P\, 0$ and $Q_2 \Vdash\!\!= val\, P\, n \Rightarrow val\, P\,(Sn)$ mean that $Q_1 \triangleright^* (D_1, (\lambda[\underline{y}].T_1)) \not\triangleright$ and $Q_2 \triangleright^* (D_2, (\lambda[\underline{z}].T_2)) \not\triangleright$, where

- $DT(D_1, \llbracket(\underline{y}) \to B\left[\begin{smallmatrix}0\\x\end{smallmatrix}\right](T_1\, nil)\rrbracket) \triangleright^* true$
- $D_1 : rt$ and $T_1 : map(pi(nil), \Pi)$
- $DT(D_2, \llbracket(q, \underline{z})\, B\left[\begin{smallmatrix}n\\x\end{smallmatrix}\right] q \to B\left[\begin{smallmatrix}Sn\\x\end{smallmatrix}\right](T_2[q])\rrbracket) \triangleright^* true$
- $D_2 : rt$ and $T_2 : map(pi[\Pi], \Pi)$.

Recall from the Expanded Term Language the recursion operator $reco_0$, satisfying $reco_0(a, g) \triangleright^* T_0$, $T_0(0) \triangleright^* a$ and $T_0(Sn) \triangleright^* g(n, T_0 n)$. Let A be $T_1\, nil$, G be $(\lambda(n, q).T_2[q])$, C be $(\lambda n.Fn(reco_0(A, G)n))$, and T be $(\lambda nil.reco_0(A, G)n)$; then let D_0 be the code of the protological derivation

$$
\dfrac{
\dfrac{
\dfrac{\dfrac{(\underline{y}) \to B\left[\begin{smallmatrix}0\\x\end{smallmatrix}\right](T_1\, nil)\ (\text{p0})}{(\underline{y}) \to C0}\,\text{conv}
\qquad
\dfrac{\dfrac{\dfrac{(q, \underline{z})\, B\left[\begin{smallmatrix}n\\x\end{smallmatrix}\right] q \to B\left[\begin{smallmatrix}Sn\\x\end{smallmatrix}\right](T_2[q])\ (\text{p1})}{(q, \underline{z})\, Fnq \to F(Sn)(G(n, q))}\,\text{red}}{(\underline{z})\, Fn(reco_0(A, G)n) \to F(Sn)(G(n, reco_0(A, G)n))}\,\text{inst}}{(\underline{z})\, Cn \to C(Sn)}\,\text{conv}}
{(n, \underline{y})\, num\, n \to Cn}\,\text{ind}}
{(q, \underline{z})\, num\, n \to Cn}\,\text{thin}}
{(q, \underline{z})\, (\lambda nil.num\, n)q \to B\left[\begin{smallmatrix}n\\x\end{smallmatrix}\right](T[q])}\,\text{red}
$$

Then by the *join*-lemma

$$DT(join(D_0)[D_1, D_2], [\![(q, \underline{z})\ (\lambda nil.num\, n)q \rightarrow B\begin{bmatrix} n \\ x \end{bmatrix}(T[q])]\!])\ \triangleright^*\ true$$

which implies that $join(D_0)[D_1, D_2]\ \triangleright^*\ D$ for some construction D. So define a recursive function Pr_{23} such that

$$Pr_{23}((D_1, (\lambda[\underline{y}].T_1)), (D_2, (\lambda[\underline{z}].T_2)), \{\Rightarrow val\, P\, 0\}, \{val\, P\, n \Rightarrow val\, P\, (Sn)\},$$
$$\{\ulcorner num\, n \urcorner \Rightarrow val\, P\, n\})\ \triangleright^*\ (D, (\lambda[\underline{z}].T)),$$

thus satisfying the decidable part of the proof relation.

To check the well-foundedness conditions,

$D\ \triangleleft^*\ join(D_0)[D_1, D_2]\ :\ rt$, by the *join*-lemma, as required.

$nil\ :\ pi(nil)$, by the *pi*-rule

A is $T_1\, nil\ :\ \Pi$.

For any construction $W\ :\ product(leaf, \Pi)$,

W is (N, Q), for some constructions $N\ :\ leaf$ and $Q\ :\ \Pi$

$T_2\begin{bmatrix} N \\ n \end{bmatrix}\ :\ map(pi[\Pi], \Pi)$, by the well-foundedness instantiation rule

$GW\ \triangleright^*\ T_2\begin{bmatrix} N \\ n \end{bmatrix}[Q]\ :\ \Pi$.

$G\ :\ map(product(leaf, \Pi), \Pi)$.

$rec_0(A, G)\ :\ map(leaf, \Pi)$, by the recursion well-foundedness rule.

For any construction $Z\ :\ pi[leaf]$,

$TZ\ \triangleright^*\ rec_0(A, G)n\ :\ \Pi$.

$T\ :\ map(pi[leaf], \Pi)$, as required.

▮

FORMAL INTERPRETATION OF CPF IN PROTOLOGIC

The previous section has shown that every axiom of CPF has a proof (in the $\Vdash$ sense) and every rule of inference preserves provability. It follows that every CPF theorem is provable. In this section I shall show explicitly how to generate a proof of a logical sequent from its CPF derivation. First we need a way of coding CPF derivations as constructions.

DEFINITION. A coding of CPF derivations as constructions is defined as follows. Number the axiom schemata and rules of CPF, 1,2,... 23, in the order in which they are listed above. A derivation with conclusion $\Gamma \Rightarrow I$, from a list of premises numbered 0,1,2,..., is coded as

$(premise(i), \{\Gamma \Rightarrow I\})$	if $\Gamma \Rightarrow I$ is premise number i
$(none(n), \{\Gamma \Rightarrow I\})$	if $\Gamma \Rightarrow I$ is an instance of axiom schema n
$(one(n, P), \{\Gamma \Rightarrow I\})$	if $\Gamma \Rightarrow I$ is derived by rule n from one subderivation, P
$(two(n, P, Q), \{\Gamma \Rightarrow I\})$	if $\Gamma \Rightarrow I$ is derived by rule n from two subderivations, P and Q.

(Note that the conclusion sequent is part of the code of the derivation. Thus this code is analogous to the code of a protological derivation *tree*, rather than the code of a protological derivation – see the beginning of Chapter 21.)

DEFINITION. Define a construction *spr* by

$$spr(none(1), a) \triangleq Pr_1(a)$$

$$\vdots$$

$$spr(none(13), a) \triangleq Pr_{13}(a)$$

$$spr(one(14, (u, b)), a) \triangleq Pr_{14}(spr(u, b), b, a)$$

$$\vdots$$

$$spr(one(20, (u, b)), a) \triangleq Pr_{20}(spr(u, b), b, a)$$

$$spr(two(21, (u, b), (v, c)), a) \triangleq Pr_{21}(spr(u, b), spr(v, c), b, c, a)$$

$$spr(two(22, (u, b), (v, c)), a) \triangleq Pr_{22}(spr(u, b), spr(v, c), b, c, a)$$

$$spr(two(23, (u, b), (v, c)), a) \triangleq Pr_{23}(spr(u, b), spr(v, c), b, c, a)$$

where a, u, b, v, c are five variables.

THEOREM 28. If D is the code of a CPF derivation (with no premises) of a logical sequent $\Gamma \Rightarrow I$ then $spr(D) \models\!\!\!\models \Gamma \Rightarrow I$.

Proof. By structural induction on the CPF derivation using the extensionality of $\models\!\!\!\models$ (Theorem 1 of this chapter). ∎

THEOREM 29. (CPF interpretation theorem.) If D is the code of a CPF derivation (with no premises) of a logical sequent $\Rightarrow I$, where I has no free variables, then $pr(spr(D)) \vdash I$.

Proof. By the previous theorem, $spr(D) \models\!\!\!\models \Rightarrow I$ so, by the soundness theorem for $\models\!\!\!\models$ with respect to $\vdash$ (Theorem 2 of this chapter), $pr(spr(D)) \vdash I$.
∎

Future chapters will apply this interpretation theorem to CPF derivations compiled from derivations in higher-level formal systems. The compilation process requires the following function, *glue*, which is used to attach CPF derivations to the premises of other CPF derivations. It is a variation of the function *join* defined for protological derivations, and can be used on derivations in a variety of formal systems, provided they are coded in a similar way to CPF derivations.

DEFINITION. Define a construction *glue* by

$$\begin{aligned}
glue((premise(0), a), ((d, a), rest)) &\triangleq (d, a) \\
glue((premise(Si), a), (first, rest)) &\triangleq glue((premise(i), a), rest) \\
glue((none(n), a), dlist) &\triangleq (none(n), a) \\
glue((one(n, p), a), dlist) &\triangleq (one(n, glue(p, dlist)), a) \\
glue((two(n, p, q), a), dlist) &\triangleq \\
&\qquad (two(n, glue(p, dlist), glue(q, dlist)), a).
\end{aligned}$$

where $a, d, i, first, rest, n, p, q, dlist$ are nine variables.

THEOREM 30. (The *glue*-lemma for CPF.) If D is the code of a CPF derivation of a logical sequent S from premise sequents $T_0, \ldots T_k$, and $D_0, \ldots D_k$ are the codes of CPF derivations (with no premises) of $T_0, \ldots T_k$ respectively, then $glue(D, [D_0, \ldots D_k]) \rhd^*$ the code of a CPF derivation (with no premises) of S.

Proof. By a straightforward structural induction on the derivation encoded by D. ∎

EXAMPLE. Let D be the coded CPF derivation

$$(two(21, (one(16, (premise(0), W)), X), (premise(1), Y)), Z),$$

where W, X, Y, Z are coded logical sequents; note that D is the code of a derivation of Z from the premises W and Y. Let D_0 be $(one(17, (none(5), U)), W)$, a coded CPF derivation of W; and let D_1 be $(none(10), Y)$, a coded CPF derivation of Y. Then

$$\begin{aligned}
&glue(D, [D_0, D_1]) \\
&\quad \rhd^* (two(21, (one(16, \underbrace{(one(17, (none(5), U)), W)}_{D_0}), X), \underbrace{(none(10), Y)}_{D_1}), Z)
\end{aligned}$$

which is the code of a CPF derivation of Z without premises.

CHAPTER 28

CALCULUS OF PROOF FUNCTIONS

LOGICAL SEQUENTS

A *logical sequent* is a sentence in the following language.

- The alphabet is that of the Expanded Term Language plus ',' and '$\Rightarrow$'.
- The tokens are those of the Expanded Term Language plus ',' and '$\Rightarrow$'.
- Lexical analysis consists of recognising constants and replacing them by the token '*con*', recognising variables and replacing them by the token '*vbl*', and recognising every other token.
- The grammar of the language is as follows.
  - The terminals are the tokens.
  - The non-terminals are *S*, *LHS*, *Ps*, *P*, *T*; the start symbol is *S*.
  - The production rules are

$$S \rightarrow LHS \Rightarrow P$$
$$LHS \rightarrow Ps \mid \varepsilon$$
$$Ps \rightarrow P \, , \, Ps \mid P$$
$$P \rightarrow T$$
$$T \rightarrow con \mid (\, vbl \,) \mid (\, T\, T \,) \mid (\, \lambda\, T \,.\, T \,) \mid (\, T \left[{}^{T}_{vbl} \right])$$

In addition there is a context-sensitive constraint, that all the substrings matching *P* must be proof functions.

METANOTATION

All the metanotation of the Expanded Term Language will be allowed in the terms in logical sequents. The letters '*I*', '*J*' and '*K*' will be used as metavariables denoting proof functions, the letters 'Γ' and 'Δ' as metavariables denoting (possibly empty) sequences of proof functions, and the letters '*P*', 'P_1' and 'P_2' as metavariables denoting predicates. *Metasequents* and their *instances* are defined by analogy with metaterms and their instances in the Expanded Term Language. The typical metanotation for a sequent will be '$\Gamma \Rightarrow I$'.

THE PROOF RELATION, $\models\!\!\models$, FOR LOGICAL SEQUENTS

A proof relation $\models\!\!\models$ is defined between terms with no free variables and logical sequents. A construction *pr* is defined.

THEOREM 1. (Extensionality of $\models\!\!\models$.) If $Q \vartriangleright^* \vartriangleleft Q'$ then $Q \models\!\!\models \Gamma \rightarrow I$ iff $Q' \models\!\!\models \Gamma \rightarrow I$.

THEOREM 2. (Soundness of $\models\!\!\models$ with respect to $\vdash$.) If $Q \models\!\!\models \Rightarrow I$, where I has no free variables, then $pr(Q) \vdash I$.

THEOREM 3. (Completeness of $\models\!\!\models$ with respect to $\vdash$.) If $Q \vdash I$ then $in(\lambda[\underline{z}].(Q,I)) \models\!\!\models I$, where $\underline{z}$ are the free variables of I and Q, for a certain construction *in*.

THEOREM 4. For any proof function I and predicate P, there is no term Q such that $Q \models\!\!\models \Box(I,P) \Rightarrow \ulcorner false \urcorner$.

CALCULUS OF PROOF FUNCTIONS (CPF)

The axioms and rules of CPF are all instances of the following schemata.

Truth introduction (*true*-in): $\Rightarrow \ulcorner true \urcorner$

Falsity elimination (*false*-el): $\ulcorner false \urcorner \Rightarrow I$

Reduction (red): $I \Rightarrow J$ where $I \vartriangleright J$ or $J \vartriangleright I$;
$\ulcorner X \urcorner \Rightarrow \ulcorner Y \urcorner$ where $X \vartriangleright Y$ or $Y \vartriangleright X$

$\wedge$-introduction ($\wedge$-in): $I,\ J \Rightarrow \wedge(I,J)$

$\wedge$-elimination ($\wedge$-el): $\wedge(I,J) \Rightarrow I \qquad \wedge(I,J) \Rightarrow J$

$\exists$-introduction ($\exists$-in): $val\,P\,x \Rightarrow \exists P$

$\Box$-expansion ($\Box$-exp): $I \Rightarrow \Box(J, con\,I)$

$\Box$-compression ($\Box$-comp): $\Box(\ulcorner true \urcorner, \Delta(I,P)) \Rightarrow \Box(I,P)$

Term existence (te): $\ulcorner T\left[\begin{smallmatrix}X\\x\end{smallmatrix}\right] \urcorner \Rightarrow \exists(predify(\lambda x.\ulcorner x = X \urcorner))$
where $x \in T$ but $x \notin X$

Equality (eq): $\ulcorner X = Y \urcorner,\ val\,P\,X \Rightarrow val\,P\,Y$ where $X, Y \not\vartriangleright$;
$\ulcorner X = Y \urcorner,\ \ulcorner T\left[\begin{smallmatrix}X\\u\end{smallmatrix}\right] \urcorner \Rightarrow \ulcorner T\left[\begin{smallmatrix}Y\\u\end{smallmatrix}\right] \urcorner$

Thinning (thin): $\dfrac{\Gamma \Rightarrow J}{I,\ \Gamma \Rightarrow J}$

Exchange (exch): $\dfrac{\Gamma,\ I,\ J,\ \Delta \Rightarrow K}{\Gamma,\ J,\ I,\ \Delta \Rightarrow K}$

Contraction (con): $\dfrac{I,\ I,\ \Gamma \Rightarrow J}{I,\ \Gamma \Rightarrow J}$

$\exists$-elimination ($\exists$-el): $\dfrac{val\,P\,x,\ \Gamma \Rightarrow I}{\exists P,\ \Gamma \Rightarrow I}$ where $x \notin P, \Gamma, I$

□-introduction & elimination (□-in & □-el): $\dfrac{I \Rightarrow val\,P\,x}{\Rightarrow \Box(I,P)}$ (double line)

where $x \notin I, P$

□-transitivity (□-tr): $\dfrac{I,\ val\,P_1\,X \Rightarrow J}{\Box(I,P_1),\ \Box(J,P_2) \Rightarrow \Box(I,P_2)}$ where $X \not\in$

Cut: $\dfrac{\Gamma \Rightarrow I \qquad I,\ \Delta \Rightarrow J}{\Gamma,\ \Delta \Rightarrow J}$

Boolean rule (bool): $\dfrac{\ulcorner x = true \urcorner,\ \Gamma \Rightarrow I \qquad \ulcorner x = false \urcorner,\ \Gamma \Rightarrow I}{\ulcorner boolean\,x \urcorner,\ \Gamma \Rightarrow I}$

Induction (ind): $\dfrac{\Rightarrow val\,P\,0 \qquad val\,P\,n \Rightarrow val\,P\,(Sn)}{\ulcorner num\,n \urcorner \Rightarrow val\,P\,n}$ where $n \notin P$

FORMAL INTERPRETATION OF CPF IN PROTOLOGIC

A coding of CPF derivations as constructions is defined. A construction *spr* is defined.

THEOREM 28. If D is the code of a CPF derivation (with no premises) of a logical sequent $\Gamma \Rightarrow I$ then $spr(D) \Vdash \Gamma \Rightarrow I$.

THEOREM 29. (CPF interpretation theorem.) If D is the code of a CPF derivation (with no premises) of a logical sequent $\Rightarrow I$, where I has no free variables, then $pr(spr(D)) \vdash I$.

DEFINITION. Define a construction *glue* by

$$
\begin{aligned}
glue((premise(0),a),((d,a),rest)) &\triangleq (d,a)\\
glue((premise(Si),a),(first,rest)) &\triangleq glue((premise(i),a),rest)\\
glue((none(n),a),dlist) &\triangleq (none(n),a)\\
glue((one(n,p),a),dlist) &\triangleq (one(n,glue(p,dlist)),a)\\
glue((two(n,p,q),a),dlist) &\triangleq\\
&\qquad (two(n,glue(p,dlist),glue(q,dlist)),a).
\end{aligned}
$$

where $a, d, i, first, rest, n, p, q, dlist$ are nine variables.

THEOREM 30. (The *glue*-lemma for CPF.) If D is the code of a CPF derivation of a logical sequent S from premise sequents $T_0, \ldots T_k$, and $D_0, \ldots D_k$ are the codes of CPF derivations (with no premises) of $T_0, \ldots T_k$ respectively, then $glue(D, [D_0, \ldots D_k]) \rhd^*$ the code of a CPF derivation (with no premises) of S.

CHAPTER 29

FROM CALCULUS OF PROOF FUNCTIONS TO THE LOGIC OF PARTIAL TERMS

In this chapter some derived logical constants, $\bigwedge$, $\bullet$, $\forall$, $>$ and $\supset$, are defined and their CPF properties are established. Next, *formulae* are introduced: these are expressions in the style of predicate calculus, built up from atomic formulae (which are simply terms) using the logical constants $\wedge$, $\vee$, $\supset$, $\exists$ and $\forall$. Every formula is interpreted as a proof function and thus inherits a notion of intuitionistic proof. The Calculus of Proof Functions (CPF), when restricted to interpreted formulae, turns into *Logic of Partial Terms* (LPT). LPT is an axiomatisation of first-order intuitionistic predicate calculus with equality, reduction and induction, adapted to take account of the fact that not all terms have values. Every theorem of LPT (without free variables) has an intuitionistic proof.

LPT is similar to Beeson's (1985, Chapter VI, §1) system of the same name, except that Beeson includes, for any term t, an atomic formula $t\downarrow$, which says that t has a value. Since $t\downarrow$ is provably equivalent to $t = t$ in Beeson's system, the former is redundant and I shall not use it. Another difference between my system and Beeson's is that I shall not use any substitution notation in LPT formulae; substitution is more trouble than it is worth, and its introduction will be delayed until Peano Arithmetic (see Chapter 33).

LEXICAL CONVENTIONS

In this chapter, the identifiers 'I', 'J' and 'K', sometimes with subscripts, will be used to denote proof functions, the identifiers 'P' and 'Q' will be used to denote predicates, and the identifiers 'Γ' and 'Δ' will be used to denote (possibly empty) sequences of proof functions.

NEW LOGICAL CONSTANTS

Let i, g, p, j be four variables.

DEFINITION. Define the construction $\bigwedge$ by

$$\bigwedge nil \triangleq \ulcorner true \urcorner$$
$$\bigwedge(i, g) \triangleq \wedge(i, \bigwedge g).$$

THEOREM 1. If $I_1, \ldots I_n$ are proof functions (where $n \geq 1$) then

$$\bigwedge[I_1, \ldots I_n] \ \rhd^* \lhd \ \wedge(I_1, \ldots \wedge (I_n, \ulcorner \mathit{true} \urcorner) \ldots).$$

THEOREM 2. If Γ is a sequence of proof functions then $\bigwedge[\Gamma]$ is a proof function.

DEFINITION. Define the construction $\bullet$ as $(\lambda i.\Box(\ulcorner \mathit{true} \urcorner, \mathit{con}\, i))$.

THEOREM 3. If I is a proof function then so is $\bullet I$.

DEFINITION. Define the construction $\forall$ as $(\lambda p.\Box(\ulcorner \mathit{true} \urcorner, p))$.

THEOREM 4. If P is a predicate then $\forall P$ is a proof function.

DEFINITION. Define the construction $>$ (pronounced 'super-implication') as $(\lambda(i,j).\Box(i, \mathit{con}\, j))$.

DEFINITION. Define the construction $\supset$ ('implication') as $(\lambda(i,j).>(i, \bullet j))$.

THEOREM 5. If I and J are proof functions then so are $>(I, J)$ and $\supset(I, J)$.

The logical constants $>$ and $\forall$ are special cases of $\Box$, and $\bullet$ may be regarded as a special case either of $>$ or of $\forall$.

To understand the meanings of these logical constants, recall first how the CPF sequent arrow $\Rightarrow$ works. A proof (in the sense of $\models\!\!\!\mid\,$) of $I \Rightarrow J$ involves a function T such that, for any construction Q: (i) if Q satisfies the decidable part of the proof relation $Q \vdash I$ then $T[Q]$ satisfies the decidable part of the proof relation $T[Q] \vdash J$; and (ii) if Q satisfies the well-foundedness part of $Q \vdash I$ then $T[Q]$ satisfies the well-foundedness part of $T[Q] \vdash J$.

Now, a proof of $\bullet I$ is (loosely speaking, ignoring some technicalities) a construction X and a derivation of the protological sequent $\rightarrow AX$, where A is the decidable part of I. If we have a construction X such that $AX \rhd^*$ *true* then we can derive $\rightarrow AX$ using the special derivation $\mathcal{E}$. Thus given something satisfying the decidable part of the requirements for a proof of I we obtain something satisfying the decidable part of the requirements for a proof of $\bullet I$. The converse, however, fails. If we have a derivation (not necessarily with a well-founded reflection tree) of $\rightarrow AX$ we cannot conclude that $AX \rhd^*$ *true*. Consequently we shall find that $I \Rightarrow \bullet I$ is a theorem of CPF but $\bullet I \Rightarrow I$ is not, except in special cases. If $\bullet \ulcorner U \urcorner \Rightarrow \ulcorner U \urcorner$ were a theorem, for arbitrary terms U, this would amount to saying that all reflection principles were sound, irrespective of the well-foundedness of the reflection tree. (In particular, Theorem 4 in the Calculus of Proof Functions shows that $\bullet\ulcorner false\urcorner \Rightarrow \ulcorner false\urcorner$ has no proof and hence is not derivable in CPF.) Thus, within CPF, $\bullet I$ is a slightly weakened form of I.

A proof of $>(I,J)$, where $I \rhd^* pfn(A,\Pi) \not\rhd$ and $J \rhd^* pfn(B,\Sigma) \not\rhd$, is (loosely speaking) a function T and a protological derivation of the sequent $(q)\,Aq \rightarrow B(Tq)$. To prove $>(I,J)$, it is necessary to map every construction Q satisfying the test A to a construction TQ satisfying the test B; the mapping must work on all such Q, *even those that are not well-founded according to* Π. In addition, T must map everything well-founded according to Π to something well-founded according to Σ. Thus super-implication ($>$) treats the decidable and well-foundedness parts of proof separately and is analogous to the CPF sequent arrow $\Rightarrow$; as such it is stronger than intuitionistic implication in the usual sense, which simply involves mapping proofs of I to proofs of J.

To obtain intuitionistic implication in the intended sense we need to form the proof function $>(I,\bullet J)$. A proof of $>(I,\bullet J)$ is (loosely speaking) a function T, a function D, and a protological derivation D' of $(q)\,Aq \rightarrow DT(Dq, [\![\rightarrow B(Tq)]\!])$. The well-foundedness conditions are: $D' : rt$; $DQ : rt$ if $Q : \Pi$; and $TQ : \Sigma$ if $Q : \Pi$. These conditions guarantee that if $AQ \rhd^* true$ and $Q : \Pi$ then $B(TQ) \rhd^* true$ and $TQ : \Sigma$. That is, if Q is a proof of I then TQ is a proof of J. Hence $>(I,\bullet J)$ may be interpreted as 'I implies J'. This explains the way implication ($\supset$) is defined above. We shall see below that $\supset$, defined in this way, satisfies the correct introduction and elimination rules.

A logical sequent $\Gamma \Rightarrow I$ may be read informally as 'Γ super-implies I', whereas one of the form $\Gamma \Rightarrow \bullet I$ may be read informally as 'Γ implies I'.

SOME DERIVATIONS IN CPF

THEOREM 6. Tautology (taut): $I \Rightarrow I$

Proof. Let J be a proof function such that $I \rhd J$ or $J \rhd I$. (If $I \not\rhd$, take J as $id\,I$; otherwise take J such that $I \rhd J$.) Then the CPF derivation is as follows.

$$\frac{I \Rightarrow J \qquad J \Rightarrow I \text{ (red)}}{I \Rightarrow I}\text{cut}$$

∎

THEOREM 7. Reduction properties (red): $\Rightarrow \ulcorner T\urcorner$ if $T \rhd^* true$,

$$\frac{\Gamma \Rightarrow I}{\Gamma \Rightarrow J} \qquad \frac{\Gamma,\, I,\, \Delta \Rightarrow K}{\Gamma,\, J,\, \Delta \Rightarrow K} \qquad \text{if } I \rhd^* \lhd J.$$

Proof. By Cut, Exchange, *true*-introduction, and the Reduction axioms. ∎

THEOREM 8. Symmetry (symm): $\ulcorner X = Y \urcorner \Rightarrow \ulcorner Y = X \urcorner$

Proof. Let x be a fresh variable and P be the predicate $predify(\lambda x. \ulcorner x = X \urcorner)$.

$$\dfrac{\ulcorner X = Y \urcorner \Rightarrow \exists P \;(\text{te}) \qquad \dfrac{\dfrac{\dfrac{\dfrac{\ulcorner x = X \urcorner, \ulcorner x = X \urcorner \Rightarrow \ulcorner X = X \urcorner \;(\text{eq})}{\ulcorner x = X \urcorner \Rightarrow \ulcorner X = X \urcorner}\text{con}}{val\, P\, x \Rightarrow \ulcorner X = X \urcorner}\text{red}}{\exists P \Rightarrow \ulcorner X = X \urcorner}\exists\text{-el} \qquad \dfrac{\ulcorner X = Y \urcorner, \ulcorner X = X \urcorner \Rightarrow \ulcorner Y = X \urcorner \;(\text{eq})}{\ulcorner X = X \urcorner, \ulcorner X = Y \urcorner \Rightarrow \ulcorner Y = X \urcorner}\text{exch}}{\exists P, \ulcorner X = Y \urcorner \Rightarrow \ulcorner Y = X \urcorner}\text{cut}}{\dfrac{\ulcorner X = Y \urcorner, \ulcorner X = Y \urcorner \Rightarrow \ulcorner Y = X \urcorner}{\ulcorner X = Y \urcorner \Rightarrow \ulcorner Y = X \urcorner}\text{con}}\text{cut}$$

∎

THEOREM 9. $\bigwedge$-theorems and rule ($\bigwedge$):

(a) $\Rightarrow \bigwedge nil$

(b) $I, \bigwedge[\Gamma] \Rightarrow \bigwedge[I, \Gamma]$

(c) $\Gamma \Rightarrow \bigwedge[\Gamma]$

(d) $\bigwedge[I, \Gamma] \Rightarrow I$

(e) $\bigwedge[I, \Gamma] \Rightarrow \bigwedge[\Gamma]$

(f) $\bigwedge[K_1, \ldots K_n] \Rightarrow K_i$, for $i = 1, \ldots n$

(g) $\dfrac{\Gamma, \Delta \Rightarrow I}{\bigwedge[\Gamma], \Delta \Rightarrow I}$ (double line)

Proof.

(a): by Reduction from the CPF axiom $\Rightarrow \ulcorner true \urcorner$.

(b): by Reduction from CPF $\wedge$-introduction.

(c): from (a) and (b) using Exchange and Cut.

(d), (e): by Reduction from CPF $\wedge$-elimination.

(f): from (d) and (e) using Cut.

(g): from (c) and (f) using Cut, Exchange and Contraction. ∎

THEOREM 10. $\bullet$-introduction ($\bullet$-in): $I \Rightarrow \bullet I$

Proof.

$$\dfrac{I \Rightarrow \Box(\ulcorner true \urcorner, con\, I) \;(\Box\text{-exp})}{I \Rightarrow \bullet I}\text{red}$$

∎

THEOREM 11. Shift rule (shift): $\dfrac{\Gamma \Rightarrow \Box(\wedge(I, J), P)}{\Gamma, I \Rightarrow \Box(J, P)}$

Proof. Let x be a fresh variable.

$$
\dfrac{
I \Rightarrow \Box(J, con\, I)\ (\Box\text{-exp})
\qquad
\dfrac{
\dfrac{
\dfrac{
\dfrac{I,\ J \Rightarrow \wedge(I,J)\ (\wedge\text{-in})}{J,\ I \Rightarrow \wedge(I,J)}\ \text{exch}
}{J,\ val(con\, I)x \Rightarrow \wedge(I,J)}\ \text{red}
}{\Box(J, con\, I),\ \Box(\wedge(I,J),P) \Rightarrow \Box(J,P)}\ \Box\text{-tr}
}{}
}{I,\ \Box(\wedge(I,J),P) \Rightarrow \Box(J,P)}\ \text{cut}
$$

$$
\dfrac{
\Gamma \Rightarrow \Box(\wedge(I,J),P)
\qquad
\dfrac{I,\ \Box(\wedge(I,J),P) \Rightarrow \Box(J,P)}{\Box(\wedge(I,J),P),\ I \Rightarrow \Box(J,P)}\ \text{exch}
}{\Gamma,\ I \Rightarrow \Box(J,P)}\ \text{cut}
$$

∎

THEOREM 12. Extended $\Box$-introduction rule ($\Box$-in+): $\dfrac{\Gamma,\ I \Rightarrow val\, P\, x}{\Gamma \Rightarrow \Box(I,P)}$

where $x \notin \Gamma, I, P$.

Proof.

$$
\dfrac{
\dfrac{
\dfrac{
\dfrac{
\dfrac{\Gamma,\ I \Rightarrow val\, P\, x}{\bigwedge[\Gamma],\ I \Rightarrow val\, P\, x}\ \bigwedge
}{\wedge(\bigwedge[\Gamma], I) \Rightarrow val\, P\, x}\ \wedge\text{-el, cut, con}
}{\Rightarrow \Box(\wedge(\bigwedge[\Gamma], I), P)}\ \Box\text{-in}
}{\bigwedge[\Gamma] \Rightarrow \Box(I,P)}\ \text{shift}
}{\Gamma \Rightarrow \Box(I,P)}\ \bigwedge
$$

∎

THEOREM 13. $\Box$-instantiation rule ($\Box$-inst): $\dfrac{I,\ val\, P\, X \Rightarrow val\, Q\, x}{\Box(I,P) \Rightarrow \Box(I,Q)}$

where $x \notin I, P, Q, X$ and $X \not\triangleright$.

Proof. Let J be $\wedge(I, val\, P\, X)$.

$$
\dfrac{
\dfrac{
\dfrac{I,\ val\, P\, X \Rightarrow val\, Q\, x}{J \Rightarrow val\, Q\, x}\ \wedge\text{-el, cut, con}
}{\Rightarrow \Box(J,Q)}\ \Box\text{-in}
\qquad
\dfrac{I,\ val\, P\, X \Rightarrow J\ (\wedge\text{-in})}{\Box(I,P),\ \Box(J,Q) \Rightarrow \Box(I,Q)}\ \Box\text{-tr}
}{\Box(I,P) \Rightarrow \Box(I,Q)}\ \text{cut}
$$

∎

THEOREM 14. $\Box$-monotonicity rule ($\Box$-mono): $\dfrac{I,\ val\, P\, x \Rightarrow val\, Q\, x}{\Box(I,P) \Rightarrow \Box(I,Q)}$

where $x \notin I, P, Q$.

Proof. Recall from Logic that $val(\Delta(I,Q))x \ \triangleright^{*}\!\triangleleft\ \Box(I, con(val\, Q\, x))$. Let y be a fresh variable.

$$\dfrac{\dfrac{\dfrac{\dfrac{\dfrac{\dfrac{I,\ val\,P\,x \Rightarrow val\,Q\,x}{I,\ val\,P\,x \Rightarrow val(con(val\,Q\,x))y}}{\Box(I,P) \Rightarrow \Box(I, con(val\,Q\,x))}\,\Box\text{-inst}}{\Box(I,P) \Rightarrow val(\Delta(I,Q))x}\,\text{red}}{\Box(I,P),\ \ulcorner true \urcorner \Rightarrow val(\Delta(I,Q))x}\,\text{thin}}{\Box(I,P) \Rightarrow \Box(\ulcorner true \urcorner, \Delta(I,Q))}\,\Box\text{-in+} \qquad \Box(\ulcorner true \urcorner, \Delta(I,Q)) \Rightarrow \Box(I,Q)\ (\Box\text{-comp})}{\Box(I,P) \Rightarrow \Box(I,Q)}\,\text{cut}$$

∎

THEOREM 15. •-compression (•-comp): $\bullet(\Box(I,P)) \Rightarrow \Box(I,P)$

Proof. Let x and y be two fresh variables, and let Q be $\Delta(I,P)$.

$$\dfrac{\dfrac{\dfrac{\dfrac{\dfrac{\dfrac{\dfrac{val\,P\,x \Rightarrow val\,P\,x\ (\text{taut})}{I,\ val\,P\,x \Rightarrow val\,P\,x}\,\text{thin}}{I,\ val\,P\,x \Rightarrow val(con(val\,P\,x))y}\,\text{red}}{\Box(I,P) \Rightarrow \Box(I, con(val\,P\,x))}\,\Box\text{-inst}}{\ulcorner true \urcorner,\ \Box(I,P) \Rightarrow \Box(I, con(val\,P\,x))}\,\text{thin}}{\ulcorner true \urcorner,\ val(con(\Box(I,P)))x \Rightarrow val\,Q\,x}\,\text{red}}{\Box(\ulcorner true \urcorner, con(\Box(I,P))) \Rightarrow \Box(\ulcorner true \urcorner, Q)}\,\Box\text{-mono}$$

$$\dfrac{\dfrac{\Box(\ulcorner true \urcorner, con(\Box(I,P))) \Rightarrow \Box(\ulcorner true \urcorner, Q)}{\bullet(\Box(I,P)) \Rightarrow \Box(\ulcorner true \urcorner, Q)}\,\text{red} \qquad \Box(\ulcorner true \urcorner, Q) \Rightarrow \Box(I,P)\ (\Box\text{-comp})}{\bullet(\Box(I,P)) \Rightarrow \Box(I,P)}\,\text{cut}$$

∎

THEOREM 16. >-introduction rule (>-in): $\dfrac{\Gamma,\ I \Rightarrow J}{\Gamma \Rightarrow >(I,J)}$

Proof. Let x be a fresh variable.

$$\dfrac{\dfrac{\dfrac{\Gamma,\ I \Rightarrow J}{\Gamma,\ I \Rightarrow val(con\,J)x}\,\text{red}}{\Gamma \Rightarrow \Box(I, con\,J)}\,\Box\text{-in+}}{\Gamma \Rightarrow >(I,J)}\,\text{red}$$

∎

THEOREM 17. >-transitivity (>-tr): $>(I,J),\ >(J,K) \Rightarrow >(I,K)$

Proof. Let x be a fresh variable.

$$\dfrac{\dfrac{\dfrac{\dfrac{J \Rightarrow J\ (\text{taut})}{I,\ J \Rightarrow J}\ \text{thin}}{I,\ val(con\,J)x \Rightarrow J}\ \text{red}}{\square(I, con\,J),\ \square(J, con\,K) \Rightarrow \square(I, con\,K)}\ \square\text{-tr}}{>(I,J),\ >(J,K) \Rightarrow >(I,K)}\ \text{red}$$

∎

THEOREM 18. •-transitivity (•-tr): $\dfrac{\Gamma,\ I \Rightarrow J}{\Gamma,\ \bullet I \Rightarrow \bullet J}$

Proof.

$$\dfrac{\dfrac{\Gamma,\ I \Rightarrow J}{\Gamma \Rightarrow >(I,J)}\ >\text{-in} \qquad \dfrac{>(\ulcorner true \urcorner, I),\ >(I,J) \Rightarrow >(\ulcorner true \urcorner, J)\ (>\text{-tr})}{\bullet I,\ >(I,J) \Rightarrow \bullet J}\ \text{red}}{\Gamma,\ \bullet I \Rightarrow \bullet J}\ \text{cut}$$

∎

THEOREM 19. •-elimination (•-el): $\bullet\bullet I \Rightarrow \bullet I \qquad \dfrac{\Rightarrow \bullet I}{\Rightarrow I} \qquad \dfrac{\Gamma,\ I \Rightarrow \bullet J}{\Gamma,\ \bullet I \Rightarrow \bullet J}$

Proof. Let x be a fresh variable.

$$\dfrac{\bullet(\square(\ulcorner true \urcorner, con\,I)) \Rightarrow \square(\ulcorner true \urcorner, con\,I)\ (\bullet\text{-comp})}{\bullet\bullet I \Rightarrow \bullet I}\ \text{red}$$

$$\dfrac{\Rightarrow \ulcorner true \urcorner\ (\mathit{true}\text{-in}) \qquad \dfrac{\dfrac{\dfrac{\Rightarrow \bullet I}{\Rightarrow \square(\ulcorner true \urcorner, con\,I)}\ \text{red}}{\ulcorner true \urcorner \Rightarrow val(con\,I)x}\ \square\text{-el}}{\ulcorner true \urcorner \Rightarrow I}\ \text{red}}{\Rightarrow I}\ \text{cut}$$

$$\dfrac{\dfrac{\Gamma,\ I \Rightarrow \bullet J}{\Gamma,\ \bullet I \Rightarrow \bullet\bullet J}\ \bullet\text{-tr} \qquad \bullet\bullet J \Rightarrow \bullet J\ (\bullet\text{-el, above})}{\Gamma,\ \bullet I \Rightarrow \bullet J}\ \text{cut}$$

∎

THEOREM 20. Cut rule with • (•-cut): $\dfrac{\Gamma \Rightarrow \bullet I \qquad I,\ \Delta \Rightarrow \bullet J}{\Gamma,\ \Delta \Rightarrow \bullet J}$

Proof.

$$\dfrac{\Gamma \Rightarrow \bullet I \qquad \dfrac{\dfrac{I,\ \Delta \Rightarrow \bullet J}{\Delta,\ I \Rightarrow \bullet J}\,\text{exch}}{\Delta,\ \bullet I \Rightarrow \bullet J}\,\bullet\text{-el}}{\Gamma,\ \Delta \Rightarrow \bullet J}\,\text{cut}$$

∎

THEOREM 21. $\forall$-elimination ($\forall$-el): $\quad \forall P \Rightarrow \bullet(val\,P\,x)$

Proof. Let y be a fresh variable.

$$\dfrac{\dfrac{\dfrac{\dfrac{val\,P\,x \Rightarrow val\,P\,x \ \text{(taut)}}{\ulcorner true \urcorner,\ val\,P\,x \Rightarrow val\,P\,x}\,\text{thin}}{\ulcorner true \urcorner,\ val\,P\,x \Rightarrow val(con(val\,P\,x))y}\,\text{red}}{\Box(\ulcorner true \urcorner, P) \Rightarrow \Box(\ulcorner true \urcorner, con(val\,P\,x))}\,\Box\text{-inst}}{\forall P \Rightarrow \bullet(val\,P\,x)}\,\text{red}$$

∎

THEOREM 22. $\forall$-introduction rule ($\forall$-in): $\quad \dfrac{\Gamma \Rightarrow \bullet(val\,P\,x)}{\Gamma \Rightarrow \forall P}$

where $x \notin \Gamma, P$.

Proof. Let Q be $\Delta(\ulcorner true \urcorner, P)$, giving $val\,Q\,x \ \rhd^* \lhd\ \Box(\ulcorner true \urcorner, con(val\,P\,x))$ $\rhd^* \lhd\ \bullet(val\,P\,x)$.

$$\dfrac{\dfrac{\dfrac{\dfrac{\Gamma \Rightarrow \bullet(val\,P\,x)}{\Gamma \Rightarrow val\,Q\,x}\,\text{red}}{\Gamma,\ \ulcorner true \urcorner \Rightarrow val\,Q\,x}\,\text{thin}}{\Gamma \Rightarrow \Box(\ulcorner true \urcorner, Q)}\,\Box\text{-in+} \qquad \dfrac{\Box(\ulcorner true \urcorner, Q) \Rightarrow \Box(\ulcorner true \urcorner, P)\ (\Box\text{-comp})}{\Box(\ulcorner true \urcorner, Q) \Rightarrow \forall P}\,\text{red}}{\Gamma \Rightarrow \forall P}\,\text{cut}$$

∎

EXERCISES. Derive the following logical sequents and rules in CPF.

$$\dfrac{\Rightarrow\ >(I, J)}{I \Rightarrow J} \qquad \dfrac{I \Rightarrow J}{\Box(J, P) \Rightarrow \Box(I, P)} \qquad \dfrac{\bullet I \Rightarrow I \qquad \bullet J \Rightarrow J}{\bullet(\wedge(I, J)) \Rightarrow \wedge(I, J)}$$

$$>(\bigwedge[\Delta], J) \Rightarrow\ >(\bigwedge[I, \Delta], J)$$

$$>(\bigwedge[\Gamma, I, J, \Delta], K) \Rightarrow\ >(\bigwedge[\Gamma, J, I, \Delta], K)$$

$$>(\bigwedge[I, I, \Gamma], J) \Rightarrow >(\bigwedge[I, \Gamma], J)$$

$$>(\bigwedge[\Gamma], I),\ >(\bigwedge[I, \Delta], J) \Rightarrow >([\Gamma, \Delta], J)$$

Define a 'new' version of implication, $\ni$, as $(\lambda(i, j).>(i, >(i, j)))$. Investigate the CPF properties of $\ni$.

EXERCISES.

1. Find proof functions I and J for which $\supset(I, J) \Rightarrow >(I, J)$ is not a CPF theorem.
2. Find a predicate P for which $\bullet\exists P \Rightarrow \exists P$ is not a CPF theorem.

LPT FORMULAE

A *formula* of LPT is a sentence in the following language.

- The alphabet is that of the Expanded Term Language (ET) plus '$\wedge$', '$\vee$', '$\supset$', '$\exists$', '$\forall$'.
- The tokens are those of ET plus '$\wedge$', '$\vee$', '$\supset$', '$\exists$', '$\forall$'.
- Lexical analysis consists of recognising constants and replacing them by the token '*con*', recognising variables and replacing them by the token '*vbl*', and recognising every other token.
- The grammar of the language is as follows.
  - The terminals are the tokens.
  - The non-terminals are F, T; the start symbol is F.
  - The production rules are

$$F \rightarrow T \mid (\,F \wedge F\,) \mid (\,F \vee F\,) \mid (\,F \supset F\,) \mid \exists\, vbl\ F \mid \forall\, vbl\ F$$

$$T \rightarrow con \mid (\,vbl\,) \mid (\,T\,T\,) \mid (\,\lambda\, T\,.\,T\,) \mid (\,T \left[\tfrac{T}{vbl} \right])$$

(Note that this grammar contains that of ET: a term of ET is any string matching the non-terminal T. A formula that is a term will be called an *atomic formula*.)

METAFORMULAE

The letters 'A', 'B' and 'C' will be used as metavariables to denote formulae. A *metaformula* is an expression that is like a formula except that it may contain metaconstants, metavariables and ET metanotation, some brackets may be omitted, and optional brackets and spaces may be added. To state it more precisely, the alphabet is that of ET metaterms plus '$\wedge$', '$\vee$', '$\supset$', '$\exists$', '$\forall$'; the tokens are those of ET metaterms plus '$\wedge$', '$\vee$', '$\supset$', '$\exists$', '$\forall$', and a new one '*formvbl*'; lexical analysis is as for ET metaterms except that metavariables denoting formulae are replaced by '*formvbl*'; and the grammar is

$$
\begin{array}{l}
F \to P \wedge F \mid P \vee F \mid P \supset F \mid P \\
P \to \exists\, vbl\, P \mid \forall\, vbl\, P \mid (\,F\,) \mid T \mid formvbl \\
T \to T\,L \mid T\left[\begin{smallmatrix} T \\ vbl \end{smallmatrix}\right] \mid L \\
L \to con \mid vbl \mid (\,T\,) \mid (\lambda\, T\,.\,T\,) \mid termcon \mid vblvbl \mid termvbl \mid \ldots
\end{array}
$$

where the start symbol is F and '...' represents production rules for ET metanotation.

EXAMPLE. Let 'x' and 'y' be metavariables denoting variables, and 'A', 'B' and 'C' be metavariables denoting formulae. Then the lexical analysis and parsing of the metaformula '$\forall x \exists y\, A \supset B \supset C$' are as follows.

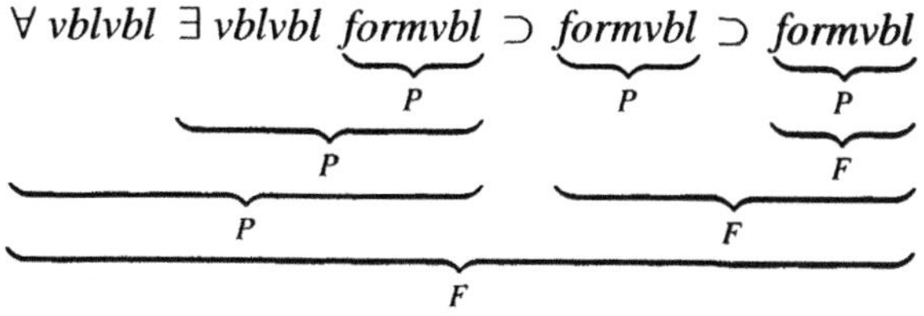

Note that quantifiers have higher syntactic precedence than propositional connectives and the precedence of propositional connectives increases from left to right: '$A \supset B \vee C \wedge D$' is parsed as if it were '$A \supset (B \vee (C \wedge D))$'.

An *instance* of a metaformula is a formula obtained from the metaformula by choosing a term, variable or formula (as required) for each metavariable in the metaformula, replacing each occurrence of each metavariable by the chosen term, variable or formula, replacing each metaconstant by the term it denotes, adding and removing brackets as required, removing spaces, and rewriting all metanotation. A metaformula may be used in two ways: as a schema, standing for any of its instances, or to denote a particular one of its instances. The latter will be implied when particular formulae, terms or variables have previously been chosen for the metavariables.

FREE VARIABLES IN FORMULAE

For any variable v and LPT formula A, the relation $v \in A$ ('v occurs free in A') is defined by:

- if A is a term T, $v \in A$ iff $v \in T$ in the sense of ET;
- $v \in A \wedge B$ iff $v \in A \vee B$ iff $v \in A \supset B$ iff $v \in A$ or $v \in B$;
- $v \in \exists x A$ iff $v \in \forall x A$ iff v isn't x and $v \in A$.

The *free variables* of A are the variables v such that $v \in A$.

THEOREM 23. Any LPT formula has finitely many free variables.

INTERPRETATION OF LPT FORMULAE AS PROOF FUNCTIONS

Each formula A is interpreted as a proof function $\ulcorner A \urcorner$, as follows.

$$\begin{aligned}
\ulcorner T \urcorner & \text{ (where } T \text{ is a term) has already been defined in Logic} \\
\ulcorner A \wedge B \urcorner & \text{ is } \wedge(\ulcorner A \urcorner, \ulcorner B \urcorner) \\
\ulcorner A \vee B \urcorner & \text{ is } \ulcorner \exists x\, (\mathit{boolean}\, x \wedge (x = \mathit{true} \supset A) \wedge (x = \mathit{false} \supset B)) \urcorner \\
& \qquad \text{for a variable } x \notin A, B \\
\ulcorner A \supset B \urcorner & \text{ is } \supset(\ulcorner A \urcorner, \ulcorner B \urcorner) \\
\ulcorner \exists x A \urcorner & \text{ is } \exists(\mathit{predify}(\lambda x. \ulcorner A \urcorner)) \\
\ulcorner \forall x A \urcorner & \text{ is } \forall(\mathit{predify}(\lambda x. \ulcorner A \urcorner))
\end{aligned}$$

THEOREM 24. (Free variables in formulae.) If A is an LPT formula then $v \in A$ iff $v \in \ulcorner A \urcorner$.

Note that the interpretation of $A \vee B$ is independent (up to compilation) of the choice of x, since $\ulcorner A \vee B \urcorner$ is, more explicitly,

$$\begin{aligned}
&\exists(\mathit{predify}(\lambda x. \wedge (\mathit{pfn}((\lambda \mathit{nil}.\mathit{boolean}\, x), \mathit{leaf}), \\
&\qquad\qquad \wedge(\supset (\mathit{pfn}((\lambda \mathit{nil}.x = \mathit{true}), \mathit{leaf}), \ulcorner A \urcorner), \\
&\qquad\qquad\quad \supset (\mathit{pfn}((\lambda \mathit{nil}.x = \mathit{false}), \mathit{leaf}), \ulcorner B \urcorner)))))
\end{aligned}$$

(recall the binding-independence property in the Expanded Term Language).

THEOREM 25. For any formulae A, B, term Q, variables $\underline{y}$, and irreducible terms $\underline{Y}$, we have $Q \vdash \ulcorner A \wedge B \urcorner \left[\frac{\underline{Y}}{\underline{y}}\right]$ iff $Q \rhd^* (R_1, R_2) \not\rhd$, where $R_1 \vdash \ulcorner A \urcorner \left[\frac{\underline{Y}}{\underline{y}}\right]$ and $R_2 \vdash \ulcorner B \urcorner \left[\frac{\underline{Y}}{\underline{y}}\right]$.

Proof. This follows immediately from Theorem 8 of Logic, since $\ulcorner A \wedge B \urcorner \left[\frac{\underline{Y}}{\underline{y}}\right] \overset{*}{\leftrightarrow} \wedge(\ulcorner A \urcorner \left[\frac{\underline{Y}}{\underline{y}}\right], \ulcorner B \urcorner \left[\frac{\underline{Y}}{\underline{y}}\right])$. ▮

THEOREM 26. For any formula A, variable x, term Q, variables $\underline{y}$ (not including x), and irreducible terms $\underline{Y}$, we have $Q \vdash \ulcorner \exists x A \urcorner \left[\frac{\underline{Y}}{\underline{y}}\right]$ iff $Q \rhd^* (X, R) \not\rhd$, where $R \vdash \ulcorner A \urcorner \left[\frac{\underline{Y}}{\underline{y}}\right]\left[\frac{X}{x}\right]$.

Proof. This follows from Theorem 10 of Logic, since, if we take P as the predicate $\mathit{predify}(\lambda x. \ulcorner A \urcorner)\left[\frac{\underline{Y}}{\underline{y}}\right]$, we have $\ulcorner \exists x A \urcorner \left[\frac{\underline{Y}}{\underline{y}}\right] \overset{*}{\leftrightarrow} \exists P$ and $\ulcorner A \urcorner \left[\frac{\underline{Y}}{\underline{y}}\right]\left[\frac{X}{x}\right] \lhd^* \mathit{val}\, P\, X$. ▮

THEOREM 27. For any formulae A, B (with free variables $\underline{y}$ between them), any terms Q, R (with no free variables), and any constructions $\underline{Y}$, if $Q \vdash \ulcorner A \supset B \urcorner \left[\frac{\underline{Y}}{\underline{y}}\right]$ and $R \vdash \ulcorner A \urcorner \left[\frac{\underline{Y}}{\underline{y}}\right]$ then $el_{\supset}(Q,R) \vdash \ulcorner B \urcorner \left[\frac{\underline{Y}}{\underline{y}}\right]$ (for a certain construction $el_{\supset}$).

Proof. Define $el_{\supset}$ as $(\lambda(q,r).el_{\Box}(el_{\Box}\, q\, r\, 0)\, nil\, 0)$. Now,

$$\ulcorner A \supset B \urcorner \left[\frac{\underline{Y}}{\underline{y}}\right] \overset{*}{\leftrightarrow} \supset(\ulcorner A \urcorner \left[\frac{\underline{Y}}{\underline{y}}\right], \ulcorner B \urcorner \left[\frac{\underline{Y}}{\underline{y}}\right]) \rhd^{*}\lhd >(\ulcorner A \urcorner \left[\frac{\underline{Y}}{\underline{y}}\right], \bullet \ulcorner B \urcorner \left[\frac{\underline{Y}}{\underline{y}}\right])$$
$$\rhd^{*} \Box(\ulcorner A \urcorner \left[\frac{\underline{Y}}{\underline{y}}\right], con(\Box(\ulcorner true \urcorner, con(\ulcorner B \urcorner \left[\frac{\underline{Y}}{\underline{y}}\right])))),$$

so applying Theorem 12 of Logic gives

$$el_{\Box}\, Q\, R\, 0 \vdash val(con(\Box(\ulcorner true \urcorner, con(\ulcorner B \urcorner \left[\frac{\underline{Y}}{\underline{y}}\right]))))0 \rhd^{*} \Box(\ulcorner true \urcorner, con(\ulcorner B \urcorner \left[\frac{\underline{Y}}{\underline{y}}\right])),$$

and, applying the same theorem again (since $nil \vdash \ulcorner true \urcorner$ and Q, R are evaluable),

$$el_{\supset}(Q,R) \rhd^{*}\lhd el_{\Box}(el_{\Box}\, Q\, R\, 0)\, nil\, 0 \vdash val(con(\ulcorner B \urcorner \left[\frac{\underline{Y}}{\underline{y}}\right]))0 \rhd^{*} \ulcorner B \urcorner \left[\frac{\underline{Y}}{\underline{y}}\right],$$

as required. ▮

THEOREM 28. For any formulae A, B (with free variables $\underline{y}$ between them), any term Q (with no free variables), and any constructions $\underline{Y}$, if $Q \vdash \ulcorner A \vee B \urcorner \left[\frac{\underline{Y}}{\underline{y}}\right]$ then either $select_{\vee}(Q) \rhd^{*}$ *true* and $el_{\vee}(Q) \vdash \ulcorner A \urcorner \left[\frac{\underline{Y}}{\underline{y}}\right]$ or $select_{\vee}(Q) \rhd^{*}$ *false* and $el_{\vee}(Q) \vdash \ulcorner B \urcorner \left[\frac{\underline{Y}}{\underline{y}}\right]$ (for certain constructions $select_{\vee}$ and $el_{\vee}$).

Proof. Let x, r_1, r_2, r_3 be four fresh variables, define $select_{\vee}$ as *left*, define $el_{\vee}$ as $(\lambda(x,(r_1,(r_2,r_3))).el_{\supset}(if\, x\, r_2\, r_3, nil))$, and let C be the formula $boolean\, x \wedge (x = true \supset A) \wedge (x = false \supset B)$. Now, $\ulcorner A \vee B \urcorner \left[\frac{\underline{Y}}{\underline{y}}\right] \overset{*}{\leftrightarrow} \exists(predify(\lambda x. \ulcorner C \urcorner)\left[\frac{\underline{Y}}{\underline{y}}\right])$; and our hypothesis that $Q \vdash \ulcorner A \vee B \urcorner \left[\frac{\underline{Y}}{\underline{y}}\right]$ implies by Theorem 10 of Logic that $Q \rhd^{*} (X,R) \not\rhd$, where $R \vdash val(predify(\lambda x. \ulcorner C \urcorner)\left[\frac{\underline{Y}}{\underline{y}}\right])X \rhd^{*} \ulcorner C \urcorner \left[\frac{\underline{Y}}{\underline{y}}\right]\left[\frac{X}{x}\right]$. Applying Theorem 25 twice, $R \overset{*}{\leftrightarrow} (R_1,(R_2,R_3))$, where $R_1 \vdash \ulcorner boolean\, x \urcorner \left[\frac{X}{x}\right]$, $R_2 \vdash \ulcorner x = true \supset A \urcorner \left[\frac{\underline{Y}}{\underline{y}}\right]\left[\frac{X}{x}\right]$, and $R_3 \vdash \ulcorner x = false \supset B \urcorner \left[\frac{\underline{Y}}{\underline{y}}\right]\left[\frac{X}{x}\right]$. Hence, by Theorem 4 of Logic, $(boolean\, x)\left[\frac{X}{x}\right] \overset{*}{\leftrightarrow} boolean\, X \rhd^{*}$ *true*, which implies that $X \overset{*}{\leftrightarrow}$ *true* or *false*. Hence $select_{\vee}(Q) \rhd^{*}$ *true* or *false*.

In the first case, in which $select_{\vee}(Q) \;\rhd^*\; X \overset{*}{\leftrightarrow} true$, we have $R_2 \vdash \ulcorner x = true \supset A\urcorner\left[\frac{\underline{Y}}{\underline{y}}\right]\left[\frac{true}{x}\right]$ and $nil \vdash \ulcorner x{=}true\urcorner\left[\frac{true}{x}\right]$, so Theorem 27 gives $el_{\vee}(Q) \;\rhd^*\; el_{\supset}(R_2, nil) \vdash \ulcorner A\urcorner\left[\frac{\underline{Y}}{\underline{y}}\right]$, as required.

In the second case, where $select_{\vee}(Q) \;\rhd^*\; X \overset{*}{\leftrightarrow} false$, we have $R_3 \vdash \ulcorner x = false \supset B\urcorner\left[\frac{\underline{Y}}{\underline{y}}\right]\left[\frac{false}{x}\right]$ and $nil \vdash \ulcorner x = false\urcorner\left[\frac{false}{x}\right]$, so Theorem 27 gives $el_{\vee}(Q) \;\rhd^*\; el_{\supset}(R_3, nil) \vdash \ulcorner B\urcorner\left[\frac{\underline{Y}}{\underline{y}}\right]$, as required. ∎

EXERCISE. State and prove a converse to Theorem 28.

THEOREM 29. For any formula of the form $\forall x A$ (with free variables $\underline{y}$), any term Q (with no free variables), and any constructions $\underline{Y}, X$, if $Q \vdash \ulcorner \forall x A\urcorner\left[\frac{\underline{Y}}{\underline{y}}\right]$ then $el_{\forall}(Q, X) \vdash \ulcorner A\urcorner\left[\frac{\underline{Y}}{\underline{y}}\right]\left[\frac{X}{x}\right]$ (for a certain construction $el_{\forall}$).

Proof. Let $el_{\forall}$ be $(\lambda(q,x).el_{\Box}\, q\, nil\, x)$, where q is a fresh variable. Now, $nil \vdash \ulcorner true\urcorner$ and $Q \vdash \forall(predify(\lambda x.\ulcorner A\urcorner)\left[\frac{\underline{Y}}{\underline{y}}\right])$, so Q is evaluable and by Theorem 12 of Logic $el_{\forall}(Q, X) \;\rhd^*\!\lhd\; el_{\Box}\, Q\, nil\, X \vdash val(predify(\lambda x.\ulcorner A\urcorner)\left[\frac{\underline{Y}}{\underline{y}}\right])X \;\rhd^*\; \ulcorner A\urcorner\left[\frac{\underline{Y}}{\underline{y}}\right]\left[\frac{X}{x}\right]$.
∎

SOME CPF DERIVATIONS INVOLVING INTERPRETED FORMULAE

THEOREM 30. $\wedge$ properties ($\wedge$):

$$\ulcorner A\urcorner,\ \ulcorner B\urcorner \Rightarrow \ulcorner A \wedge B\urcorner \qquad \ulcorner A \wedge B\urcorner \Rightarrow \ulcorner A\urcorner \qquad \ulcorner A \wedge B\urcorner \Rightarrow \ulcorner B\urcorner$$

$$\frac{\ulcorner A\urcorner,\ \ulcorner B\urcorner \Rightarrow K}{\overline{\ulcorner A \wedge B\urcorner \Rightarrow K}} \qquad \frac{\Gamma \Rightarrow \ulcorner A\urcorner \quad \Delta \Rightarrow \ulcorner B\urcorner}{\Gamma,\ \Delta \Rightarrow \ulcorner A \wedge B\urcorner}$$

Proof. Since $\ulcorner A \wedge B\urcorner$ is $\wedge(\ulcorner A\urcorner, \ulcorner B\urcorner)$ it follows that $\ulcorner A\urcorner,\ \ulcorner B\urcorner \Rightarrow \ulcorner A \wedge B\urcorner$, $\ulcorner A \wedge B\urcorner \Rightarrow \ulcorner A\urcorner$, and $\ulcorner A \wedge B\urcorner \Rightarrow \ulcorner B\urcorner$ are CPF axioms. The rules then follow by Cut, Exchange and Contraction. ∎

THEOREM 31. $\supset$ properties ($\supset$):

$$\ulcorner A\urcorner,\ \ulcorner A \supset B\urcorner \Rightarrow \bullet\ulcorner B\urcorner \qquad \frac{\Gamma,\ \ulcorner A\urcorner \Rightarrow \bullet\ulcorner B\urcorner}{\overline{\Gamma \Rightarrow \ulcorner A \supset B\urcorner}} \qquad \frac{\Gamma,\ \ulcorner A\urcorner \Rightarrow \ulcorner B\urcorner}{\Gamma \Rightarrow \ulcorner A \supset B\urcorner}$$

Proof. The CPF derivations are as follows; bear in mind that $\ulcorner A \supset B\urcorner$ is $\supset(\ulcorner A\urcorner, \ulcorner B\urcorner) \;\rhd^*\!\lhd\; {>}(\ulcorner A\urcorner, \bullet\ulcorner B\urcorner)$.

$$\dfrac{\ulcorner A\urcorner \Rightarrow \bullet\ulcorner A\urcorner\ (\bullet\text{-in}) \qquad \dfrac{\dfrac{>(\ulcorner true\urcorner, \ulcorner A\urcorner),\ >(\ulcorner A\urcorner, \bullet\ulcorner B\urcorner) \Rightarrow\ >(\ulcorner true\urcorner, \bullet\ulcorner B\urcorner)\ (>\text{-tr})}{\bullet\ulcorner A\urcorner,\ \ulcorner A\supset B\urcorner \Rightarrow \bullet\bullet\ulcorner B\urcorner}\text{red} \qquad \bullet\bullet\ulcorner B\urcorner \Rightarrow \bullet\ulcorner B\urcorner\ (\bullet\text{-el})}{\bullet\ulcorner A\urcorner,\ \ulcorner A\supset B\urcorner \Rightarrow \bullet\ulcorner B\urcorner}\text{cut}}{\ulcorner A\urcorner,\ \ulcorner A\supset B\urcorner \Rightarrow \bullet\ulcorner B\urcorner}\text{cut}$$

$$\dfrac{\dfrac{\Gamma,\ \ulcorner A\urcorner \Rightarrow \bullet\ulcorner B\urcorner}{\Gamma \Rightarrow\ >(\ulcorner A\urcorner, \bullet\ulcorner B\urcorner)}>\text{-in}}{\Gamma \Rightarrow \ulcorner A\supset B\urcorner}\text{red} \qquad \dfrac{\Gamma \to \ulcorner A\supset B\urcorner \qquad \dfrac{\ulcorner A\urcorner,\ \ulcorner A\supset B\urcorner \Rightarrow \bullet\ulcorner B\urcorner\ (\supset\text{, above})}{\ulcorner A\supset B\urcorner,\ \ulcorner A\urcorner \Rightarrow \bullet\ulcorner B\urcorner}\text{exch}}{\Gamma,\ \ulcorner A\urcorner \Rightarrow \bullet\ulcorner B\urcorner}\text{cut}$$

$$\dfrac{\dfrac{\Gamma,\ \ulcorner A\urcorner \Rightarrow \ulcorner B\urcorner \qquad \ulcorner B\urcorner \Rightarrow \bullet\ulcorner B\urcorner\ (\bullet\text{-in})}{\Gamma,\ \ulcorner A\urcorner \Rightarrow \bullet\ulcorner B\urcorner}\text{cut}}{\Gamma \Rightarrow \ulcorner A\supset B\urcorner}\supset\text{, above}$$

∎

THEOREM 32. $\exists$ properties ($\exists$): $\quad \ulcorner A\urcorner \Rightarrow \ulcorner\exists x A\urcorner \quad \dfrac{\ulcorner A\urcorner,\ \Gamma \Rightarrow J}{\ulcorner\exists x A\urcorner,\ \Gamma \Rightarrow J}$
where $x \notin \Gamma, J$.

Proof. Let P be the predicate *predify*($\lambda x.\ulcorner A\urcorner$); thus $\exists P$ is $\ulcorner\exists x A\urcorner$ and $val\, P\, x$ $\triangleright^*\triangleleft$ $\ulcorner A\urcorner$.

$$\dfrac{val\, P\, x \Rightarrow \exists P\ (\exists\text{-in})}{\ulcorner A\urcorner \Rightarrow \ulcorner\exists x A\urcorner}\text{red} \qquad \dfrac{\dfrac{\ulcorner A\urcorner,\ \Gamma \Rightarrow J}{val\, P\, x,\ \Gamma \Rightarrow J}\text{red}}{\ulcorner\exists x A\urcorner,\ \Gamma \Rightarrow J}\exists\text{-el}$$

∎

THEOREM 33. $\forall$ properties ($\forall$): $\quad \ulcorner\forall x A\urcorner \Rightarrow \bullet\ulcorner A\urcorner \quad \dfrac{\Gamma \Rightarrow \bullet\ulcorner A\urcorner}{\Gamma \Rightarrow \ulcorner\forall x A\urcorner}$
where $x \notin \Gamma$.

Proof. Let P be the predicate *predify*($\lambda x.\ulcorner A\urcorner$); thus $\forall P$ is $\ulcorner\forall x A\urcorner$ and $val\, P\, x$ $\triangleright^*\triangleleft$ $\ulcorner A\urcorner$.

$$\dfrac{\forall P \Rightarrow \bullet(val\, P\, x)\ (\forall\text{-el})}{\ulcorner\forall x A\urcorner \Rightarrow \bullet\ulcorner A\urcorner}\text{red} \qquad \dfrac{\dfrac{\Gamma \Rightarrow \bullet\ulcorner A\urcorner}{\Gamma \Rightarrow \bullet(val\, P\, x)}\text{red}}{\Gamma \Rightarrow \forall P}\forall\text{-in}$$

∎

THEOREM 34. Specification rule (spec): $\dfrac{\ulcorner x=X\urcorner,\ \Gamma\Rightarrow J}{\Gamma\Rightarrow J}$
where $X \not\triangleright$ and $x\notin X,\Gamma,J$.

Proof.

$$\dfrac{\Rightarrow \ulcorner X=X\urcorner\ (\text{red}) \qquad \ulcorner X=X\urcorner \Rightarrow \ulcorner \exists x\, x=X\urcorner\ (\text{te}) \qquad \dfrac{\ulcorner x=X\urcorner,\ \Gamma\Rightarrow J}{\ulcorner\exists x\,x=X\urcorner,\ \Gamma\Rightarrow J}\,\exists}{\Gamma\Rightarrow J}\,\text{cut}$$

∎

THEOREM 35. $\vee$-introduction ($\vee$-in): $\ulcorner A\urcorner \Rightarrow \ulcorner A\vee B\urcorner \qquad \ulcorner B\urcorner \Rightarrow \ulcorner A\vee B\urcorner$

Proof. I shall only give the first derivation; the second is similar. Let x be a fresh variable, let T be the formula $x = \mathit{true}$, and let F be the formula $x=\mathit{false}$. First we need the following two CPF derivations.

$$\dfrac{\Rightarrow \ulcorner \mathit{boolean\ true}\urcorner\ (\text{red}) \qquad \dfrac{\dfrac{\ulcorner T\urcorner \Rightarrow \ulcorner \mathit{true}=x\urcorner\ (\text{symm}) \qquad \ulcorner \mathit{true}=x\urcorner,\ \ulcorner \mathit{boolean\ true}\urcorner \Rightarrow \ulcorner \mathit{boolean}\,x\urcorner\ (\text{eq})}{\ulcorner T\urcorner,\ \ulcorner \mathit{boolean\ true}\urcorner\Rightarrow\ulcorner \mathit{boolean}\,x\urcorner}\,\text{cut}}{\ulcorner \mathit{boolean\ true}\urcorner,\ \ulcorner T\urcorner \Rightarrow \ulcorner \mathit{boolean}\,x\urcorner}\,\text{exch}}{\ulcorner T\urcorner\Rightarrow\ulcorner \mathit{boolean}\,x\urcorner \quad (1)}\,\text{cut}$$

$$\dfrac{\dfrac{\ulcorner T\urcorner,\ \ulcorner F\urcorner \Rightarrow \ulcorner \mathit{true}=\mathit{false}\urcorner\ (\text{eq}) \qquad \ulcorner \mathit{true}=\mathit{false}\urcorner \Rightarrow \ulcorner \mathit{false}\urcorner\ (\text{red})}{\ulcorner T\urcorner,\ \ulcorner F\urcorner \Rightarrow \ulcorner \mathit{false}\urcorner}\,\text{cut} \qquad \ulcorner \mathit{false}\urcorner \Rightarrow \ulcorner B\urcorner\ (\mathit{false}\text{-el})}{\ulcorner T\urcorner,\ \ulcorner F\urcorner \Rightarrow \ulcorner B\urcorner \quad (2)}\,\text{cut}$$

Then the main CPF derivation is as follows.

$$\dfrac{\dfrac{\ulcorner T\urcorner \Rightarrow \ulcorner \mathit{boolean}\,x\urcorner\ (1) \qquad \dfrac{\dfrac{\dfrac{\dfrac{\ulcorner A\urcorner \Rightarrow \ulcorner A\urcorner\ (\text{taut})}{\ulcorner A\urcorner,\ \ulcorner T\urcorner \Rightarrow \ulcorner A\urcorner}\,\text{thin}}{\ulcorner A\urcorner \Rightarrow \ulcorner T\supset A\urcorner}\,\supset \qquad \dfrac{\ulcorner T\urcorner,\ \ulcorner F\urcorner \Rightarrow \ulcorner B\urcorner\ (2)}{\ulcorner T\urcorner \Rightarrow \ulcorner F\supset B\urcorner}\,\supset}{\ulcorner A\urcorner,\ \ulcorner T\urcorner \Rightarrow \ulcorner (T\supset A)\wedge(F\supset B)\urcorner}\,\wedge}{\ulcorner T\urcorner,\ \ulcorner A\urcorner \Rightarrow \ulcorner (T\supset A)\wedge(F\supset B)\urcorner}\,\text{exch}}{\dfrac{\ulcorner T\urcorner,\ \ulcorner A\urcorner \Rightarrow \ulcorner \mathit{boolean}\,x\wedge(T\supset A)\wedge(F\supset B)\urcorner}{\ulcorner T\urcorner,\ \ulcorner A\urcorner \Rightarrow \ulcorner A\vee B\urcorner}\,\exists,\text{cut}}\,\wedge,\text{con}}{\ulcorner A\urcorner \Rightarrow \ulcorner A\vee B\urcorner}\,\text{spec}$$

∎

THEOREM 36. $\vee$-elimination rule ($\vee$-el): $\dfrac{\ulcorner A\urcorner,\ \Gamma\Rightarrow \bullet K \qquad \ulcorner B\urcorner,\ \Gamma\Rightarrow \bullet K}{\ulcorner A\vee B\urcorner,\ \Gamma\Rightarrow \bullet K}$

Proof. Again, let x be a fresh variable, let T be the formula $x = true$, and let F be the formula $x = false$. First we need the following two CPF derivations.

$$\dfrac{\ulcorner T\urcorner,\ \ulcorner T\supset A\urcorner \Rightarrow \bullet\ulcorner A\urcorner\ (\supset) \qquad \ulcorner A\urcorner,\ \Gamma\Rightarrow \bullet K}{\ulcorner T\urcorner,\ \ulcorner T\supset A\urcorner,\ \Gamma\Rightarrow \bullet K\quad (1)}\ \bullet\text{-cut}$$

$$\dfrac{\ulcorner F\urcorner,\ \ulcorner F\supset B\urcorner \Rightarrow \bullet\ulcorner B\urcorner\ (\supset) \qquad \ulcorner B\urcorner,\ \Gamma\Rightarrow \bullet K}{\ulcorner F\urcorner,\ \ulcorner F\supset B\urcorner,\ \Gamma\Rightarrow \bullet K\quad (2)}\ \bullet\text{-cut}$$

Now the main CPF derivation is as follows.

$$\dfrac{\dfrac{\dfrac{\ulcorner T\urcorner,\ \ulcorner T\supset A\urcorner,\ \Gamma\Rightarrow \bullet K\ (1) \qquad \ulcorner F\urcorner,\ \ulcorner F\supset B\urcorner,\ \Gamma\Rightarrow \bullet K\ (2)}{\ulcorner boolean\, x\urcorner,\ \ulcorner T\supset A\urcorner,\ \ulcorner F\supset B\urcorner,\ \Gamma\Rightarrow \bullet K}\ \text{thin,bool}}{\ulcorner boolean\, x\wedge (T\supset A)\wedge(F\supset B)\urcorner,\ \Gamma\Rightarrow \bullet K}\ \wedge}{\ulcorner A\vee B\urcorner,\ \Gamma\Rightarrow \bullet K}\ \exists$$

∎

THEOREM 37. Extraction theorems (ext):

$$\ulcorner\forall x\,(x = X\supset A)\urcorner \Rightarrow \bullet(val(predify(\lambda x.\ulcorner A\urcorner))X)$$

$$\bullet(val(predify(\lambda x.\ulcorner A\urcorner))X) \Rightarrow \ulcorner\forall x\,(x = X\supset A)\urcorner$$

where $x\notin X$ and $X \not>$.

Proof. Let B be the formula $x{=}X\supset A$ and P be the predicate $predify(\lambda x.\ulcorner A\urcorner)$.

$$\dfrac{\dfrac{\ulcorner\forall x B\urcorner \Rightarrow \bullet\ulcorner B\urcorner\ (\forall) \qquad \dfrac{\dfrac{\ulcorner x=X\urcorner,\ \ulcorner B\urcorner \Rightarrow \bullet\ulcorner A\urcorner\ (\supset) \qquad \dfrac{\dfrac{\dfrac{\ulcorner x=X\urcorner,\ val\,P\,x\Rightarrow val\,P\,X\ (\text{eq})}{\ulcorner x=X\urcorner,\ \ulcorner A\urcorner\Rightarrow val\,P\,X}\ \text{red}}{\ulcorner x=X\urcorner,\ \bullet\ulcorner A\urcorner\Rightarrow \bullet(val\,P\,X)}\ \bullet\text{-tr}}{\bullet\ulcorner A\urcorner,\ \ulcorner x=X\urcorner\Rightarrow \bullet(val\,P\,X)}\ \text{exch}}{\ulcorner x=X\urcorner,\ \ulcorner B\urcorner\Rightarrow \bullet(val\,P\,X)}\ \text{cut, con}}{\ulcorner B\urcorner,\ \ulcorner x=X\urcorner\Rightarrow \bullet(val\,P\,X)}\ \text{exch}}{\ulcorner\forall x B\urcorner,\ \ulcorner x=X\urcorner\Rightarrow \bullet(val\,P\,X)}\ \bullet\text{-cut}}{\ulcorner\forall x B\urcorner\Rightarrow \bullet(val\,P\,X)}\ \text{spec, exch}$$

$$
\dfrac{
\ulcorner x = X \urcorner \Rightarrow \ulcorner X = x \urcorner \ \text{(symm)}
\qquad
\dfrac{\ulcorner X = x \urcorner,\ val\,P\,X \Rightarrow val\,P\,x \ \text{(eq)}}{\ulcorner X = x \urcorner,\ val\,P\,X \Rightarrow \ulcorner A \urcorner}\,\text{red}
}{
\dfrac{\ulcorner x = X \urcorner,\ val\,P\,X \Rightarrow \ulcorner A \urcorner}{
\dfrac{val\,P\,X \Rightarrow \ulcorner B \urcorner}{
\dfrac{\bullet(val\,P\,X) \Rightarrow \bullet\ulcorner B \urcorner}{\bullet(val\,P\,X) \Rightarrow \ulcorner \forall x\, B \urcorner}\,\forall
}\,\bullet\text{-tr}
}\,\text{exch}, \supset
}\,\text{cut}
$$

▮

EXERCISE. Derive in CPF the logical sequents

$$\ulcorner \exists x\,(x = X \wedge A) \urcorner \Rightarrow val(predify(\lambda x.\ulcorner A \urcorner))X$$
$$val(predify(\lambda x.\ulcorner A \urcorner))X \Rightarrow \ulcorner \exists x\,(x = X \wedge A) \urcorner$$

where $x \notin X$ and $X \not\triangleright$

LOGIC OF PARTIAL TERMS

The axioms and rules of LPT are all instances of the following schemata.

$\wedge$-axioms: $A \supset B \supset (A \wedge B) \quad (A \wedge B) \supset A \quad (A \wedge B) \supset B$

$\vee$-axioms: $A \supset (A \vee B) \quad B \supset (A \vee B) \quad (A \vee B) \supset (A \supset C) \supset (B \supset C) \supset C$

$\supset$-axioms: $A \supset B \supset A \quad (A \supset B \supset C) \supset (A \supset B) \supset A \supset C$

Truth and falsity (*true*-in, *false*-el): *true* $\quad$ *false* $\supset A$

Quantifier axioms ($\exists$-in, $\forall$-el): $A \supset \exists x A \quad \forall x A \supset A$

Term existence (te): $T\begin{bmatrix} X \\ x \end{bmatrix} \supset \exists x\, x = X$ where $x \in T$ but $x \notin X$

Reduction (red): $X \supset Y$ where $X \triangleright Y$ or $Y \triangleright X$

Equality (eq): $X = Y \supset T\begin{bmatrix} X \\ u \end{bmatrix} \supset T\begin{bmatrix} Y \\ u \end{bmatrix}$

Modus ponens: $\dfrac{A \quad A \supset B}{B}$

Quantifier rules ($\exists$-el, $\forall$-in): $\dfrac{A \supset B}{\exists x A \supset B} \quad \dfrac{B \supset A}{B \supset \forall x A}$ where $x \notin B$

Induction (ind): $\dfrac{\forall x\,(x = 0 \supset A) \quad \forall x\,(x = n \supset A) \supset \forall x\,(x = Sn \supset A)}{num\,n \supset \forall x\,(x = n \supset A)}$

where $n \notin x, A$

CPF DERIVATIONS CORRESPONDING TO THE LPT AXIOMS AND RULES

I shall show that for each axiom A of LPT the logical sequent $\Rightarrow \ulcorner A \urcorner$ is derivable in CPF, and that for each rule $\dfrac{A \ldots B}{C}$ of LPT the rule $\dfrac{\Rightarrow \ulcorner A \urcorner \ \ldots \ \Rightarrow \ulcorner B \urcorner}{\Rightarrow \ulcorner C \urcorner}$ is derivable in CPF.

THEOREM 38. LPT $\wedge$-axioms: $A \supset B \supset (A \wedge B)$ $\quad (A \wedge B) \supset A \quad (A \wedge B) \supset B$

Proof. The corresponding CPF derivations are as follows.

$$\dfrac{\dfrac{\ulcorner A \urcorner,\ \ulcorner B \urcorner \Rightarrow \ulcorner A \wedge B \urcorner\ (\wedge)}{\ulcorner A \urcorner \Rightarrow \ulcorner B \supset (A \wedge B) \urcorner}\supset}{\Rightarrow \ulcorner A \supset B \supset (A \wedge B) \urcorner}\supset \qquad \dfrac{\ulcorner A \wedge B \urcorner \Rightarrow \ulcorner A \urcorner\ (\wedge)}{\Rightarrow \ulcorner (A \wedge B) \supset A \urcorner}\supset \qquad \dfrac{\ulcorner A \wedge B \urcorner \Rightarrow \ulcorner B \urcorner\ (\wedge)}{\Rightarrow \ulcorner (A \wedge B) \supset B \urcorner}\supset$$

∎

THEOREM 39. LPT $\vee$-axioms:
$A \supset (A \vee B) \qquad B \supset (A \vee B) \qquad (A \vee B) \supset (A \supset C) \supset (B \supset C) \supset C$

Proof. The CPF derivations are as follows.

$$\dfrac{\ulcorner A \urcorner \Rightarrow \ulcorner A \vee B \urcorner\ (\vee)}{\Rightarrow \ulcorner A \supset (A \vee B) \urcorner}\supset \qquad \dfrac{\ulcorner B \urcorner \Rightarrow \ulcorner A \vee B \urcorner\ (\vee)}{\Rightarrow \ulcorner B \supset (A \vee B) \urcorner}\supset$$

$$\dfrac{\dfrac{\dfrac{\dfrac{\dfrac{\ulcorner A \urcorner,\ \ulcorner A \supset C \urcorner \Rightarrow \bullet \ulcorner C \urcorner\ (\supset)}{\ulcorner A \urcorner,\ \ulcorner A \supset C \urcorner,\ \ulcorner B \supset C \urcorner \Rightarrow \bullet \ulcorner C \urcorner}\text{thin} \qquad \dfrac{\ulcorner B \urcorner,\ \ulcorner B \supset C \urcorner \Rightarrow \bullet \ulcorner C \urcorner\ (\supset)}{\ulcorner B \urcorner,\ \ulcorner A \supset C \urcorner,\ \ulcorner B \supset C \urcorner \Rightarrow \bullet \ulcorner C \urcorner}\text{thin}}{\ulcorner A \vee B \urcorner,\ \ulcorner A \supset C \urcorner,\ \ulcorner B \supset C \urcorner \Rightarrow \bullet \ulcorner C \urcorner}\vee}{\ulcorner A \vee B \urcorner,\ \ulcorner A \supset C \urcorner \Rightarrow \ulcorner (B \supset C) \supset C \urcorner}\supset}{\ulcorner A \vee B \urcorner \Rightarrow \ulcorner (A \supset C) \supset (B \supset C) \supset C \urcorner}\supset}{\Rightarrow \ulcorner (A \vee B) \supset (A \supset C) \supset (B \supset C) \supset C \urcorner}\supset$$

∎

THEOREM 40. LPT $\supset$-axioms: $A \supset B \supset A \quad (A \supset B \supset C) \supset (A \supset B) \supset A \supset C$

Proof. The CPF derivations are as follows.

$$\dfrac{\dfrac{\dfrac{\ulcorner A \urcorner \Rightarrow \ulcorner A \urcorner\ \text{(taut)}}{\ulcorner A \urcorner,\ \ulcorner B \urcorner \Rightarrow \ulcorner A \urcorner}\text{thin}}{\ulcorner A \urcorner \Rightarrow \ulcorner B \supset A \urcorner}\supset}{\Rightarrow \ulcorner A \supset B \supset A \urcorner}\supset$$

$$
\dfrac{\dfrac{\ulcorner A\urcorner,\ \ulcorner A \supset B\urcorner \Rightarrow \bullet\ulcorner B\urcorner\ (\supset) \qquad \dfrac{\dfrac{\ulcorner A\urcorner,\ \ulcorner A\supset B\supset C\urcorner \Rightarrow \bullet\ulcorner B\supset C\urcorner\ (\supset) \qquad \dfrac{\ulcorner B\urcorner,\ \ulcorner B\supset C\urcorner \Rightarrow \bullet\ulcorner C\urcorner\ (\supset)}{\ulcorner B\supset C\urcorner,\ \ulcorner B\urcorner \Rightarrow \bullet\ulcorner C\urcorner}\,\text{exch}}{\ulcorner A\urcorner,\ \ulcorner A\supset B\supset C\urcorner,\ \ulcorner B\urcorner \Rightarrow \bullet\ulcorner C\urcorner}\,\bullet\text{-cut}}{\ulcorner B\urcorner,\ \ulcorner A\urcorner,\ \ulcorner A\supset B\supset C\urcorner \Rightarrow \bullet\ulcorner C\urcorner}\,\text{exch}}{\ulcorner A\urcorner,\ \ulcorner A\supset B\urcorner,\ \ulcorner A\urcorner,\ \ulcorner A\supset B\supset C\urcorner \Rightarrow \bullet\ulcorner C\urcorner}\,\bullet\text{-cut}
$$

$$
\dfrac{\dfrac{\dfrac{\dfrac{\ulcorner A\urcorner,\ \ulcorner A\supset B\urcorner,\ \ulcorner A\urcorner,\ \ulcorner A\supset B\supset C\urcorner \Rightarrow \bullet\ulcorner C\urcorner}{\ulcorner A\supset B\supset C\urcorner,\ \ulcorner A\supset B\urcorner,\ \ulcorner A\urcorner \Rightarrow \bullet\ulcorner C\urcorner}\,\text{exch, con}}{\ulcorner A\supset B\supset C\urcorner,\ \ulcorner A\supset B\urcorner \Rightarrow \ulcorner A\supset C\urcorner}\,\supset}{\ulcorner A\supset B\supset C\urcorner \Rightarrow \ulcorner (A\supset B)\supset A\supset C\urcorner}\,\supset}{\Rightarrow \ulcorner (A\supset B\supset C)\supset (A\supset B)\supset A\supset C\urcorner}\,\supset
$$

▮

THEOREM 41. LPT truth and falsity axioms (*true*-in, *false*-el):

$$
true \qquad false \supset A
$$

Proof. $\Rightarrow \ulcorner true\urcorner$ is a CPF axiom, while for $false \supset A$ the CPF derivation is as follows.

$$
\dfrac{\ulcorner false\urcorner \Rightarrow \ulcorner A\urcorner\ (false\text{-el})}{\Rightarrow \ulcorner false \supset A\urcorner}\,\supset
$$

▮

THEOREM 42. LPT quantifier axioms ($\exists$-in, $\forall$-el): $A \supset \exists x A \qquad \forall x A \supset A$

Proof. The CPF derivations are as follows.

$$
\dfrac{\ulcorner A\urcorner \Rightarrow \ulcorner \exists x A\urcorner\ (\exists)}{\Rightarrow \ulcorner A \supset \exists x A\urcorner}\,\supset \qquad \dfrac{\ulcorner \forall x A\urcorner \Rightarrow \bullet\ulcorner A\urcorner\ (\forall)}{\Rightarrow \ulcorner \forall x A \supset A\urcorner}\,\supset
$$

▮

THEOREM 43. LPT term existence axiom (te): $T\left[\begin{smallmatrix}X\\x\end{smallmatrix}\right] \supset \exists x\, x = X$, where $x \in T$ but $x \notin X$.

Proof. The CPF derivation is as follows.

$$
\dfrac{\ulcorner T\left[\begin{smallmatrix}X\\x\end{smallmatrix}\right]\urcorner \Rightarrow \ulcorner \exists x\, x = X\urcorner\ (\text{te})}{\Rightarrow \ulcorner T\left[\begin{smallmatrix}X\\x\end{smallmatrix}\right] \supset \exists x\, x = X\urcorner}\,\supset
$$

▮

THEOREM 44. LPT reduction axiom (red): $X \supset Y$ where $X \rhd Y$ or $Y \rhd X$.

Proof. The CPF derivation is as follows.

$$\dfrac{\ulcorner X \urcorner \Rightarrow \ulcorner Y \urcorner \ (\text{red})}{\Rightarrow \ulcorner X \supset Y \urcorner}\supset$$

∎

THEOREM 45. LPT equality axiom (eq): $X = Y \supset T\begin{bmatrix} X \\ u \end{bmatrix} \supset T\begin{bmatrix} Y \\ u \end{bmatrix}$

Proof. The CPF derivation is as follows.

$$\dfrac{\dfrac{\ulcorner X = Y \urcorner, \ulcorner T\begin{bmatrix} X \\ u \end{bmatrix} \urcorner \Rightarrow \ulcorner T\begin{bmatrix} Y \\ u \end{bmatrix} \urcorner \ (\text{eq})}{\ulcorner X = Y \urcorner \Rightarrow \ulcorner T\begin{bmatrix} X \\ u \end{bmatrix} \supset T\begin{bmatrix} Y \\ u \end{bmatrix} \urcorner}\supset}{\Rightarrow \ulcorner X = Y \supset T\begin{bmatrix} X \\ u \end{bmatrix} \supset T\begin{bmatrix} Y \\ u \end{bmatrix} \urcorner}\supset$$

∎

THEOREM 46. LPT modus ponens rule: $\dfrac{A \quad A \supset B}{B}$

Proof. The CPF derivation is as follows.

$$\dfrac{\dfrac{\Rightarrow \ulcorner A \urcorner \qquad \dfrac{\Rightarrow \ulcorner A \supset B \urcorner}{\ulcorner A \urcorner \Rightarrow \bullet \ulcorner B \urcorner}\supset}{\Rightarrow \bullet \ulcorner B \urcorner}\text{cut}}{\Rightarrow \ulcorner B \urcorner}\bullet\text{-el}$$

∎

THEOREM 47. LPT quantifier rules ($\exists$-el, $\forall$-in): $\dfrac{A \supset B}{\exists x A \supset B}$ $\dfrac{B \supset A}{B \supset \forall x A}$ where $x \notin B$.

Proof. The CPF derivations are as follows.

$$\dfrac{\dfrac{\dfrac{\Rightarrow \ulcorner A \supset B \urcorner}{\ulcorner A \urcorner \Rightarrow \bullet \ulcorner B \urcorner}\supset}{\ulcorner \exists x A \urcorner \Rightarrow \bullet \ulcorner B \urcorner}\exists}{\Rightarrow \ulcorner \exists x A \supset B \urcorner}\supset \qquad \dfrac{\dfrac{\dfrac{\Rightarrow \ulcorner B \supset A \urcorner}{\ulcorner B \urcorner \Rightarrow \bullet \ulcorner A \urcorner}\supset}{\ulcorner B \urcorner \Rightarrow \ulcorner \forall x A \urcorner}\forall}{\Rightarrow \ulcorner B \supset \forall x A \urcorner}\supset$$

∎

THEOREM 48. LPT induction rule (ind):

$$\frac{\forall x\,(x=0\supset A) \quad \forall x\,(x=n\supset A)\supset\forall x\,(x=Sn\supset A)}{\mathit{num}\,n\supset\forall x\,(x=n\supset A)} \quad \text{where } n\notin x, A$$

Proof. Let P be the predicate $\mathit{predify}(\lambda x.\ulcorner A\urcorner)$, and let Q be $\triangle(\ulcorner \mathit{true}\urcorner, P)$. Recall from Logic that $\mathit{val}\,Q\,x \;\triangleright^{*}\!\triangleleft\; \Box(\ulcorner \mathit{true}\urcorner, \mathit{con}(\mathit{val}\,P\,x)) \;\triangleright^{*}\!\triangleleft\; \bullet(\mathit{val}\,P\,x)$. The CPF derivation is as follows.

$$\dfrac{\dfrac{\dfrac{\dfrac{\dfrac{\Rightarrow \ulcorner\forall x\,(x=0\supset A)\urcorner}{\Rightarrow \bullet(\mathit{val}\,P\,0)}\,\text{ext, cut}}{\Rightarrow \mathit{val}\,Q\,0}\,\text{red} \qquad \dfrac{\dfrac{\dfrac{\dfrac{\Rightarrow \ulcorner\forall x\,(x=n\supset A)\supset\forall x\,(x=Sn\supset A)\urcorner}{\ulcorner\forall x\,(x=n\supset A)\urcorner \Rightarrow \bullet\ulcorner\forall x\,(x=Sn\supset A)\urcorner}\,\supset}{\bullet(\mathit{val}\,P\,n)\Rightarrow \bullet\ulcorner\forall x\,(x=Sn\supset A)\urcorner}\,\text{ext, cut}}{\bullet(\mathit{val}\,P\,n)\Rightarrow\bullet(\mathit{val}\,P\,(Sn))}\,\text{ext, }\bullet\text{-cut}}{\mathit{val}\,Q\,n\Rightarrow \mathit{val}\,Q\,(Sn)}\,\text{red}}{\ulcorner \mathit{num}\,n\urcorner \Rightarrow \mathit{val}\,Q\,n}\,\text{ind}}{\ulcorner \mathit{num}\,n\urcorner \Rightarrow \bullet(\mathit{val}\,P\,n)}\,\text{red}}{\ulcorner \mathit{num}\,n\urcorner \Rightarrow \ulcorner\forall x\,(x=n\supset A)\urcorner}\,\text{ext, cut}}{\Rightarrow \ulcorner \mathit{num}\,n\supset\forall x\,(x=n\supset A)\urcorner}\,\supset$$

∎

FORMAL INTERPRETATION OF LPT

DEFINITION. A coding of formulae as constructions is defined as follows. Let *atomic*, *and*, *or*, *implies*, *exists* and *all* be six fresh 1-ary constructors. If A is a formula with free variables $\underline{z}$ (listed in standard order), the code of A, $\langle\!\langle A\rangle\!\rangle$, is defined as $(\lambda[\underline{z}].\langle A\rangle)$, where

$$\begin{aligned}
\langle T\rangle &\text{ is } \mathit{atomic}(T) \quad \text{if } T \text{ is atomic}\\
\langle A\wedge B\rangle &\text{ is } \mathit{and}(\langle A\rangle, \langle B\rangle)\\
\langle A\vee B\rangle &\text{ is } \mathit{or}(\langle A\rangle, \langle B\rangle)\\
\langle A\supset B\rangle &\text{ is } \mathit{implies}(\langle A\rangle, \langle B\rangle)\\
\langle \exists x A\rangle &\text{ is } \mathit{exists}(\lambda[x].\langle A\rangle)\\
\langle \forall x A\rangle &\text{ is } \mathit{all}(\lambda[x].\langle A\rangle).
\end{aligned}$$

DEFINITION. A coding of LPT derivations as constructions is defined as follows. Number the axiom schemata and rules of inference of LPT, 1,2,... 19. A derivation with conclusion A, from a list of premises numbered 0,1,2..., is coded as

$(premise(i), \langle\langle A\rangle\rangle)$	if A is premise number i
$(none(n), \langle\langle A\rangle\rangle)$	if A is an instance of axiom schema n
$(one(n, X), \langle\langle A\rangle\rangle)$	if A is derived by rule n from the subderivation X
$(two(n, X, Y), \langle\langle A\rangle\rangle)$	if A is derived by rule n from subderivations X, Y.

DEFINITION. Define recursive functions $T_1, \ldots T_{19}$ such that

- for $n = 1, \ldots 15$, if A is an instance of LPT axiom schema n then $T_n(\langle\langle A\rangle\rangle)$ $\triangleright^*$ the code of the corresponding CPF derivation (given above) of $\Rightarrow \ulcorner A \urcorner$ from no premises;
- for $n = 17, 18$, if $\dfrac{B}{A}$ is an instance of LPT rule n then $T_n(\langle\langle B\rangle\rangle, \langle\langle A\rangle\rangle)$ $\triangleright^*$ the code of the corresponding CPF derivation (given above) of $\Rightarrow \ulcorner A \urcorner$ from the premise $\Rightarrow \ulcorner B \urcorner$;
- for $n = 16, 19$, if $\dfrac{B\ C}{A}$ is an instance of LPT rule n then $T_n(\langle\langle B\rangle\rangle, \langle\langle C\rangle\rangle, \langle\langle A\rangle\rangle)$ $\triangleright^*$ the code of the corresponding CPF derivation (given above) of $\Rightarrow \ulcorner A \urcorner$ from the premises $\Rightarrow \ulcorner B \urcorner$ and $\Rightarrow \ulcorner C \urcorner$.

DEFINITION. Define a construction *cpf* by

$$
\begin{aligned}
cpf(none(1), a) &\triangleq T_1(a) \\
&\vdots \\
cpf(none(15), a) &\triangleq T_{15}(a) \\
cpf(two(16, (u, b), (v, c)), a) &\triangleq glue(T_{16}(b, c, a), [cpf(u, b), cpf(v, c)]) \\
cpf(one(17, (u, b)), a) &\triangleq glue(T_{17}(b, a), [cpf(u, b)]) \\
cpf(one(18, (u, b)), a) &\triangleq glue(T_{18}(b, a), [cpf(u, b)]) \\
cpf(two(19, (u, b), (v, c)), a) &\triangleq glue(T_{19}(b, c, a), [cpf(u, b), cpf(v, c)])
\end{aligned}
$$

where a, b, c, u, v are five variables. (This uses the *glue* function defined in Chapter 28.)

THEOREM 49. If D is the code of an LPT derivation (with no premises) of a formula A then $cpf(D)$ $\triangleright^*$ the code of a CPF derivation (with no premises) of the logical sequent $\Rightarrow \ulcorner A \urcorner$.

Proof. The argument is by structural induction on the LPT derivation encoded by D. Since the derivation has no premises there are only three cases.

Case 1: D is $(none(n), \langle\langle A\rangle\rangle)$, where A is an instance of LPT axiom schema n. Then $cpf(D) \rhd^* T_n(\langle\langle A\rangle\rangle) \rhd^*$ a CPF derivation of $\Rightarrow \ulcorner A\urcorner$ from no premises, as required.

Case 2: D is $(one(n, (U, \langle\langle B\rangle\rangle)), \langle\langle A\rangle\rangle)$, where $\frac{B}{A}$ is an instance of LPT rule of inference n. Then $T_n(\langle\langle B\rangle\rangle, \langle\langle A\rangle\rangle) \rhd^*$ a CPF derivation of $\Rightarrow \ulcorner A\urcorner$ from the premise $\Rightarrow \ulcorner B\urcorner$. By the inductive hypothesis, $cpf(U, \langle\langle B\rangle\rangle) \rhd^*$ the code of a CPF derivation of $\Rightarrow \ulcorner B\urcorner$ from no premises. Hence by the *glue*-lemma for CPF $cpf(D) \rhd^* glue(T_n(\langle\langle B\rangle\rangle, \langle\langle A\rangle\rangle), [cpf(U, \langle\langle B\rangle\rangle)]) \rhd^*$ a CPF derivation of $\Rightarrow \ulcorner A\urcorner$ from no premises, as required.

Case 3: D is $(two(n, (U, \langle\langle B\rangle\rangle), (V, \langle\langle C\rangle\rangle)), \langle\langle A\rangle\rangle)$, where $\frac{B\ C}{A}$ is an instance of LPT rule of inference n. Then $T_n(\langle\langle B\rangle\rangle, \langle\langle C\rangle\rangle, \langle\langle A\rangle\rangle) \rhd^*$ a CPF derivation of $\Rightarrow \ulcorner A\urcorner$ from the premises $\Rightarrow \ulcorner B\urcorner$ and $\Rightarrow \ulcorner C\urcorner$. By the inductive hypothesis, $cpf(U, \langle\langle B\rangle\rangle)$ and $cpf(V, \langle\langle C\rangle\rangle)$ reduce to the codes of CPF derivations of $\Rightarrow \ulcorner B\urcorner$ and $\Rightarrow \ulcorner C\urcorner$, respectively, from no premises. Therefore

$$cpf(D) \rhd^* glue(T_n(\langle\langle B\rangle\rangle, \langle\langle C\rangle\rangle, \langle\langle A\rangle\rangle), [cpf(U, \langle\langle B\rangle\rangle), cpf(V, \langle\langle C\rangle\rangle)])$$

which by the *glue*-lemma for CPF reduces to the code of a CPF derivation of $\Rightarrow \ulcorner A\urcorner$ from no premises, as required. ▮

THEOREM 50. (LPT interpretation theorem.) If D is the code of an LPT derivation (with no premises) of a formula A with no free variables, then $pr(spr(cpf(D))) \vdash \ulcorner A\urcorner$.

Proof. By the previous theorem, the extensionality theorem for $\vdash$, and the interpretation theorem for CPF. ▮

THEOREM 51. (The *glue*-lemma for LPT.) If D is the code of an LPT derivation of an LPT formula A from premises $B_0, \ldots B_k$, and $D_0, \ldots D_k$ are the codes of LPT derivations (with no premises) of $B_0, \ldots B_k$ respectively, then $glue(D, [D_0, \ldots D_k]) \rhd^*$ the code of an LPT derivation (with no premises) of A.

Proof. By structural induction on the derivation encoded by D. ▮

EXERCISE. Define a substitution notation $A\binom{X}{x}$, meaning the formula obtained by substituting the term X for the variable x in the formula A, provided no variable clashes occur, in such a way that the logical sequents

$$\ulcorner A\tbinom{X}{x}\urcorner \Rightarrow \ulcorner A\urcorner\begin{bmatrix}X\\x\end{bmatrix} \qquad \ulcorner A\urcorner\begin{bmatrix}X\\x\end{bmatrix} \Rightarrow \ulcorner A\tbinom{X}{x}\urcorner$$

are derivable in CPF, assuming $x \notin X$ and $X \not\rhd$. (Hint: see the end of Chapter 33.) Why may these sequents be underivable if X is reducible?

CHAPTER 30

LOGIC OF PARTIAL TERMS

LPT FORMULAE

A *formula* of LPT is a sentence in the following language.

- The alphabet is that of the Expanded Term Language (ET) plus '$\wedge$', '$\vee$', '$\supset$', '$\exists$', '$\forall$'.
- The tokens are those of ET plus '$\wedge$', '$\vee$', '$\supset$', '$\exists$', '$\forall$'.
- Lexical analysis consists of recognising constants and replacing them by the token '*con*', recognising variables and replacing them by the token '*vbl*', and recognising every other token.
- The grammar of the language is as follows.
  - The terminals are the tokens.
  - The non-terminals are *F*, *T*; the start symbol is *F*.
  - The production rules are

$$F \rightarrow T \mid (F \wedge F) \mid (F \vee F) \mid (F \supset F) \mid \exists\, vbl\, F \mid \forall\, vbl\, F$$
$$T \rightarrow con \mid (vbl) \mid (T\,T) \mid (\lambda\, T\,.\,T) \mid (T\left[{}^{T}_{vbl}\right])$$

(Note that this grammar contains that of ET: a term of ET is any string matching the non-terminal *T*. A formula that is a term will be called an *atomic formula*.)

METAFORMULAE

The letters '*A*', '*B*' and '*C*' will be used as metavariables to denote formulae. A *metaformula* is an expression that is like a formula except that it may contain metaconstants, metavariables and ET metanotation, some brackets may be omitted, and optional brackets and spaces may be added. To state it more precisely, the alphabet is that of ET metaterms plus '$\wedge$', '$\vee$', '$\supset$', '$\exists$', '$\forall$'; the tokens are those of ET metaterms plus '$\wedge$', '$\vee$', '$\supset$', '$\exists$', '$\forall$', and a new one '*formvbl*'; lexical analysis is as for ET metaterms except that metavariables denoting formulae are replaced by '*formvbl*'; and the grammar is

$$\begin{aligned}
F &\to P \wedge F \mid P \vee F \mid P \supset F \mid P \\
P &\to \exists\, vbl\, P \mid \forall\, vbl\, P \mid (\,F\,) \mid T \mid formvbl \\
T &\to T\,L \mid T\begin{bmatrix} T \\ vbl \end{bmatrix} \mid L \\
L &\to con \mid vbl \mid (\,T\,) \mid (\lambda\, T\,.\,T\,) \mid termcon \mid vblvbl \mid termvbl \mid \ldots
\end{aligned}$$

where the start symbol is F and '...' represents production rules for ET metanotation.

An *instance* of a metaformula is a formula obtained from the metaformula by choosing a term, variable or formula (as required) for each metavariable in the metaformula, replacing each occurrence of each metavariable by the chosen term, variable or formula, replacing each metaconstant by the term it denotes, adding and removing brackets as required, removing spaces, and rewriting all metanotation. A metaformula may be used in two ways: as a schema, standing for any of its instances, or to denote a particular one of its instances. The latter will be implied when particular formulae, terms or variables have previously been chosen for the metavariables.

FREE VARIABLES IN FORMULAE

For any variable v and LPT formula A, the relation $v \in A$ ('v occurs free in A') is defined by:

- if A is a term T, $v \in A$ iff $v \in T$ in the sense of ET;
- $v \in A \wedge B$ iff $v \in A \vee B$ iff $v \in A \supset B$ iff $v \in A$ or $v \in B$;
- $v \in \exists x A$ iff $v \in \forall x A$ iff v isn't x and $v \in A$.

The *free variables* of A are the variables v such that $v \in A$.

THEOREM 23. Any LPT formula has finitely many free variables.

INTERPRETATION OF LPT FORMULAE AS PROOF FUNCTIONS

Each formula A is interpreted as a proof function $\ulcorner A \urcorner$. This interpretation extends the interpretation of terms T as proof functions $\ulcorner T \urcorner$, introduced in Logic.

THEOREM 25. For any formulae A, B, term Q, variables $\underline{y}$, and irreducible terms $\underline{Y}$, we have $Q \vdash \ulcorner A \wedge B \urcorner \begin{bmatrix} \underline{Y} \\ \underline{y} \end{bmatrix}$ iff $Q \ \triangleright^* \ (R_1, R_2) \not\triangleright$, where $R_1 \vdash \ulcorner A \urcorner \begin{bmatrix} \underline{Y} \\ \underline{y} \end{bmatrix}$ and $R_2 \vdash \ulcorner B \urcorner \begin{bmatrix} \underline{Y} \\ \underline{y} \end{bmatrix}$.

THEOREM 26. For any formula A, variable x, term Q, variables $\underline{y}$ (not including x), and irreducible terms $\underline{Y}$, we have $Q \vdash \ulcorner \exists x A \urcorner \begin{bmatrix} \underline{Y} \\ \underline{y} \end{bmatrix}$ iff $Q \ \triangleright^*$ $(X, R) \not\triangleright$, where $R \vdash \ulcorner A \urcorner \begin{bmatrix} \underline{Y} \\ \underline{y} \end{bmatrix} \begin{bmatrix} X \\ x \end{bmatrix}$.

THEOREM 27. For any formulae A, B (with free variables $\underline{y}$ between them), any terms Q, R (with no free variables), and any constructions $\underline{Y}$, if $Q \vdash \ulcorner A \supset B \urcorner \left[\frac{\underline{Y}}{\underline{y}}\right]$ and $R \vdash \ulcorner A \urcorner \left[\frac{\underline{Y}}{\underline{y}}\right]$ then $el_{\supset}(Q, R) \vdash \ulcorner B \urcorner \left[\frac{\underline{Y}}{\underline{y}}\right]$ (for a certain construction $el_{\supset}$).

THEOREM 28. For any formulae A, B (with free variables $\underline{y}$ between them), any term Q (with no free variables), and any constructions $\underline{Y}$, if $Q \vdash \ulcorner A \vee B \urcorner \left[\frac{\underline{Y}}{\underline{y}}\right]$ then either $select_{\vee}(Q) \;\triangleright^*\; true$ and $el_{\vee}(Q) \vdash \ulcorner A \urcorner \left[\frac{\underline{Y}}{\underline{y}}\right]$ or $select_{\vee}(Q) \;\triangleright^*\; false$ and $el_{\vee}(Q) \vdash \ulcorner B \urcorner \left[\frac{\underline{Y}}{\underline{y}}\right]$ (for certain constructions $select_{\vee}$ and $el_{\vee}$).

THEOREM 29. For any formula of the form $\forall x A$ (with free variables $\underline{y}$), any term Q (with no free variables), and any constructions $\underline{Y}, X$, if $Q \vdash \ulcorner \forall x A \urcorner \left[\frac{\underline{Y}}{\underline{y}}\right]$ then $el_{\forall}(Q, X) \vdash \ulcorner A \urcorner \left[\frac{\underline{Y}}{\underline{y}}\right]\left[\frac{X}{x}\right]$ (for a certain construction $el_{\forall}$).

LOGIC OF PARTIAL TERMS

The axioms and rules of LPT are all instances of the following schemata.

$\wedge$-axioms: $A \supset B \supset (A \wedge B)$ $\quad (A \wedge B) \supset A$ $\quad (A \wedge B) \supset B$
$\vee$-axioms: $A \supset (A \vee B)$ $\quad B \supset (A \vee B)$ $\quad (A \vee B) \supset (A \supset C) \supset (B \supset C) \supset C$
$\supset$-axioms: $A \supset B \supset A$ $\quad (A \supset B \supset C) \supset (A \supset B) \supset A \supset C$
Truth and falsity (*true*-in, *false*-el): *true* $\quad$ *false* $\supset A$
Quantifier axioms ($\exists$-in, $\forall$-el): $A \supset \exists x A$ $\quad \forall x A \supset A$
Term existence (te): $T\left[\frac{X}{x}\right] \supset \exists x\, x = X$ where $x \in T$ but $x \notin X$
Reduction (red): $X \supset Y$ where $X \triangleright Y$ or $Y \triangleright X$
Equality (eq): $X = Y \supset T\left[\frac{X}{u}\right] \supset T\left[\frac{Y}{u}\right]$

Modus ponens: $\dfrac{A \quad A \supset B}{B}$

Quantifier rules ($\exists$-el, $\forall$-in): $\dfrac{A \supset B}{\exists x A \supset B}$ $\quad \dfrac{B \supset A}{B \supset \forall x A}$ $\quad$ where $x \notin B$

Induction (ind): $\dfrac{\forall x (x = 0 \supset A) \quad \forall x (x = n \supset A) \supset \forall x (x = Sn \supset A)}{num\, n \supset \forall x (x = n \supset A)}$

where $n \notin x, A$

FORMAL INTERPRETATION OF LPT

A coding of LPT derivations as constructions is defined, using six 1-ary constructors *atomic*, *and*, *or*, *implies*, *exists* and *all*. A construction *cpf* is defined.

THEOREM 49. If D is the code of an LPT derivation (with no premises) of a formula A then $cpf(D) \triangleright^*$ the code of a CPF derivation (with no premises) of the logical sequent $\Rightarrow \ulcorner A \urcorner$.

THEOREM 50. (LPT interpretation theorem.) If D is the code of an LPT derivation (with no premises) of a formula A with no free variables, then $pr(spr(cpf(D))) \vdash \ulcorner A \urcorner$.

THEOREM 51. (The *glue*-lemma for LPT.) If D is the code of an LPT derivation of an LPT formula A from premises $B_0, \dots B_k$, and $D_0, \dots D_k$ are the codes of LPT derivations (with no premises) of $B_0, \dots B_k$ respectively, then $glue(D, [D_0, \dots D_k]) \triangleright^*$ the code of an LPT derivation (with no premises) of A.

CHAPTER 31

FROM LOGIC OF PARTIAL TERMS TO HEYTING ARITHMETIC

Heyting Arithmetic (HA) is first-order intuitionistic number theory. It will be obtained from the Logic of Partial Terms (LPT) by restricting the range of the variables to numbers. Every HA derivation can be transformed into an LPT derivation. Therefore every theorem of HA (without free variables) has an intuitionistic proof.

LEXICAL CONVENTIONS

The identifier 'A' will denote an LPT formula; all other upper-case letters will denote terms. These conventions will apply until HA formulae are introduced.

'$\underline{n}$' will be used as an abbreviation for a (possibly empty) sequence of variables $n_1, \ldots n_k$; and '$num\, \underline{n} \supset A$' will be used as a metanotation for $num\, n_1 \supset \ldots num\, n_k \supset A$. Other underlined metavariables will be used in a similar way.

SOME THEOREMS AND DERIVED RULES OF LPT

First note that LPT contains intuitionistic propositional calculus, so I shall use the latter freely in the LPT derivations that follow, marking each use with the label 'pc'.

THEOREM 1. LPT reduction theorems (red): $U \supset V$, where $U \vartriangleright^{*} \vartriangleleft V$;
T, where $T \vartriangleright^{*}$ *true*.

Proof. $U \supset V$ follows from LPT reduction axioms using intuitionistic propositional calculus; T then follows from *true* $\supset T$ and the axiom *true*. ∎

THEOREM 2. LPT self-equality theorem (se): $T\begin{bmatrix} X \\ x \end{bmatrix} \supset X = X$,
where $x \in T$ but $x \notin X$.

Proof.

$$\dfrac{T\begin{bmatrix}X\\x\end{bmatrix} \supset \exists x\, x=X \ \text{(te)} \qquad \dfrac{\dfrac{x=X\supset x=X\supset X=X \ \text{(eq)}}{x=X\supset X=X}\,\text{pc}}{\exists x\, x=X\supset X=X}\,\exists\text{-el}}{T\begin{bmatrix}X\\x\end{bmatrix}\supset X=X}\,\text{pc}$$

∎

THEOREM 3. LPT symmetry theorem (symm): $X=Y\supset Y=X.$

Proof.

$$\dfrac{X=Y\supset X=X \ \text{(se)} \qquad X=Y\supset X=X\supset Y=X \ \text{(eq)}}{X=Y\supset Y=X}\,\text{pc}$$

∎

THEOREM 4. LPT reversed equality theorem (r.eq): $X=Y\supset T\begin{bmatrix}Y\\u\end{bmatrix}\supset T\begin{bmatrix}X\\u\end{bmatrix}$

Proof.

$$\dfrac{X=Y\supset Y=X \ \text{(symm)} \qquad Y=X\supset T\begin{bmatrix}Y\\u\end{bmatrix}\supset T\begin{bmatrix}X\\u\end{bmatrix} \ \text{(eq)}}{X=Y\supset T\begin{bmatrix}Y\\u\end{bmatrix}\supset T\begin{bmatrix}X\\u\end{bmatrix}}\,\text{pc}$$

∎

THEOREM 5. LPT specification rule (spec): $\dfrac{x=X\supset A}{T\begin{bmatrix}X\\x\end{bmatrix}\supset A}$

where $x\in T$ but $x\notin X, A$.

Proof.

$$\dfrac{T\begin{bmatrix}X\\x\end{bmatrix} \supset \exists x\, x=X \ \text{(te)} \qquad \dfrac{x=X\supset A}{\exists x\, x=X\supset A}\,\exists\text{-el}}{T\begin{bmatrix}X\\x\end{bmatrix}\supset A}\,\text{pc}$$

∎

THEOREM 6. LPT simplification rule (simp): $\dfrac{num\, n\supset A}{A}$ where $n\notin A$.

Proof.

$$\dfrac{\dfrac{num\,0 \text{ (red)} \qquad n=0 \supset num\,0 \supset num\,n \text{ (r.eq)} \qquad num\,n \supset A}{\dfrac{n=0\supset A}{num\,0 \supset A}\,\text{spec}}\,\text{pc}}{}$$

$$\dfrac{num\,0 \text{ (red)} \qquad num\,0 \supset A}{A}\,\text{pc}$$

∎

THEOREM 7. LPT instantiation rule (inst): $\dfrac{P \supset \ldots Q \supset R}{P\begin{bmatrix}X\\x\end{bmatrix} \supset \ldots Q\begin{bmatrix}X\\x\end{bmatrix} \supset R\begin{bmatrix}X\\x\end{bmatrix}}$

where $P, \ldots Q, R$ are terms, $x \notin X$, and $x \in P$ or $\ldots$ or $x \in Q$.

Proof. Let T be one of the terms $P, \ldots Q$ such that $x \in T$.

$$\dfrac{\dfrac{\dfrac{\dfrac{P \supset \ldots Q \supset R}{x = X \supset P \supset \ldots Q \supset R}\,\text{pc}}{x = X \supset P\begin{bmatrix}X\\x\end{bmatrix} \supset \ldots Q\begin{bmatrix}X\\x\end{bmatrix} \supset R\begin{bmatrix}X\\x\end{bmatrix}}\,\text{r.eq, eq, pc}}{T\begin{bmatrix}X\\x\end{bmatrix} \supset P\begin{bmatrix}X\\x\end{bmatrix} \supset \ldots Q\begin{bmatrix}X\\x\end{bmatrix} \supset R\begin{bmatrix}X\\x\end{bmatrix}}\,\text{spec}}{P\begin{bmatrix}X\\x\end{bmatrix} \supset \ldots Q\begin{bmatrix}X\\x\end{bmatrix} \supset R\begin{bmatrix}X\\x\end{bmatrix}}\,\text{pc}$$

∎

THEOREM 8. LPT primitive recursion theorem (pr): $num\,\underline{n} \supset num\,(F(\underline{n}))$ where $\underline{n}$ are k variables, $n_1, \ldots n_k$, and F is a k-ary primitive recursive function.

Proof. The argument is by induction on the structure of F. There are five cases; uses of the inductive hypothesis are marked by 'ih'.

Case 1. F is 0. Then the formula to be derived is $num\,0$, which is an instance of the LPT reduction theorem, since $num\,0 \;\triangleright^*\; true$.

Case 2. F is S. Then the formula to be derived is $num\,n_1 \supset num\,(S(n_1))$, which is an instance of the LPT reduction theorem, since $num\,(S(n_1)) \;\triangleright^*\; num\,n_1$.

Case 3: F is $proj_i^k$. Then the required formula is derived in LPT by

$$\dfrac{num\,n_i \supset num\,(proj_i^k(\underline{n}))\ \text{(red)}}{num\,\underline{n} \supset num\,(proj_i^k(\underline{n}))}\,\text{pc}$$

Case 4: F is $comp_l(H,(F_1,\ldots F_l))$, where H is l-ary primitive recursive and $F_1,\ldots F_l$ are k-ary primitive recursive (with $k,l \geq 1$). Let $\underline{x}$ be l fresh variables, $x_1,\ldots x_l$. Then the required formula is derived in LPT by

$$\dfrac{\begin{array}{c}(\text{for } i=1,\ldots l)\\ num\,\underline{n} \supset num\,(F_i(\underline{n}))\ \text{(ih)}\end{array} \qquad \dfrac{\dfrac{num\,\underline{x} \supset num\,(H(\underline{x}))\ \text{(ih)}}{num\,(F_1(\underline{n})) \supset \ldots \supset num\,(H(F_1(\underline{n}),\ldots F_l(\underline{n})))}\,\text{inst}}{num\,(F_1(\underline{n})) \supset \ldots \supset num\,(F(\underline{n}))}\,\text{red}}{num\,\underline{n} \supset num\,(F(\underline{n}))}\,\text{pc}$$

Case 5: F is $rec_{k-1}(G,H)$, where G is $(k-1)$-ary primitive recursive and H is $(k+1)$-ary primitive recursive. Let $\underline{n}$ be $m,\underline{y}$ and let x,v be two fresh variables. Let A be the formula $num\,\underline{y} \supset (num\,x \wedge num\,(F(x,\underline{y})))$. For any term X such that $X \not\succ$ and $x \notin X$, let A_X be the formula $num\,\underline{y} \supset (num\,X \wedge num\,(F(X,\underline{y})))$. First we need the following three LPT derivations.

$$\dfrac{X=X\ \text{(red)} \qquad \dfrac{\dfrac{\dfrac{\dfrac{x=X \supset num\,x \supset num\,X\ \text{(eq)} \qquad x=X \supset num\,(F(x,\underline{y})) \supset num\,(F(X,\underline{y}))\ \text{(eq)}}{x=X \supset A \supset A_X}\,\text{pc}}{x=X \supset (x=X \supset A) \supset A_X}\,\text{pc}}{x=X \supset \forall x\,(x=X \supset A) \supset A_X}\,\forall\text{-el, pc}}{X=X \supset \forall x\,(x=X \supset A) \supset A_X}\,\text{spec}}{\forall x\,(x=X \supset A) \supset A_X \quad (1)}\,\text{pc}$$

$$\dfrac{\dfrac{\dfrac{x=X \supset num\,X \supset num\,x\ \text{(r.eq)} \qquad x=X \supset num\,(F(X,\underline{y})) \supset num\,(F(x,\underline{y}))\ \text{(r.eq)}}{x=X \supset A_X \supset A}\,\text{pc}}{A_X \supset x=X \supset A}\,\text{pc}}{A_X \supset \forall x\,(x=X \supset A) \quad (2)}\,\forall\text{-in}$$

$$\dfrac{\dfrac{num\,0\ \text{(red)} \qquad \dfrac{num\,\underline{y} \supset num\,(G(\underline{y}))\ \text{(ih)}}{num\,\underline{y} \supset num\,(F(0,\underline{y}))}\,\text{red, pc}}{A_0}\,\text{pc}}{\forall x\,(x=0 \supset A) \quad (3)}\,\text{2, pc}$$

Now the required formula is derived in LPT as follows.

$$
\cfrac{\cfrac{\forall x\,(x=0\supset A)\ \text{(3)} \qquad \cfrac{\cfrac{\cfrac{\cfrac{\cfrac{num\, m\supset num\, \underline{y}\supset num\, v\supset num\,(H(m,\underline{y},v))\ \text{(ih)}}{num\, m\supset num\, \underline{y}\supset num\,(F(m,\underline{y}))\supset num\,(H(m,\underline{y},F(m,\underline{y})))}\,\text{inst}}{num\, m\supset num\, \underline{y}\supset num\,(F(m,\underline{y}))\supset num\,(F(Sm,\underline{y}))}\,\text{red}}{num\, m\supset num\, \underline{y}\supset num\,(F(m,\underline{y}))\supset (num\,(Sm)\wedge num\,(F(Sm,\underline{y})))}\,\text{red, pc}}{A_m\supset A_{Sm}}\,\text{pc}}{\forall x\,(x=m\supset A)\supset \forall x\,(x=Sm\supset A)}\,\text{1, 2, pc}}{num\, m\supset \forall x\,(x=m\supset A)}\,\text{ind}}{\cfrac{num\, m\supset A_m}{num\, m\supset num\, \underline{y}\supset num\,(F(m,\underline{y}))}\,\text{pc}}\,\text{1, pc}
$$

∎

THEOREM 9. LPT numeric terms theorem (num): $num\, \underline{n}\supset num\, N$, where N is a numeric term with free variables $\underline{n}$.

Proof. Recall from the Expanded Term Language that a numeric term is 0, a variable, or $F(N_1,\ldots N_k)$, where $k\geq 1$, F is a k-ary primitive recursive function and $N_1,\ldots N_k$ are numeric terms. The proof is by induction on the structure of the numeric terms. There are three cases.

Case 1: N is 0. Then the formula to be derived is $num\, 0$, which is an LPT reduction theorem.

Case 2: N is a variable, n. Then the formula to be derived is $num\, n\supset num\, n$, which is a theorem of LPT by propositional calculus.

Case 3: N is $F(N_1,\ldots N_k)$, where $k\geq 1$, F is a k-ary primitive recursive function, and $N_1,\ldots N_k$ are numeric terms. Let $\underline{z_i}$ be the free variables of N_i, for $i=1,\ldots k$, and let $\underline{x}$ be k fresh variables. The required formula is derived by

$$
\cfrac{\begin{matrix}(\text{for } i=1,\ldots k)\\ num\, \underline{z_i}\supset num\, N_i\ \text{(ih)}\end{matrix} \qquad \cfrac{num\, \underline{x}\supset num\,(F(\underline{x}))\ \text{(pr)}}{num\, N_1\supset\ldots\supset num\, N_k\supset num\, N}\,\text{inst}}{num\, \underline{n}\supset num\, N}\,\text{pc}
$$

where use of the inductive hypothesis is marked by ‘ih’. ∎

THEOREM 10. LPT conversion theorem (conv): $num\, \underline{x}\supset M=N$ where M and N are numeric terms with free variables $\underline{x}$, and $M \rhd^{*}\lhd N$.

Proof. Let $\underline{n}$ be the free variables of N.

$$\cfrac{\cfrac{num\,\underline{n} \supset num\,N \text{ (num)} \qquad num\,N \supset N = N \text{ (se)} \qquad N = N \supset M = N \text{ (red)}}{num\,\underline{n} \supset M = N}\,\text{pc}}{num\,\underline{x} \supset M = N}\,\text{pc}$$

▮

EXERCISE. Recall the substitution notation $A\binom{X}{x}$ for LPT formulae introduced in the exercise at the end of Chapter 29. Derive in LPT the formulae

$$\exists x\, x = X \qquad U = V \supset A\binom{U}{x} \supset A\binom{V}{x}$$

$$A\binom{X}{x} \supset \exists x\,(x = X \wedge A) \qquad \exists x\,(x = X \wedge A) \supset A\binom{X}{x}$$

$$\exists x\, x = U \supset A\binom{U}{x} \supset \exists x A \qquad \exists x\, x = U \supset \forall x A \supset A\binom{U}{x}$$

assuming that $x \notin X, U, V$ and $X \not\succ$.

FORMULAE OF HEYTING ARITHMETIC

An *HA formula* is a sentence in the following language.

- The alphabet is that of the Expanded Term Language plus '$\wedge$', '$\vee$', '$\supset$', '$\exists^{\mathrm{N}}$', '$\forall^{\mathrm{N}}$'.
- The tokens are those of the Expanded Term Language plus '$\wedge$', '$\vee$', '$\supset$', '$\exists^{\mathrm{N}}$', '$\forall^{\mathrm{N}}$'.
- Lexical analysis consists of recognising constants and replacing them by the token '*con*', recognising variables and replacing them by the token '*vbl*', and recognising every other token.
- The grammar of the language is as follows.
  - The terminals are the tokens.
  - The non-terminals are F, N, T; the start symbol is F.
  - The production rules are

$$F \to true \mid false \mid N = N \mid (\,F \wedge F\,) \mid (\,F \vee F\,) \mid (\,F \supset F\,) \mid \exists^{\mathrm{N}}\, vbl\, F \mid \forall^{\mathrm{N}}\, vbl\, F$$

$$N \to T$$

$$T \to con \mid (\,vbl\,) \mid (\,T\,T\,) \mid (\,\lambda\, T\,.\,T\,) \mid (\,T \left[\tfrac{T}{vbl}\right])$$

In addition, there is a context-sensitive constraint that strings matching N must be *numeric* terms. Formulae matching *true*, *false* or $N = N$ are called *atomic* formulae.

HA METAFORMULAE

From now on, the identifiers 'A', 'B' and 'C' will be used as metavariables to denote HA formulae. *Metaformulae* and their *instances* are defined as for LPT formulae, except that metaformulae may contain the metanotation $\neg A$ for $A \supset \mathit{false}$; '$\neg$' has the same syntactic precedence as the quantifiers (formally, we add a production rule $P \to \neg\, P$ to the grammar of metaformulae).

The identifiers 'F', 'G' and 'H', possibly with subscripts, will denote primitive recursive functions. All other identifiers beginning with capital letters will denote numeric terms.

FREE VARIABLES IN HA FORMULAE

For any variable v and HA formula A, the relation $v \in A$ ('v occurs free in A') is defined by:

- $v \notin 0$, $v \in n$ iff v is n,
- $v \in F(N_1, \ldots N_k)$ iff $v \in N_1$ or $\ldots$ or $v \in N_k$,
- $v \notin \mathit{true}$, $v \notin \mathit{false}$, $v \in N_1 = N_2$ iff $v \in N_1$ or $v \in N_2$,
- $v \in A \wedge B$ iff $v \in A \vee B$ iff $v \in A \supset B$ iff $v \in A$ or $v \in B$,
- $v \in \exists^{\mathrm{N}} n\, A$ iff $v \in \forall^{\mathrm{N}} n\, A$ iff v isn't n and $v \in A$.

The *free variables* of A are the variables v such that $v \in A$, listed in standard order.

THEOREM 11. Any formula has finitely many free variables.

INTERPRETATION OF HA FORMULAE IN LPT

For any HA formula A, with free variables $\underline{z}$,

- A^{N} is defined by
  A^{N} is A, for an atomic formula A;
  $(A \wedge B)^{\mathrm{N}}$ is $A^{\mathrm{N}} \wedge B^{\mathrm{N}}$; $(A \vee B)^{\mathrm{N}}$ is $A^{\mathrm{N}} \vee B^{\mathrm{N}}$; $(A \supset B)^{\mathrm{N}}$ is $A^{\mathrm{N}} \supset B^{\mathrm{N}}$;
  $(\exists^{\mathrm{N}} n\, A)^{\mathrm{N}}$ is $\exists n\, (\mathit{num}\, n \wedge A^{\mathrm{N}})$; $(\forall^{\mathrm{N}} n\, A)^{\mathrm{N}}$ is $\forall n\, (\mathit{num}\, n \supset A^{\mathrm{N}})$.
- A^{LPT} is $\mathit{num}\, \underline{z} \supset A^{\mathrm{N}}$.

A^{LPT} is the LPT interpretation of the HA formula A. Note that A^{N} and A^{LPT} have the same free variables as A. For an atomic formula A with no free variables, A^{LPT} is A.

EXERCISE. Suppose we define a substitution notation $A\binom{N}{n}$, meaning the HA formula obtained by substituting the numeric term N for the variable n in the HA formula A, assuming no variable clashes occur, similar to the substitution notation for LPT formulae considered in previous exercises. Show that any formula of the form

$$A^{\mathrm{LPT}}\binom{N}{n} \supset A\binom{N}{n}^{\mathrm{LPT}}$$

is derivable in LPT. Why does the converse fail?

AXIOMS AND RULES OF INFERENCE OF HA

$\wedge$-axioms: $A \supset B \supset (A \wedge B) \quad (A \wedge B) \supset A \quad (A \wedge B) \supset B$
$\vee$-axioms: $A \supset (A \vee B) \quad B \supset (A \vee B) \quad (A \vee B) \supset (A \supset C) \supset (B \supset C) \supset C$
$\supset$-axioms: $A \supset B \supset A \quad (A \supset B \supset C) \supset (A \supset B) \supset A \supset C$
Truth and falsity (*true*-in, *false*-el): $\mathit{true} \quad \mathit{false} \supset A$
Quantifier axioms ($\exists^{\mathrm{N}}$-in, $\forall^{\mathrm{N}}$-el): $A \supset \exists^{\mathrm{N}} n A \quad \forall^{\mathrm{N}} n A \supset A$
Term existence (te): $\exists^{\mathrm{N}} n\, n = N$ where $n \notin N$
Equality (eq): $N = N \quad M = N \supset N = M \quad U = V \supset V = W \supset U = W$
$M = N \supset F(X_1, \ldots X_{i-1}, M, X_{i+1}, \ldots X_k) = F(X_1, \ldots X_{i-1}, N, X_{i+1}, \ldots X_k)$
Peano axioms (P): $\neg Sn = 0 \quad Sm = Sn \supset m = n$
Axioms for primitive recursive functions (pr):
$\mathit{proj}_i^k(\underline{n}) = n_i$ where $1 \leq i \leq k$
$\mathit{comp}_l(H, (F_1, \ldots F_l))(\underline{n}) = H(F_1(\underline{n}), \ldots F_l(\underline{n}))$
$\mathit{rec}_k(F, G)(0, \underline{n}) = F(\underline{n}) \quad \mathit{rec}_k(F, G)(Sm, \underline{n}) = G(m, \underline{n}, \mathit{rec}_k(F, G)(m, \underline{n}))$

where $\underline{n}$ is $n_1, \ldots n_k$, H is l-ary primitive recursive, $F, F_1, \ldots F_l$ are k-ary primitive recursive, and G is $(k+2)$-ary primitive recursive

Modus ponens: $\dfrac{A \quad A \supset B}{B}$

Quantifier rules ($\exists^{\mathrm{N}}$-el, $\forall^{\mathrm{N}}$-in): $\dfrac{A \supset B}{\exists^{\mathrm{N}} n A \supset B} \quad \dfrac{B \supset A}{B \supset \forall^{\mathrm{N}} n A}$ where $n \notin B$

Induction (ind): $\dfrac{\forall^{\mathrm{N}} x (x = 0 \supset A) \quad \forall^{\mathrm{N}} x (x = n \supset A) \supset \forall^{\mathrm{N}} x (x = Sn \supset A)}{\forall^{\mathrm{N}} x (x = n \supset A)}$

where $n \notin x, A$

LPT DERIVATIONS CORRESPONDING TO THE HA AXIOMS AND RULES

I shall show that, for each axiom A of HA, A^{LPT} is a theorem of LPT, and that, for each rule of inference $\dfrac{A \ldots B}{C}$ of HA, $\dfrac{A^{\mathrm{LPT}} \ldots B^{\mathrm{LPT}}}{C^{\mathrm{LPT}}}$ is a derived rule of LPT.

THEOREM 12. The HA $\wedge$, $\vee$, $\supset$, truth and falsity axioms all translate to theorems of LPT.

Proof. For example, the interpretation of $A \supset B \supset A$ is $num\,\underline{z} \supset A^{\mathrm{N}} \supset B^{\mathrm{N}} \supset A^{\mathrm{N}}$, where $\underline{z}$ are the free variables of A and B, which is derived from the LPT axiom $A^{\mathrm{N}} \supset B^{\mathrm{N}} \supset A^{\mathrm{N}}$. Similarly for the others. ∎

THEOREM 13. The HA quantifier axioms: $\quad A \supset \exists^{\mathrm{N}} n\, A \qquad \forall^{\mathrm{N}} n\, A \supset A$.

Proof. Let $\underline{z}$ be the free variables of A. The LPT derivations of $(A \supset \exists^{\mathrm{N}} n\, A)^{\mathrm{LPT}}$ and $(\forall^{\mathrm{N}} n\, A \supset A)^{\mathrm{LPT}}$ are

$$\dfrac{\dfrac{\forall n\,(num\,n \supset A^{\mathrm{N}}) \supset (num\,n \supset A^{\mathrm{N}})\ (\forall\text{-el})}{num\,n \supset \forall n\,(num\,n \supset A^{\mathrm{N}}) \supset A^{\mathrm{N}}}\,\text{pc}}{num\,\underline{z} \supset \forall n\,(num\,n \supset A^{\mathrm{N}}) \supset A^{\mathrm{N}}}$$

$$\dfrac{\dfrac{(num\,n \wedge A^{\mathrm{N}}) \supset \exists n\,(num\,n \wedge A^{\mathrm{N}})\ (\exists\text{-in})}{num\,n \supset A^{\mathrm{N}} \supset \exists n\,(num\,n \wedge A^{\mathrm{N}})}\,\text{pc}}{num\,\underline{z} \supset A^{\mathrm{N}} \supset \exists n\,(num\,n \wedge A^{\mathrm{N}})}$$

In each case the last step is sound because if $n \in A$ then n is one of the $\underline{z}$ variables, while if $n \notin A$ then we may remove the $num\,n$, using Simplification, and add $num\,\underline{z}$ in its place. ∎

THEOREM 14. The HA term existence axiom: $\quad \exists^{\mathrm{N}} n\, n = N$, where $n \notin N$.

Proof. Let $\underline{z}$ be the free variables of N. The LPT derivation of $(\exists^{\mathrm{N}} n\, n = N)^{\mathrm{LPT}}$ is

$$\dfrac{num\,\underline{z} \supset num\,N\ (\text{num}) \qquad \dfrac{\dfrac{n = N \supset num\,N \supset num\,n\ (\text{r.eq}) \qquad (num\,n \wedge n = N) \supset \exists n\,(num\,n \wedge n = N)\ (\exists\text{-in})}{n = N \supset num\,N \supset \exists n\,(num\,n \wedge n = N)}\,\text{pc}}{num\,N \supset num\,N \supset \exists n\,(num\,n \wedge n = N)}\,\text{spec}}{num\,\underline{z} \supset \exists n\,(num\,n \wedge n = N)}\,\text{pc}$$

∎

THEOREM 15. The first three HA equality axioms:
$$N = N \qquad M = N \supset N = M \qquad U = V \supset V = W \supset U = W.$$

Proof. Let $\underline{x}$ be the free variables of N, $\underline{y}$ be the free variables of M and N, and $\underline{z}$ be the free variables of U, V and W. The LPT derivations of the interpretations of the axioms are as follows.

$$\frac{num\,\underline{x} \supset num\,N \text{ (num)} \qquad num\,N \supset N = N \text{ (se)}}{num\,\underline{x} \supset N = N}\,\text{pc}$$

$$\frac{M = N \supset N = M \text{ (symm)}}{num\,\underline{y} \supset M = N \supset N = M}\,\text{pc}$$

$$\frac{U = V \supset V = W \supset U = W \text{ (r.eq)}}{num\,\underline{z} \supset U = V \supset V = W \supset U = W}\,\text{pc}$$

∎

THEOREM 16. The fourth HA equality axiom:
$M = N \supset F(X_1, \ldots X_{i-1}, M, X_{i+1}, \ldots X_k) = F(X_1, \ldots X_{i-1}, N, X_{i+1}, \ldots X_k)$.

Proof. Let U be the term $F(X_1, \ldots X_{i-1}, M, X_{i+1}, \ldots X_k)$ and let V be the term $F(X_1, \ldots X_{i-1}, N, X_{i+1}, \ldots X_k)$. Let $\underline{x}$ be the free variables of U, and $\underline{y}$ be the free variables of U and V. The LPT derivation of the interpretation of the axiom is as follows.

$$\cfrac{\cfrac{\cfrac{num\,\underline{x} \supset num\,U \text{ (num)} \qquad num\,U \supset U = U \text{ (se)}}{num\,\underline{x} \supset U = U}\,\text{pc} \qquad M = N \supset U = U \supset U = V \text{ (eq)}}{num\,\underline{x} \supset M = N \supset U = V}\,\text{pc}}{num\,\underline{y} \supset M = N \supset U = V}\,\text{pc}$$

∎

THEOREM 17. The HA Peano axioms $\neg Sn = 0$ and $Sm = Sn \supset m = n$.

Proof. Recall the construction *pre* from the Expanded Term Language and recall (Theorem 53) that $pre(Sn) \vartriangleright n$ and $Sn = 0 \vartriangleright false$. Let $\underline{x}$ be the variables m, n in standard order. The LPT derivations of the interpretations of the axioms are

$$\frac{Sn = 0 \supset false \text{ (red)}}{num\,n \supset Sn = 0 \supset false}\,\text{pc}$$

$$\cfrac{\cfrac{m = m \text{ (red)} \qquad \cfrac{Sm = Sn \supset pre(Sm) = pre(Sm) \supset pre(Sm) = pre(Sn) \text{ (eq)}}{Sm = Sn \supset m = m \supset m = n}\,\text{red, pc}}{Sm = Sn \supset m = n}\,\text{pc}}{num\,\underline{x} \supset Sm = Sn \supset m = n}\,\text{pc}$$

∎

THEOREM 18. The HA axioms for primitive recursive functions:

$proj_i^k(\underline{n}) = n_i$ where $1 \leq i \leq k$

$comp_l(H,(F_1,\ldots F_l))(\underline{n}) = H(F_1(\underline{n}),\ldots F_l(\underline{n}))$

$rec_k(F,G)(0,\underline{n}) = F(\underline{n})$ $rec_k(F,G)(Sm,\underline{n}) = G(m,\underline{n},rec_k(F,G)(m,\underline{n}))$

where $\underline{n}$ is $n_1,\ldots n_k$, H is l-ary primitive recursive, $F,F_1,\ldots F_l,F$ are k-ary primitive recursive, and G is $(k+2)$-ary primitive recursive.

Proof. The LPT interpretations of these formulae are all instances of the LPT conversion theorem. ∎

THEOREM 19. The HA modus ponens rule: $\dfrac{A \quad A \supset B}{B}$.

Proof. Let $\underline{x}$ be the free variables of A, $\underline{y}$ be the free variables of B, and $\underline{z}$ be the free variables of $A \supset B$. Then $\dfrac{A^{\mathrm{LPT}} \quad (A \supset B)^{\mathrm{LPT}}}{B^{\mathrm{LPT}}}$ is derived in LPT by

$$\dfrac{\dfrac{num\,\underline{x} \supset A^{\mathrm{N}} \qquad num\,\underline{z} \supset A^{\mathrm{N}} \supset B^{\mathrm{N}}}{num\,\underline{z} \supset B^{\mathrm{N}}}\,\text{pc}}{num\,\underline{y} \supset B^{\mathrm{N}}}\,\text{simp, pc}$$

∎

THEOREM 20. The HA quantifier rules: $\dfrac{A \supset B}{\exists^{\mathrm{N}} n\, A \supset B} \qquad \dfrac{B \supset A}{B \supset \forall^{\mathrm{N}} n\, A}$

where $n \notin B$.

Proof. Let $\underline{y}$ be the free variables of $A \supset B$ (which are also the free variables of $B \supset A$) and $\underline{z}$ be the free variables of $\exists^{\mathrm{N}} n\, A \supset B$ (which are also the free variables of $B \supset \forall^{\mathrm{N}} n\, A$). Then $\underline{y}$ is $\underline{z}$ with the possible addition of n. The LPT derivations of the interpretations of the rules are

$$\dfrac{\dfrac{\dfrac{num\,\underline{y} \supset A^{\mathrm{N}} \supset B^{\mathrm{N}}}{(num\,n \wedge A^{\mathrm{N}}) \supset num\,\underline{z} \supset B^{\mathrm{N}}}\,\text{pc}}{\exists n\,(num\,n \wedge A^{\mathrm{N}}) \supset num\,\underline{z} \supset B^{\mathrm{N}}}\,\exists\text{-el}}{num\,\underline{z} \supset \exists n\,(num\,n \wedge A^{\mathrm{N}}) \supset B^{\mathrm{N}}}\,\text{pc} \qquad \dfrac{\dfrac{num\,\underline{y} \supset B^{\mathrm{N}} \supset A^{\mathrm{N}}}{num\,\underline{z} \supset B^{\mathrm{N}} \supset num\,n \supset A^{\mathrm{N}}}\,\text{pc}}{num\,\underline{z} \supset B^{\mathrm{N}} \supset \forall n\,(num\,n \supset A^{\mathrm{N}})}\,\forall\text{-in, pc}$$

where in each case the use of the quantifier rule is justified since $n \notin num\,\underline{z}, B^{\mathrm{N}}$. ∎

THEOREM 21. The HA induction rule:

$$\dfrac{\forall^{\mathrm{N}} x\,(x = 0 \supset A) \qquad \forall^{\mathrm{N}} x\,(x = n \supset A) \supset \forall^{\mathrm{N}} x\,(x = Sn \supset A)}{\forall^{\mathrm{N}} x\,(x = n \supset A)}$$

where $n \notin x, A$.

Proof. Let $\underline{y}$ be the free variables of $\forall^{\mathrm{N}}x\,(x=n\supset A)$ and let $\underline{z}$ be the free variables of $\forall^{\mathrm{N}}x\,(x=0\supset A)$. (Thus $\underline{y}$ is $\underline{z}$ with the addition of n.) Let B be $\mathit{num}\,\underline{z}\supset \mathit{num}\,x\supset A^{\mathrm{N}}$. First we need the following LPT derivations, where N is any numeric term.

$$\dfrac{\dfrac{(\forall^{\mathrm{N}}x\,(x=N\supset A))^{\mathrm{N}}\supset \mathit{num}\,x\supset x=N\supset A^{\mathrm{N}}\ \ (\forall\text{-el})}{(\mathit{num}\,\underline{z}\supset(\forall^{\mathrm{N}}x\,(x=N\supset A))^{\mathrm{N}})\supset x=N\supset B}\,\text{pc}}{(\mathit{num}\,\underline{z}\supset(\forall^{\mathrm{N}}x\,(x=N\supset A))^{\mathrm{N}})\supset \forall x\,(x=N\supset B)\quad(1)}\,\forall\text{-in}$$

$$\dfrac{\dfrac{\forall x\,(x=N\supset B)\supset(x=N\supset B)\ \ (\forall\text{-el})}{\forall x\,(x=N\supset B)\supset \mathit{num}\,\underline{z}\supset \mathit{num}\,x\supset x=N\supset A^{\mathrm{N}}}\,\text{pc}}{\forall x\,(x=N\supset B)\supset \mathit{num}\,\underline{z}\supset(\forall^{\mathrm{N}}x\,(x=N\supset A))^{\mathrm{N}}\quad(2)}\,\forall\text{-in, pc}$$

$$\dfrac{\dfrac{\dfrac{x=Sn\supset \mathit{num}\,x\supset \mathit{num}\,(Sn)\ \ (\text{eq})}{x=Sn\supset \mathit{num}\,x\supset \mathit{num}\,n}\,\text{red, pc}\qquad \forall x\,(x=Sn\supset B)\supset(x=Sn\supset B)\ \ (\forall\text{-el})}{(\mathit{num}\,n\supset\forall x\,(x=Sn\supset B))\supset x=Sn\supset B}\,\text{pc}}{(\mathit{num}\,n\supset\forall x\,(x=Sn\supset B))\supset\forall x\,(x=Sn\supset B)\quad(3)}\,\forall\text{-in}$$

$$\dfrac{\mathit{num}\,\underline{z}\supset(\forall^{\mathrm{N}}x\,(x=0\supset A))^{\mathrm{N}}\ \ (\text{premise})}{\forall x\,(x=0\supset B)\quad(4)}\,1,\text{pc}$$

Then the LPT derivation of the interpretation of the induction rule is as follows.

$$\dfrac{\dfrac{\forall x\,(x=0\supset B)\ (4)\qquad \dfrac{\dfrac{\dfrac{\mathit{num}\,\underline{y}\supset(\forall^{\mathrm{N}}x\,(x=n\supset A))^{\mathrm{N}}\supset(\forall^{\mathrm{N}}x\,(x=Sn\supset A))^{\mathrm{N}}\ \ (\text{premise})}{\mathit{num}\,n\supset(\mathit{num}\,\underline{z}\supset(\forall^{\mathrm{N}}x\,(x=n\supset A))^{\mathrm{N}})\supset \mathit{num}\,\underline{z}\supset(\forall^{\mathrm{N}}x\,(x=Sn\supset A)^{\mathrm{N}})}\,\text{pc}}{\mathit{num}\,n\supset\forall x\,(x=n\supset B)\supset\forall x\,(x=Sn\supset B)}\,1,2,\text{pc}}{\forall x\,(x=n\supset B)\supset\forall x\,(x=Sn\supset B)}\,3,\text{pc}}{\mathit{num}\,n\supset\forall x\,(x=n\supset B)}\,\text{ind}}{\mathit{num}\,\underline{y}\supset(\forall^{\mathrm{N}}x\,(x=n\supset A))^{\mathrm{N}}}\,2,\text{pc}$$

∎

FORMAL INTERPRETATION OF HA

DEFINITION. A coding of HA formulae as constructions is defined as follows. If A is a formula with free variables $\underline{z}$, the code of A, $\langle\!\langle A\rangle\!\rangle$, is defined as

$(\lambda[\underline{z}].\langle A\rangle)$, where

$$\begin{aligned}
\langle T\rangle &\text{ is } \mathit{atomic}(T) \quad \text{if } T \text{ is atomic}\\
\langle A \wedge B\rangle &\text{ is } \mathit{and}(\langle A\rangle, \langle B\rangle)\\
\langle A \vee B\rangle &\text{ is } \mathit{or}(\langle A\rangle, \langle B\rangle)\\
\langle A \supset B\rangle &\text{ is } \mathit{implies}(\langle A\rangle, \langle B\rangle)\\
\langle \exists^{\mathrm{N}} x A\rangle &\text{ is } \mathit{exists}(\lambda[x].\langle A\rangle)\\
\langle \forall^{\mathrm{N}} x A\rangle &\text{ is } \mathit{all}(\lambda[x].\langle A\rangle).
\end{aligned}$$

DEFINITION. A coding of HA derivations as constructions is defined as follows. Number the axiom schemata and rules of inference of HA, 1,2,... 27. A derivation with conclusion A, from a list of premises numbered 0,1,2..., is coded as

$(\mathit{premise}(i), \langle\!\langle A\rangle\!\rangle)$	if A is premise number i
$(\mathit{none}(n), \langle\!\langle A\rangle\!\rangle)$	if A is an instance of axiom schema n
$(\mathit{one}(n, X), \langle\!\langle A\rangle\!\rangle)$	if A is derived by rule n from the subderivation X
$(\mathit{two}(n, X, Y), \langle\!\langle A\rangle\!\rangle)$	if A is derived by rule n from subderivations X, Y.

DEFINITION. Define recursive functions $T_1, \ldots T_{27}$ such that

- for $n=1,\ldots 23$, if A is an instance of HA axiom schema n then $T_n(\langle\!\langle A\rangle\!\rangle)$ $\triangleright^*$ the code of the corresponding LPT derivation (given above) of A^{LPT} from no premises;
- for $n = 25, 26$, if $\dfrac{B}{A}$ is an instance of HA rule n then $T_n(\langle\!\langle B\rangle\!\rangle, \langle\!\langle A\rangle\!\rangle)$ $\triangleright^*$ the code of the corresponding LPT derivation (given above) of A^{LPT} from the premise B^{LPT};
- for $n = 24, 27$, if $\dfrac{B \quad C}{A}$ is an instance of HA rule n then $T_n(\langle\!\langle B\rangle\!\rangle, \langle\!\langle C\rangle\!\rangle,$ $\langle\!\langle A\rangle\!\rangle)$ $\triangleright^*$ the code of the corresponding LPT derivation (given above) of A^{LPT} from the premises B^{LPT} and C^{LPT}.

DEFINITION. Define a construction *lpt* by

$$\begin{aligned}
\mathit{lpt}(\mathit{none}(1), a) &\triangleq T_1(a)\\
&\vdots\\
\mathit{lpt}(\mathit{none}(23), a) &\triangleq T_{23}(a)
\end{aligned}$$

$$lpt(two(24,(u,b),(v,c)),a) \triangleq glue(T_{24}(b,c,a),[lpt(u,b),lpt(v,c)])$$
$$lpt(one(25,(u,b)),a) \triangleq glue(T_{25}(b,a),[lpt(u,b)])$$
$$lpt(one(26,(u,b)),a) \triangleq glue(T_{26}(b,a),[lpt(u,b)])$$
$$lpt(two(27,(u,b),(v,c)),a) \triangleq glue(T_{27}(b,c,a),[lpt(u,b),lpt(v,c)])$$

where a, b, c, u, v are five variables.

THEOREM 22. If D is the code of an HA derivation (with no premises) of a formula A then $lpt(D) \triangleright^*$ the code of an LPT derivation (with no premises) of A^{LPT}.

Proof. By structural induction on the derivation encoded by D. ∎

THEOREM 23. (HA interpretation theorem.) If D is the code of an HA derivation (with no premises) of a formula A with no free variables, then $pr(spr(cpf(lpt(D)))) \vdash \ulcorner A^{\mathrm{LPT}} \urcorner$.

Proof. By the previous theorem, the extensionality theorem for $\vdash$, and the interpretation theorem for LPT. ∎

THEOREM 24. (The *glue*-lemma for HA.) If D is the code of an HA derivation of A from premises $B_1, \ldots B_k$, and $D_1, \ldots D_k$ are the codes of HA derivations (with no premises) of $B_1, \ldots B_k$ respectively, then $glue(D, [D_1, \ldots D_k]) \triangleright^*$ the code of an HA derivation (with no premises) of A.

Proof. By structural induction on the derivation encoded by D. ∎

UNPROVABILITY OF THE EXCLUDED MIDDLE

The theory of proof developed in this book is meant to provide the *intended* interpretation of intuitionistic logic and arithmetic. Hence it is necessary to check not merely that all theorems of HA have proofs but that intuitionistically unsound principles do not have proofs; preferably this should be done by arguments that follow our informal reasons for believing that the principles in question are unsound, rather than by roundabout methods that happen to produce results coinciding with our informal expectations. I shall construct an instance of the principle of excluded middle that has no proof and hence is not a theorem of HA; and I shall follow as closely as possible the informal motivation for disbelieving the excluded middle, namely that if it were sound we would have a general method for deciding Π_1 problems and hence for solving the halting problem.

Let $\{M\}$ be the Mth function in an enumeration of the partial recursive functions and let F be a primitive recursive function such that, for any numbers M, N, $F(M,N) \rhd^* 0$ iff the evaluation of $\{M\}(M)$ halts in exactly N steps. Let A be the HA atomic formula $F(m,n) = 0$ (where m and n are two variables) and let $\overline{A}$ be the contrary atomic formula $minus(1, F(m,n)) = 0$. Thus, for any numbers M, N,

$$A\begin{bmatrix} M,N \\ m,n \end{bmatrix} \rhd^* \ true \quad \text{iff } \{M\}(M) \text{ halts in exactly } N \text{ steps;}$$

$$\overline{A}\begin{bmatrix} M,N \\ m,n \end{bmatrix} \rhd^* \ true \quad \text{iff } \{M\}(M) \text{ does not halt in exactly } N \text{ steps.}$$

THEOREM 25. The proof function $\ulcorner(\forall^{\mathrm{N}} m\,(\exists^{\mathrm{N}} n\, A \vee \forall^{\mathrm{N}} n\, \overline{A}))^{\mathrm{LPT}}\urcorner$ has no proof and hence $\forall^{\mathrm{N}} m\,(\exists^{\mathrm{N}} n\, A \vee \forall^{\mathrm{N}} n\, \overline{A})$ is not a theorem of HA.

Proof. Suppose $P \vdash \ulcorner(\forall^{\mathrm{N}} m\,(\exists^{\mathrm{N}} n\, A \vee \forall^{\mathrm{N}} n\, \overline{A}))^{\mathrm{LPT}}\urcorner$. Without loss of generality, assume P has no free variables, since otherwise we can instantiate the free variables by 0, using Theorem 3 of Logic. Then, for any number M, by Theorem 29 of the Logic of Partial Terms (LPT),

$$el_{\forall}(P, M) \vdash \ulcorner num\, m \supset ((\exists^{\mathrm{N}} n\, A)^{\mathrm{N}} \vee (\forall^{\mathrm{N}} n\, \overline{A})^{\mathrm{N}})\urcorner \begin{bmatrix} M \\ m \end{bmatrix}.$$

Now, we certainly have $nil \vdash \ulcorner num\, m\urcorner \begin{bmatrix} M \\ m \end{bmatrix}$, by Theorem 4 of Logic, so, applying Theorem 27 of LPT,

$$el_{\supset}(el_{\forall}(P, M), nil) \vdash \ulcorner((\exists^{\mathrm{N}} n\, A)^{\mathrm{N}} \vee \forall^{\mathrm{N}} n\, \overline{A})^{\mathrm{N}}\urcorner \begin{bmatrix} M \\ m \end{bmatrix}$$

and hence, by Theorem 28 of LPT, $select_{\vee}(el_{\supset}(el_{\forall}(P, M), nil)) \rhd^* \ true$ or $false$.

In the first case, where $select_{\vee}(el_{\supset}(el_{\forall}(P, M), nil)) \rhd^* \ true$, we have

$$el_{\vee}(el_{\supset}(el_{\forall}(P, M), nil)) \vdash \ulcorner \exists n\,(num\, n \wedge A)\urcorner \begin{bmatrix} M \\ m \end{bmatrix}$$

and hence, by Theorem 26 of LPT, $el_{\vee}(el_{\supset}(el_{\forall}(P, M), nil)) \rhd^* \ (N, R) \not\rhd$, where

$$R \vdash \ulcorner num\, n \wedge A\urcorner \begin{bmatrix} M,N \\ m,n \end{bmatrix}.$$

Hence, by Theorem 25 of LPT, $R \overset{*}{\leftrightarrow} (R_1, R_2)$, where

$$R_1 \vdash \ulcorner num\, n\urcorner \begin{bmatrix} N \\ n \end{bmatrix} \qquad \text{and} \qquad R_2 \vdash \ulcorner A\urcorner \begin{bmatrix} M,N \\ m,n \end{bmatrix}$$

giving $(num\, n)\begin{bmatrix} N \\ n \end{bmatrix} \rhd^* \ true$ and $A\begin{bmatrix} M,N \\ m,n \end{bmatrix} \rhd^* \ true$. Thus N is a number and the evaluation of $\{M\}(M)$ halts in N steps.

In the second case, where $select_\vee(el_\supset(el_\forall(P,M),nil)) \rhd^* false$, we have

$$el_\vee(el_\supset(el_\forall(P,M),nil)) \vdash \ulcorner\forall n\,(num\,n \supset \overline{A})\urcorner\begin{bmatrix}M\\m\end{bmatrix}$$

and hence, by Theorem 29 of LPT, for any number N,

$$el_\forall(el_\vee(el_\supset(el_\forall(P,M),nil)),N) \vdash \ulcorner num\,n \supset \overline{A}\urcorner\begin{bmatrix}M,N\\m,n\end{bmatrix}$$

and by Theorem 27 of LPT, since $nil \vdash \ulcorner num\,n\urcorner\begin{bmatrix}N\\n\end{bmatrix}$,

$$el_\supset(el_\forall(el_\vee(el_\supset(el_\forall(P,M),nil)),N),nil) \vdash \ulcorner\overline{A}\urcorner\begin{bmatrix}M,N\\m,n\end{bmatrix}.$$

This implies $\overline{A}\begin{bmatrix}M,N\\m,n\end{bmatrix} \rhd^* true$, and hence the evaluation of $\{M\}(M)$ does not halt in exactly N steps. Since N was an arbitrary number, $\{M\}(M)$ is undefined.

To summarise, $select_\vee(el_\supset(el_\forall(P,M),nil))$ reduces to *true* or *false* according to whether $\{M\}(M)$ is defined or not. This of course leads to a contradiction by a familiar argument: we can define a total recursive function H on numbers such that

$$H(x) = \begin{cases} \{x\}(x)+1 & \text{if } select_\vee(el_\supset(el_\forall(P,x),nil)) \rhd^* true \\ 0 & \text{if } select_\vee(el_\supset(el_\forall(P,x),nil)) \rhd^* false \end{cases}$$

and then, if k is the code number of H in the enumeration of partial recursive functions, $\{k\}(k)$ is defined, so

$$\{k\}(k) = \{k\}(k)+1$$

which is the desired contradiction. ∎

Some variations on this result, which are provable in the same way, are:

- there is no general method of proving $\ulcorner(\exists^{\mathrm{N}} n A \vee \forall^{\mathrm{N}} n \overline{A})^{\mathrm{LPT}}\urcorner\begin{bmatrix}M\\m\end{bmatrix}$;
- there is no general method of deriving

  $$\exists^{\mathrm{N}} n\, F(M,n)=0 \vee \forall^{\mathrm{N}} n\, minus(1,F(M,n))=0$$

  in HA.

CHAPTER 32

HEYTING ARITHMETIC

FORMULAE OF HEYTING ARITHMETIC

An *HA formula* is a sentence in the following language.

- The alphabet is that of the Expanded Term Language plus '$\wedge$', '$\vee$', '$\supset$', '$\exists^{\mathrm{N}}$', '$\forall^{\mathrm{N}}$'.
- The tokens are those of the Expanded Term Language plus '$\wedge$', '$\vee$', '$\supset$', '$\exists^{\mathrm{N}}$', '$\forall^{\mathrm{N}}$'.
- Lexical analysis consists of recognising constants and replacing them by the token '*con*', recognising variables and replacing them by the token '*vbl*', and recognising every other token.
- The grammar of the language is as follows.
  - The terminals are the tokens.
  - The non-terminals are *F*, *N*, *T*; the start symbol is *F*.
  - The production rules are

$$F \to \mathit{true} \mid \mathit{false} \mid N = N \mid (\, F \wedge F \,) \mid (\, F \vee F \,) \mid (\, F \supset F \,) \mid \exists^{\mathrm{N}}\, \mathit{vbl}\, F \mid \forall^{\mathrm{N}}\, \mathit{vbl}\, F$$

$$N \to T$$

$$T \to \mathit{con} \mid (\, \mathit{vbl} \,) \mid (\, T\, T \,) \mid (\, \lambda\, T \,.\, T \,) \mid (\, T \left[{}^{T}_{\mathit{vbl}} \right])$$

In addition, there is a context-sensitive constraint that strings matching *N* must be *numeric* terms. Formulae matching *true*, *false* or $N = N$ are called *atomic* formulae.

HA METAFORMULAE

From now on, the identifiers '*A*', '*B*' and '*C*' will be used as metavariables to denote HA formulae. *Metaformulae* and their *instances* are defined as for LPT formulae, except that metaformulae may contain the metanotation $\neg A$ for $A \supset \mathit{false}$; '$\neg$' has the same syntactic precedence as the quantifiers (formally, we add a production rule $P \to \neg\, P$ to the grammar of metaformulae).

The identifiers '*F*', '*G*' and '*H*', possibly with subscripts, will denote primitive recursive functions. All other identifiers beginning with capital letters will denote numeric terms.

FREE VARIABLES IN HA FORMULAE

For any variable v and HA formula A, the relation $v \in A$ ('v occurs free in A') is defined by:

- $v \notin 0, \quad v \in n$ iff v is n,
- $v \in F(N_1, \ldots N_k)$ iff $v \in N_1$ or $\ldots$ or $v \in N_k$,
- $v \notin \mathit{true}, \quad v \notin \mathit{false}, \quad v \in N_1 = N_2$ iff $v \in N_1$ or $v \in N_2$,
- $v \in A \wedge B$ iff $v \in A \vee B$ iff $v \in A \supset B$ iff $v \in A$ or $v \in B$,
- $v \in \exists^{\mathrm{N}} n A$ iff $v \in \forall^{\mathrm{N}} n A$ iff v isn't n and $v \in A$.

The *free variables* of A are the variables v such that $v \in A$, listed in standard order.

THEOREM 11. Any formula has finitely many free variables.

INTERPRETATION OF HA FORMULAE IN LPT

A mapping from HA formulae to LPT formulae is defined: if A is an HA formula then A^{LPT} is the corresponding formula. For an atomic formula A with no free variables, A^{LPT} is A.

AXIOMS AND RULES OF INFERENCE OF HA

$\wedge$-axioms: $A \supset B \supset (A \wedge B) \quad (A \wedge B) \supset A \quad (A \wedge B) \supset B$
$\vee$-axioms: $A \supset (A \vee B) \quad B \supset (A \vee B) \quad (A \vee B) \supset (A \supset C) \supset (B \supset C) \supset C$
$\supset$-axioms: $A \supset B \supset A \quad (A \supset B \supset C) \supset (A \supset B) \supset A \supset C$
Truth and falsity (*true*-in, *false*-el): $\mathit{true} \quad \mathit{false} \supset A$
Quantifier axioms ($\exists^{\mathrm{N}}$-in, $\forall^{\mathrm{N}}$-el): $A \supset \exists^{\mathrm{N}} n A \quad \forall^{\mathrm{N}} n A \supset A$
Term existence (te): $\exists^{\mathrm{N}} n\, n = N$ where $n \notin N$
Equality (eq): $N = N \quad M = N \supset N = M \quad U = V \supset V = W \supset U = W$
$M = N \supset F(X_1, \ldots X_{i-1}, M, X_{i+1}, \ldots X_k) = F(X_1, \ldots X_{i-1}, N, X_{i+1}, \ldots X_k)$
Peano axioms (P): $\neg Sn = 0 \quad Sm = Sn \supset m = n$
Axioms for primitive recursive functions (pr):
$\mathit{proj}_i^k(\underline{n}) = n_i$ where $1 \leq i \leq k$
$\mathit{comp}_l(H, (F_1, \ldots F_l))(\underline{n}) = H(F_1(\underline{n}), \ldots F_l(\underline{n}))$
$\mathit{rec}_k(F, G)(0, \underline{n}) = F(\underline{n}) \quad \mathit{rec}_k(F, G)(Sm, \underline{n}) = G(m, \underline{n}, \mathit{rec}_k(F, G)(m, \underline{n}))$

where $\underline{n}$ is $n_1, \ldots n_k$, H is l-ary primitive recursive, $F, F_1, \ldots F_l$ are k-ary primitive recursive, and G is $(k+2)$-ary primitive recursive

Modus ponens: $\dfrac{A \quad A \supset B}{B}$

Quantifier rules ($\exists^{\mathrm{N}}$-el, $\forall^{\mathrm{N}}$-in): $\dfrac{A \supset B}{\exists^{\mathrm{N}} n A \supset B} \quad \dfrac{B \supset A}{B \supset \forall^{\mathrm{N}} n A}$ where $n \notin B$

Induction (ind): $$\frac{\forall^{\mathsf{N}} x\,(x=0 \supset A) \quad \forall^{\mathsf{N}} x\,(x=n \supset A) \supset \forall^{\mathsf{N}} x\,(x=Sn \supset A)}{\forall^{\mathsf{N}} x\,(x=n \supset A)}$$

where $n \notin x, A$

FORMAL INTERPRETATION OF HA

A coding of HA derivations as constructions is defined. A construction *lpt* is defined.

THEOREM 22. If D is the code of an HA derivation (with no premises) of a formula A then $lpt(D) \triangleright^{*}$ the code of an LPT derivation (with no premises) of A^{LPT}.

THEOREM 23. (HA interpretation theorem.) If D is the code of an HA derivation (with no premises) of a formula A with no free variables, then $pr(spr(cpf(lpt(D)))) \vdash \ulcorner A^{\mathrm{LPT}} \urcorner$.

THEOREM 24. (The *glue*-lemma for HA.) If D is the code of an HA derivation of A from premises $B_1, \ldots B_k$, and $D_1, \ldots D_k$ are the codes of HA derivations (with no premises) of $B_1, \ldots B_k$ respectively, then $glue(D, [D_1, \ldots D_k]) \triangleright^{*}$ the code of an HA derivation (with no premises) of A.

UNPROVABILITY OF THE EXCLUDED MIDDLE

Let $\{M\}$ be the Mth function in an enumeration of the partial recursive functions and let F be a primitive recursive function such that, for any numbers M, N, $F(M,N) \triangleright^{*} 0$ iff the evaluation of $\{M\}(M)$ halts in exactly N steps. Let A be the HA atomic formula $F(m,n)=0$ (where m and n are two variables) and let $\overline{A}$ be the contrary atomic formula $minus(1, F(m,n)) = 0$.

THEOREM 25. The proof function $\ulcorner(\forall^{\mathsf{N}} m\,(\exists^{\mathsf{N}} n A \vee \forall^{\mathsf{N}} n \overline{A}))^{\mathrm{LPT}}\urcorner$ has no proof and hence $\forall^{\mathsf{N}} m\,(\exists^{\mathsf{N}} n A \vee \forall^{\mathsf{N}} n \overline{A})$ is not a theorem of HA.

CHAPTER 33

FROM HEYTING ARITHMETIC TO PEANO ARITHMETIC

Peano Arithmetic (PA) is classical first-order number theory; it differs from Heyting Arithmetic (HA) only in including the Excluded Middle, $A \vee \neg A$. I shall obtain PA from HA by a variation, due to Gentzen (1933), of Gödel's (1933a) double-negation interpretation. It follows that every theorem of PA (without free variables) has, via its HA, LPT and CPF interpretations, an intuitionistic proof.

My formulation of PA differs from the usual versions in that it lacks any notion of substitution, which makes the quantifier and equality axioms appear weaker. This appearance however is illusory: the usual axioms and rules of PA can be derived from my PA, as I shall show at the end of this chapter.

LEXICAL CONVENTIONS

The letters 'A', 'B' and 'C', sometimes with subscripts, will denote HA formulae; the letters 'F', 'G' and 'H', possibly with subscripts, will denote primitive recursive functions; all other metavariables beginning with a capital letter denote numeric terms.

SOME HA DERIVATIONS

Since HA contains intuitionistic propositional calculus I shall use the latter freely in the HA derivations that follow, marking all such uses with the label 'pc'.

THEOREM 1. HA reversed equality theorem (r.eq): $M=N \supset M=U \supset N=U$

Proof.

$$\frac{M=N \supset N=M \ (\text{eq}) \qquad N=M \supset M=U \supset N=U \ (\text{eq})}{M=N \supset M=U \supset N=U}\,\text{pc}$$

∎

THEOREM 2. HA specification rule (spec): $\dfrac{n=N \supset A}{A}$ where $n \notin N, A$.

Proof.

$$\frac{\exists^{\mathsf{N}} n\, n=N \ \text{(te)} \qquad \dfrac{n=N \supset A}{\exists^{\mathsf{N}} n\, n=N \supset A}\ \exists^{\mathsf{N}}\text{-el}}{A}\ \text{pc}$$

∎

THEOREM 3. HA bivalence theorem (biv): $m=n \vee \neg m=n$,
where m and n are two variables.

Proof. Let x be a fresh variable. For any numeric terms P and Q, let A_{PQ} be the formula $P=Q \vee \neg P=Q$, let B_P be the formula $\forall^{\mathsf{N}} n\, A_{Pn}$, and let C_Q be the formula $B_m \supset A_{Sm,Q}$. Then we have the following HA derivations, where where M, N, U, V and W are any numeric terms such that $x, n \notin M$ and $x \notin N$.

$$\frac{U=V \supset V=U \ \text{(eq)} \qquad V=U \supset U=V \ \text{(eq)}}{A_{UV} \supset A_{VU} \quad (1)}\ \text{pc}$$

$$\frac{U=V \supset U=W \supset V=W \ \text{(r.eq)} \qquad U=V \supset V=W \supset U=W \ \text{(eq)}}{U=V \supset A_{UW} \supset A_{VW} \quad (2)}\ \text{pc}$$

$$\frac{\dfrac{A_{0N} \supset A_{N0}\ (1) \quad x=N \supset N=x\ \text{(eq)} \quad N=x \supset A_{N0} \supset A_{x0}\ (2) \quad A_{x0} \supset A_{0x}\ (1)}{A_{0N} \supset x=N \supset A_{0x}}\ \text{pc}}{A_{0N} \supset \forall^{\mathsf{N}} x\, (x=N \supset A_{0x}) \quad (3)}\ \forall^{\mathsf{N}}\text{-in}$$

$$\frac{\dfrac{\forall^{\mathsf{N}} x\, (x=n \supset A_{0x}) \supset (x=n \supset A_{0x})\ (\forall^{\mathsf{N}}\text{-el}) \qquad x=n \supset A_{0x} \supset A_{0n}\ (2{,}1{,}\text{pc})}{x=n \supset \forall^{\mathsf{N}} x\, (x=n \supset A_{0x}) \supset A_{0n}}\ \text{pc}}{\forall^{\mathsf{N}} x\, (x=n \supset A_{0x}) \supset A_{0n} \quad (4)}\ \text{spec}$$

$$\frac{\dfrac{x=M \supset M=x\ \text{(eq)} \qquad \dfrac{\dfrac{B_M \supset A_{Mn}\ (\forall^{\mathsf{N}}\text{-el}) \qquad M=x \supset A_{Mn} \supset A_{xn}\ (2)}{M=x \supset B_M \supset A_{xn}}\ \text{pc}}{M=x \supset B_M \supset B_x}\ \forall^{\mathsf{N}}\text{-in,pc}}{B_M \supset x=M \supset B_x}\ \text{pc}}{B_M \supset \forall^{\mathsf{N}} x\, (x=M \supset B_x) \quad (5)}\ \forall^{\mathsf{N}}\text{-in}$$

$$\dfrac{\forall^{\mathrm{N}}x\,(x=m\supset B_x)\supset(x=m\supset B_x)\ (\forall^{\mathrm{N}}\text{-el}) \qquad \dfrac{\dfrac{B_x\supset A_{xn}\ (\forall^{\mathrm{N}}\text{-el}) \qquad x=m\supset A_{xn}\supset A_{mn}\ (2)}{x=m\supset B_x\supset A_{mn}}\,\text{pc}}{x=m\supset B_x\supset B_m}\,\forall^{\mathrm{N}}\text{-in,pc}}{\dfrac{x=m\supset\forall^{\mathrm{N}}x\,(x=m\supset B_x)\supset B_m}{\forall^{\mathrm{N}}x\,(x=m\supset B_x)\supset B_m\quad(6)}\,\text{spec}}\,\text{pc}$$

$$\dfrac{\dfrac{\dfrac{x=N\supset N=x\ (\text{eq}) \qquad N=x\supset A_{Sm,N}\supset A_{Sm,x}\ (2,1,\text{pc})}{A_{Sm,N}\supset x=N\supset A_{Sm,x}}\,\text{pc}}{C_N\supset x=N\supset C_x}\,\text{pc}}{C_N\supset\forall^{\mathrm{N}}x\,(x=N\supset C_x)\quad(7)}\,\forall^{\mathrm{N}}\text{-in}$$

$$\dfrac{\forall^{\mathrm{N}}x\,(x=n\supset C_x)\supset(x=n\supset C_x)\ (\forall^{\mathrm{N}}\text{-el}) \qquad \dfrac{x=n\supset A_{Sm,x}\supset A_{Sm,n}\ (2,1,\text{pc})}{x=n\supset C_x\supset C_n}\,\text{pc}}{\dfrac{x=n\supset\forall^{\mathrm{N}}x\,(x=n\supset C_x)\supset C_n}{\forall^{\mathrm{N}}x\,(x=n\supset C_x)\supset C_n\quad(8)}\,\text{spec}}\,\text{pc}$$

$$\dfrac{true\ (true\text{-in}) \qquad \dfrac{\dfrac{\dfrac{\dfrac{\dfrac{\dfrac{0=0\ (\text{eq})}{A_{00}}\,\text{pc}}{\forall^{\mathrm{N}}x\,(x=0\supset A_{0x})}\,3,\text{pc} \qquad \dfrac{\dfrac{\dfrac{\dfrac{0=Sn\supset Sn=0\ (\text{eq}) \qquad \neg Sn=0\ (\text{P})}{\neg 0=Sn}\,\text{pc}}{A_{0,Sn}}\,\text{pc}}{\forall^{\mathrm{N}}x\,(x=Sn\supset A_{0x})}\,3,\text{pc}}{\forall^{\mathrm{N}}x\,(x=n\supset A_{0x})\supset\forall^{\mathrm{N}}x\,(x=Sn\supset A_{0x})}\,\text{pc}}{\forall^{\mathrm{N}}x\,(x=n\supset A_{0x})}\,\text{ind}}{A_{0n}}\,4,\text{pc}}{true\supset A_{0n}}\,\text{pc}}{true\supset B_0}\,\forall^{\mathrm{N}}\text{-in}}{B_0\quad(9)}\,\text{pc}$$

$$\dfrac{B_m\supset A_{mn}\ (\forall^{\mathrm{N}}\text{-el}) \qquad \dfrac{m=n\supset Sm=Sn\ (\text{eq}) \qquad Sm=Sn\supset m=n\ (\text{P})}{A_{mn}\supset A_{Sm,Sn}}\,\text{pc}}{C_{Sn}\quad(10)}\,\text{pc}$$

$$
\dfrac{\dfrac{\dfrac{\dfrac{\dfrac{\dfrac{\neg Sm = 0 \;\text{(P)}}{A_{Sm,0}}\,\text{pc}}{C_0}\,\text{pc}}{\forall^{\mathrm{N}} x\,(x = 0 \supset C_x)}\,\text{7,pc} \quad \dfrac{\dfrac{C_{Sn}\;\text{(10)}}{\forall^{\mathrm{N}} x\,(x = Sn \supset C_x)}\,\text{7,pc}}{\forall^{\mathrm{N}} x\,(x = n \supset C_x) \supset \forall^{\mathrm{N}} x\,(x = Sn \supset C_x)}\,\text{pc}}{\forall^{\mathrm{N}} x\,(x = n \supset C_x)}\,\text{ind}}{C_n}\,\text{8,pc}
$$

$$
\dfrac{\dfrac{\dfrac{\dfrac{B_0\;\text{(9)}}{\forall^{\mathrm{N}} x\,(x = 0 \supset B_x)}\,\text{5,pc} \quad \dfrac{\dfrac{C_n}{B_m \supset B_{Sm}}\,\forall^{\mathrm{N}}\text{-in}}{\forall^{\mathrm{N}} x\,(x = m \supset B_x) \supset \forall^{\mathrm{N}} x\,(x = Sm \supset B_x)}\,\text{5,6,pc}}{\forall^{\mathrm{N}} x\,(x = m \supset B_x)}\,\text{ind}}{B_m}\,\text{6,pc} \quad B_m \supset A_{mn}\;(\forall^{\mathrm{N}}\text{-el})}{A_{mn}}\,\text{pc}
$$

The conclusion, A_{mn}, is $m = n \vee \neg m = n$, as required. ∎

PA FORMULAE

The formulae, metaformulae and free variables of Peano Arithmetic are the same as those of Heyting Arithmetic, except that metaformulae may contain the substitution metanotation introduced below. The letters 'A', 'B' and 'C' will from now on be used as metavariables denoting PA formulae; 'F', 'G' and 'H', possibly with subscripts, denote primitive recursive functions; all other metavariables starting with a capital letter denote numeric terms.

INTERPRETATION OF PA FORMULAE IN HA

For any PA formula A, the corresponding HA formula, A^{H}, is defined as follows.

A^{H} is A if A is an atomic formula
$(B \wedge C)^{\mathrm{H}}$ is $B^{\mathrm{H}} \wedge C^{\mathrm{H}}$
$(B \vee C)^{\mathrm{H}}$ is $\neg(\neg B^{\mathrm{H}} \wedge \neg C^{\mathrm{H}})$
$(B \supset C)^{\mathrm{H}}$ is $B^{\mathrm{H}} \supset C^{\mathrm{H}}$
$(\exists^{\mathrm{N}} n\, B)^{\mathrm{H}}$ is $\neg \forall^{\mathrm{N}} n \neg B^{\mathrm{H}}$
$(\forall^{\mathrm{N}} n\, B)^{\mathrm{H}}$ is $\forall^{\mathrm{N}} n\, B^{\mathrm{H}}$

THEOREM 4. $v \in A$ iff $v \in A^{\mathrm{H}}$.

THEOREM 5. PA double negation theorem ($\neg\neg$): for any PA formula A, $\neg\neg A^{\mathrm{H}} \supset A^{\mathrm{H}}$ is a theorem of HA.

Proof. By structural induction on A, as follows. There are eight cases; uses of the inductive hypothesis are marked by 'ih'.

Case 1: A is *true*. Then $\neg\neg A^{\mathrm{H}} \supset A^{\mathrm{H}}$ is derived in HA by

$$\dfrac{\mathit{true} \quad (\mathit{true}\text{-in})}{\neg\neg\mathit{true} \supset \mathit{true}}\,\text{pc}$$

Case 2: A is *false*. Then $\neg\neg A^{\mathrm{H}} \supset A^{\mathrm{H}}$ is $((\mathit{false} \supset \mathit{false}) \supset \mathit{false}) \supset \mathit{false}$, which is a theorem of HA by intuitionistic propositional calculus.

Case 3: A is $M = N$. Then $\neg\neg A^{\mathrm{H}} \supset A^{\mathrm{H}}$ is derived by

$$\dfrac{\dfrac{\dfrac{m = n \vee \neg m = n \ (\text{biv})}{m = M \supset n = N \supset (M = N \vee \neg M = N)}\,\text{eq, pc}}{M = N \vee \neg M = N}\,\text{spec}}{\neg\neg M = N \supset M = N}\,\text{pc}$$

where m, n are two fresh variables.

Case 4: A is $B \wedge C$. Then $\neg\neg A^{\mathrm{H}} \supset A^{\mathrm{H}}$ is derived in HA by

$$\dfrac{\dfrac{A^{\mathrm{H}} \supset B^{\mathrm{H}} \ (\text{pc})}{\neg\neg A^{\mathrm{H}} \supset \neg\neg B^{\mathrm{H}}}\,\text{pc} \quad \neg\neg B^{\mathrm{H}} \supset B^{\mathrm{H}} \ (\text{ih}) \quad \dfrac{A^{\mathrm{H}} \supset C^{\mathrm{H}} \ (\text{pc})}{\neg\neg A^{\mathrm{H}} \supset \neg\neg C^{\mathrm{H}}}\,\text{pc} \quad \neg\neg C^{\mathrm{H}} \supset C^{\mathrm{H}} \ (\text{ih})}{\neg\neg A^{\mathrm{H}} \supset A^{\mathrm{H}}}\,\text{pc}$$

Case 5: A is $B \vee C$. Then $\neg\neg A^{\mathrm{H}} \supset A^{\mathrm{H}}$ is

$$\neg\neg\neg(\neg B^{\mathrm{H}} \wedge \neg C^{\mathrm{H}}) \supset \neg(\neg B^{\mathrm{H}} \wedge \neg C^{\mathrm{H}}),$$

which is a theorem of HA by intuitionistic propositional calculus.

Case 6: A is $B \supset C$. Then $\neg\neg A^{\mathrm{H}} \supset A^{\mathrm{H}}$ is derived in HA by

$$\dfrac{\dfrac{B^{\mathrm{H}} \supset (B^{\mathrm{H}} \supset C^{\mathrm{H}}) \supset C^{\mathrm{H}} \ (\text{pc})}{B^{\mathrm{H}} \supset \neg\neg(B^{\mathrm{H}} \supset C^{\mathrm{H}}) \supset \neg\neg C^{\mathrm{H}}}\,\text{pc} \qquad \neg\neg C^{\mathrm{H}} \supset C^{\mathrm{H}} \ (\text{ih})}{\neg\neg(B^{\mathrm{H}} \supset C^{\mathrm{H}}) \supset (B^{\mathrm{H}} \supset C^{\mathrm{H}})}\,\text{pc}$$

Case 7: A is $\exists^{\mathrm{N}} n\, B$. Then $\neg\neg A^{\mathrm{H}} \supset A^{\mathrm{H}}$ is $\neg\neg\neg\forall^{\mathrm{N}} n \neg B^{\mathrm{H}} \supset \neg\forall^{\mathrm{N}} n \neg B^{\mathrm{H}}$, which is a theorem of HA by intuitionistic propositional calculus.

Case 8: A is $\forall^{N} n\, B$. Then $\neg\neg A^{H} \supset A^{H}$ is derived by

$$\dfrac{\dfrac{\dfrac{\forall^{N} n\, B^{H} \supset B^{H}\ (\forall^{N}\text{-el})}{\neg\neg\forall^{N} n\, B^{H} \supset \neg\neg B^{H}}\ \text{pc} \qquad \neg\neg B^{H} \supset B^{H}\ \text{(ih)}}{\neg\neg\forall^{N} n\, B^{H} \supset B^{H}}\ \text{pc}}{\neg\neg\forall^{N} n\, B^{H} \supset \forall^{N} n\, B^{H}}\ \forall^{N}\text{-in}$$

This completes the proof. ∎

AXIOMS AND RULES OF INFERENCE OF PEANO ARITHMETIC

$\wedge$-axioms: $A \supset B \supset (A \wedge B)$ $\quad (A \wedge B) \supset A \quad (A \wedge B) \supset B$
$\vee$-axioms: $A \supset (A \vee B) \quad B \supset (A \vee B) \quad (A \vee B) \supset (A \supset C) \supset (B \supset C) \supset C$
$\supset$-axioms: $A \supset B \supset A \quad (A \supset B \supset C) \supset (A \supset B) \supset A \supset C$
Excluded middle: $A \vee \neg A$
Truth and falsity (*true*-in, *false*-el): $\mathit{true} \quad \mathit{false} \supset A$
Quantifier axioms ($\exists^{N}$-in, $\forall^{N}$-el): $A \supset \exists^{N} n A \quad \forall^{N} n A \supset A$
Term existence (te): $\exists^{N} n\, n = N$ where $n \notin N$
Equality (eq): $N = N \quad M = N \supset N = M \quad U = V \supset V = W \supset U = W$
$M = N \supset F(X_1, \ldots X_{i-1}, M, X_{i+1}, \ldots X_k) = F(X_1, \ldots X_{i-1}, N, X_{i+1}, \ldots X_k)$
Peano axioms (P): $\neg Sn = 0 \quad Sm = Sn \supset m = n$
Axioms for primitive recursive functions (pr):
$\mathit{proj}_i^k(\underline{n}) = n_i$ where $1 \leq i \leq k$
$\mathit{comp}_l(H, (F_1, \ldots F_l))(\underline{n}) = H(F_1(\underline{n}), \ldots F_l(\underline{n}))$
$\mathit{rec}_k(F, G)(0, \underline{n}) = F(\underline{n}) \quad \mathit{rec}_k(F, G)(Sm, \underline{n}) = G(m, \underline{n}, \mathit{rec}_k(F, G)(m, \underline{n}))$

where $\underline{n}$ is $n_1, \ldots n_k$, H is l-ary primitive recursive, $F_1, \ldots F_l$ and F are k-ary primitive recursive, and G is $(k+2)$-ary primitive recursive

Modus ponens: $\dfrac{A \quad A \supset B}{B}$

Quantifier rules ($\exists^{N}$-el, $\forall^{N}$-in): $\dfrac{A \supset B}{\exists^{N} n A \supset B} \quad \dfrac{B \supset A}{B \supset \forall^{N} n A}$ where $n \notin B$

Induction (ind): $\dfrac{\forall^{N} x\,(x = 0 \supset A) \quad \forall^{N} x\,(x = n \supset A) \supset \forall^{N} x\,(x = Sn \supset A)}{\forall^{N} x\,(x = n \supset A)}$

where $n \notin x, A$

HA DERIVATIONS CORRESPONDING TO THE PA AXIOMS AND RULES OF INFERENCE

I shall show that, for any axiom A of PA, A^{H} is a theorem of HA, and that, for any rule of inference $\dfrac{A \ldots B}{C}$ of PA, $\dfrac{A^{H} \ldots B^{H}}{C^{H}}$ is a derived rule of HA.

THEOREM 6. The interpretations of the PA axiom schemata and rules that involve only atomic formulae, $\wedge$, $\supset$ and $\forall^{\mathrm{N}}$ are also axioms and rules of HA.

Proof. For example, the PA axiom $(A \wedge B) \supset A$ is interpreted as the HA axiom $(A^{\mathrm{H}} \wedge B^{\mathrm{H}}) \supset A^{\mathrm{H}}$. Similarly for the others. ∎

THEOREM 7. PA $\vee$-axioms:
$A \supset (A \vee B) \quad B \supset (A \vee B) \quad (A \vee B) \supset (A \supset C) \supset (B \supset C) \supset C.$

Proof. The corresponding HA derivations are

$$\frac{A^{\mathrm{H}} \supset (A^{\mathrm{H}} \vee B^{\mathrm{H}}) \text{ (pc)}}{A^{\mathrm{H}} \supset \neg(\neg A^{\mathrm{H}} \wedge \neg B^{\mathrm{H}})} \text{pc} \qquad \frac{B^{\mathrm{H}} \supset (A^{\mathrm{H}} \vee B^{\mathrm{H}}) \text{ (pc)}}{B^{\mathrm{H}} \supset \neg(\neg A^{\mathrm{H}} \wedge \neg B^{\mathrm{H}})} \text{pc}$$

$$\dfrac{\dfrac{\dfrac{\dfrac{A^{\mathrm{H}} \supset (A^{\mathrm{H}} \supset C^{\mathrm{H}}) \supset C^{\mathrm{H}} \text{ (pc)}}{\neg C^{\mathrm{H}} \supset (A^{\mathrm{H}} \supset C^{\mathrm{H}}) \supset \neg A^{\mathrm{H}}} \text{pc} \quad \dfrac{B^{\mathrm{H}} \supset (B^{\mathrm{H}} \supset C^{\mathrm{H}}) \supset C^{\mathrm{H}} \text{ (pc)}}{\neg C^{\mathrm{H}} \supset (B^{\mathrm{H}} \supset C^{\mathrm{H}}) \supset \neg B^{\mathrm{H}}} \text{pc}}{\neg C^{\mathrm{H}} \supset (A^{\mathrm{H}} \supset C^{\mathrm{H}}) \supset (B^{\mathrm{H}} \supset C^{\mathrm{H}}) \supset (\neg A^{\mathrm{H}} \wedge \neg B^{\mathrm{H}})} \text{pc}}{\neg(\neg A^{\mathrm{H}} \wedge \neg B^{\mathrm{H}}) \supset (A^{\mathrm{H}} \supset C^{\mathrm{H}}) \supset (B^{\mathrm{H}} \supset C^{\mathrm{H}}) \supset \neg\neg C^{\mathrm{H}}} \text{pc} \quad \neg\neg C^{\mathrm{H}} \supset C^{\mathrm{H}} \text{ } (\neg\neg)}{\neg(\neg A^{\mathrm{H}} \wedge \neg B^{\mathrm{H}}) \supset (A^{\mathrm{H}} \supset C^{\mathrm{H}}) \supset (B^{\mathrm{H}} \supset C^{\mathrm{H}}) \supset C^{\mathrm{H}}} \text{pc}$$

∎

THEOREM 8. PA excluded middle axiom: $A \vee \neg A$.

Proof. This is interpreted as $\neg(\neg A^{\mathrm{H}} \wedge \neg\neg A^{\mathrm{H}})$, which is a theorem of HA by intuitionistic propositional calculus. ∎

THEOREM 9. PA $\exists^{\mathrm{N}}$-in axiom: $A \supset \exists^{\mathrm{N}} n A$.

Proof. The corresponding HA derivation is

$$\frac{\forall^{\mathrm{N}} n \neg A^{\mathrm{H}} \supset \neg A^{\mathrm{H}} \text{ } (\forall^{\mathrm{N}}\text{-el})}{A^{\mathrm{H}} \supset \neg \forall^{\mathrm{N}} n \neg A^{\mathrm{H}}} \text{pc}$$

∎

THEOREM 10. PA term existence axiom: $\exists^{\mathrm{N}} n\, n = N$ where $n \notin N$.

Proof. The corresponding HA derivation is

$$\dfrac{\dfrac{\forall^{\mathrm{N}} n \neg n = N \supset \neg n = N \text{ } (\forall^{\mathrm{N}}\text{-el})}{n = N \supset \neg \forall^{\mathrm{N}} n \neg n = N} \text{pc}}{\neg \forall^{\mathrm{N}} n \neg n = N} \text{spec}$$

∎

THEOREM 11. PA $\exists^{\mathrm{N}}$-el rule: $$\frac{A \supset B}{\exists^{\mathrm{N}} n A \supset B} \quad \text{where } n \notin B.$$

Proof. The rule's interpretation is derived as follows.

$$\dfrac{\dfrac{\dfrac{\dfrac{A^{\mathrm{H}} \supset B^{\mathrm{H}}}{\neg B^{\mathrm{H}} \supset \neg A^{\mathrm{H}}}\,\text{pc}}{\neg B^{\mathrm{H}} \supset \forall^{\mathrm{N}} n \neg A^{\mathrm{H}}}\,\forall\text{-in}}{\neg \forall^{\mathrm{N}} n \neg A^{\mathrm{H}} \supset \neg\neg B^{\mathrm{H}}}\,\text{pc} \qquad \neg\neg B^{\mathrm{H}} \supset B^{\mathrm{H}}\ (\neg\neg)}{\neg \forall^{\mathrm{N}} n \neg A^{\mathrm{H}} \supset B^{\mathrm{H}}}\,\text{pc}$$

∎

FORMAL INTERPRETATION OF PA

DEFINITION. A coding of PA formulae as constructions is defined as follows. If A is a formula with free variables $\underline{z}$ (listed in standard order), the code of A, $\langle\!\langle A \rangle\!\rangle$, is defined as $(\lambda[\underline{z}].\langle A \rangle)$, where

$\langle T \rangle$ is *atomic*(T) if T is atomic
$\langle A \wedge B \rangle$ is *and*$(\langle A \rangle, \langle B \rangle)$
$\langle A \vee B \rangle$ is *or*$(\langle A \rangle, \langle B \rangle)$
$\langle A \supset B \rangle$ is *implies*$(\langle A \rangle, \langle B \rangle)$
$\langle \exists^{\mathrm{N}} x A \rangle$ is *exists*$(\lambda[x].\langle A \rangle)$
$\langle \forall^{\mathrm{N}} x A \rangle$ is *all*$(\lambda[x].\langle A \rangle)$.

DEFINITION. A coding of PA derivations as constructions is defined as follows. Number the axiom schemata and rules of inference of PA, 1,2,... 28. A derivation with conclusion A, from a list of premises numbered 0,1,2..., is coded as

(*premise*(i), $\langle\!\langle A \rangle\!\rangle$)	if A is premise number i
(*none*(n), $\langle\!\langle A \rangle\!\rangle$)	if A is an instance of axiom schema n
(*one*(n, X), $\langle\!\langle A \rangle\!\rangle$)	if A is derived by rule n from the subderivation X
(*two*(n, X, Y), $\langle\!\langle A \rangle\!\rangle$)	if A is derived by rule n from subderivations X, Y.

DEFINITION. Define recursive functions $T_1, \ldots T_{28}$ such that

- for $n = 1, \ldots 24$, if A is an instance of PA axiom schema n then $T_n(\langle\langle A \rangle\rangle)$ $\triangleright^*$ the code of the corresponding HA derivation (given above) of A^{H} from no premises;
- for $n = 26, 27$, if $\dfrac{B}{A}$ is an instance of PA rule n then $T_n(\langle\langle B \rangle\rangle, \langle\langle A \rangle\rangle)$ $\triangleright^*$ the code of the corresponding HA derivation (given above) of A^{H} from the premise B^{H};
- for $n{=}25, 28$, if $\dfrac{B\ C}{A}$ is an instance of PA rule n then $T_n(\langle\langle B \rangle\rangle, \langle\langle C \rangle\rangle, \langle\langle A \rangle\rangle)$ $\triangleright^*$ the code of the corresponding HA derivation (given above) of A^{H} from the premises B^{H} and C^{H}.

DEFINITION. Define a construction *ha* by

$$
\begin{aligned}
ha(none(1), a) &\triangleq T_1(a) \\
&\vdots \\
ha(none(24), a) &\triangleq T_{24}(a) \\
ha(two(25, (u, b), (v, c)), a) &\triangleq glue(T_{25}(b, c, a), [ha(u, b), ha(v, c)]) \\
ha(one(26, (u, b)), a) &\triangleq glue(T_{26}(b, a), [ha(u, b)]) \\
ha(one(27, (u, b)), a) &\triangleq glue(T_{27}(b, a), [ha(u, b)]) \\
ha(two(28, (u, b), (v, c)), a) &\triangleq glue(T_{28}(b, c, a), [ha(u, b), ha(v, c)])
\end{aligned}
$$

where a, b, c, u, v are five variables.

THEOREM 12. If D is the code of a PA derivation (with no premises) of a formula A then $ha(D)$ $\triangleright^*$ the code of an HA derivation (with no premises) of A^{H}.

Proof. By structural induction on the derivation encoded by D. ∎

THEOREM 13. (PA interpretation theorem.) If D is the code of a PA derivation (with no premises) of a formula A with no free variables, then $pr(spr(cpf(lpt(ha(D))))) \vdash \ulcorner (A^{\mathrm{H}})^{\mathrm{LPT}} \urcorner$.

Proof. By the previous theorem, the extensionality theorem for $\vdash$, and the interpretation theorem for HA. ∎

SUBSTITUTION IN PA FORMULAE

Peano Arithmetic is often stated in an apparently stronger form than the version I have given. For example, the elimination axiom for $\forall^{\mathrm{N}}$ is usually stated not as $\forall^{\mathrm{N}} n A \supset A$ but as $\forall^{\mathrm{N}} n A \supset A\binom{N}{n}$, where $A\binom{N}{n}$ is the formula obtained by substituting a term N for all free occurrences of n in A, assuming there are no 'variable clashes'. Here I shall show that the stronger forms of the axioms and rules are derivable in my formulation of PA. (The same can be done in HA, since the derivations do not require the Excluded Middle, or even in a modified form in LPT; but the notion of substitution is such a nuisance that I prefer to delay its introduction to the very end.)

The first step is to define when substitution is possible. A notation $N \overset{n}{\hookrightarrow} A$, meaning that the numeric term N *is substitutable for* the variable n in the formula A, is defined as follows.

- $N \overset{n}{\hookrightarrow} A$ if A is an atomic formula;
- $N \overset{n}{\hookrightarrow} A \wedge B$ iff $N \overset{n}{\hookrightarrow} A \vee B$ iff $N \overset{n}{\hookrightarrow} A \supset B$ iff $N \overset{n}{\hookrightarrow} A$ and $N \overset{n}{\hookrightarrow} B$;
- $N \overset{n}{\hookrightarrow} \exists^{\mathrm{N}} m A$ iff $N \overset{n}{\hookrightarrow} \forall^{\mathrm{N}} m A$ iff $n \notin \exists^{\mathrm{N}} m A$ or ($m \notin N$ and $N \overset{n}{\hookrightarrow} A$).

If $N \overset{n}{\hookrightarrow} A$, then the formula $A\binom{N}{n}$ is defined by:

- $0\binom{N}{n}$ is 0, $n\binom{N}{n}$ is N, $m\binom{N}{n}$ is m if m isn't n;
- $F(N_1, \ldots N_k)\binom{N}{n}$ is $F(N_1\binom{N}{n}, \ldots N_k\binom{N}{n})$, for $k \geq 1$;
- $\mathit{true}\binom{N}{n}$ is true, $\mathit{false}\binom{N}{n}$ is false, $(N_1 = N_2)\binom{N}{n}$ is $N_1\binom{N}{n} = N_2\binom{N}{n}$;
- $(A \wedge B)\binom{N}{n}$ is $A\binom{N}{n} \wedge B\binom{N}{n}$, $(A \vee B)\binom{N}{n}$ is $A\binom{N}{n} \vee B\binom{N}{n}$,
- $(A \supset B)\binom{N}{n}$ is $A\binom{N}{n} \supset B\binom{N}{n}$;
- $(\exists^{\mathrm{N}} m A)\binom{N}{n}$ is $\exists^{\mathrm{N}} m A$ and $(\forall^{\mathrm{N}} m A)\binom{N}{n}$ is $\forall^{\mathrm{N}} m A$, if $n \notin \exists^{\mathrm{N}} m A$;
- $(\exists^{\mathrm{N}} m A)\binom{N}{n}$ is $\exists^{\mathrm{N}} m (A\binom{N}{n})$ and $(\forall^{\mathrm{N}} m A)\binom{N}{n}$ is $\forall^{\mathrm{N}} m (A\binom{N}{n})$, if $n \in \exists^{\mathrm{N}} m A$ and $m \notin N$.

THEOREM 14. Properties of substitution:

- $0 \overset{n}{\hookrightarrow} A$;
- if M and N have the same free variables, then $M \overset{n}{\hookrightarrow} A$ iff $N \overset{n}{\hookrightarrow} A$;
- $n \overset{n}{\hookrightarrow} A$ and $A\binom{n}{n}$ is A;
- if $n \notin A$ then $N \overset{n}{\hookrightarrow} A$ and $A\binom{N}{n}$ is A;
- if $N \overset{n}{\hookrightarrow} A$ then $m \in A\binom{N}{n}$ iff ($m \in N$ and $n \in A$) or ($m \in A$ and n isn't m).

EXERCISE. Show that if $N \overset{n}{\hookrightarrow} A$ then $N \overset{n}{\hookrightarrow} A^{\mathrm{H}}$ and $A\binom{N}{n}^{\mathrm{H}}$ is $A^{\mathrm{H}}\binom{N}{n}$, using the same definition of substitution for HA formulae as for PA formulae.

PA THEOREMS AND DERIVED RULES INVOLVING SUBSTITUTION

Note that PA contains classical propositional calculus, so I shall use the latter freely in the derivations that follow, marking such uses with the label 'pc'.

THEOREM 15. PA equality theorem for terms (eq.term):

$$M = N \supset T\binom{M}{x} = T\binom{N}{x}.$$

Proof. By structural induction on T using propositional calculus and the equality axioms

$$N = N \qquad U = V \supset V = W \supset U = W$$

$$M = N \supset F(X_1, \ldots X_{i-1}, M, X_{i+1}, \ldots X_k) = F(X_1, \ldots X_{i-1}, N, X_{i+1}, \ldots X_k).$$

∎

THEOREM 16. PA equality theorem for formulae (eq.form.):

$$M = N \supset A\binom{M}{x} \supset A\binom{N}{x},$$

where $M \overset{x}{\hookrightarrow} A$ and $N \overset{x}{\hookrightarrow} A$.

Proof. By structural induction on A. There are eight cases; uses of the inductive hypothesis are marked by 'ih'.

Case 1: A is *true*. Then the formula to be derived is $M = N \supset \mathit{true} \supset \mathit{true}$, which is a theorem of PA, by propositional calculus.

Case 2: A is *false*. Then the formula to be derived is $M = N \supset \mathit{false} \supset \mathit{false}$, which is also a theorem of PA, by propositional calculus.

Case 3: A is $U = V$, where U and V are numeric terms. Then the desired formula, $M = N \supset U\binom{M}{x} = V\binom{M}{x} \supset U\binom{N}{x} = V\binom{N}{x}$, follows by propositional calculus from the PA theorems

$$\begin{array}{ll} M = N \supset N = M & \text{(eq axiom)} \\ N = M \supset U\binom{N}{x} = U\binom{M}{x} & \text{(eq.term)} \\ U\binom{N}{x} = U\binom{M}{x} \supset U\binom{M}{x} = V\binom{M}{x} \supset U\binom{N}{x} = V\binom{M}{x} & \text{(eq axiom)} \\ M = N \supset V\binom{M}{x} = V\binom{N}{x} & \text{(eq.term)} \\ U\binom{N}{x} = V\binom{M}{x} \supset V\binom{M}{x} = V\binom{N}{x} \supset U\binom{N}{x} = V\binom{N}{x} & \text{(eq axiom).} \end{array}$$

Case 4: A is $B \wedge C$. Then the desired formula is derived in PA as follows.

$$\frac{M = N \supset B\binom{M}{x} \supset B\binom{N}{x} \ \text{(ih)} \qquad M = N \supset C\binom{M}{x} \supset C\binom{N}{x} \ \text{(ih)}}{M = N \supset A\binom{M}{x} \supset A\binom{N}{x}} \ \text{pc}$$

Case 5: A is $B \vee C$. This is similar to Case 4.

Case 6: A is $B \supset C$. Then the desired formula is derived in PA as follows.

$$\dfrac{M=N\supset C\binom{M}{x}\supset C\binom{N}{x}\ \text{(ih)} \qquad \dfrac{M=N\supset N=M\ \text{(eq)}\quad N=M\supset B\binom{N}{x}\supset B\binom{M}{x}\ \text{(ih)}}{M=N\supset B\binom{N}{x}\supset B\binom{M}{x}}\,\text{ih}}{M=N\supset A\binom{M}{x}\supset A\binom{N}{x}}\,\text{pc}$$

Case 7: A is $\exists^{\mathrm{N}} n\, B$. The substitutability conditions tell us that $x \notin \exists^{\mathrm{N}} n\, B$ or ($n \notin M, N$ and $M \overset{x}{\hookrightarrow} B$ and $N \overset{x}{\hookrightarrow} B$). In the former subcase ($x \notin \exists^{\mathrm{N}} n\, B$) the formula to be derived is $M = N \supset \exists^{\mathrm{N}} n\, B \supset \exists^{\mathrm{N}} n\, B$, which is a theorem of PA by propositional calculus. In the latter subcase the PA derivation is as follows.

$$\dfrac{\dfrac{M=N\supset B\binom{M}{x}\supset B\binom{N}{x}\ \text{(ih)} \qquad B\binom{N}{x}\supset \exists^{\mathrm{N}} n\, B\binom{N}{x}\ (\exists^{\mathrm{N}}\text{-in})}{M=N\supset B\binom{M}{x}\supset \exists^{\mathrm{N}} n\, B\binom{N}{x}}\,\text{pc}}{M=N\supset \exists^{\mathrm{N}} n\, B\binom{M}{x}\supset \exists^{\mathrm{N}} n\, B\binom{N}{x}}\,\exists^{\mathrm{N}}\text{-el, pc}$$

Case 8: A is $\forall^{\mathrm{N}} n\, B$. This is similar to Case 7. ∎

THEOREM 17. PA specification rule (spec): $\dfrac{n=N\supset A}{A}$ where $n \notin N, A$.

Proof.

$$\dfrac{\exists^{\mathrm{N}} n\, n=N\ \text{(te)} \qquad \dfrac{n=N\supset A}{\exists^{\mathrm{N}} n\, n=N\supset A}\,\exists^{\mathrm{N}}\text{-el}}{A}\,\text{pc}$$

∎

THEOREM 18. PA substitution rule (sub): $\dfrac{B}{B\binom{N}{n}}$ where $N \overset{n}{\hookrightarrow} B$.

Proof. Let m be a fresh variable (and thus $m \overset{n}{\hookrightarrow} B$). The PA derivation is as follows.

$$\dfrac{\dfrac{\dfrac{\dfrac{B \qquad n=m\supset B\supset B\binom{m}{n}\ \text{(eq.form.)}}{n=m\supset B\binom{m}{n}}\,\text{pc}}{B\binom{m}{n}}\,\text{spec, since } n\notin m, B\binom{m}{n} \qquad m=N\supset B\binom{m}{n}\supset B\binom{N}{n}\ \text{(eq.form.)}}{m=N\supset B\binom{N}{n}}\,\text{pc}}{B\binom{N}{n}}\,\text{spec, since } m\notin N, B\binom{N}{n}$$

∎

THEOREM 19. PA quantifier theorems with substitution:

$$A\binom{N}{n} \supset \exists^{N} n A, \qquad \forall^{N} n A \supset A\binom{N}{n}, \qquad \text{where } N \overset{n}{\hookrightarrow} A.$$

Proof.

$$\dfrac{A \supset \exists^{N} n A \;(\exists^{N}\text{-in})}{A\binom{N}{n} \supset \exists^{N} n A}\,\text{sub} \qquad\qquad \dfrac{\forall^{N} n A \supset A \;(\forall^{N}\text{-el})}{\forall^{N} n A \supset A\binom{N}{n}}\,\text{sub}$$

∎

THEOREM 20. PA variable change rule (vch): $\dfrac{B\binom{n}{m}}{B}$

where $n \notin B$ and $n \overset{m}{\hookrightarrow} B$.

Proof. If n is m then there is nothing to prove, so assume n is not m. The PA derivation is as follows.

$$\dfrac{\dfrac{B\binom{n}{m} \qquad\qquad n = m \supset B\binom{n}{m} \supset B \;(\text{eq.form.})}{n = m \supset B}\,\text{pc}}{B}\,\text{spec, since } n \notin m, B$$

∎

THEOREM 21. PA quantifier rules with substitution:

$$\dfrac{A\binom{n}{m} \supset B}{\exists^{N} m A \supset B} \qquad \dfrac{B \supset A\binom{n}{m}}{B \supset \forall^{N} m A} \qquad \text{where } n \overset{m}{\hookrightarrow} A \text{ and } n \notin \exists^{N} m A, B.$$

Proof. If n is m then the rules are simply the $\exists^{N}$-el and $\forall^{N}$-in rules of PA, so assume n is not m. Then $n \notin A, B$ and the PA derivations are as follows.

$$\dfrac{\dfrac{\dfrac{A\binom{n}{m} \supset \exists^{N} n A\binom{n}{m} \;(\exists^{N}\text{-in})}{A \supset \exists^{N} n A\binom{n}{m}}\,\text{vch}}{\exists^{N} m A \supset \exists^{N} n A\binom{n}{m}}\,\exists^{N}\text{-el} \qquad \dfrac{A\binom{n}{m} \supset B}{\exists^{N} n A\binom{n}{m} \supset B}\,\exists^{N}\text{-el}}{\exists^{N} m A \supset B}\,\text{pc}$$

$$\dfrac{\dfrac{B \supset A\binom{n}{m}}{B \supset \forall^{N} n A\binom{n}{m}}\,\forall^{N}\text{-in} \qquad \dfrac{\dfrac{\forall^{N} n A\binom{n}{m} \supset A\binom{n}{m} \;(\forall^{N}\text{-el})}{\forall^{N} n A\binom{n}{m} \supset A}\,\text{vch}}{\forall^{N} n A\binom{n}{m} \supset \forall^{N} m A}\,\forall^{N}\text{-in}}{B \supset \forall^{N} m A}\,\text{pc}$$

∎

THEOREM 22. PA extraction theorems (ext):

$$A\binom{N}{m} \supset \forall^{\mathrm{N}} m\,(m = N \supset A) \qquad \forall^{\mathrm{N}} m\,(m = N \supset A) \supset A\binom{N}{m},$$

where $m \notin N$ and $N \overset{m}{\hookrightarrow} A$.

Proof.

$$\dfrac{\dfrac{\dfrac{m = N \supset N = m \ \text{(eq)} \qquad N = m \supset A\binom{N}{m} \supset A \ \text{(eq.form.)}}{m = N \supset A\binom{N}{m} \supset A}\,\text{pc}}{A\binom{N}{m} \supset m = N \supset A}\,\text{pc}}{A\binom{N}{m} \supset \forall^{\mathrm{N}} m\,(m = N \supset A)}\,\forall^{\mathrm{N}}\text{-in, since } m \notin A\binom{N}{m}$$

$$\dfrac{\dfrac{\forall^{\mathrm{N}} m\,(m = N \supset A) \supset (m = N \supset A)\ (\forall^{\mathrm{N}}\text{-el}) \qquad m = N \supset A \supset A\binom{N}{m}\ \text{(eq.form.)}}{m = N \supset \forall^{\mathrm{N}} m\,(m = N \supset A) \supset A\binom{N}{m}}\,\text{pc}}{\forall^{\mathrm{N}} m\,(m = N \supset A) \supset A\binom{N}{m}}\,\text{spec, since } m \notin N, A\binom{N}{m}$$

∎

THEOREM 23. PA induction rule with substitution: $\dfrac{A\binom{0}{m} \qquad A\binom{n}{m} \supset A\binom{Sn}{m}}{A\binom{n}{m}}$

where $n \notin A$ and $n \overset{m}{\hookrightarrow} A$.

Proof. If m is n then $m \notin A$, so the rule is $\dfrac{A \quad A \supset A}{A}$, which is trivial. If m is not n then the PA derivation is as follows. (Note that $0 \overset{m}{\hookrightarrow} A$ and $Sn \overset{m}{\hookrightarrow} A$ by Theorem 14.)

$$\dfrac{\dfrac{\dfrac{A\binom{0}{m}}{\forall^{\mathrm{N}} m\,(m = 0 \supset A)}\,\text{ext, pc} \qquad \dfrac{A\binom{n}{m} \supset A\binom{Sn}{m}}{\forall^{\mathrm{N}} m\,(m = n \supset A) \supset \forall^{\mathrm{N}} m\,(m = Sn \supset A)}\,\text{ext, pc}}{\forall^{\mathrm{N}} m\,(m = n \supset A)}\,\text{ind}}{A\binom{n}{m}}\,\text{ext, pc}$$

∎

CHAPTER 34

PEANO ARITHMETIC

PA FORMULAE

The formulae, metaformulae and free variables of Peano Arithmetic are the same as those of Heyting Arithmetic, except that metaformulae may contain the substitution metanotation introduced below. The letters 'A', 'B' and 'C' will from now on be used as metavariables denoting PA formulae; 'F', 'G' and 'H', possibly with subscripts, denote primitive recursive functions; all other metavariables starting with a capital letter denote numeric terms.

A mapping from PA to HA is defined: if A is a PA formula then A^{H} is the corresponding PA formula. For atomic A, A^{H} is A.

AXIOMS AND RULES OF INFERENCE OF PEANO ARITHMETIC

$\wedge$-axioms: $A \supset B \supset (A \wedge B) \quad (A \wedge B) \supset A \quad (A \wedge B) \supset B$
$\vee$-axioms: $A \supset (A \vee B) \quad B \supset (A \vee B) \quad (A \vee B) \supset (A \supset C) \supset (B \supset C) \supset C$
$\supset$-axioms: $A \supset B \supset A \quad (A \supset B \supset C) \supset (A \supset B) \supset A \supset C$
Excluded middle: $A \vee \neg A$
Truth and falsity (*true*-in, *false*-el): $\mathit{true} \quad \mathit{false} \supset A$
Quantifier axioms ($\exists^{\mathrm{N}}$-in, $\forall^{\mathrm{N}}$-el): $A \supset \exists^{\mathrm{N}} n\, A \quad \forall^{\mathrm{N}} n\, A \supset A$
Term existence (te): $\exists^{\mathrm{N}} n\, n = N$ where $n \notin N$
Equality (eq): $N = N \quad M = N \supset N = M \quad U = V \supset V = W \supset U = W$
$M = N \supset F(X_1, \ldots X_{i-1}, M, X_{i+1}, \ldots X_k) = F(X_1, \ldots X_{i-1}, N, X_{i+1}, \ldots X_k)$
Peano axioms (P): $\neg Sn = 0 \quad Sm = Sn \supset m = n$
Axioms for primitive recursive functions (pr):
$\mathit{proj}_i^k(\underline{n}) = n_i$ where $1 \leq i \leq k$
$\mathit{comp}_l(H, (F_1, \ldots F_l))(\underline{n}) = H(F_1(\underline{n}), \ldots F_l(\underline{n}))$
$\mathit{rec}_k(F, G)(0, \underline{n}) = F(\underline{n}) \quad \mathit{rec}_k(F, G)(Sm, \underline{n}) = G(m, \underline{n}, \mathit{rec}_k(F, G)(m, \underline{n}))$

where $\underline{n}$ is $n_1, \ldots n_k$, H is l-ary primitive recursive, $F_1, \ldots F_l$ and F are k-ary primitive recursive, and G is $(k+2)$-ary primitive recursive

Modus ponens: $\dfrac{A \qquad A \supset B}{B}$

Quantifier rules ($\exists^{\mathrm{N}}$-el, $\forall^{\mathrm{N}}$-in): $\dfrac{A \supset B}{\exists^{\mathrm{N}} n\, A \supset B} \qquad \dfrac{B \supset A}{B \supset \forall^{\mathrm{N}} n\, A}$ where $n \notin B$

Induction (ind): $\dfrac{\forall^{\mathrm{N}} x\,(x=0\supset A) \quad \forall^{\mathrm{N}} x\,(x=n\supset A)\supset \forall^{\mathrm{N}} x\,(x=Sn\supset A)}{\forall^{\mathrm{N}} x\,(x=n\supset A)}$

where $n \notin x, A$

FORMAL INTERPRETATION OF PA

A coding of PA derivations as constructions is defined. A construction *ha* is defined.

THEOREM 12. If D is the code of a PA derivation (with no premises) of a formula A then $ha(D) \rhd^*$ the code of an HA derivation (with no premises) of A^{H}.

THEOREM 13. (PA interpretation theorem.) If D is the code of a PA derivation (with no premises) of a formula A with no free variables, then $pr(spr(cpf(lpt(ha(D))))) \vdash \ulcorner (A^{\mathrm{H}})^{\mathrm{LPT}} \urcorner$.

SUBSTITUTION IN PA FORMULAE

A notation $N \overset{n}{\hookrightarrow} A$, meaning that the numeric term N *is substitutable for* the variable n in the formula A, is defined as follows.

- $N \overset{n}{\hookrightarrow} A$ if A is an atomic formula;
- $N \overset{n}{\hookrightarrow} A \wedge B$ iff $N \overset{n}{\hookrightarrow} A \vee B$ iff $N \overset{n}{\hookrightarrow} A \supset B$ iff $N \overset{n}{\hookrightarrow} A$ and $N \overset{n}{\hookrightarrow} B$;
- $N \overset{n}{\hookrightarrow} \exists^{\mathrm{N}} m A$ iff $N \overset{n}{\hookrightarrow} \forall^{\mathrm{N}} m A$ iff $n \notin \exists^{\mathrm{N}} m A$ or ($m \notin N$ and $N \overset{n}{\hookrightarrow} A$).

If $N \overset{n}{\hookrightarrow} A$, then the formula $A\binom{N}{n}$ is defined by:

- $0\binom{N}{n}$ is 0, $\quad n\binom{N}{n}$ is N, $\quad m\binom{N}{n}$ is m if m isn't n;
- $F(N_1, \ldots N_k)\binom{N}{n}$ is $F(N_1\binom{N}{n}, \ldots N_k\binom{N}{n})$, for $k \geq 1$;
- $true\binom{N}{n}$ is $true$, $\quad false\binom{N}{n}$ is $false$, $\quad (N_1 = N_2)\binom{N}{n}$ is $N_1\binom{N}{n} = N_2\binom{N}{n}$;
- $(A \wedge B)\binom{N}{n}$ is $A\binom{N}{n} \wedge B\binom{N}{n}$, $\quad (A \vee B)\binom{N}{n}$ is $A\binom{N}{n} \vee B\binom{N}{n}$,
- $(A \supset B)\binom{N}{n}$ is $A\binom{N}{n} \supset B\binom{N}{n}$;
- $(\exists^{\mathrm{N}} m A)\binom{N}{n}$ is $\exists^{\mathrm{N}} m A$ and $(\forall^{\mathrm{N}} m A)\binom{N}{n}$ is $\forall^{\mathrm{N}} m A$, if $n \notin \exists^{\mathrm{N}} m A$;
- $(\exists^{\mathrm{N}} m A)\binom{N}{n}$ is $\exists^{\mathrm{N}} m\,(A\binom{N}{n})$ and $(\forall^{\mathrm{N}} m A)\binom{N}{n}$ is $\forall^{\mathrm{N}} m\,(A\binom{N}{n})$, if $n \in \exists^{\mathrm{N}} m A$ and $m \notin N$.

PA THEOREMS AND DERIVED RULES INVOLVING SUBSTITUTION

THEOREM 16. PA equality theorem for formulae (eq.form.):

$$M = N \supset A\binom{M}{x} \supset A\binom{N}{x},$$

where $M \overset{x}{\hookrightarrow} A$ and $N \overset{x}{\hookrightarrow} A$.

THEOREM 19. PA quantifier theorems with substitution:

$$A\binom{N}{n} \supset \exists^{\mathrm{N}} n A, \quad \forall^{\mathrm{N}} n A \supset A\binom{N}{n}, \quad \text{where } N \overset{n}{\hookrightarrow} A.$$

THEOREM 21. PA quantifier rules with substitution:

$$\frac{A\binom{n}{m} \supset B}{\exists^{\mathrm{N}} m A \supset B} \qquad \frac{B \supset A\binom{n}{m}}{B \supset \forall^{\mathrm{N}} m A} \qquad \text{where } n \overset{m}{\hookrightarrow} A \text{ and } n \notin \exists^{\mathrm{N}} m A, B.$$

THEOREM 23. PA induction rule with substitution: $$\frac{A\binom{0}{m} \qquad A\binom{n}{m} \supset A\binom{Sn}{m}}{A\binom{n}{m}}$$

where $n \notin A$ and $n \overset{m}{\hookrightarrow} A$.

CHAPTER 35

CONCLUSIONS ON ARITHMETIC

The interpretation of arithmetic is now complete. What does it accomplish?

- It demonstrates that any formula of Peano Arithmetic may be interpreted as a proof function, that is, as a demand for a construction of a certain sort. The PA formula A is translated into a Heyting Arithmetic formula A^{H}, which is in turn translated into a formula of the Logic of Partial Terms, $(A^{\mathrm{H}})^{\mathrm{LPT}}$, which is in turn interpreted as a proof function, $\ulcorner (A^{\mathrm{H}})^{\mathrm{LPT}} \urcorner$.
- It shows that any formal derivation in Peano Arithmetic of the formula A may be understood as a high-level description of an intuitionistic proof of a proof function. The (coded) Peano Arithmetic derivation D is compiled into a (coded) Heyting Arithmetic derivation $ha(D)$, which is in turn compiled into a (coded) derivation in Logic of Partial Terms, $lpt(ha(D))$, which is in turn compiled into a (coded) derivation in the Calculus of Proof Functions, $cpf(lpt(ha(D)))$, which is in turn compiled into a proof $spr(cpf(lpt(ha(D))))$ (in the sense of $\Vdash$), which is finally compiled into a proof $pr(spr(cpf(lpt(ha(D)))))$ (in the sense of $\vdash$) of the proof function $\ulcorner (A^{\mathrm{H}})^{\mathrm{LPT}} \urcorner$.
- Consequently, formulae of Peano Arithmetic are meaningful, not as propositions but as demands for constructions. Formal derivations within Peano Arithmetic are meaningful, not as arguments establishing truths but as systematic ways of building constructions.
- Assuming the Central Dogma of Formalism (see Chapter 9), that all mathematics can be captured in formal systems (and in particular that ordinary arithmetic is captured in Peano Arithmetic), this shows that our grasp of arithmetic can be understood entirely in terms of constructions and protologic, and hence ultimately in terms of our abstract understanding of algorithms and well-foundedness.
- A grasp of 'the infinite set of natural numbers', or even 'all the natural numbers at once', is therefore *not* essential to our understanding and use of numbers. This completes the proof (Chapter 3) that these phrases are meaningless. Thus, as explained in Chapter 3, not only does the intuitionistic account of arithmetic succeed but also the platonist account fails.

- The interpretation of arithmetic may also be regarded as an implementation of Hilbert's programme, if one is prepared to accept protologic as finitary reasoning (as advocated in Chapter 9). For, let T be a term with no free variables, considered as a PA atomic formula. Then T^{H} is T (see Peano Arithmetic, Chapter 34), and hence $(T^{\mathrm{H}})^{\mathrm{LPT}}$ is T (see Heyting Arithmetic, Chapter 32). Thus, by the conservativeness theorem of $\vdash$ over reduction (see Logic, Chapter 26),

$$X \vdash \ulcorner (T^{\mathrm{H}})^{\mathrm{LPT}} \urcorner \quad \text{iff} \quad X \vdash \ulcorner T \urcorner \quad \text{iff} \quad T \rhd^{*} \mathit{true}$$

for any construction X. That is, $\ulcorner (T^{\mathrm{H}})^{\mathrm{LPT}} \urcorner$ has an intuitionistic proof iff $T \rhd^{*}$ *true*. In particular, $\ulcorner (\mathit{false}^{\mathrm{H}})^{\mathrm{LPT}} \urcorner$ has no intuitionistic proof, and hence (by the interpretation theorem for PA) *false* is not a theorem of PA. Thus PA is consistent.

- Someone who didn't accept the soundness theorem for protologic (see the Coding of Trees, Chapter 22) would not accept the conclusion that PA is consistent. They would accept that PA derivations can be transformed into proofs of logical sequents but would not accept the soundness of the mapping *pr* from proofs of logical sequents to proofs of proof functions (see the Calculus of Proof Functions, Chapter 28). Their conclusion would be that if D is the code of a PA derivation of *false* then $spr(cpf(lpt(ha(D)))) \not\models\; \Rightarrow \ulcorner \mathit{false} \urcorner$, which implies that there is a protological derivation with well-founded reflection tree for the sequent $\rightarrow \mathit{false}$. In short, PA is consistent relative to protologic.

- Let us return briefly to the question of logicism. Opponents of logicism claim that PA consists of 'logic' (by which they mean the axioms and rules for $\wedge$, $\vee$, $\supset$, $\exists$ and $\forall$, and perhaps equality) plus special axioms and rules about the 'mathematical' subject matter (the Peano axioms, the axioms for the primitive recursive functions, the term existence axiom, and the induction rule). Tracing this distinction back through HA, LPT and CPF to Expanded Protologic and Protologic, they would seem committed to the position that there is something special about the Fxpt Rules and primitive recursive function definitions that sets them apart from the rest of Protologic and the Term Language as 'mathematical'. Since Protologic simply consists of structural axioms and rules plus elimination rules for various features of the Term Language, it is hard to see why the Fxpt Rules should be considered mathematical while the If Rule and Decomposition Rules are not. Equally, it is hard to see anything especially 'mathematical' about primitive recursive definitions compared with the other metanotation in the Expanded Term Language. Alternatively the anti-logicist might take the view that the whole Theory of Constructions is mathematical, in which case the conclusion is that mathematics is prior to logic, as Brouwer said. In any case, the distinction between the 'logical' and 'mathematical' aspects of PA is superficial: it evaporates as one delves into the protological underpinning of PA.

PART IV: THE INTERPRETATION OF ANALYSIS

CHAPTER 36

INTRODUCTION TO PART IV

By *analysis* I shall mean a system of first-order predicate calculus with variables for real numbers and the usual axioms for a complete ordered field. The completeness property is expressed in an axiom schema

$$(\exists x A \wedge \exists y \forall x (A \supset x \leq y)) \supset \exists z \forall y (z \leq y \Leftrightarrow \forall x (A \supset x \leq y))$$

for any formula A in which y and z do not occur free. Note that variables for functions or sets of reals are not included. I shall not make use of the continuity principles that are common in intuitionistic analysis, since, as explained in Chapter 11, their correct formulation is not yet clear.

Analysis may be interpreted in second-order number theory, if we regard a real number as a Dedekind cut in the rationals, a rational as a pair of integers, and an integer as a pair of natural numbers. Thus the problem of finding an interpretation of classical analysis reduces to the problem of interpreting second-order Peano Arithmetic (2PA). 2PA is just like first-order Peano Arithmetic except that in addition to the *first-order* variables (ranging over the natural numbers) it has *second-order* variables (ranging over predicates or sets of natural numbers), with second-order quantifiers ($\exists^2$ and $\forall^2$) to bind them, and corresponding axioms and rules. The atomic formula $N \models \alpha$ means that the numeric term N satisfies the predicate α. The Comprehension Axiom Schema states that any formula A defines a predicate α:

$$\exists^2 \alpha \forall^{\mathrm{N}} n (n \models \alpha \Leftrightarrow A)$$

where A may contain n as a free variable but not α.

Recall that the interpretation of arithmetic (Part III) started from the Coding of Trees (the final theory of Part II) and proceeded through a sequence of theories, Logic, Calculus of Proof Functions, Logic of Partial Terms, Heyting Arithmetic and finally to Peano Arithmetic. In Part IV I shall extend each of the aforementioned theories to 'second-order' versions, as follows.

- The Coding of Trees (CT, Chapter 22) will be extended to the Second-Order Coding of Trees (2CT, Chapter 38) by the addition of type variables representing arbitrary tree codings.

- Logic (L, Chapter 26) will be extended to Second-Order Logic (2L, Chapter 40) by the addition of type predicates and some new logical constants, *apply*, *val'*, *predify'* and $\forall'$.
- Calculus of Proof Functions (CPF, Chapter 28) will be extended to Second-Order Calculus of Proof Functions (2CPF, Chapter 42) by the addition of introduction and elimination rules for *apply* and $\forall'$, and instantiation rules.
- Logic of Partial Terms (LPT, Chapter 30) will be extended to Second-Order Logic of Partial Terms (2LPT, Chapter 44) by the addition of a new atomic formula $x \models \alpha$ and a second-order universal quantifier $\forall^2\alpha$, with appropriate axioms and rules. Note that no second-order existential quantifier can be defined at this stage.
- Heyting Arithmetic (HA, Chapter 32) will be extended to Second-Order Heyting Arithmetic (2HA, Chapter 46) by the addition of a new atomic formula $N \models \alpha$ (where N is a numeric term) and a second-order universal quantifier.
- Peano Arithmetic (PA, Chapter 34) will be extended to Second-Order Peano Arithmetic (2PA, Chapter 48) by the addition of the atomic formula $N \models \alpha$ and the second-order universal and existential quantifiers.

In addition, the intermediate chapters (Chapters 21, 25, 27, 29, 31 and 33) will be extended accordingly. For example, Chapter 31, which interprets HA in LPT, will be extended to an interpretation of 2HA in 2LPT (Chapter 45). The theories and intermediate chapters are shown in Figure 4.

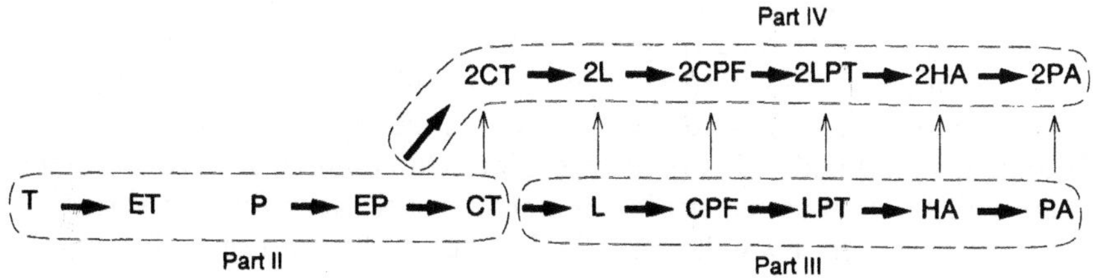

Figure 4: the theories in Parts II, III and IV; a thick arrow represents an intermediate chapter, where one theory is interpreted in another (the arrow points from the interpreting theory to the interpreted theory), and a thin arrow shows where one theory is an extension of another (the arrow points towards the extended theory).

The final result will be an interpretation theorem for 2PA similar to the interpretation theorem for PA. This gives an intuitionistic meaning to classical analysis.

Note the absence of a second-order existential quantifier in all theories except 2PA; even in 2PA it is only present in a formal sense, and is interpreted as $\neg\forall^2\neg$ in 2HA, by analogy with the interpretation of the first-order existential quantifier of PA in HA. This is because it seems to be impossible to define a constructive notion of existence for predicates of numbers; that is, one cannot specify convincingly what it means to *construct* a predicate of numbers in general (see Chapter 11). This is no obstacle to the interpretation of analysis, however.

THE FOUR SORTS OF VARIABLES

In Parts II and III all the variables were considered to be of the same sort. In Part IV these variables are divided into several sorts: *object* variables, *type* variables, and *second-order* variables. Object variables are further subdivided into *first-order* variables and *function* variables. Each of these sorts of variable is given as an infinite sequence, arranged in a standard order.

For the sake of definiteness, let us say that a first-order variable is a variable beginning with '$\mathcal{A}$', a function variable is a variable beginning with '$\mathcal{B}$', a type variable is a variable beginning with '$\mathcal{C}$', and a second-order variable is a variable beginning with '$\mathcal{D}$'–'$\mathcal{Z}$'. Let us say that the standard order for each sort of variable is the restriction to those variables of the standard order on all the variables.

An *object term* is a term all of whose free variables are object variables.

All the metavariables that denote variables used in Parts II and III began with lower-case roman letters and ranged over all variables. When these theories are incorporated in second-order theories the metavariables in the Term Language, the Expanded Term Language, Protologic and Expanded Protologic will still range over all variables, but the metavariables in the Coding of Trees, Logic and Calculus of Proof Functions will be reinterpreted as ranging only over object variables, and the metavariables in Logic of Partial Terms, Heyting Arithmetic and Peano Arithmetic will be reinterpreted as ranging only over first-order variables. All the arguments of Parts II and III remain sound under this reinterpretation. In addition, lower-case greek letters will be used as metavariables ranging over other sorts of variables. The conventions on this will be explicitly stated at the start of each chapter.

CHAPTER 37

FROM EXPANDED PROTOLOGIC TO THE SECOND-ORDER CODING OF TREES

Recall that the Theory of Constructions (Part II) ended with the Coding of Trees (CT). The first step in the interpretation of analysis is to extend CT to a second-order version (2CT), incorporating variables for tree codings; this is the purpose of the present chapter, which is to be viewed as extending the arguments in Chapter 21 ('From Expanded Protologic to the Coding of Trees').

LEXICAL CONVENTIONS

Two sorts of variables will be used in this chapter: object variables and type variables. The metavariables that denote variables will be interpreted as follows.

- Those beginning with lower-case roman letters range over object variables.
- The letters 'π', 'σ', 'ϕ', 'χ', 'ψ' and 'ω' range over type variables.

These conventions are also imposed on all the definitions and theorems inherited from Chapter 21. All the metavariables denoting variables in Chapter 21 began with lower-case roman letters, and consequently they will now be interpreted as ranging over object variables (this of course does not affect the soundness of the proofs).

TYPE VARIABLES AS TREE CODINGS

A type variable is interpreted as standing for a 'socket', in the sense of Chapter 11. That is, it represents a connection to a black box that, given an input construction, returns an output construction obtained in an unknown but deterministic way from the input construction. Chapter 11 describes how a socket may be interpreted as a tree coding. Consequently, a type variable represents a tree coding, and hence may be used as a type symbol and as a component in composite type symbols, such as $map(\pi, product(\sigma, \phi))$.

In Chapter 21, all type symbols were constructions; now, type variables are allowed to occur free in type symbols. A type variable π may occur in a type symbol Σ in any position where it would be meaningful to insert *any* type symbol; this is expressed in the following theorem.

THEOREM 1. If Π and Σ are type symbols then so is $\Sigma\begin{bmatrix}\Pi\\ \pi\end{bmatrix}$.

If Π contains free type variables $\pi, \ldots \sigma$ then a statement such as 'T is a well-founded tree of type Π' is meaningless in isolation, since we have not specified what black boxes are plugged into the sockets $\pi, \ldots \sigma$. However, as explained in Chapter 11, it is possible to make general statements of the form 'for any input-output behaviour of $\pi, \ldots \sigma$, A', where A is a well-foundedness statement or argument involving free type variables $\pi, \ldots \sigma$. This will be represented by the notation ${}_{\pi,\ldots\sigma}\{A\}$; the variables $\pi, \ldots \sigma$ are to be regarded as bound in ${}_{\pi,\ldots\sigma}\{A\}$. As an additional piece of notation, if $\phi, \ldots \psi$ are *all* the free type variables of A then ${}_{\phi,\ldots\psi}\{A\}$ is abbreviated to $\{A\}$.

TYPE MAPPINGS AND *anytype*

DEFINITION. A *type mapping* is an irreducible term M such that $M\pi$ reduces to a type symbol for any type variable π.

DEFINITION. Let *anytype* be a fresh 1-ary constructor. For any type mapping M, choose $\pi \notin M$ and let $M\pi \;\triangleright^*\; \Sigma \not\triangleright$. Now let *anytype*($M$) represent the following coding of trees. To interpret a construction X as a tree of type *anytype*(M), connect a new black box to the socket corresponding to π and then interpret X as a tree of type Σ.

This definition uses the indeterministic 'connect a new black box' operation introduced in Chapter 11; hence whether a construction is a tree of type *anytype*(M), and if so whether it is well-founded, will depend on the black box chosen. The interpretation of X does not, however, depend on the choice of the variable π, since one socket is as good as another.

Note the difference between *anytype*(M) and $M\pi$; *anytype*(M) is *indeterministic* (that is, if we interpret X as a tree of this type repeatedly we will get differing results), whereas $M\pi$ is *indeterminate* (that is, we cannot interpret X as a tree of this type until someone tells us what π represents).

EXAMPLE. *map*(*anytype*($\lambda\pi$.*product*(σ, π)), σ) is a type symbol, which is indeterminate until a black box is connected to the socket represented by σ.

THE WELL-FOUNDEDNESS RELATION

Due to the indeterminism introduced by *anytype*(M) the basic well-foundedness statement $T : \Pi$, defined in Chapter 21, needs to be amended as follows (the new features are in italics).

DEFINITION. The relation $T : \Pi$ is defined as follows.

- Let T be a construction and Π be a type symbol; then $T : \Pi$ means that T is a well-founded tree, according to the coding Π, *for any input-output behaviour of the black boxes connected in the course of interpreting T and its subtrees.*
- Let T be an *object* term and Π be a term that reduces to a type symbol; then $T : \Pi$ is obtained from the previous clause as in Chapter 21. That is, let $\underline{x}$ be the free variables of T and let $\Pi \rhd^* \Pi' \not\rhd$; then $T : \Pi$ means that, for any constructions $\underline{X}$, if $T\left[\frac{\underline{X}}{\underline{x}}\right] \rhd^* T' \not\rhd$ then T' is a well-founded tree of type Π', for any behaviour of the black boxes chosen in interpreting T' and its subtrees.

(Observe that the relation $T : \Pi$ is only defined if T is an object term and if black boxes have already been chosen for all the free type variables in Π.) All the theorems in CT concerning the $T : \Pi$ relation remain true under this extended definition, with type variables allowed in the type symbols and the terms T restricted to object terms, provided we enclose each entire argument in $\{\!\!\{ \ldots \}\!\!\}$.

If a type variable π occurs free in a type symbol Σ then the meaning of Σ as a tree coding depends on the meaning of π as a tree coding. However, this dependence will always satisfy the property that the well-foundedness condition $X : \Sigma$ (for an arbitrary construction X) is a function of the well-foundedness condition $Y : \pi$ (for arbitrary Y). Hence we may replace π in Σ by any other type symbol having the same well-foundedness condition without disturbing the meaning of the well-foundedness condition for Σ. This property is a consequence of the way that type symbols are built up from their components.

EXAMPLE. I shall show that $(\lambda x.(x,x))$: *anytype*($\lambda\pi$.*map*(π, *product*(π,π))). To interpret $(\lambda x.(x,x))$ as a tree we must begin by connecting a black box to the socket π. (We will get different trees depending on the black box chosen.) Now, the subtrees of $(\lambda x.(x,x))$ are of the form (X,X), where $X : \pi$, and they are interpreted using the *product*(π,π) coding; hence they each have two subtrees, X and X, both interpreted according to the π coding. Thus we see that $(\lambda x.(x,x))$ is interpreted as a well-founded tree.

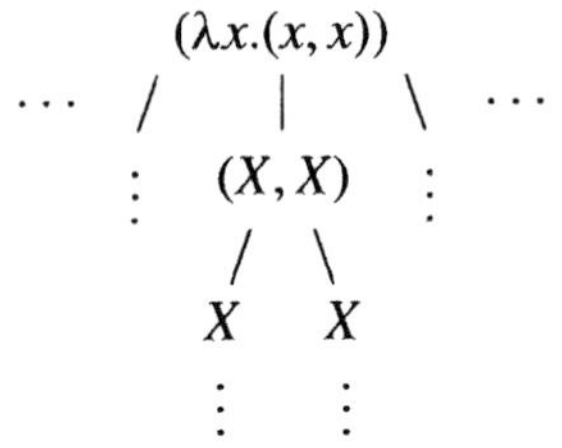

EXAMPLE. I shall show that

$$\{\!\{\, id \;:\; map(anytype(\lambda\pi.map(\pi, product(\pi,\pi))), map(\sigma, product(\sigma,\sigma)))\,\}\!\}.$$

There is a single free variable, σ, here, so we must choose a black box to connect to the socket σ. Now, whatever the behaviour of this black box, we can interpret *id* as a tree according to the *map* definition. One subtree of *id* is $(\lambda x.(x,x))$, since, as shown in the previous example, $(\lambda x.(x,x)) : anytype(\lambda\pi.map(\pi, product(\pi,\pi)))$, and $id(\lambda x.(x,x)) \rhd (\lambda x.(x,x)) \not\rhd$. The subtree $(\lambda x.(x,x))$ is interpreted as a tree by the coding $map(\sigma, product(\sigma,\sigma))$ and is found to be well-founded. A similar argument applies to the other subtrees of *id*. Hence the whole tree is well-founded.

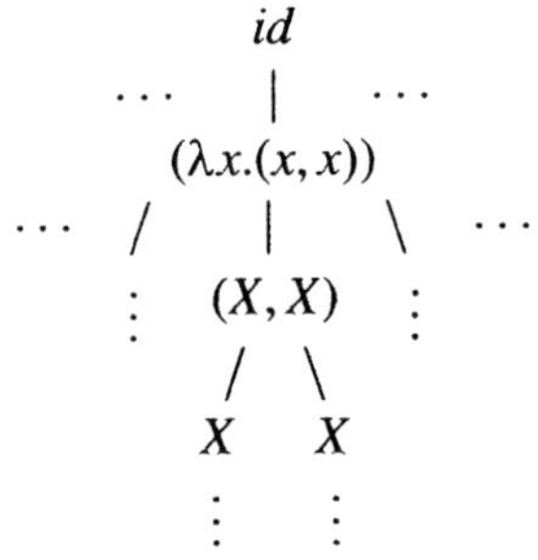

THEOREM 2. (Well-foundedness type instantiation rule.) If Π and Σ are type symbols and T is an object term then $\{\!\{\, \text{if } {}_\pi\{\!\{\, T : \Sigma \,\}\!\} \text{ then } T : \Sigma\left[\begin{smallmatrix}\Pi\\ \pi\end{smallmatrix}\right] \,\}\!\}$.

Proof. Assume $\pi \in \Sigma$, since otherwise the conclusion follows immediately. Let $\pi, \phi, \ldots \psi$ be the free variables of Σ (which must all be type variables, since Σ is a type symbol), and let $\chi, \ldots \omega$ be the free variables of Π (which are also all type variables). Now, regardless of the input-output behaviour of the sockets $\phi, \ldots \psi, \chi, \ldots \omega$, the following holds. Suppose ${}_\pi\{\!\{\, T : \Sigma \,\}\!\}$, let $\underline{x}$ be the free variables of T (which must all be object variables) and let $\underline{X}$ be any constructions. If $T\left[\frac{X}{\underline{x}}\right]$ reduces to a construction T' then, by hypothesis, $T' : \Sigma$ for any input-output behaviour of the socket π.

We wish to infer from this that $T' : \Sigma\left[\frac{\Pi}{\pi}\right]$. The simplest way to do this would be to imagine that connected to the socket π is a black box that behaves, as a tree coding, like Π. Unfortunately this may not be possible, since Π may be indeterministic while black boxes are always deterministic. We can, however, consider a black box that, when connected the socket π, gives $X : \pi$ iff $X : \Pi$ for any construction X. This black box would work as follows: it would accept the branch $[X]$, for any construction X, and would accept branches of the form $[X, X, \ldots X]$, with more than one X, iff $X : \Pi$ does not hold. Then X would represent a well-founded tree iff $X : \Pi$ holds, as required. It follows that replacing π by Π in Σ would not alter the well-foundedness condition of Σ; this gives $T' : \Sigma\left[\frac{\Pi}{\pi}\right]$, as required. ∎

THEOREM 3. (*anytype* rule.) Let M be a type mapping and T be an object term.

- If $\pi \notin M$ then $\{\!\![T : \mathit{anytype}(M)$ iff ${}_\pi\{\!\![T : M\pi]\!\!\}]\!\!\}$.
- If Π is a type symbol then $\{\!\![$ if $T : \mathit{anytype}(M)$ then $T : M\Pi]\!\!\}$.

Proof. For the first part, let $M\pi \rhd^* \Sigma \not\rhd$, let $\phi, \ldots \psi$ be the free variables of M, and let $\underline{x}$ be the free variables of T. Then, regardless of the input-output behaviour of the sockets $\phi, \ldots \psi$, the following two paragraphs hold.

Suppose $T : \mathit{anytype}(M)$. Now, regardless of the input-output behaviour of the socket π, the following holds. Let $\underline{X}$ be any constructions and suppose $T\left[\frac{X}{\underline{x}}\right]$ reduces to a construction T'. This means that $T' : \mathit{anytype}(M)$. Then by the definition of $\mathit{anytype}(M)$, $T' : \Sigma$, which verifies that $T : M\pi$, as required.

Conversely, suppose ${}_\pi\{\!\![T : M\pi]\!\!\}$, let $\underline{X}$ be any constructions, and suppose $T\left[\frac{X}{\underline{x}}\right]$ reduces to a construction T'. This means that, regardless of the input-output behaviour of π, $T' : \Sigma$. By the definition of $\mathit{anytype}(M)$ this shows that $T' : \mathit{anytype}(M)$. This establishes that $T : \mathit{anytype}(M)$, and hence completes the proof of the first part of the theorem.

For the second part, let π be a fresh type variable and let $M\pi \rhd^* \Sigma \not\rhd$. By the first part of this theorem and Theorem 2,

$$\{\!\![\text{if } T : \mathit{anytype}(M) \text{ then } {}_\pi\{\!\![T : M\pi \rhd^* \Sigma]\!\!\} \text{ and so } T : \Sigma\left[\tfrac{\Pi}{\pi}\right] \lhd^* M\Pi]\!\!\}.$$

∎

CHAPTER 38

THE SECOND-ORDER CODING OF TREES

This theory consists of the Coding of Trees (Chapter 22), with the metavariables that denote variables restricted to range only over object variables, plus the following definitions and theorems. In this chapter, metavariables that denote variables are interpreted as follows.

- Those beginning with lower-case roman letters range over object variables.
- The letters 'π', 'σ', 'ϕ', 'χ', 'ψ' and 'ω' range over type variables.

Any type variable is a type symbol, interpreted as a socket connected to a black box. If A is a well-foundedness statement or argument then ${}_{\pi,\ldots\sigma}\{\!\!\{ A \}\!\!\}$ means 'for any input-output behaviour of the sockets $\pi, \ldots \sigma$, A'; and if $\phi, \ldots \psi$ are all the free type variables of A then $\{\!\!\{ A \}\!\!\}$ means ${}_{\phi,\ldots\psi}\{\!\!\{ A \}\!\!\}$.

THEOREM 1. If Π and Σ are type symbols then so is $\Sigma\left[\begin{smallmatrix}\Pi\\ \pi\end{smallmatrix}\right]$.

TYPE MAPPINGS AND *anytype*

DEFINITION. A *type mapping* is an irreducible term M such that $M\pi$ reduces to a type symbol for any type variable π.

A fresh 1-ary constructor, *anytype*, is introduced, and a new type symbol *anytype*(M) is defined, for any type mapping M.

THE WELL-FOUNDEDNESS RELATION

The well-foundedness relation $T : \Pi$ is amended; it is only defined when T is an object term and when black boxes have been chosen for all the free type variables of Π. The well-foundedness theorems of Chapter 22 remain sound, when enclosed in $\{\!\!\{ \ldots \}\!\!\}$.

THEOREM 2. (Well-foundedness type instantiation rule.) If Π and Σ are type symbols and T is an object term then $\{\!\!\{$ if ${}_{\pi}\{\!\!\{ T : \Sigma \}\!\!\}$ then $T : \Sigma\left[\begin{smallmatrix}\Pi\\ \pi\end{smallmatrix}\right] \}\!\!\}$.

THEOREM 3. (*anytype* rule.) Let M be a type mapping and T be an object term.

- If $\pi \notin M$ then $\{\!\!\{ T : \mathit{anytype}(M)$ iff $_\pi\{\!\!\{ T : M\pi \}\!\!\} \}\!\!\}$.
- If Π is a type symbol then $\{\!\!\{$ if $T : \mathit{anytype}(M)$ then $T : M\Pi \}\!\!\}$.

CHAPTER 39

FROM THE SECOND-ORDER CODING OF TREES TO SECOND-ORDER LOGIC

Second-order logic (2L) is obtained from first-order logic (L) by adding type predicates and the logical constants *apply*, *val*′, *predify*′ and $\forall'$. In this chapter I shall extend the arguments of Chapter 25 ('From the Coding of Trees to Logic') to include these new logical constants and their properties. Some notions defined in Chapter 25 need to be modified slightly to take account of the difference between object and type variables.

VARIABLE RESTRICTIONS

The variable conventions are as in the previous chapter; that is, metavariables that denote variables are interpreted as follows.

- Those beginning with lower-case roman letters range over object variables.
- The letters 'π', 'σ', 'ϕ', 'χ', 'ψ' and 'ω' range over type variables.

These conventions are also imposed on all the definitions and theorems inherited from Chapter 25 (hence all the metavariables denoting variables there now range over object variables).

Recall that a proof function was defined in Chapter 25 as a term I such that $I \triangleright^* \mathit{pfn}(A,\Pi) \not\triangleright$ where Π is a type symbol; and a predicate was defined as a term P such that $P \triangleright^* \mathit{pred}(F,\Pi) \not\triangleright$ where Fx is evaluable for any variable x and Π is a type symbol. I now impose the additional requirement that A and F be object terms (that is, all their free variables must be object variables). Similarly, in the proof function $\ulcorner T \urcorner$, I require that T be an object term; and Theorem 1 of Chapter 25 is replaced by the following theorem.

THEOREM 1. Let Π be a type symbol and X be an irreducible object term.

- If I is a proof function then so are $I\begin{bmatrix}\Pi\\ \pi\end{bmatrix}$ and $I\begin{bmatrix}X\\ x\end{bmatrix}$.
- If P is a predicate then so are $P\begin{bmatrix}\Pi\\ \pi\end{bmatrix}$ and $P\begin{bmatrix}X\\ x\end{bmatrix}$.

Proof. For the first part, let $I \rhd^* pfn(A,\Sigma) \not\rhd$. Then $I\begin{bmatrix}\Pi\\ \pi\end{bmatrix} \rhd^* pfn(A,\Sigma\begin{bmatrix}\Pi\\ \pi\end{bmatrix}) \not\rhd$, which is a proof function since $\Sigma\begin{bmatrix}\Pi\\ \pi\end{bmatrix}$ is a type symbol. Also, $I\begin{bmatrix}X\\ x\end{bmatrix} \rhd^* pfn(A\begin{bmatrix}X\\ x\end{bmatrix},\Sigma) \not\rhd$, which is a proof function since $A\begin{bmatrix}X\\ x\end{bmatrix}$ is an object term.

For the second part, let $P \rhd^* pred(F,\Phi) \not\rhd$ and let u be a fresh variable; thus Fu is evaluable. Now, $P\begin{bmatrix}\Pi\\ \pi\end{bmatrix} \rhd^* pred(F,\Phi\begin{bmatrix}\Pi\\ \pi\end{bmatrix}) \not\rhd$, and $\Phi\begin{bmatrix}\Pi\\ \pi\end{bmatrix}$ is a type symbol; thus $P\begin{bmatrix}\Pi\\ \pi\end{bmatrix}$ is a predicate. Moreover, $P\begin{bmatrix}X\\ x\end{bmatrix} \rhd^* pred(F\begin{bmatrix}X\\ x\end{bmatrix},\Phi) \not\rhd$, where, for any variable v, $F\begin{bmatrix}X\\ x\end{bmatrix} v \overset{*}{\leftrightarrow} (Fu)\begin{bmatrix}X\\ x\end{bmatrix}\begin{bmatrix}v\\ u\end{bmatrix}$, which is evaluable since Fu is. This establishes that $P\begin{bmatrix}X\\ x\end{bmatrix}$ is a predicate. ∎

THE PROOF RELATION

In Chapter 25 the proof relation $\vdash$ was defined by

$$X \vdash I \qquad \text{iff} \qquad I \rhd^* pfn(A,\Pi) \not\rhd\,, \quad AX \rhd^* true \quad \text{and} \quad X : \Pi.$$

I now modify this by requiring that X be an object term and replacing $X : \Pi$ by $\{\!\!\{ X : \Pi \}\!\!\}$.

THE LOGICAL CONSTANT *apply*

It is my intention in Chapter 43 to expand the Logic of Partial Terms to include second-order variables $\alpha, \beta, \ldots$ ranging over predicates. A new atomic formula $x \models \alpha$ will be introduced, meaning 'the construction x satisfies the predicate α', and it will be necessary to interpret this atomic formula as a proof function. The obvious interpretation of $x \models \alpha$ is $val\,\alpha\,x$, but unfortunately this is not a proof function since it is not evaluable. No other way of interpreting $x \models \alpha$ is apparent, and this technical difficulty threatens to prevent us from developing a second-order logic. The problem can be overcome by the following awkward but workable procedure, which is best explained in two steps.

(1) Interpret $x \models \alpha$ as $val(pred(f,\pi))x$ rather than $val\,\alpha\,x$, where f and π are suitably chosen variables. This is an improvement since we can make some progress in reducing $val(pred(f,\pi))x$:

$$val(pred(f,\pi))x \rhd^* pfn(fx,\pi) \rhd \cdots.$$

Unfortunately the reduction gets stuck at this point, since fx reduces only to itself.

(2) Replace $val(pred(f,\pi))x$ by $apply(pred(f,\pi))x$, where *apply* is a new logical constant defined in such a way that $apply(pred(f,\pi))x$ is evaluable but *apply* is otherwise as similar as possible to *val*.

I shall define *apply* now and carry out the rest of the procedure in Chapter 43. In what follows let π be a type variable and f, x be two object variables.

DEFINITION. Define the construction *apply* as

$$(\lambda(pred(f,\pi)).(\lambda x.pfn(s(s(kf)(kx))id,\pi))).$$

THEOREM 2. For any predicate P, $apply\,P\,x$ is a proof function. Also, $apply(pred(f,\pi))x$ is a proof function.

Proof. Let $P \;\triangleright^*\; pred(F,\Pi) \not\triangleright$. Then

$$apply\,P\,x \;\triangleright^*\; apply(pred(F,\Pi))x \;\triangleright^*\; pfn(s(s(kF)(kx))id,\Pi) \not\triangleright ,$$

which is a proof function. Similarly,

$$apply(pred(f,\pi))x \;\triangleright^*\; pfn(s(s(kf)(kx))id,\pi) \not\triangleright ,$$

which is a proof function. ∎

TYPE PREDICATES AND ASSOCIATED LOGICAL CONSTANTS

Recall that a *predicate* was defined as a certain kind of term P that, when applied to a variable x, produces a proof function $val\,P\,x$ whose decidable part depends on x. Similarly I shall define a *type predicate* as a certain kind of term P' that, when applied to a type variable π, produces a proof function $val'\,P'\,\pi$ whose well-foundedness part depends on π. Type predicates are used in conjunction with a quantifier $\forall'$ that ranges over types.

In the following definitions and theorems, let π be a type variable and a, f, m, x be four object variables.

DEFINITION. Let $pred'$ be a fresh 1-ary constructor. A *type predicate* is a term P' such that $P' \;\triangleright^*\; pred'(A,M) \not\triangleright$, where A is an object term and M is a type mapping.

DEFINITION. Define the construction val' as $(\lambda pred'(a,m).(\lambda\pi.pfn(a,m\pi)))$.

THEOREM 3. If P' is a type predicate then $val'\,P'\,\pi$ is a proof function.

Proof. Let $P' \;\triangleright^*\; pred'(A,M) \not\triangleright$, where M is a type mapping. Then $M\pi \;\triangleright^*\; \Sigma \not\triangleright$ for some type symbol Σ, and $val'\,P'\,\pi \;\triangleright^*\; pfn(A,M\pi) \;\triangleright^*\; pfn(A,\Sigma) \not\triangleright$, which is a proof function, as required. ∎

DEFINITION. Define a construction *predify'* as

$$(\lambda f.pred'((\lambda pfn(a,\pi).a)(f\ nil), s(k(\lambda pfn(a,\pi).\pi))f)).$$

THEOREM 4. If I is a proof function and π is a type variable then $predify'(\lambda\pi.I)$ is a type predicate and $val'(predify'(\lambda\pi.I))\pi \ \triangleright^{*}\!\triangleleft\ I$.

Proof. Let $I \ \triangleright^{*}\ pfn(A,\Pi) \not\triangleright$. Then

$$\begin{aligned} predify'(\lambda\pi.I) \ &\triangleright^{*}\ pred'((\lambda pfn(a,\pi).a)((\lambda\pi.I)\ nil), s(k(\lambda pfn(a,\pi).\pi))(\lambda\pi.I)) \\ &\triangleright^{*}\ pred'(A, s(k(\lambda pfn(a,\pi).\pi))(\lambda\pi.I)) \not\triangleright \end{aligned}$$

Now $s(k(\lambda pfn(a,\pi).\pi))(\lambda\pi.I)$ is a type mapping since, for any type variable σ,

$$s(k(\lambda pfn(a,\pi).\pi))(\lambda\pi.I)\sigma \ \triangleright^{*}\ (\lambda pfn(a,\pi).\pi)(I\begin{bmatrix}\sigma\\ \pi\end{bmatrix}) \ \triangleright^{*}\ \Pi\begin{bmatrix}\sigma\\ \pi\end{bmatrix}$$

which is a type symbol. This establishes that $predify'(\lambda\pi.I)$ is a type predicate.

Moreover, $val'(predify'(\lambda\pi.I))\pi \ \triangleright^{*}\ pfn(A,\Pi) \ \triangleleft^{*}\ I$, as required. ∎

DEFINITION. Define the construction $\forall'$ as $(\lambda pred'(a,m).pfn(a, anytype(m)))$.

THEOREM 5. If P' is a type predicate then $\forall' P'$ is a proof function.

Proof. Let $P' \ \triangleright^{*}\ pred'(A,M) \not\triangleright$. Then $\forall' P' \ \triangleright^{*}\ pfn(A, anytype(M)) \not\triangleright$, which is a proof function since $anytype(M)$ is a type symbol. ∎

CHAPTER 40

SECOND-ORDER LOGIC

Second-order logic (2L) consists of first-order logic (L, Chapter 26), with the metavariables that denote variables restricted to range only over object variables, plus the following definitions and theorems. In this chapter, metavariables that denote variables are interpreted as follows.

- Those beginning with lower-case roman letters range over object variables.
- The letters 'π', 'σ', 'ϕ', 'χ', 'ψ' and 'ω' range over type variables.

The definitions of proof functions, predicates and $\ulcorner T \urcorner$, inherited from Chapter 26, are now restricted by requiring that the terms A, F and T, respectively, be object terms.

Theorem 1 of Chapter 26 is replaced by the following theorem.

THEOREM 1. Let Π be a type symbol and X be an irreducible object term.

- If I is a proof function then so are $I\begin{bmatrix}\Pi\\ \pi\end{bmatrix}$ and $I\begin{bmatrix}X\\ x\end{bmatrix}$.
- If P is a predicate then so are $P\begin{bmatrix}\Pi\\ \pi\end{bmatrix}$ and $P\begin{bmatrix}X\\ x\end{bmatrix}$.

THE PROOF RELATION

The proof relation $\vdash$ is defined by

$$X \vdash I \qquad \text{iff} \qquad I \rhd^* \mathit{pfn}(A, \Pi) \not\rhd, \; AX \rhd^* \mathit{true} \quad \text{and} \quad \{\!\!\{ X : \Pi \}\!\!\}$$

where X is an object term and I is a proof function.

THE LOGICAL CONSTANT *apply*

Let π be a type variable and f, x be two object variables.

DEFINITION. Define the construction *apply* as

$$(\lambda(\mathit{pred}(f,\pi)).(\lambda x.\mathit{pfn}(s(s(kf)(kx))id,\pi))).$$

THEOREM 2. For any predicate P, $\mathit{apply}\,P\,x$ is a proof function. Also, $\mathit{apply}(\mathit{pred}(f,\pi))x$ is a proof function.

TYPE PREDICATES AND ASSOCIATED LOGICAL CONSTANTS

Let π be a type variable and a, f, m, x be four object variables.

DEFINITION. Let *pred'* be a fresh 1-ary constructor. A *type predicate* is a term P' such that $P' \triangleright^* \mathit{pred}'(A, M) \not\triangleright$, where A is an object term and M is a type mapping.

DEFINITION. Define the construction *val'* as $(\lambda \mathit{pred}'(a,m).(\lambda\pi.\mathit{pfn}(a, m\pi)))$.

THEOREM 3. If P' is a type predicate then $\mathit{val}'\ P'\ \pi$ is a proof function.

DEFINITION. A construction *predify'* is defined.

THEOREM 4. If I is a proof function and π is a type variable then $\mathit{predify}'(\lambda\pi.I)$ is a type predicate and $\mathit{val}'(\mathit{predify}'(\lambda\pi.I))\pi \ \triangleright^*\triangleleft\ I.$

DEFINITION. Define the construction $\forall'$ as $(\lambda \mathit{pred}'(a,m).\mathit{pfn}(a, \mathit{anytype}(m)))$.

THEOREM 5. If P' is a type predicate then $\forall' P'$ is a proof function.

CHAPTER 41

FROM SECOND-ORDER LOGIC TO SECOND-ORDER CALCULUS OF PROOF FUNCTIONS

Second-Order Calculus of Proof Functions (2CPF) is obtained from Calculus of Proof Functions (CPF) by adding axioms and rules for the logical constants introduced in Second-Order Logic. In this chapter I shall extend the arguments of Chapter 27 ('From Logic to Calculus of Proof Functions') to incorporate these new axioms and rules.

LEXICAL CONVENTIONS

In addition to the conventions of Chapter 27, 'P'' will be used to denote a type predicate. As in the previous chapter, metavariables that denote variables will be interpreted as follows.

- Those beginning with lower-case roman letters range over object variables.
- The letters 'π', 'σ', 'ϕ', 'χ', 'ψ' and 'ω' range over type variables.

LOGICAL SEQUENTS AND $\Vdash$

The notion of a logical sequent is carried over unchanged from Chapter 27. Note that in the definition of the code of a sequent the variables $\underline{z}$ are *all* the free variables, of any sort. Recall that the proof relation $\Vdash$ for sequents was defined in Chapter 27 as follows.

DEFINITION. The relation $Q \Vdash I_1, \ldots I_k \Rightarrow J$, where $I_1, \ldots I_k \Rightarrow J$ is a logical sequent, Q is a term with no free variables, and the proof functions $I_1, \ldots I_k, J$ have free variables $\underline{x}$, means that

- $Q \rhd^* (D, (\lambda[\underline{x}].T)) \not\rhd$, for some D and T,
- $I_1 \rhd^* \mathit{pfn}(A_1, \Pi_1) \not\rhd, \ldots I_k \rhd^* \mathit{pfn}(A_k, \Pi_k) \not\rhd, J \rhd^* \mathit{pfn}(B, \Sigma) \not\rhd$,
- $\mathit{DT}(D, [\![(q_1, \ldots q_k, \underline{x})\, A_1 q_1, \ldots A_k q_k \to B(T[q_1, \ldots q_k])]\!]) \rhd^* \mathit{true}$,
- $D : \mathit{rt}$ and $T : \mathit{map}(\mathit{pi}[\Pi_1, \ldots \Pi_k], \Sigma)$,

where $q_1, \ldots q_k$ are k variables different from $\underline{x}$.

This definition now needs to be modified in two ways to take account of the presence of type variables: the variables $\underline{x}$ are now merely all the free *object* variables; and secondly T : $map(pi[\Pi_1, \ldots \Pi_k], \Sigma)$ is replaced by $\{\!\!\{ T : map(pi[\Pi_1, \ldots \Pi_k], \Sigma) \}\!\!\}$. (There would be no point in enclosing $D : rt$ in $\{\!\!\{ \ldots \}\!\!\}$ as rt has no free type variables.)

The theorems about $\Vert\!\!\models$ (Theorems 1–4) continue to hold.

SECOND-ORDER CALCULUS OF PROOF FUNCTIONS

The axiom schemata and rules of inference of 2CPF are those of CPF (with the metavariables that denote variables restricted to object variables and the metavariables that denote terms restricted to object terms) plus the following.

apply-introduction (*apply*-in): $val\,P\,x \Rightarrow apply\,P\,x$

apply-elimination (*apply*-el): $apply\,P\,x \Rightarrow val\,P\,x$

$\forall'$-elimination ($\forall'$-el): $\forall' P' \Rightarrow val'\,P'\,\pi$

$\forall'$-introduction rule ($\forall'$-in): $$\frac{I \Rightarrow val'\,P'\,\pi}{I \Rightarrow \forall' P'} \quad \text{where } \pi \notin I, P'$$

Instantiation rules (inst): $$\frac{I \Rightarrow J}{I\begin{bmatrix}F\\f\end{bmatrix} \Rightarrow J\begin{bmatrix}F\\f\end{bmatrix}} \qquad \frac{I \Rightarrow J}{I\begin{bmatrix}\Pi\\\pi\end{bmatrix} \Rightarrow J\begin{bmatrix}\Pi\\\pi\end{bmatrix}}$$

where F is an irreducible object term and Π is a type symbol.

Note that the *apply* axioms show that $apply\,P\,x$ is equivalent to $val\,P\,x$. This is in accordance with the purpose of introducing *apply* in Chapter 39: *apply* was to be a variation of *val* that allows second-order variables to be introduced into LPT formulae in such a way that those formulae can be interpreted as proof functions – see Chapter 43. Note also that the $\forall'$ axiom and rule differ from the $\forall$ theorem and derived rule in CPF in that they do not involve the logical constant $\bullet$.

INTERPRETATIONS OF THE 2CPF AXIOMS AND RULES

For each 2CPF axiom I shall produce a proof of it, and for each 2CPF rule of inference I shall show how to obtain a proof of the conclusion sequent from proofs of the premise sequents. The arguments for the axiom schemata and rules inherited from CPF carry over unchanged from Chapter 27, except that all variables referred to there are object variables, all terms are object terms, and all well-foundedness arguments must now be enclosed in $\{\!\!\{ \ldots \}\!\!\}$. Hence it is only necessary to consider the new 2CPF axiom schemata and rules.

THEOREM 1. (2CPF *apply*-introduction axiom.)
$Pr_{24}(\{val\,P\,x \Rightarrow apply\,P\,x\}) \Vdash val\,P\,x \Rightarrow apply\,P\,x,$
where the construction Pr_{24} is defined below.

Proof. Let $P \rhd^* pred(F,\Pi) \not\rhd$. Then

$$val\,P\,x \rhd^* pfn(Fx,\Pi) \rhd^* pfn(A,\Pi) \not\rhd$$
$$apply\,P\,x \rhd^* pfn(B,\Pi) \not\rhd$$

for some term A, where B is $s(s(kF)(kx))id$.

Now let $\underline{z}$ be the free object variables of P and x, let q be a fresh object variable, and let T be the term $(\lambda[q].q)$. Then $B(T[q]) \rhd^* Fxq \rhd^* Aq$, so let D_1 be the code of the protological derivation of the sequent $(q,\underline{z})\,Aq \to B(T[q])$ by Reduction. Then by the *join*-lemma

$$DT(join(D_1)nil, [\![(q,\underline{z})\,Aq \to B(T[q])]\!]) \rhd^* true$$

which implies that $join(D_1)nil \rhd^* D$ for some construction D. So define a recursive function Pr_{24} such that $Pr_{24}(\{val\,P\,x \Rightarrow apply\,P\,x\}) \rhd^* (D,(\lambda[\underline{z}].T))$, thus satisfying the decidable part of the proof relation.

To check the well-foundedness conditions,
$D \lhd^* join(D_1)nil : rt$, by the *join*-lemma, as required.
$\{$ For any construction $X : pi[\Pi]$,
X is $[Q]$ for some construction $Q : \Pi$
$TX \rhd^* Q : \Pi$.
$T : map(pi[\Pi],\Pi)$, as required. $\}$
∎

THEOREM 2. (2CPF *apply*-elimination axiom.)
$Pr_{25}(\{apply\,P\,x \Rightarrow val\,P\,x\}) \Vdash apply\,P\,x \Rightarrow val\,P\,x,$
where the construction Pr_{25} is defined below.

Proof. Similar to the previous theorem. ∎

THEOREM 3. (2CPF $\forall'$-elimination axiom.)
$Pr_{26}(\{\forall' P' \Rightarrow val'\,P'\,\pi\}) \Vdash \forall' P' \Rightarrow val'\,P'\,\pi,$
where the construction Pr_{26} is defined below.

Proof. Let $P' \rhd^* pred'(A,M) \not\rhd$. Then

$$\forall' P' \rhd^* pfn(A, anytype(M)) \not\rhd$$
$$val'\,P'\,\pi \rhd^* pfn(A,M\pi) \rhd^* pfn(A,\Sigma) \not\rhd$$

for some type symbol Σ. Let $\underline{x}$ be the free object variables of P', let q be a fresh object variable, let T be $(\lambda[q].q)$, and let D_1 be the code of the protological derivation of the sequent $(q,\underline{x})\ Aq \rightarrow A(T[q])$ by Reduction. Then by the *join*-lemma

$$DT(join(D_1)nil, [\![(q,\underline{x})\ Aq \rightarrow A(T[q])]\!]) \triangleright^* true$$

which implies that $join(D_1)nil \triangleright^* D$ for some construction D. So define a recursive function Pr_{26} such that $Pr_{26}(\{\forall' P' \Rightarrow val'\ P'\ \pi\}) \triangleright^* (D, (\lambda[\underline{x}].T))$, thus satisfying the decidable part of the proof relation.

To check the well-foundedness conditions,
$D \triangleleft^* join(D_1)nil : rt$, by the *join*-lemma, as required.
$\{\!|$ For any construction $X : pi[anytype(M)]$,
X is $[Q]$, for some construction $Q : anytype(M)$
$Q : M\pi \triangleright^* \Sigma$, by the *anytype* rule (second part)
$TX \triangleright^* Q : \Sigma$.
$T : map(pi[anytype(M)], \Sigma)$, as required. $|\!\}$
∎

THEOREM 4. (2CPF $\forall'$-introduction rule.) If $Q \models I \Rightarrow val'\ P'\ \pi$ then $Pr_{27}(Q, \{I \Rightarrow val'\ P'\ \pi\}, \{I \Rightarrow \forall' P'\}) \models I \Rightarrow \forall' P'$, where $\pi \notin I, P'$, and the construction Pr_{27} is defined below.

Proof. Let $I \triangleright^* pfn(B, \Phi) \not\triangleright$ and $P' \triangleright^* pred'(A, M) \not\triangleright$. Then

$$\forall' P' \triangleright^* pfn(A, anytype(M)) \not\triangleright$$
$$val'\ P'\ \pi \triangleright^* pfn(A, M\pi) \triangleright^* pfn(A, \Sigma) \not\triangleright$$

for some type symbol Σ. Let $\underline{x}$ be the free object variables of I, P', and let q, y, z be three fresh object variables. The hypothesis that $Q \models I \Rightarrow val'\ P'\ \pi$ means that $Q \triangleright^* (D, (\lambda[\underline{x}].T)) \not\triangleright$ where

- $DT(D, [\![(q,\underline{x})\ Bq \rightarrow A(T[q])]\!]) \triangleright^* true$,
- $D : rt$ and $\{\!| T : map(pi[\Phi], \Sigma) |\!\}$.

This shows that Q also satisfies the decidable part of the requirements for a proof of $I \Rightarrow \forall' P'$, so define Pr_{27} by

$$Pr_{27}(q, y, z) \triangleq q.$$

To check the well-foundedness conditions,
$D : rt$, by hypothesis, as required.
$\{\!|\ {}_\pi\{\!| T : map(pi[\Phi], \Sigma) |\!\}$, by hypothesis (separating the π quantifier from the others).
For any construction $X : pi[\Phi]$,
${}_\pi\{\!| TX : \Sigma \triangleleft^* M\pi |\!\}$,
$TX : anytype(M)$, by the *anytype* rule.
$T : map(pi[\Phi], anytype(M))$, as required. $|\!\}$
∎

THEOREM 5. (First 2CPF instantiation rule.)
If $Q \VDash I \Rightarrow J$ then $Pr_{28}(Q, \{I \Rightarrow J\}, \{I\left[\begin{matrix}F\\ f\end{matrix}\right] \Rightarrow J\left[\begin{matrix}F\\ f\end{matrix}\right]\}) \VDash I\left[\begin{matrix}F\\ f\end{matrix}\right] \Rightarrow J\left[\begin{matrix}F\\ f\end{matrix}\right]$, where F is an irreducible object term and the construction Pr_{28} is defined below.

Proof. Let $I \triangleright^{*} pfn(A,\Sigma) \not\triangleright$ and $J \triangleright^{*} pfn(B,\Phi) \not\triangleright$. Then

$$\begin{aligned} I\left[\begin{matrix}F\\ f\end{matrix}\right] &\triangleright^{*} pfn(A\left[\begin{matrix}F\\ f\end{matrix}\right],\Sigma) \not\triangleright \\ J\left[\begin{matrix}F\\ f\end{matrix}\right] &\triangleright^{*} pfn(B\left[\begin{matrix}F\\ f\end{matrix}\right],\Phi) \not\triangleright \ . \end{aligned}$$

Let $\underline{y}$ be the free object variables of I, J, let $\underline{z}$ be the free object variables of $I\left[\begin{matrix}F\\ f\end{matrix}\right], J\left[\begin{matrix}F\\ f\end{matrix}\right]$, and let q be a fresh object variable. The hypothesis that $Q \VDash I \Rightarrow J$ means that $Q \triangleright^{*} (D, (\lambda[\underline{y}].T)) \not\triangleright$ where

- $DT(D, [\![(q,\underline{y})\ Aq \to B(T[q])]\!]) \triangleright^{*} true,$
- $D : rt$ and $\{T : map(pi[\Sigma],\Phi)\}$.

Let D_1 be the code of the protological derivation

$$\dfrac{\dfrac{(q,\underline{y})\ Aq \to B(T[q])\ \ (\mathrm{p0})}{(q,f,\underline{z})\ Aq \to B(T[q])}\ \text{thin}}{(q,\underline{z})\ A\left[\begin{matrix}F\\ f\end{matrix}\right] q \to B\left[\begin{matrix}F\\ f\end{matrix}\right] (T\left[\begin{matrix}F\\ f\end{matrix}\right] [q])}\ \text{inst}$$

Then by the *join*-lemma

$$DT(join(D_1)[D], [\![(q,\underline{z})\ A\left[\begin{matrix}F\\ f\end{matrix}\right] q \to B\left[\begin{matrix}F\\ f\end{matrix}\right] (T\left[\begin{matrix}F\\ f\end{matrix}\right] [q])]\!]) \triangleright^{*} true$$

which implies that $join(D_1)D \triangleright^{*} D'$ for some construction D'. So define a recursive function Pr_{28} such that $Pr_{28}(Q, \{I \Rightarrow J\}, \{I\left[\begin{matrix}F\\ f\end{matrix}\right] \Rightarrow J\left[\begin{matrix}F\\ f\end{matrix}\right]\}) \triangleright^{*} (D', (\lambda[\underline{z}].T\left[\begin{matrix}F\\ f\end{matrix}\right]))$, thus satisfying the decidable part of the proof relation.

To check the well-foundedness conditions,
$D' \triangleleft^{*} join(D_1)[D] : rt$, by the *join*-lemma, as required.
$\{ T : map(pi[\Sigma],\Phi)$, by hypothesis
$T\left[\begin{matrix}F\\ f\end{matrix}\right] : map(pi[\Sigma],\Phi)$, by instantiation (Theorem 6 of the Coding of Trees), as required. $\}$

∎

THEOREM 6. (Second 2CPF instantiation rule.)

If $Q \models I \Rightarrow J$ then $Pr_{29}(Q, \{I \Rightarrow J\}, \{I\begin{bmatrix}\Pi\\ \pi\end{bmatrix} \Rightarrow J\begin{bmatrix}\Pi\\ \pi\end{bmatrix}\}) \models I\begin{bmatrix}\Pi\\ \pi\end{bmatrix} \Rightarrow J\begin{bmatrix}\Pi\\ \pi\end{bmatrix}$,

where Π is a type symbol and the construction Pr_{29} is defined below.

Proof. Let $I \rhd^* pfn(A, \Sigma) \not\rhd$ and $J \rhd^* pfn(B, \Phi) \not\rhd$. Then

$$I\begin{bmatrix}\Pi\\ \pi\end{bmatrix} \rhd^* pfn(A, \Sigma\begin{bmatrix}\Pi\\ \pi\end{bmatrix}) \not\rhd$$
$$J\begin{bmatrix}\Pi\\ \pi\end{bmatrix} \rhd^* pfn(B, \Phi\begin{bmatrix}\Pi\\ \pi\end{bmatrix}) \not\rhd .$$

Let $\underline{x}$ be the free object variables of I, J, and let q, y, z be three fresh object variables. The hypothesis that $Q \models I \Rightarrow J$ means that $Q \rhd^* (D, (\lambda[\underline{x}].T)) \not\rhd$ where

- $DT(D, [\![(q, \underline{x})\ Aq \rightarrow B(T[q])]\!]) \rhd^* true$,
- $D : rt$ and $\{\!\!\{ T : map(pi[\Sigma], \Phi) \}\!\!\}$.

This shows that Q also satisfies the decidable part of the requirements for a proof of $I\begin{bmatrix}\Pi\\ \pi\end{bmatrix} \Rightarrow J\begin{bmatrix}\Pi\\ \pi\end{bmatrix}$, so define Pr_{29} by

$$Pr_{29}(q, y, z) \triangleq q.$$

To check the well-foundedness conditions,

$D : rt$, by hypothesis, as required.

$\{\!\!\{\ {}_{\pi}\{\!\!\{ T : map(pi[\Sigma], \Phi) \}\!\!\}$, by hypothesis (separating the π quantifier from the others),

$T : map(pi[\Sigma], \Phi)\begin{bmatrix}\Pi\\ \pi\end{bmatrix} \overset{*}{\mapsto} map(pi[\Sigma\begin{bmatrix}\Pi\\ \pi\end{bmatrix}], \Phi\begin{bmatrix}\Pi\\ \pi\end{bmatrix})$, by instantiation (Theorem 2 of the Second-Order Coding of Trees), as required. $\}\!\!\}$

∎

FORMAL INTERPRETATION OF 2CPF IN PROTOLOGIC

DEFINITION. The coding of CPF derivations as constructions is extended to a coding of 2CPF derivations by including the three new axiom schemata and the three new rules, numbered 24,... 29.

DEFINITION. The definition of the construction *spr* is augmented by the clauses

$$spr(none(24), a) \triangleq Pr_{24}(a)$$

$$spr(none(25), a) \triangleq Pr_{25}(a)$$

$$spr(none(26), a) \triangleq Pr_{26}(a)$$

$$spr(one(27, (u, b)), a) \triangleq Pr_{27}(spr(u, b), b, a)$$

$$spr(one(28, (u, b)), a) \triangleq Pr_{28}(spr(u, b), b, a)$$

$$spr(one(29, (u, b)), a) \triangleq Pr_{29}(spr(u, b), b, a).$$

THEOREM 7. If D is the code of a 2CPF derivation (with no premises) of a logical sequent $\Gamma \Rightarrow I$ then $spr(D) \Vdash \Gamma \Rightarrow I$.

Proof. By structural induction on the 2CPF derivation, using the extensionality of $\Vdash$ (Theorem 1 of Chapter 27). ∎

THEOREM 8. (2CPF interpretation theorem.) If D is the code of a 2CPF derivation (with no premises) of a logical sequent $\Rightarrow I$, where I has no free variables, then $pr(spr(D)) \vdash I$.

Proof. By the previous theorem and the soundness theorem for $\Vdash$ with respect to $\vdash$ (Theorem 2 of Chapter 27). ∎

THEOREM 9. (The *glue*-lemma for 2CPF.) If D is the code of a 2CPF derivation of a logical sequent S from premise sequents $T_0, \ldots T_k$, and $D_0, \ldots D_k$ are the codes of 2CPF derivations (with no premises) of $T_0, \ldots T_k$ respectively, then $glue(D, [D_0, \ldots D_k]) \triangleright^*$ the code of a 2CPF derivation (with no premises) of S.

Proof. By a straightforward structural induction on the derivation encoded by D. ∎

CHAPTER 42

SECOND-ORDER CALCULUS OF PROOF FUNCTIONS

The Second-Order Calculus of Proof Functions (2CPF) is obtained from the Calculus of Proof Functions (CPF) in Chapter 28 as follows.

LEXICAL CONVENTIONS

In addition to the conventions of Chapter 28, 'P'' will be used to denote a type predicate. As in the previous chapter, metavariables that denote variables will be interpreted as follows.

- Those beginning with lower-case roman letters range over object variables.
- The letters 'π', 'σ', 'ϕ', 'χ', 'ψ' and 'ω' range over type variables.

LOGICAL SEQUENTS AND $\Vdash$

The notion of a logical sequent is carried over unchanged from CPF. The proof relation $\Vdash$ is modified slightly to take account of the presence of type variables. The theorems about $\Vdash$ (Theorems 1–4) continue to hold.

SECOND-ORDER CALCULUS OF PROOF FUNCTIONS

The axiom schemata and rules of inference of 2CPF are those of CPF (with the metavariables that denote variables restricted to object variables and the metavariables that denote terms restricted to object terms) plus the following.

apply-introduction (*apply*-in): $val\, P\, x \Rightarrow apply\, P\, x$

apply-elimination (*apply*-el): $apply\, P\, x \Rightarrow val\, P\, x$

$\forall'$-elimination ($\forall'$-el): $\forall' P' \Rightarrow val'\, P'\, \pi$

$\forall'$-introduction rule ($\forall'$-in): $\dfrac{I \Rightarrow val'\, P'\, \pi}{I \Rightarrow \forall' P'}$ where $\pi \notin I, P'$

Instantiation rules (inst): $\dfrac{I \Rightarrow J}{I\begin{bmatrix}F\\f\end{bmatrix} \Rightarrow J\begin{bmatrix}F\\f\end{bmatrix}}$ $\dfrac{I \Rightarrow J}{I\begin{bmatrix}\Pi\\\pi\end{bmatrix} \Rightarrow J\begin{bmatrix}\Pi\\\pi\end{bmatrix}}$

where F is an irreducible object term and Π is a type symbol.

FORMAL INTERPRETATION OF 2CPF IN PROTOLOGIC

The coding of CPF derivations as constructions and the definition of *spr* are extended to 2CPF derivations.

THEOREM 7. If D is the code of a 2CPF derivation (with no premises) of a logical sequent $\Gamma \Rightarrow I$ then $spr(D) \Vdash\!\!= \Gamma \Rightarrow I$.

THEOREM 8. (2CPF interpretation theorem.) If D is the code of a 2CPF derivation (with no premises) of a logical sequent $\Rightarrow I$, where I has no free variables, then $pr(spr(D)) \vdash I$.

THEOREM 9. (The *glue*-lemma for 2CPF.) If D is the code of a 2CPF derivation of a logical sequent S from premise sequents $T_0, \ldots T_k$, and $D_0, \ldots D_k$ are the codes of 2CPF derivations (with no premises) of $T_0, \ldots T_k$ respectively, then $glue(D, [D_0, \ldots D_k]) \triangleright^*$ the code of a 2CPF derivation (with no premises) of S.

CHAPTER 43

FROM SECOND-ORDER CALCULUS OF PROOF FUNCTIONS TO SECOND-ORDER LOGIC OF PARTIAL TERMS

In this chapter, the syntax of formulae in the Logic of Partial Terms (LPT) is extended to include the atomic formula $x \models \alpha$ and the second-order universal quantifier $\forall^2$, interpreted as a combination of $\forall'$ and $\forall$. Axioms and rules for these new constructs are added to LPT, giving Second-Order Logic of Partial Terms (2LPT). In this chapter I shall extend the interpretation of LPT in CPF given in Chapter 29 to an interpretation of 2LPT in 2CPF.

Recall from Chapter 36 that there are several sorts of variables: *object* variables (consisting of *first-order* variables and *function* variables), *type* variables, and *second-order* variables. Metavariables that denote variables are interpreted in this chapter as follows.

- Those beginning with lower-case roman letters range over object variables (unless otherwise specified).
- The letters 'π', 'σ', 'ϕ', 'χ', 'ψ' and 'ω' range over type variables.
- The letters 'α', 'β' and 'γ' range over second-order variables.

Informally, the first-order variables are to be thought of as ranging over the universe of 'individuals' of 2LPT. A predicate of individuals is represented by a second-order variable, or sometimes by $pred(f, \pi)$, where f is a function variable and π is a type variable.

2LPT FORMULAE

The language of 2LPT formulae is obtained by augmenting the language of LPT formulae as follows.

- The characters '$\models$' and '$\forall^2$' are added to the alphabet.
- The tokens 'vbl_2', '$\models$' and '$\forall^2$' are added to the lexicon.
- Lexical analysis is modified so that all first-order variables are replaced by 'vbl' and all second-order variables are replaced by 'vbl_2' (no function or type variables are allowed).
- The production rules $F \rightarrow vbl \models vbl_2 \mid \forall^2\, vbl_2\, F$ are added to the grammar.

(Note that, as a consequence of this grammar, strings matching T are terms containing only first-order variables.)

EXAMPLE. $\forall^2\alpha\, \exists x\, (x \models \alpha \wedge x = y) \supset y \models \beta$ is a formula, provided x, y are first-order variables and α, β are second-order variables.

The metavariables 'A', 'B' and 'C' denote 2LPT formulae; 'I', 'J' and 'K' denote proof functions; 'Γ' and 'Δ' denote sequences of proof functions; 'P' and 'Q' denote predicates; 'P'' denotes a type predicate.

Metaformulae and their instances are defined by the obvious generalisation of the LPT case, except that metaformulae may contain the metanotation $A \Leftrightarrow B$ for $(A \supset B) \wedge (B \supset A)$ and $\neg A$ for $A \supset \mathit{false}$.

FREE VARIABLES IN 2LPT FORMULAE

The relation $v \in A$, defined for variables v and LPT formulae A, is extended to 2LPT formulae by letting v be first- or second-order and adding the clauses

- $v \in x \models \alpha$ iff v is x or v is α,
- $v \in \forall^2\alpha\, A$ iff v isn't α and $v \in A$.

The *free variables* of A are the (finitely many) first- and second-order variables v such that $v \in A$, listed in standard order.

INTERPRETATION OF 2LPT FORMULAE AS PROOF FUNCTIONS

Establish a one-to-one correspondence between the second-order variables and the function variables (since both supplies of variables are given as infinite sequences), and likewise establish a one-to-one correspondence between the second-order variables and the type variables. When I want to make use of these two one-to-one correspondences I shall say something like 'let f and π be the function and type variables corresponding to α'.

The interpretation of LPT formulae is extended to 2LPT formulae by
$\ulcorner x \models \alpha \urcorner$ is $\mathit{apply}(\mathit{pred}(f, \pi))x$,
$\ulcorner \forall^2\alpha\, A \urcorner$ is $\forall'(\mathit{predify}'(\lambda\pi.\forall(\mathit{predify}(\lambda f.\ulcorner A \urcorner))))$;
where f and π are the function and type variables corresponding to α.

THEOREM 1. Let A be a 2LPT formula.

- If x is a first-order variable then $x \in A$ iff $x \in \ulcorner A \urcorner$.
- If α is a second-order variable, and f and π are the function and type variables corresponding to α, then $\alpha \in A$ iff $f \in \ulcorner A \urcorner$ iff $\pi \in \ulcorner A \urcorner$.

THEOREM 2. Let A be a 2LPT formula.

- $\ulcorner A \urcorner$ is a proof function.
- If x is a first-order variable then $\mathit{predify}(\lambda x.\ulcorner A \urcorner)$ is a predicate.
- If π is a type variable then $\mathit{predify}'(\lambda\pi.\ulcorner A \urcorner)$ is a type predicate.

EXAMPLE. To return to the example of $\forall^2\alpha\, \exists x\,(x \models \alpha \wedge x = y) \supset y \models \beta$, its interpretation as a proof function is

$$\supset(\forall'(\mathit{predify}'(\lambda\pi.\forall(\mathit{predify}(\lambda f.\exists(\mathit{predify}(\lambda x. \wedge(\mathit{apply}(\mathit{pred}(f,\pi))x, \ulcorner x = y\urcorner))))))), \mathit{apply}(\mathit{pred}(g,\sigma))y)$$

where f and π are the function and type variables corresponding to α, and g and σ are the function and type variables corresponding to β. We can check that this is a proof function as follows:

- $\mathit{apply}(\mathit{pred}(f,\pi))x$ and $\mathit{apply}(\mathit{pred}(g,\sigma))y$ are proof functions, by Theorem 2 of Second-Order Logic;
- $\ulcorner x = y\urcorner$ is a proof function, by definition (see Logic);
- $\wedge(\mathit{apply}(\mathit{pred}(f,\pi))x, \ulcorner x = y\urcorner)$ is a proof function, by Theorem 7 of Logic;
- $\exists(\mathit{predify}(\lambda x. \wedge(\mathit{apply}(\mathit{pred}(f,\pi))x, \ulcorner x = y\urcorner)))$ is a proof function (call it I), by Theorems 6 and 9 of Logic;
- $\forall(\mathit{predify}(\lambda f.I))$ is a proof function, by Theorem 6 of Logic and Theorem 4 of Chapter 29;
- $\forall'(\mathit{predify}'(\lambda\pi.\forall(\mathit{predify}(\lambda f.I))))$ is a proof function, by Theorems 4 and 5 of Second-Order Logic;
- $\supset(\forall'(\mathit{predify}'(\lambda\pi.\forall(\mathit{predify}(\lambda f.I)))), \mathit{apply}(\mathit{pred}(g,\sigma))y)$ is a proof function, by Theorem 5 of Chapter 29.

SOME 2CPF DERIVATIONS INVOLVING INTERPRETED FORMULAE

All the CPF derivations given in Chapter 29 also apply in 2CPF, with the metavariables denoting variables restricted to object variables (or to first-order variables if the metavariable occurs in a metaformula), the metavariables denoting terms restricted to object terms (or to terms with only first-order free variables if the metavariable occurs in a metaformula), and the metavariables denoting formulae ranging over 2LPT formulae.

THEOREM 3. 2CPF $\forall^2$ properties ($\forall^2$-el & $\forall^2$-in):

$$\ulcorner\forall^2\alpha\, A\urcorner \Rightarrow \bullet\ulcorner A\urcorner \qquad \frac{\ulcorner B\urcorner \Rightarrow \bullet\ulcorner A\urcorner}{\ulcorner B\urcorner \Rightarrow \ulcorner\forall^2\alpha\, A\urcorner} \quad \text{where } \alpha \notin B.$$

Proof. Let f and π be the function and type variables corresponding to α. Let P be the predicate $\mathit{predify}(\lambda f.\ulcorner A\urcorner)$ and P' be the type predicate $\mathit{predify}'(\lambda\pi.\forall P)$; note that $\ulcorner\forall^2\alpha\, A\urcorner$ is $\forall' P'$. The required 2CPF derivations are as follows.

$$\dfrac{\dfrac{\forall' P' \Rightarrow val'\, P'\, \pi \ (\forall'\text{-el})}{\forall' P' \Rightarrow \forall P}\,\text{red} \qquad \dfrac{\forall P \Rightarrow \bullet(val\, P f) \ (\forall\text{-el})}{\forall P \Rightarrow \bullet \ulcorner A \urcorner}\,\text{red}}{\forall' P' \Rightarrow \bullet \ulcorner A \urcorner}\,\text{cut} \qquad\qquad \dfrac{\dfrac{\dfrac{\dfrac{\ulcorner B \urcorner \Rightarrow \bullet \ulcorner A \urcorner}{\ulcorner B \urcorner \Rightarrow \bullet(val\, P f)}\,\text{red}}{\ulcorner B \urcorner \Rightarrow \forall P}\,\forall\text{-in}}{\ulcorner B \urcorner \Rightarrow val'\, P'\, \pi}\,\text{red}}{\ulcorner B \urcorner \Rightarrow \forall' P'}\,\forall'\text{-in}$$

∎

THEOREM 4. 2CPF $\neg\forall^2\neg$ theorem: $\ulcorner A \urcorner \Rightarrow \ulcorner \neg \forall^2 \alpha\, \neg A \urcorner$

Proof. The 2CPF derivation is as follows.

$$\dfrac{\dfrac{\dfrac{\ulcorner \forall^2 \alpha\, \neg A \urcorner \Rightarrow \bullet \ulcorner \neg A \urcorner \ (\forall^2\text{-el}) \qquad \dfrac{\ulcorner A \urcorner,\ \ulcorner A \supset false \urcorner \Rightarrow \bullet \ulcorner false \urcorner \ (\supset)}{\ulcorner \neg A \urcorner,\ \ulcorner A \urcorner \Rightarrow \bullet \ulcorner false \urcorner}\,\text{exch}}{\ulcorner \forall^2 \alpha\, \neg A \urcorner,\ \ulcorner A \urcorner \Rightarrow \bullet \ulcorner false \urcorner}\,\bullet\text{-cut}}{\ulcorner A \urcorner \Rightarrow \ulcorner \forall^2 \alpha\, \neg A \supset false \urcorner}\,\supset}{\ulcorner A \urcorner \Rightarrow \ulcorner \neg \forall^2 \alpha\, \neg A \urcorner}\,\text{rewriting}$$

∎

SECOND-ORDER LOGIC OF PARTIAL TERMS

Second-Order Logic of Partial Terms (2LPT) consists of the axiom schemata and rules of inference of LPT (with the formula metavariables ranging over 2LPT formulae, the term metavariables ranging over terms whose free variables are all first-order, and the variable metavariables ranging over first-order variables) plus the following.

Equality (eq_2): $x = y \supset x \models \alpha \supset y \models \alpha$

Comprehension (comp): $\neg\forall^2 \alpha\, \neg \forall x\, (x \models \alpha \Leftrightarrow A)$ where $\alpha \notin A$

$\forall^2$-elimination ($\forall^2$-el): $\forall^2 \alpha\, A \supset A$

$\forall^2$-introduction rule ($\forall^2$-in): $\dfrac{B \supset A}{B \supset \forall^2 \alpha\, A}$ where $\alpha \notin B$

Note that the Comprehension schema says '$\neg\forall^2\alpha\,\neg$' instead of '$\exists^2\alpha$' because 2LPT lacks an $\exists^2$ quantifier.

EXAMPLE. '$(A \wedge B) \supset B$' is an axiom schema of LPT and therefore also an axiom schema of 2LPT, but its instances are different in the two theories. In LPT, $(x = y \wedge w = z) \supset x = y$ is an instance for any variables x, y, w, z, whereas in 2LPT it is only an instance (indeed, only a formula) if x, y, w, z are first-order variables. On the other hand, 2LPT has instances that LPT lacks, such as $(x \models \alpha \wedge y \models \beta) \supset x \models \alpha$ (where x, y are first-order and α, β are second-order).

2CPF DERIVATIONS CORRESPONDING TO THE 2LPT AXIOMS AND RULES

I shall show that for each axiom A of 2LPT the logical sequent $\Rightarrow \ulcorner A \urcorner$ is derivable in 2CPF, and that for each rule of inference $\dfrac{A \ \ldots \ B}{C}$ of 2LPT the rule of inference $\dfrac{\Rightarrow \ulcorner A \urcorner \ \ldots \ \Rightarrow \ulcorner B \urcorner}{\Rightarrow \ulcorner C \urcorner}$ is derivable in 2CPF. The arguments for the axiom schemata and rules of LPT still work here (with the metavariables reinterpreted as above), so it is sufficient to consider the new axiom schemata and rules.

THEOREM 5. 2LPT equality axiom: $\quad x = y \supset x \models \alpha \supset y \models \alpha$

Proof. Let f and π be the function and type variables corresponding to α, and let P be the predicate *predify*($\lambda x.$*apply*(*pred*(f, π))x). Then $val\, P x \ \triangleright^* \ \ulcorner x \models \alpha \urcorner$, and $val\, P y \ \triangleright^* \ \ulcorner y \models \alpha \urcorner$. The required 2CPF derivation is as follows.

$$\dfrac{\dfrac{\dfrac{\ulcorner x = y \urcorner,\ val\, P x \Rightarrow val\, P y \ \text{(eq)}}{\ulcorner x = y \urcorner,\ \ulcorner x \models \alpha \urcorner \Rightarrow \ulcorner y \models \alpha \urcorner}\,\text{red}}{\ulcorner x = y \urcorner \Rightarrow \ulcorner x \models \alpha \supset y \models \alpha \urcorner}\,\supset}{\Rightarrow \ulcorner x = y \supset x \models \alpha \supset y \models \alpha \urcorner}\,\supset$$

∎

THEOREM 6. 2LPT comprehension axiom: $\quad \neg \forall^2 \alpha \, \neg \forall x \, (x \models \alpha \Leftrightarrow A)$ where $\alpha \notin A$.

Proof. Let f and π be the function and type variables corresponding to α.

Let P be the predicate *predify*($\lambda x. \ulcorner A \urcorner$), and let $P \ \triangleright^* \ pred(F, \Pi) \not\triangleright$.

Let Q be the predicate *predify*($\lambda x. \wedge(\supset(\ulcorner x \models \alpha \urcorner, \ulcorner A \urcorner), \supset(\ulcorner A \urcorner, \ulcorner x \models \alpha \urcorner)))$.

Let B be the formula $\forall x \, (x \models \alpha \Leftrightarrow A)$.

By Theorem 1 of Second-Order Logic, $Q\begin{bmatrix} F, \Pi \\ f, \pi \end{bmatrix}$ is a predicate and $\ulcorner B \urcorner \begin{bmatrix} F, \Pi \\ f, \pi \end{bmatrix}$ is a proof function. The 2CPF derivation is as follows.

$$\dfrac{\dfrac{\dfrac{\dfrac{apply\,P\,x \Rightarrow val\,P\,x \ (apply\text{-el})}{apply\,P\,x \Rightarrow \bullet(val\,P\,x)}\,\bullet\text{-in, cut}}{\Rightarrow\ >(apply\,P\,x, \bullet(val\,P\,x))}\,>\text{-in}}{\Rightarrow\ \supset(apply\,P\,x, \ulcorner A \urcorner)}\,\text{red} \qquad \dfrac{\dfrac{\dfrac{val\,P\,x \Rightarrow apply\,P\,x \ (apply\text{-in})}{val\,P\,x \Rightarrow \bullet(apply\,P\,x)}\,\bullet\text{-in, cut}}{\Rightarrow\ >(val\,P\,x, \bullet(apply\,P\,x))}\,>\text{-in}}{\Rightarrow\ \supset(\ulcorner A \urcorner, apply\,P\,x)}\,\text{red}}{\Rightarrow\ \wedge(\supset(apply\,P\,x, \ulcorner A \urcorner), \supset(\ulcorner A \urcorner, apply\,P\,x))}\,\wedge\text{-in, cut}$$

$$\dfrac{\dfrac{\dfrac{\dfrac{\dfrac{\Rightarrow\ \wedge(\supset(apply\,P\,x, \ulcorner A \urcorner), \supset(\ulcorner A \urcorner, apply\,P\,x))}{\Rightarrow\ val\,(Q\begin{bmatrix} F,\Pi \\ f,\pi \end{bmatrix})\,x}\,\text{red}}{\Rightarrow\ \bullet(val\,(Q\begin{bmatrix} F,\Pi \\ f,\pi \end{bmatrix})\,x)}\,\bullet\text{-in, cut}}{\Rightarrow\ \forall Q\begin{bmatrix} F,\Pi \\ f,\pi \end{bmatrix}}\,\forall\text{-in}}{\Rightarrow\ \ulcorner B \urcorner\begin{bmatrix} F,\Pi \\ f,\pi \end{bmatrix}}\,\text{rewriting} \qquad \dfrac{\ulcorner B \urcorner \Rightarrow \ulcorner \neg\forall^2\alpha\,\neg B \urcorner \ (\neg\forall^2\neg)}{\ulcorner B \urcorner\begin{bmatrix} F,\Pi \\ f,\pi \end{bmatrix} \Rightarrow \ulcorner \neg\forall^2\alpha\,\neg B \urcorner}\,\text{inst}}{\Rightarrow\ \ulcorner \neg\forall^2\alpha\,\neg B \urcorner}\,\text{cut}$$

▮

THEOREM 7. 2LPT $\forall^2$-el and $\forall^2$-in: $\forall^2\alpha\,A \supset A \qquad \dfrac{B \supset A}{B \supset \forall^2\alpha\,A}$
where $\alpha \notin B$.

Proof. The required 2CPF derivations are as follows.

$$\dfrac{\ulcorner \forall^2\alpha\,A \urcorner \Rightarrow \bullet\ulcorner A \urcorner \ (\forall^2\text{-el})}{\Rightarrow\ \ulcorner \forall^2\alpha\,A \supset A \urcorner}\,\supset \qquad\qquad \dfrac{\dfrac{\dfrac{\Rightarrow\ \ulcorner B \supset A \urcorner}{\ulcorner B \urcorner \Rightarrow \bullet\ulcorner A \urcorner}\,\supset}{\ulcorner B \urcorner \Rightarrow \ulcorner \forall^2\alpha\,A \urcorner}\,\forall^2\text{-in}}{\Rightarrow\ \ulcorner B \supset \forall^2\alpha\,A \urcorner}\,\supset$$

▮

FORMAL INTERPRETATION OF 2LPT

DEFINITION. Let $atomic_2$ and all_2 be two fresh 1-ary constructors. The coding, $\langle\langle A \rangle\rangle$, of LPT formulae, A, as constructions is extended to 2LPT formulae by adding the clauses

$$\langle x \models \alpha \rangle \text{ is } atomic_2(x, \alpha)$$
$$\langle \forall^2\alpha\,A \rangle \text{ is } all_2(\lambda[\alpha].\langle A \rangle).$$

DEFINITION. The coding of LPT derivations as constructions is extended to 2LPT derivations by including the three new axioms and the new rule, numbered 20,...23.

DEFINITION. Define recursive functions $T_{20}, \ldots T_{23}$ such that

- for $n = 20, 21, 22$, if A is an instance of 2LPT axiom schema n then $T_n(\langle\!\langle A \rangle\!\rangle) \rhd^*$ the code of the corresponding 2CPF derivation (given above) of $\Rightarrow \ulcorner A \urcorner$ from no premises;
- if $\dfrac{B}{A}$ is an instance of 2LPT rule 23 then $T_{23}(\langle\!\langle B \rangle\!\rangle, \langle\!\langle A \rangle\!\rangle) \rhd^*$ the code of the corresponding 2CPF derivation (given above) of $\Rightarrow \ulcorner A \urcorner$ from the premise $\Rightarrow \ulcorner B \urcorner$.

DEFINITION. The definition of the construction *cpf* is extended by the clauses

$$\begin{aligned} cpf(none(20), a) &\triangleq T_{20}(a) \\ cpf(none(21), a) &\triangleq T_{21}(a) \\ cpf(none(22), a) &\triangleq T_{22}(a) \\ cpf(one(23, (u, b)), a) &\triangleq glue(T_{23}(b, a), [cpf(u, b)]). \end{aligned}$$

THEOREM 8. If D is the code of a 2LPT derivation (with no premises) of a formula A then $cpf(D) \rhd^*$ the code of a 2CPF derivation (with no premises) of the logical sequent $\Rightarrow \ulcorner A \urcorner$.

Proof. By structural induction on the derivation encoded by D. ∎

THEOREM 9. (2LPT interpretation theorem.) If D is the code of a 2LPT derivation (with no premises) of a formula A with no free variables, then $pr(spr(cpf(D))) \vdash \ulcorner A \urcorner$.

Proof. By the previous theorem, the extensionality theorem for $\vdash$, and the interpretation theorem for 2CPF. ∎

THEOREM 10. (The *glue*-lemma for 2LPT.) If D is the code of a 2LPT derivation of a 2LPT formula A from premises $B_0, \ldots B_k$, and $D_0, \ldots D_k$ are the codes of 2LPT derivations (with no premises) of $B_0, \ldots B_k$ respectively, then $glue(D, [D_0, \ldots D_k]) \rhd^*$ the code of a 2LPT derivation (with no premises) of A.

Proof. By structural induction on the derivation encoded by D. ∎

CHAPTER 44

SECOND-ORDER LOGIC OF PARTIAL TERMS

Second-Order Logic of Partial Terms (2LPT) is obtained from Logic of Partial Terms (LPT, Chapter 30) as follows. In this chapter, metavariables that denote variables are interpreted as follows.

- Those beginning with lower-case roman letters range over first-order variables.
- The letters 'α', 'β' and 'γ' range over second-order variables.

2LPT FORMULAE

The language of 2LPT formulae is obtained by augmenting the language of LPT formulae as follows.

- The characters '$\models$' and '$\forall^2$' are added to the alphabet.
- The tokens 'vbl_2', '$\models$' and '$\forall^2$' are added to the lexicon.
- Lexical analysis is modified so that all first-order variables are replaced by '*vbl*' and all second-order variables are replaced by 'vbl_2' (no function or type variables are allowed).
- The production rules $F \rightarrow vbl \models vbl_2 \mid \forall^2\, vbl_2\, F$ are added to the grammar.

The metavariables 'A', 'B' and 'C' denote 2LPT formulae.

Metaformulae and their instances are defined by the obvious generalisation of the LPT case, except that metaformulae may contain the metanotation $A \Leftrightarrow B$ for $(A \supset B) \wedge (B \supset A)$ and $\neg A$ for $A \supset \mathit{false}$.

FREE VARIABLES IN 2LPT FORMULAE

The relation $v \in A$, defined for variables v and LPT formulae A, is extended to 2LPT formulae by letting v be first- or second-order and adding the clauses

- $v \in x \models \alpha$ iff v is x or v is α,
- $v \in \forall^2\alpha A$ iff v isn't α and $v \in A$.

The *free variables* of A are the (finitely many) first- and second-order variables v such that $v \in A$, listed in standard order.

The interpretation of each LPT formula A as a proof function $\ulcorner A \urcorner$ is extended to 2LPT formulae.

SECOND-ORDER LOGIC OF PARTIAL TERMS

Second-Order Logic of Partial Terms (2LPT) consists of the axiom schemata and rules of inference of LPT (with the formula metavariables ranging over 2LPT formulae, the term metavariables ranging over terms whose free variables are all first-order, and the variable metavariables ranging over first-order variables) plus the following.

Equality (eq_2): $x = y \supset x \models \alpha \supset y \models \alpha$
Comprehension (comp): $\neg \forall^2 \alpha \neg \forall x (x \models \alpha \Leftrightarrow A)$ where $\alpha \notin A$
$\forall^2$-elimination ($\forall^2$-el): $\forall^2 \alpha A \supset A$

$\forall^2$-introduction rule ($\forall^2$-in): $$\frac{B \supset A}{B \supset \forall^2 \alpha A} \quad \text{where } \alpha \notin B$$

FORMAL INTERPRETATION OF 2LPT

The coding of LPT derivations as constructions and the definition of *cpf* are extended to 2LPT derivations, using two fresh 1-ary constructors, $atomic_2$ and all_2.

THEOREM 8. If D is the code of a 2LPT derivation (with no premises) of a formula A then $cpf(D) \triangleright^*$ the code of a 2CPF derivation (with no premises) of the logical sequent $\Rightarrow \ulcorner A \urcorner$.

THEOREM 9. (2LPT interpretation theorem.) If D is the code of a 2LPT derivation (with no premises) of a formula A with no free variables, then $pr(spr(cpf(D))) \vdash \ulcorner A \urcorner$.

THEOREM 10. (The *glue*-lemma for 2LPT.) If D is the code of a 2LPT derivation of a 2LPT formula A from premises $B_0, \ldots B_k$, and $D_0, \ldots D_k$ are the codes of 2LPT derivations (with no premises) of $B_0, \ldots B_k$ respectively, then $glue(D, [D_0, \ldots D_k]) \triangleright^*$ the code of a 2LPT derivation (with no premises) of A.

CHAPTER 45

FROM SECOND-ORDER LOGIC OF PARTIAL TERMS TO SECOND-ORDER HEYTING ARITHMETIC

Second-Order Heyting Arithmetic (2HA) is obtained from Heyting Arithmetic (HA) by admitting atomic formulae of the form $N \models \alpha$ and the second-order universal quantifier. In this chapter I shall extend the interpretation of HA in LPT given in Chapter 31 to an interpretation of 2HA in 2LPT. Metavariables that denote variables will be interpreted as follows.

- Those beginning with lower-case roman letters range over first-order variables.
- The letters 'α', 'β' and 'γ' range over second-order variables.

The definition of *numeric terms* (given originally in the Expanded Term Language) is now restricted by requiring that all their free variables be first-order.

SOME 2LPT THEOREMS

All the LPT derivations in Chapter 31 still apply, with 'A' now interpreted as ranging over 2LPT formulae, the metavariables denoting terms restricted to terms whose variables are all first-order, and the metavariables denoting variables restricted to first-order variables. In addition we have the following 2LPT derivations.

THEOREM 1. First 2LPT replacement theorem (rep1):
$M = N \supset (\forall z(z = M \supset A) \Leftrightarrow \forall z(z = N \supset A))$ where $z \notin M, N$.

Proof. Half of this is derived by

$$\dfrac{\forall z(z=M\supset A)\supset(z=M\supset A)\ (\forall\text{-el}) \qquad \dfrac{M=N\supset z=N\supset z=M\ \text{(r.eq)}}{M=N\supset(z=M\supset A)\supset z=N\supset A}\,\text{pc}}{\dfrac{M=N\supset\forall z(z=M\supset A)\supset z=N\supset A}{M=N\supset\forall z(z=M\supset A)\supset\forall z(z=N\supset A)}\,\forall\text{-in, pc}}\,\text{pc}$$

and the other half follows from this by using $M = N \supset N = M$ and swapping M and N. ∎

THEOREM 2. Second 2LPT replacement theorem (rep2):
$\forall u\,(u=T\supset u\models\alpha)\Leftrightarrow\forall v\,(v=T\supset v\models\alpha)$ where $u,v\notin T$.

Proof. If u is v then the theorem is an instance of $A\Leftrightarrow A$, so assume u is not v. The 2LPT derivation, in one direction, is

$$\dfrac{\dfrac{v=v\ (\text{red}) \qquad \dfrac{\dfrac{\dfrac{u=v\supset v=T\supset u=T\ (\text{r.eq}) \qquad u=v\supset u\models\alpha\supset v\models\alpha\ (\text{eq}_2)}{u=v\supset(u=T\supset u\models\alpha)\supset v=T\supset v\models\alpha}\,\text{pc}}{u=v\supset\forall u\,(u=T\supset u\models\alpha)\supset v=T\supset v\models\alpha}\,\forall\text{-el, pc}}{v=v\supset\forall u\,(u=T\supset u\models\alpha)\supset v=T\supset v\models\alpha}\,\text{spec}}{\forall u\,(u=T\supset u\models\alpha)\supset v=T\supset v\models\alpha}\,\text{pc}}{\forall u\,(u=T\supset u\models\alpha)\supset\forall v\,(v=T\supset v\models\alpha)}\,\forall\text{-in}$$

and the converse is obtained by swapping u and v. ∎

THEOREM 3. Third 2LPT replacement theorem (rep3):
$n\models\alpha\Leftrightarrow\forall x\,(x=n\supset x\models\alpha)$ where x is not n.

Proof. The implication in one direction is derived by

$$\dfrac{\dfrac{x=n\supset n=x\ (\text{symm}) \qquad n=x\supset n\models\alpha\supset x\models\alpha\ (\text{eq}_2)}{n\models\alpha\supset x=n\supset x\models\alpha}\,\text{pc}}{n\models\alpha\supset\forall x\,(x=n\supset x\models\alpha)}\,\forall\text{-in}$$

and the converse is derived by

$$\dfrac{n=n\ (\text{red}) \qquad \dfrac{\dfrac{\dfrac{x=n\supset x\models\alpha\supset n\models\alpha\ (\text{eq}_2)}{x=n\supset(x=n\supset x\models\alpha)\supset n\models\alpha}\,\text{pc}}{x=n\supset\forall x\,(x=n\supset x\models\alpha)\supset n\models\alpha}\,\forall\text{-el, pc}}{n=n\supset\forall x\,(x=n\supset x\models\alpha)\supset n\models\alpha}\,\text{spec}}{\forall x\,(x=n\supset x\models\alpha)\supset n\models\alpha}\,\text{pc}$$

∎

FORMULAE OF SECOND-ORDER HEYTING ARITHMETIC

The language of 2HA formulae is obtained by augmenting the language of HA formulae as follows.

- The characters '$\models$' and '$\forall^2$' are added to the alphabet.
- The tokens 'vbl_2', '$\models$' and '$\forall^2$' are added to the lexicon.

- Lexical analysis is modified so that all first-order variables are replaced by '*vbl*' and all second-order variables are replaced by 'vbl_2' (no function or type variables are allowed).
- The productions rules $F \rightarrow N \models vbl_2 \mid \forall^2\, vbl_2\, F$ are added to the grammar.

The strings matching N are still required to be numeric terms (in the new sense, that is, with all free variables first-order).

2HA METAFORMULAE

The metavariables 'A', 'B', and 'C' will now denote 2HA formulae. As in HA, 'F', 'G' and 'H', possibly with subscripts, denote primitive recursive functions; all other metavariables starting with a capital letter denote numeric terms.

Metaformulae and their instances are defined by an obvious generalisation of the HA case, except that metaformulae may contain the metanotation $A \Leftrightarrow B$ for $(A \supset B) \wedge (B \supset A)$.

FREE VARIABLES IN 2HA FORMULAE

The relation $v \in A$, defined for variables v and HA formulae A, is extended to 2HA by allowing v to be first- or second-order and adding the clauses

- $v \in N \models \alpha$ iff $v \in N$ or v is α,
- $v \in \forall^2 \alpha\, A$ iff v isn't α and $v \in A$.

The *free variables* of A are the (finitely many) first- or second-order variables v such that $v \in A$, listed in standard order.

INTERPRETATION OF 2HA FORMULAE IN 2LPT

The mapping from HA formulae to LPT formula is extended to a mapping from 2HA formulae to 2LPT formulae. The definition of A^{N} is extended by:

- $(N \models \alpha)^{\mathrm{N}}$ is $\forall n\, (n{=}N \supset n \models \alpha)$ (where n is the first first-order variable, in standard order, such that $n \notin N$);
- $(\forall^2 \alpha\, A)^{\mathrm{N}}$ is $\forall^2 \alpha\, (A^{\mathrm{N}})$.

The definition of A^{LPT} becomes: A^{LPT} is $num\, \underset{\sim}{z} \supset A^{\mathrm{N}}$, where $\underset{\sim}{z}$ are the free *first-order* variables of A.

Note that A^{N} and A^{LPT} have the same free variables as A.

AXIOMS AND RULES OF INFERENCE OF 2HA

The axiom schemata and rules of inference of 2HA are those of HA (with the formula metavariables ranging over 2HA formulae, the term metavariables ranging over numeric terms, and the variable metavariables ranging over first-order variables) plus the following.

Equality (eq_2): $M = N \supset (M \models \alpha \Leftrightarrow N \models \alpha)$
Comprehension (comp): $\neg\forall^2\alpha\,\neg\forall^{\mathrm{N}}n\,(n \models \alpha \Leftrightarrow A)$ where $\alpha \notin A$
$\forall^2$-elimination ($\forall^2$-el): $\forall^2\alpha\, A \supset A$

$\forall^2$-introduction rule ($\forall^2$-in): $\dfrac{B \supset A}{B \supset \forall^2\alpha\, A}$ where $\alpha \notin B$

2LPT DERIVATIONS CORRESPONDING TO THE AXIOMS AND RULES OF 2HA

I shall show that, for each axiom A of 2HA, A^{LPT} is a theorem of 2LPT, and, for each rule of inference $\dfrac{A \ldots B}{C}$ of 2HA, $\dfrac{A^{\mathrm{LPT}} \ldots B^{\mathrm{LPT}}}{C^{\mathrm{LPT}}}$ is a derived rule of 2LPT. The arguments of Chapter 31 still apply (with the metavariables reinterpreted as above), so it is only necessary to consider the new axiom schemata and rules.

THEOREM 4. 2HA equality axiom: $M = N \supset (M \models \alpha \Leftrightarrow N \models \alpha)$.

Proof. The axiom translates into 2LPT as

$$num\,\underline{z} \supset M = N \supset (\forall m\,(m = M \supset m \models \alpha) \Leftrightarrow \forall n\,(n = N \supset n \models \alpha))$$

where $m \notin M$, $n \notin N$, and $\underline{z}$ are the free first-order variables of M and N. Let x be a fresh first-order variable. The 2LPT formula follows by propositional calculus from the 2LPT theorems

$$\begin{aligned} &\forall m\,(m = M \supset m \models \alpha) \Leftrightarrow \forall x\,(x = M \supset x \models \alpha) && \text{(rep2)}\\ &M = N \supset (\forall x\,(x = M \supset x \models \alpha) \Leftrightarrow \forall x\,(x = N \supset x \models \alpha)) && \text{(rep1)}\\ &\forall x\,(x = N \supset x \models \alpha) \Leftrightarrow \forall n\,(n = N \supset n \models \alpha) && \text{(rep2).} \end{aligned}$$

∎

THEOREM 5. 2HA comprehension axiom: $\neg\forall^2\alpha\,\neg\forall^{\mathrm{N}}n\,(n \models \alpha \Leftrightarrow A)$ where $\alpha \notin A$.

Proof. Let $\underline{z}$ be the free first-order variables of A, excluding n. Now, $(n \models \alpha)^{\mathrm{N}}$ is $\forall x\,(x = n \supset x \models \alpha)$ for some $x \notin n$. The 2LPT derivation of the interpretation of the axiom is as follows.

$$\cfrac{\cfrac{\cfrac{\neg\forall^2\alpha\,\neg\forall n\,(n \models \alpha \Leftrightarrow A^{\mathrm{N}})\ \text{(comp)} \qquad n \models \alpha \Leftrightarrow \forall x\,(x = n \supset x \models \alpha)\ \text{(rep3)}}{\neg\forall^2\alpha\,\neg\forall n\,(\forall x\,(x = n \supset x \models \alpha) \Leftrightarrow A^{\mathrm{N}})}\ \forall\text{-rules,pc}}{\neg\forall^2\alpha\,\neg\forall n\,(\mathit{num}\,n \supset (\forall x\,(x = n \supset x \models \alpha) \Leftrightarrow A^{\mathrm{N}}))}\ \forall\text{-rules,pc}}{\mathit{num}\,\underline{z} \supset \neg\forall^2\alpha\,\neg\forall n\,(\mathit{num}\,n \supset (\forall x\,(x = n \supset x \models \alpha) \Leftrightarrow A^{\mathrm{N}}))}\ \text{pc}$$

where '$\forall$-rules' denotes use of the $\forall$-in, $\forall$-el, $\forall^2$-in and $\forall^2$-el axioms and rules. ∎

THEOREM 6. 2HA $\forall^2$-elimination axiom: $\quad \forall^2\alpha\, A \supset A$.

Proof. Let $\underline{z}$ be the free first-order variables of A. The formula $(\forall^2\alpha\, A \supset A)^{\mathrm{LPT}}$ is derived in 2LPT as follows.

$$\cfrac{\forall^2\alpha\, A^{\mathrm{N}} \supset A^{\mathrm{N}}\ (\forall^2\text{-el})}{\mathit{num}\,\underline{z} \supset \forall^2\alpha\, A^{\mathrm{N}} \supset A^{\mathrm{N}}}\ \text{pc}$$

∎

THEOREM 7. 2HA $\forall^2$-introduction rule: $\quad \dfrac{B \supset A}{B \supset \forall^2\alpha\, A} \quad$ where $\alpha \notin B$.

Proof. Let $\underline{z}$ be the free first-order variables of A and B. The 2LPT derivation of the 2LPT interpretation of the rule is as follows.

$$\cfrac{\mathit{num}\,\underline{z} \supset B^{\mathrm{N}} \supset A^{\mathrm{N}}}{\mathit{num}\,\underline{z} \supset B^{\mathrm{N}} \supset \forall^2\alpha\, A^{\mathrm{N}}}\ \forall^2\text{-in, pc}$$

∎

FORMAL INTERPRETATION OF 2HA

DEFINITION. The coding, $\langle\!\langle A \rangle\!\rangle$, of HA formulae, A, as constructions is extended to 2HA formula by adding the clauses

$$\langle N \models \alpha \rangle \text{ is } \mathit{atomic}_2(N, \alpha)$$
$$\langle \forall^2\alpha\, A \rangle \text{ is } \mathit{all}_2(\lambda[\alpha].\langle A \rangle).$$

DEFINITION. The coding of HA derivations as constructions is extended to 2HA derivations by including the three new axioms and the new rule, numbered 28,... 31.

DEFINITION. Define recursive functions $T_{28}, \ldots T_{31}$ such that

- for $n = 28, 29, 30$, if A is an instance of 2HA axiom schema n then $T_n(\langle\!\langle A \rangle\!\rangle) \ \triangleright^*$ the code of the corresponding 2LPT derivation (given above) of A^{LPT} from no premises;
- if $\dfrac{B}{A}$ is an instance of 2HA rule 31 then $T_{31}(\langle\!\langle B \rangle\!\rangle, \langle\!\langle A \rangle\!\rangle) \ \triangleright^*$ the code of the corresponding 2LPT derivation (given above) of A^{LPT} from the premise B^{LPT}.

DEFINITION. Extend the definition of the construction *lpt* by adding the clauses

$$\begin{aligned} lpt(none(28), a) &\triangleq T_{28}(a) \\ lpt(none(29), a) &\triangleq T_{29}(a) \\ lpt(none(30), a) &\triangleq T_{30}(a) \\ lpt(one(31, (u, b)), a) &\triangleq glue(T_{31}(b, a), [lpt(u, b)]). \end{aligned}$$

THEOREM 8. If D is the code of a 2HA derivation (with no premises) of a formula A then $lpt(D) \ \triangleright^*$ the code of a 2LPT derivation (with no premises) of A^{LPT}.

Proof. By structural induction on the derivation encoded by D. ∎

THEOREM 9. (2HA interpretation theorem.) If D is the code of a 2HA derivation (with no premises) of a formula A with no free variables, then $pr(spr(cpf(lpt(D)))) \vdash \ulcorner A^{\mathrm{LPT}} \urcorner$.

Proof. By the previous theorem, the extensionality theorem for $\vdash$, and the interpretation theorem for 2LPT. ∎

THEOREM 10. (The *glue*-lemma for 2HA.) If D is the code of a 2HA derivation of A from premises $B_1, \ldots B_k$, and $D_1, \ldots D_k$ are the codes of 2HA derivations (with no premises) of $B_1, \ldots B_k$ respectively, then $glue(D, [D_1, \ldots D_k]) \ \triangleright^*$ the code of a 2HA derivation (with no premises) of A.

Proof. By structural induction on the derivation encoded by D. ∎

CHAPTER 46

SECOND-ORDER HEYTING ARITHMETIC

Second-Order Heyting Arithmetic (2HA) is obtained from Heyting Arithmetic (HA, Chapter 32) as follows. Metavariables that denote variables will be interpreted as follows.

- Those beginning with lower-case roman letters range over first-order variables.
- The letters 'α', 'β' and 'γ' range over second-order variables.

The definition of *numeric terms* (given originally in the Expanded Term Language) is now restricted by requiring that all their free variables be first-order.

FORMULAE OF SECOND-ORDER HEYTING ARITHMETIC

The language of 2HA formulae is obtained by augmenting the language of HA formulae as follows.

- The characters '$\models$' and '$\forall^2$' are added to the alphabet.
- The tokens 'vbl_2', '$\models$' and '$\forall^2$' are added to the lexicon.
- Lexical analysis is modified so that all first-order variables are replaced by '*vbl*' and all second-order variables are replaced by 'vbl_2' (no function or type variables are allowed).
- The productions rules $F \to N \models vbl_2 \mid \forall^2\ vbl_2\ F$ are added to the grammar.

The strings matching N are still required to be numeric terms (in the new sense, that is, with all free variables first-order).

2HA METAFORMULAE

The metavariables 'A', 'B', and 'C' will now denote 2HA formulae. As in HA, 'F', 'G' and 'H', possibly with subscripts, denote primitive recursive functions; all other metavariables starting with a capital letter denote numeric terms.

Metaformulae and their instances are defined by an obvious generalisation of the HA case, except that metaformulae may contain the metanotation $A \Leftrightarrow B$ for $(A \supset B) \wedge (B \supset A)$.

FREE VARIABLES IN 2HA FORMULAE

The relation $v \in A$, defined for variables v and HA formulae A, is extended to 2HA by allowing v to be first- or second-order and adding the clauses

- $v \in N \models \alpha$ iff $v \in N$ or v is α,
- $v \in \forall^2 \alpha A$ iff v isn't α and $v \in A$.

The *free variables* of A are the (finitely many) first- or second-order variables v such that $v \in A$, listed in standard order.

The mapping from each HA formulae A to an LPT formula A^{LPT} is extended to a mapping from 2HA formulae to 2LPT formulae.

AXIOMS AND RULES OF INFERENCE OF 2HA

The axiom schemata and rules of inference of 2HA are those of HA (with the formula metavariables ranging over 2HA formulae, the term metavariables ranging over numeric terms, and the variable metavariables ranging over first-order variables) plus the following.

Equality (eq_2): $M = N \supset (M \models \alpha \Leftrightarrow N \models \alpha)$
Comprehension (comp): $\neg \forall^2 \alpha \neg \forall^{\mathrm{N}} n\, (n \models \alpha \Leftrightarrow A)$ where $\alpha \notin A$
$\forall^2$-elimination ($\forall^2$-el): $\forall^2 \alpha A \supset A$

$\forall^2$-introduction rule ($\forall^2$-in): $\dfrac{B \supset A}{B \supset \forall^2 \alpha A}$ where $\alpha \notin B$

FORMAL INTERPRETATION OF 2HA

The coding of HA derivations as constructions and the definition of *lpt* are extended to 2HA derivations.

THEOREM 8. If D is the code of a 2HA derivation (with no premises) of a formula A then $lpt(D) \rhd^*$ the code of a 2LPT derivation (with no premises) of A^{LPT}.

THEOREM 9. (2HA interpretation theorem.) If D is the code of a 2HA derivation (with no premises) of a formula A with no free variables, then $pr(spr(cpf(lpt(D)))) \vdash \ulcorner A^{\mathrm{LPT}} \urcorner$.

THEOREM 10. (The *glue*-lemma for 2HA.) If D is the code of a 2HA derivation of A from premises $B_1, \ldots B_k$, and $D_1, \ldots D_k$ are the codes of 2HA derivations (with no premises) of $B_1, \ldots B_k$ respectively, then $glue(D, [D_1, \ldots D_k]) \rhd^*$ the code of a 2HA derivation (with no premises) of A.

CHAPTER 47

FROM SECOND-ORDER HEYTING ARITHMETIC TO SECOND-ORDER PEANO ARITHMETIC

Second-Order Peano Arithmetic (2PA) is obtained from Peano Arithmetic (PA) by adding second-order quantifiers, $\forall^2$ and $\exists^2$, and atomic formulae of the form $N \models \alpha$. In this chapter I shall extend the interpretation of PA in HA given in Chapter 33 to an interpretation of 2PA in 2HA. Metavariables that denote variables will be interpreted as in 2HA, namely as follows.

- Those beginning with lower-case roman letters range over first-order variables.
- The letters 'α', 'β' and 'γ' range over second-order variables.

FORMULAE OF 2PA

The language of 2PA formulae is obtained by augmenting the language of PA formulae as follows.

- The characters '$\models$', '$\exists^2$' and '$\forall^2$' are added to the alphabet.
- The tokens 'vbl_2', '$\models$', '$\exists^2$' and '$\forall^2$' are added to the lexicon.
- Lexical analysis is modified so that all first-order variables are replaced by '*vbl*' and all second-order variables are replaced by 'vbl_2' (no function or type variables are allowed).
- The productions rules $F \to N \models vbl_2 \mid \exists^2\, vbl_2\, F \mid \forall^2\, vbl_2\, F$ are added to the grammar.

The strings matching N must be still be numeric terms.

The metavariables 'A', 'B', and 'C' will now denote 2PA formulae. As in PA, 'F', 'G' and 'H', possibly with subscripts, denote primitive recursive functions; all other metavariables starting with a capital letter denote numeric terms.

Metaformulae and their instances are defined by an obvious generalisation of the PA case, except that metaformulae may contain the extended substitution notation defined below, the metanotation $A \Leftrightarrow B$ for $(A \supset B) \wedge (B \supset A)$, and the metanotation $\alpha \simeq \beta$ for $\forall^{\mathrm{N}} n\,(n \models \alpha \Leftrightarrow n \models \beta)$, where n is the first first-order variable in standard order.

FREE VARIABLES IN 2PA FORMULAE

The relation $v \in A$ in PA is extended to 2PA by allowing v to be first- or second-order and adding the clauses

- $v \in N \models \alpha$ iff $v \in N$ or v is α,
- $v \in \exists^2 \alpha A$ iff $v \in \forall^2 \alpha A$ iff v isn't α and $v \in A$.

As before, the *free variables* of A are the (finitely many) first- and second-order variables v such that $v \in A$, listed in standard order.

INTERPRETATION OF 2PA FORMULAE IN 2HA

The interpretation of PA formulae in HA is extended to 2PA by the clauses:
$(N \models \alpha)^{\mathrm{H}}$ is $\neg\neg N \models \alpha$
$(\exists^2 \alpha A)^{\mathrm{H}}$ is $\neg \forall^2 \alpha \neg (A^{\mathrm{H}})$
$(\forall^2 \alpha A)^{\mathrm{H}}$ is $\forall^2 \alpha (A^{\mathrm{H}})$.

THEOREM 1. $v \in A$ iff $v \in A^{\mathrm{H}}$, for any first- or second-order variable v.

THEOREM 2. 2PA double negation theorem ($\neg\neg$): for any 2PA formula A, $\neg\neg A^{\mathrm{H}} \supset A^{\mathrm{H}}$ is a theorem of 2HA.

Proof. Using structural induction on A, the cases where A is *true*, *false*, $M = N$, $B \wedge C$, $B \vee C$, $B \supset C$, $\exists^{\mathrm{N}} n B$ and $\forall^{\mathrm{N}} n B$ are dealt with as in the PA double negation theorem. There are three new cases.

Case 9: A is $N \models \alpha$. Then $\neg\neg A^{\mathrm{H}} \supset A^{\mathrm{H}}$ is $\neg\neg\neg\neg N \models \alpha \supset \neg\neg N \models \alpha$, which is a theorem of 2HA by propositional calculus.

Case 10: A is $\exists^2 \alpha B$. Then $\neg\neg A^{\mathrm{H}} \supset A^{\mathrm{H}}$ is $\neg\neg\neg\forall^2 \alpha \neg B^{\mathrm{H}} \supset \neg\forall^2 \alpha \neg B^{\mathrm{H}}$, which is a theorem of 2HA by propositional calculus.

Case 11: A is $\forall^2 \alpha B$. Then $\neg\neg A^{\mathrm{H}} \supset A^{\mathrm{H}}$ is $\neg\neg\forall^2 \alpha B^{\mathrm{H}} \supset \forall^2 \alpha B^{\mathrm{H}}$, which is derived in 2HA as follows.

$$\dfrac{\dfrac{\dfrac{\forall^2 \alpha B^{\mathrm{H}} \supset B^{\mathrm{H}} \; (\forall^2\text{-el})}{\neg\neg\forall^2 \alpha B^{\mathrm{H}} \supset \neg\neg B^{\mathrm{H}}}\,\text{pc} \qquad \neg\neg B^{\mathrm{H}} \supset B^{\mathrm{H}} \; \text{(inductive hypothesis)}}{\neg\neg\forall^2 \alpha B^{\mathrm{H}} \supset B^{\mathrm{H}}}\,\text{pc}}{\neg\neg\forall^2 \alpha B^{\mathrm{H}} \supset \forall^2 \alpha B^{\mathrm{H}}}\,\forall^2\text{-in}$$

∎

AXIOMS AND RULES OF 2PA

The axiom schemata and rules of inference of Second-Order Peano Arithmetic are those of Peano Arithmetic (with the formula metavariables ranging over 2PA formulae, the term metavariables ranging over numeric terms, and the variable metavariables ranging over first-order variables) plus the following.

Equality (eq_2): $M = N \supset (M \models \alpha \Leftrightarrow N \models \alpha)$
Comprehension (comp): $\exists^2\alpha\, \forall^{\mathrm{N}} n\, (n \models \alpha \Leftrightarrow A)$ where $\alpha \notin A$
Quantifier axioms ($\exists^2$-in, $\forall^2$-el): $A \supset \exists^2\alpha\, A$ $\quad \forall^2\alpha\, A \supset A$

Quantifier rules ($\exists^2$-el, $\forall^2$-in): $\dfrac{A \supset B}{\exists^2\alpha\, A \supset B}$ $\quad \dfrac{B \supset A}{B \supset \forall^2\alpha\, A}$ where $\alpha \notin B$

2HA DERIVATIONS CORRESPONDING TO THE 2PA AXIOMS AND RULES

For each axiom A of 2PA I shall show that A^{H} is a theorem of 2HA, and for each rule of inference $\dfrac{A \ldots B}{C}$ of 2PA I shall show that $\dfrac{A^{\mathrm{H}} \ldots B^{\mathrm{H}}}{C^{\mathrm{H}}}$ is a derived rule of inference of 2HA. All the HA derivations in Chapter 33 still work in 2HA (with the metavariables reinterpreted as above), so it is sufficient to consider the new axiom schemata and rules.

THEOREM 3. 2PA equality axiom: $M = N \supset (M \models \alpha \Leftrightarrow N \models \alpha)$.

Proof. The theorem translates into 2HA as

$$M = N \supset (\neg\neg M \models \alpha \Leftrightarrow \neg\neg N \models \alpha),$$

which follows by propositional calculus from the 2HA axiom

$$M = N \supset (M \models \alpha \Leftrightarrow N \models \alpha).$$

∎

THEOREM 4. 2PA comprehension axiom: $\exists^2\alpha\, \forall^{\mathrm{N}} n\, (n \models \alpha \Leftrightarrow A)$ where $\alpha \notin A$.

Proof. Let B be the 2HA comprehension axiom, $\neg\forall^2\alpha\, \neg\forall^{\mathrm{N}} n\, (n \models \alpha \Leftrightarrow A^{\mathrm{H}})$. The 2HA derivation of $(\exists^2\alpha\, \forall^{\mathrm{N}} n\, (n \models \alpha \Leftrightarrow A))^{\mathrm{H}}$ is as follows.

$$\dfrac{B\ \text{(comp)} \qquad \dfrac{\dfrac{\dfrac{(n \models \alpha \Leftrightarrow A^{\mathrm{H}}) \supset (\neg\neg n \models \alpha \Leftrightarrow \neg\neg A^{\mathrm{H}})\ \text{(pc)} \qquad \dfrac{\neg\neg A^{\mathrm{H}} \supset A^{\mathrm{H}}\ (\neg\neg)}{\neg\neg A^{\mathrm{H}} \Leftrightarrow A^{\mathrm{H}}}\,\text{pc}}{(n \models \alpha \Leftrightarrow A^{\mathrm{H}}) \supset (\neg\neg n \models \alpha \Leftrightarrow A^{\mathrm{H}})}\,\text{pc}}{\forall^{\mathrm{N}} n\,(n \models \alpha \Leftrightarrow A^{\mathrm{H}}) \supset \forall^{\mathrm{N}} n\,(\neg\neg n \models \alpha \Leftrightarrow A^{\mathrm{H}})}\,\forall^{\mathrm{N}}\text{-in}, \forall^{\mathrm{N}}\text{-el,pc}}{B \supset \neg\forall^2\alpha\,\neg\forall^{\mathrm{N}} n\,(\neg\neg n \models \alpha \Leftrightarrow A^{\mathrm{H}})}\,\forall^2\text{-in}, \forall^2\text{-el,pc}}{\neg\forall^2\alpha\,\neg\forall^{\mathrm{N}} n\,(\neg\neg n \models \alpha \Leftrightarrow A^{\mathrm{H}})}\,\text{pc}$$

∎

THEOREM 5. 2PA quantifier axioms: $A \supset \exists^2\alpha\, A \qquad \forall^2\alpha\, A \supset A.$

Proof. The 2HA derivations of $(A \supset \exists^2\alpha\, A)^{\mathrm{H}}$ and $(\forall^2\alpha\, A \supset A)^{\mathrm{H}}$ are as follows.

$$\dfrac{\forall^2\alpha\,\neg A^{\mathrm{H}} \supset \neg A^{\mathrm{H}}\ (\forall^2\text{-el})}{A^{\mathrm{H}} \supset \neg\forall^2\alpha\,\neg A^{\mathrm{H}}}\,\text{pc} \qquad\qquad \forall^2\alpha\, A^{\mathrm{H}} \supset A^{\mathrm{H}}\ (\forall^2\text{-el})$$

∎

THEOREM 6. 2PA quantifier rules: $\dfrac{A \supset B}{\exists^2\alpha\, A \supset B} \qquad \dfrac{B \supset A}{B \supset \forall^2\alpha\, A}$
where $\alpha \notin B$.

Proof. The corresponding 2HA derivations are as follows.

$$\dfrac{\dfrac{\dfrac{\dfrac{A^{\mathrm{H}} \supset B^{\mathrm{H}}}{\neg B^{\mathrm{H}} \supset \neg A^{\mathrm{H}}}\,\text{pc}}{\neg B^{\mathrm{H}} \supset \forall^2\alpha\,\neg A^{\mathrm{H}}}\,\forall^2\text{-in}}{\neg\forall^2\alpha\,\neg A^{\mathrm{H}} \supset \neg\neg B^{\mathrm{H}}}\,\text{pc} \qquad \neg\neg B^{\mathrm{H}} \supset B^{\mathrm{H}}\ (\neg\neg)}{\neg\forall^2\alpha\,\neg A^{\mathrm{H}} \supset B^{\mathrm{H}}}\,\text{pc} \qquad\qquad \dfrac{B^{\mathrm{H}} \supset A^{\mathrm{H}}}{B^{\mathrm{H}} \supset \forall^2\alpha\, A^{\mathrm{H}}}\,\forall^2\text{-in}$$

∎

FORMAL INTERPRETATION OF 2PA

DEFINITION. Let *exists*$_2$ be a fresh 1-ary constructor. The coding, $\langle\!\langle A\rangle\!\rangle$, of PA formulae, A, as constructions is extended to 2PA formula by adding the clauses

$$\langle N \models \alpha\rangle \text{ is } atomic_2(N,\alpha)$$
$$\langle \exists^2\alpha\, A\rangle \text{ is } exists_2(\lambda[\alpha].\langle A\rangle)$$
$$\langle \forall^2\alpha\, A\rangle \text{ is } all_2(\lambda[\alpha].\langle A\rangle).$$

DEFINITION. The coding of PA derivations as constructions is extended to 2PA derivations by including the four new axioms and two new rules, numbered 29,... 34.

DEFINITION. Define recursive functions $T_{29}, \dots T_{34}$ such that

- for $n = 29, 30, 31, 32$, if A is an instance of 2PA axiom schema n then $T_n(\langle\!\langle A \rangle\!\rangle) \rhd^*$ the code of the corresponding 2HA derivation (given above) of A^{H} from no premises;
- for $n = 33, 34$, if $\dfrac{B}{A}$ is an instance of 2PA rule n then $T_n(\langle\!\langle B \rangle\!\rangle, \langle\!\langle A \rangle\!\rangle) \rhd^*$ the code of the corresponding 2HA derivation (given above) of A^{H} from the premise B^{H}.

DEFINITION. Extend the definition of the construction *ha* by adding the clauses

$$
\begin{aligned}
ha(none(29), a) &\triangleq T_{29}(a)\\
ha(none(30), a) &\triangleq T_{30}(a)\\
ha(none(31), a) &\triangleq T_{31}(a)\\
ha(none(32), a) &\triangleq T_{32}(a)\\
ha(one(33, (u, b)), a) &\triangleq glue(T_{33}(b, a), [ha(u, b)])\\
ha(one(34, (u, b)), a) &\triangleq glue(T_{34}(b, a), [ha(u, b)]).
\end{aligned}
$$

THEOREM 7. If D is the code of a 2PA derivation (with no premises) of a formula A then $ha(D) \rhd^*$ the code of a 2HA derivation (with no premises) of A^{H}.

Proof. By structural induction on the derivation encoded by D. ∎

THEOREM 8. (2PA interpretation theorem.) If D is the code of a 2PA derivation (with no premises) of a formula A with no free variables, then $pr(spr(cpf(lpt(ha(D))))) \vdash \ulcorner (A^{\mathrm{H}})^{\mathrm{LPT}} \urcorner$.

Proof. By the previous theorem, the extensionality theorem for $\vdash$, and the interpretation theorem for 2HA. ∎

SUBSTITUTION IN 2PA FORMULAE

The substitutability relation $N \overset{n}{\hookrightarrow} A$ is defined as in PA (with 'n' ranging over first-order variables) plus the following clauses:

- $N \overset{n}{\hookrightarrow} M \models \alpha$;
- $N \overset{n}{\hookrightarrow} \exists^2\gamma A$ iff $N \overset{n}{\hookrightarrow} \forall^2\gamma A$ iff $N \overset{n}{\hookrightarrow} A$;
- $\beta \overset{\alpha}{\hookrightarrow} A$, if A is atomic (*true*, *false*, $M = N$ or $N \models \alpha$);
- $\beta \overset{\alpha}{\hookrightarrow} A \wedge B$ iff $\beta \overset{\alpha}{\hookrightarrow} A \vee B$ iff $\beta \overset{\alpha}{\hookrightarrow} A \supset B$ iff $\beta \overset{\alpha}{\hookrightarrow} A$ and $\beta \overset{\alpha}{\hookrightarrow} B$;
- $\beta \overset{\alpha}{\hookrightarrow} \exists^{\mathrm{N}} m A$ iff $\beta \overset{\alpha}{\hookrightarrow} \forall^{\mathrm{N}} m A$ iff $\beta \overset{\alpha}{\hookrightarrow} A$;
- $\beta \overset{\alpha}{\hookrightarrow} \exists^2\gamma A$ iff $\beta \overset{\alpha}{\hookrightarrow} \forall^2\gamma A$ iff $\alpha \notin \exists^2\gamma A$ or (γ isn't β and $\beta \overset{\alpha}{\hookrightarrow} A$).

If $N \overset{n}{\hookrightarrow} A$, then the formula $A\binom{N}{n}$ is defined as in PA with the following additional clauses:

- $(M \models \alpha)\binom{N}{n}$ is $M\binom{N}{n} \models \alpha$;
- $(\exists^2\gamma A)\binom{N}{n}$ is $\exists^2\gamma\,(A\binom{N}{n})$ and $(\forall^2\gamma A)\binom{N}{n}$ is $\forall^2\gamma\,(A\binom{N}{n})$.

If $\beta \overset{\alpha}{\hookrightarrow} A$, then the formula $A\binom{\beta}{\alpha}$ is defined by:

- $true\binom{\beta}{\alpha}$ is *true*, $false\binom{\beta}{\alpha}$ is *false*, $(M = N)\binom{\beta}{\alpha}$ is $M = N$;
- $(N \models \alpha)\binom{\beta}{\alpha}$ is $N \models \beta$, $(N \models \gamma)\binom{\beta}{\alpha}$ is $N \models \gamma$ if γ isn't α;
- $(A \wedge B)\binom{\beta}{\alpha}$ is $A\binom{\beta}{\alpha} \wedge B\binom{\beta}{\alpha}$, $(A \vee B)\binom{\beta}{\alpha}$ is $A\binom{\beta}{\alpha} \vee B\binom{\beta}{\alpha}$;
- $(A \supset B)\binom{\beta}{\alpha}$ is $A\binom{\beta}{\alpha} \supset B\binom{\beta}{\alpha}$;
- $(\exists^{\mathrm{N}} m A)\binom{\beta}{\alpha}$ is $\exists^{\mathrm{N}} m\,(A\binom{\beta}{\alpha})$, $(\forall^{\mathrm{N}} m A)\binom{\beta}{\alpha}$ is $\forall^{\mathrm{N}} m\,(A\binom{\beta}{\alpha})$;
- $(\exists^2\gamma A)\binom{\beta}{\alpha}$ is $\exists^2\gamma A$ and $(\forall^2\gamma A)\binom{\beta}{\alpha}$ is $\forall^2\gamma A$, if $\alpha \notin \exists^2\gamma A$;
- $(\exists^2\gamma A)\binom{\beta}{\alpha}$ is $\exists^2\gamma\,(A\binom{\beta}{\alpha})$ and $(\forall^2\gamma A)\binom{\beta}{\alpha}$ is $\forall^2\gamma\,(A\binom{\beta}{\alpha})$, if $\alpha \in \exists^2\gamma A$ and γ isn't β.

THEOREM 9. (Properties of substitution.) Let v be a first- or second-order variable.

- $0 \overset{v}{\hookrightarrow} A$.
- If M and N have the same free variables, then $M \overset{v}{\hookrightarrow} A$ iff $N \overset{v}{\hookrightarrow} A$.
- $v \overset{v}{\hookrightarrow} A$ and $A\binom{v}{v}$ is A.
- If $n \notin A$ then $N \overset{n}{\hookrightarrow} A$ and $A\binom{N}{n}$ is A.
- If $\alpha \notin A$ then $\beta \overset{\alpha}{\hookrightarrow} A$ and $A\binom{\beta}{\alpha}$ is A.
- If $N \overset{n}{\hookrightarrow} A$ then $v \in A\binom{N}{n}$ iff ($v \in N$ and $n \in A$) or ($v \in A$ and n isn't v).
- If $\beta \overset{\alpha}{\hookrightarrow} A$ then $v \in A\binom{\beta}{\alpha}$ iff (v is β and $\alpha \in A$) or ($v \in A$ and α isn't v).

2PA THEOREMS AND DERIVED RULES INVOLVING SUBSTITUTION

2PA contains classical propositional calculus; I shall mark uses of the latter in 2PA derivations with the label 'pc'.

THEOREM 10. 2PA first-order equality theorem for formulae (eq.form.):

$$M = N \supset (A\binom{M}{x} \Leftrightarrow A\binom{N}{x}), \quad \text{where } M \overset{x}{\hookrightarrow} A \text{ and } N \overset{x}{\hookrightarrow} A.$$

Proof. By structural induction on A, as in the PA equality theorem for formulae. ▮

The substitution theorems and rules derived in PA still apply, with 'A', 'B' and 'C' ranging over 2PA formulae and the metavariables denoting variables ranging over first-order variables.

THEOREM 11. 2PA second-order equality theorem for formulae (eq_2.form.):

$$(\alpha \simeq \beta) \supset (A\binom{\alpha}{\gamma} \Leftrightarrow A\binom{\beta}{\gamma}), \quad \text{where } \alpha \overset{\gamma}{\hookrightarrow} A \text{ and } \beta \overset{\gamma}{\hookrightarrow} A.$$

Proof. By structural induction on A, as follows. If A is $N \models \gamma$ then the formula is $\forall^{\mathsf{N}} n\,(n \models \alpha \Leftrightarrow n \models \beta) \supset (N \models \alpha \Leftrightarrow N \models \beta)$, which is an instance of the theorem $\forall^{\mathsf{N}} n A \supset A\binom{N}{n}$. (Note that $N \overset{n}{\hookrightarrow} (n \models \alpha \Leftrightarrow n \models \beta)$.) If A is any other atomic formula then the whole formula is $(\alpha \simeq \beta) \supset (A \Leftrightarrow A)$, which is a 2PA theorem by propositional calculus.

The other cases are as in the previous theorem. ▮

THEOREM 12. 2PA second-order specification rule (spec_2): $\dfrac{\alpha \simeq \beta \supset A}{A}$ where $\alpha \notin \beta, A$.

Proof.

$$\dfrac{\exists^2\alpha\,\alpha \simeq \beta \;(\text{comp}) \qquad \dfrac{\alpha \simeq \beta \supset A}{\exists^2\alpha\,\alpha \simeq \beta \supset A}\,\exists^2\text{-el}}{A}\,\text{pc}$$

▮

THEOREM 13. 2PA second-order substitution rule (sub_2): $\dfrac{B}{B\binom{\beta}{\alpha}}$

where $\beta \overset{\alpha}{\hookrightarrow} B$.

Proof. Let γ be a fresh variable (and so $\gamma \overset{\alpha}{\hookrightarrow} B$). The 2PA derivation is as follows.

$$\dfrac{\dfrac{\dfrac{B \qquad (\alpha \simeq \gamma) \supset (B \supset B\binom{\gamma}{\alpha})\ \text{(eq}_2\text{.form.)}}{(\alpha \simeq \gamma) \supset B\binom{\gamma}{\alpha}}\,\text{pc}}{B\binom{\gamma}{\alpha}}\,\text{spec}_2 \qquad (\gamma \simeq \beta) \supset (B\binom{\gamma}{\alpha} \supset B\binom{\beta}{\alpha})\ \text{(eq}_2\text{.form.)}}{\dfrac{(\gamma \simeq \beta) \supset B\binom{\beta}{\alpha}}{B\binom{\beta}{\alpha}}\,\text{spec}_2}\,\text{pc}$$

∎

THEOREM 14. 2PA second-order quantifier theorems with substitution:

$$A\binom{\beta}{\alpha} \supset \exists^2\alpha\, A, \qquad \forall^2\alpha\, A \supset A\binom{\beta}{\alpha}, \qquad \text{where } \beta \overset{\alpha}{\hookrightarrow} A.$$

Proof. The 2PA derivations are as follows.

$$\dfrac{A \supset \exists^2\alpha\, A\ (\exists^2\text{-in})}{A\binom{\beta}{\alpha} \supset \exists^2\alpha\, A}\,\text{sub}_2 \qquad\qquad \dfrac{\forall^2\alpha\, A \supset A\ (\forall^2\text{-el})}{\forall^2\alpha\, A \supset A\binom{\beta}{\alpha}}\,\text{sub}_2$$

∎

THEOREM 15. 2PA second-order variable change rule (vch_2): $\dfrac{B\binom{\beta}{\alpha}}{B}$ where $\beta \notin B$ and $\beta \overset{\alpha}{\hookrightarrow} B$.

Proof. If β is α then there is nothing to prove, so assume β is not α. The 2PA derivation is as follows.

$$\dfrac{B\binom{\beta}{\alpha} \qquad (\beta \simeq \alpha) \supset (B\binom{\beta}{\alpha} \supset B)\ \text{(eq}_2\text{.form.)}}{\dfrac{(\beta \simeq \alpha) \supset B}{B}\,\text{spec}_2}\,\text{pc}$$

∎

THEOREM 16. 2PA second-order quantifier rules with substitution:

$$\dfrac{A\binom{\beta}{\alpha} \supset B}{\exists^2\alpha\, A \supset B} \qquad \dfrac{B \supset A\binom{\beta}{\alpha}}{B \supset \forall^2\alpha\, A} \qquad \text{where } \beta \overset{\alpha}{\hookrightarrow} A \text{ and } \beta \notin \exists^2\alpha\, A,\ B.$$

Proof. If β is α then the rules are $\exists^2$-el and $\forall^2$-in, so assume β is not α. Then $\beta \notin A, B$ and the 2PA derivations are as follows.

$$\dfrac{\dfrac{\dfrac{A\binom{\beta}{\alpha} \supset \exists^2\beta\, A\binom{\beta}{\alpha}\ (\exists^2\text{-in})}{A \supset \exists^2\beta\, A\binom{\beta}{\alpha}}\,\text{vch}_2}{\exists^2\alpha\, A \supset \exists^2\beta\, A\binom{\beta}{\alpha}}\,\exists^2\text{-el} \qquad \dfrac{A\binom{\beta}{\alpha} \supset B}{\exists^2\beta\, A\binom{\beta}{\alpha} \supset B}\,\exists^2\text{-el}}{\exists^2\alpha\, A \supset B}\,\text{pc}$$

$$\dfrac{\dfrac{B \supset A\binom{\beta}{\alpha}}{B \supset \forall^2\beta\, A\binom{\beta}{\alpha}}\,\forall^2\text{-in} \qquad \dfrac{\dfrac{\forall^2\beta\, A\binom{\beta}{\alpha} \supset A\binom{\beta}{\alpha}\ (\forall^2\text{-el})}{\forall^2\beta\, A\binom{\beta}{\alpha} \supset A}\,\text{vch}_2}{\forall^2\beta\, A\binom{\beta}{\alpha} \supset \forall^2\alpha\, A}\,\forall^2\text{-in}}{B \supset \forall^2\alpha\, A}\,\text{pc}$$

CHAPTER 48

SECOND-ORDER PEANO ARITHMETIC

Second-Order Peano Arithmetic (2PA) is obtained from Peano Arithmetic (PA, Chapter 34) as follows. Metavariables that denote variables will be interpreted as in 2HA, namely as follows.

- Those beginning with lower-case roman letters range over first-order variables.
- The letters 'α', 'β' and 'γ' range over second-order variables.

FORMULAE OF 2PA

The language of 2PA formulae is obtained by augmenting the language of PA formulae as follows.

- The characters '$\models$', '$\exists^2$' and '$\forall^2$' are added to the alphabet.
- The tokens 'vbl_2', '$\models$', '$\exists^2$' and '$\forall^2$' are added to the lexicon.
- Lexical analysis is modified so that all first-order variables are replaced by '*vbl*' and all second-order variables are replaced by 'vbl_2' (no function or type variables are allowed).
- The productions rules $F \to N \models vbl_2 \mid \exists^2\, vbl_2\, F \mid \forall^2\, vbl_2\, F$ are added to the grammar.

The strings matching N must be still be numeric terms.

The metavariables 'A', 'B', and 'C' will now denote 2PA formulae. As in PA, 'F', 'G' and 'H', possibly with subscripts, denote primitive recursive functions; all other metavariables starting with a capital letter denote numeric terms.

Metaformulae and their instances are defined by an obvious generalisation of the PA case, except that metaformulae may contain the extended substitution notation defined below, the metanotation $A \Leftrightarrow B$ for $(A \supset B) \wedge (B \supset A)$, and the metanotation $\alpha \simeq \beta$ for $\forall^{\mathrm{N}} n\,(n \models \alpha \Leftrightarrow n \models \beta)$, where n is the first first-order variable in standard order.

FREE VARIABLES IN 2PA FORMULAE

The relation $v \in A$ in PA is extended to 2PA by allowing v to be first- or second-order and adding the clauses

- $v \in N \models \alpha$ iff $v \in N$ or v is α,
- $v \in \exists^2\alpha A$ iff $v \in \forall^2\alpha A$ iff v isn't α and $v \in A$.

As before, the *free variables* of A are the (finitely many) first- and second-order variables v such that $v \in A$, listed in standard order.

The mapping from each PA formula A to an HA formula A^{H} is extended to a mapping from 2PA formulae to 2HA formulae.

AXIOMS AND RULES OF 2PA

The axiom schemata and rules of inference of Second-Order Peano Arithmetic are those of Peano Arithmetic (with the formula metavariables ranging over 2PA formulae, the term metavariables ranging over numeric terms, and the variable metavariables ranging over first-order variables) plus the following.

Equality (eq_2): $M = N \supset (M \models \alpha \Leftrightarrow N \models \alpha)$
Comprehension (comp): $\exists^2\alpha\, \forall^{\mathrm{N}} n\, (n \models \alpha \Leftrightarrow A)$ where $\alpha \notin A$
Quantifier axioms ($\exists^2$-in, $\forall^2$-el): $A \supset \exists^2\alpha A$ $\quad \forall^2\alpha A \supset A$

Quantifier rules ($\exists^2$-el, $\forall^2$-in): $\dfrac{A \supset B}{\exists^2\alpha A \supset B} \qquad \dfrac{B \supset A}{B \supset \forall^2\alpha A}$ where $\alpha \notin B$

FORMAL INTERPRETATION OF 2PA

The coding of PA derivations as constructions and the construction *ha* are extended to 2PA derivations.

THEOREM 7. If D is the code of a 2PA derivation (with no premises) of a formula A then $ha(D) \rhd^*$ the code of a 2HA derivation (with no premises) of A^{H}.

THEOREM 8. (2PA interpretation theorem.) If D is the code of a 2PA derivation (with no premises) of a formula A with no free variables, then $pr(spr(cpf(lpt(ha(D))))) \vdash \ulcorner (A^{\mathrm{H}})^{\mathrm{LPT}} \urcorner$.

SUBSTITUTION IN 2PA FORMULAE

The substitutability relation $N \overset{n}{\hookrightarrow} A$ is defined as in PA (with 'n' ranging over first-order variables) plus the following clauses:

- $N \overset{n}{\hookrightarrow} M \models \alpha$;
- $N \overset{n}{\hookrightarrow} \exists^2\gamma A$ iff $N \overset{n}{\hookrightarrow} \forall^2\gamma A$ iff $N \overset{n}{\hookrightarrow} A$;
- $\beta \overset{\alpha}{\hookrightarrow} A$, if A is atomic (*true*, *false*, $M = N$ or $N \models \alpha$);
- $\beta \overset{\alpha}{\hookrightarrow} A \wedge B$ iff $\beta \overset{\alpha}{\hookrightarrow} A \vee B$ iff $\beta \overset{\alpha}{\hookrightarrow} A \supset B$ iff $\beta \overset{\alpha}{\hookrightarrow} A$ and $\beta \overset{\alpha}{\hookrightarrow} B$;
- $\beta \overset{\alpha}{\hookrightarrow} \exists^{\mathrm{N}}m A$ iff $\beta \overset{\alpha}{\hookrightarrow} \forall^{\mathrm{N}}m A$ iff $\beta \overset{\alpha}{\hookrightarrow} A$;
- $\beta \overset{\alpha}{\hookrightarrow} \exists^2\gamma A$ iff $\beta \overset{\alpha}{\hookrightarrow} \forall^2\gamma A$ iff $\alpha \notin \exists^2\gamma A$ or (γ isn't β and $\beta \overset{\alpha}{\hookrightarrow} A$).

If $N \overset{n}{\hookrightarrow} A$, then the formula $A\binom{N}{n}$ is defined as in PA with the following additional clauses:

- $(M \models \alpha)\binom{N}{n}$ is $M\binom{N}{n} \models \alpha$;
- $(\exists^2\gamma A)\binom{N}{n}$ is $\exists^2\gamma\,(A\binom{N}{n})$ and $(\forall^2\gamma A)\binom{N}{n}$ is $\forall^2\gamma\,(A\binom{N}{n})$.

If $\beta \overset{\alpha}{\hookrightarrow} A$, then the formula $A\binom{\beta}{\alpha}$ is defined by:

- $true\binom{\beta}{\alpha}$ is *true*, $false\binom{\beta}{\alpha}$ is *false*, $(M = N)\binom{\beta}{\alpha}$ is $M = N$;
- $(N \models \alpha)\binom{\beta}{\alpha}$ is $N \models \beta$, $(N \models \gamma)\binom{\beta}{\alpha}$ is $N \models \gamma$ if γ isn't α;
- $(A \wedge B)\binom{\beta}{\alpha}$ is $A\binom{\beta}{\alpha} \wedge B\binom{\beta}{\alpha}$, $(A \vee B)\binom{\beta}{\alpha}$ is $A\binom{\beta}{\alpha} \vee B\binom{\beta}{\alpha}$;
- $(A \supset B)\binom{\beta}{\alpha}$ is $A\binom{\beta}{\alpha} \supset B\binom{\beta}{\alpha}$;
- $(\exists^{\mathrm{N}}m A)\binom{\beta}{\alpha}$ is $\exists^{\mathrm{N}}m\,(A\binom{\beta}{\alpha})$, $(\forall^{\mathrm{N}}m A)\binom{\beta}{\alpha}$ is $\forall^{\mathrm{N}}m\,(A\binom{\beta}{\alpha})$;
- $(\exists^2\gamma A)\binom{\beta}{\alpha}$ is $\exists^2\gamma A$ and $(\forall^2\gamma A)\binom{\beta}{\alpha}$ is $\forall^2\gamma A$, if $\alpha \notin \exists^2\gamma A$;
- $(\exists^2\gamma A)\binom{\beta}{\alpha}$ is $\exists^2\gamma\,(A\binom{\beta}{\alpha})$ and $(\forall^2\gamma A)\binom{\beta}{\alpha}$ is $\forall^2\gamma\,(A\binom{\beta}{\alpha})$, if $\alpha \in \exists^2\gamma A$ and γ isn't β.

2PA THEOREMS AND DERIVED RULES INVOLVING SUBSTITUTION

THEOREM 10. 2PA first-order equality theorem for formulae (eq.form.):

$$M = N \supset (A\tbinom{M}{x} \Leftrightarrow A\tbinom{N}{x}), \quad \text{where } M \overset{x}{\hookrightarrow} A \text{ and } N \overset{x}{\hookrightarrow} A.$$

The substitution theorems and rules derived in PA still apply, with 'A', 'B' and 'C' ranging over 2PA formulae and the metavariables denoting variables ranging over first-order variables.

THEOREM 11. 2PA second-order equality theorem for formulae (eq_2.form.):

$$(\alpha \simeq \beta) \supset (A\binom{\alpha}{\gamma} \Leftrightarrow A\binom{\beta}{\gamma}), \quad \text{where } \alpha \overset{\gamma}{\hookrightarrow} A \text{ and } \beta \overset{\gamma}{\hookrightarrow} A.$$

THEOREM 14. 2PA second-order quantifier theorems with substitution:

$$A\binom{\beta}{\alpha} \supset \exists^2\alpha\, A, \qquad \forall^2\alpha\, A \supset A\binom{\beta}{\alpha}, \qquad \text{where } \beta \overset{\alpha}{\hookrightarrow} A.$$

THEOREM 16. 2PA second-order quantifier rules with substitution:

$$\frac{A\binom{\beta}{\alpha} \supset B}{\exists^2\alpha\, A \supset B} \qquad \frac{B \supset A\binom{\beta}{\alpha}}{B \supset \forall^2\alpha\, A} \qquad \text{where } \beta \overset{\alpha}{\hookrightarrow} A \text{ and } \beta \notin \exists^2\alpha\, A,\, B.$$

CHAPTER 49

CONCLUSIONS ON ANALYSIS

The interpretation of Second-Order Peano Arithmetic is now complete. As pointed out in Chapter 36, this is also an interpretation of a formal theory of analysis containing variables for real numbers, the axioms for a complete ordered field, and classical predicate calculus. Real numbers are present in this system in all their impredicative glory, due to the Comprehension Axiom Schema of 2PA.

The interpretation accomplishes the same things for analysis as the interpretation of Peano Arithmetic did for arithmetic in Part III (see Chapter 35). That is, it shows that analytic formulae and formal derivations may be understood in purely constructive terms, and hence it undermines the classical account of analysis. It also accomplishes Hilbert's programme for analysis.

A project for the future is to extend the interpretation to include sets of reals or functions defined on the reals, and also to incorporate some of the continuity principles used in traditional intuitionistic analysis.

REFERENCES

Aho, A.V., Sethi, R. & Ullman, J.D. (1986) *Compilers: Principles, Techniques and Tools.* Reading, Massachusetts: Addison-Wesley.

ACM (1992) *Special Issue on the Functional Programming Language Haskell. SIGPLAN Notices*, **27**, no. 5, published as a book. New York: Association for Computing Machinery.

Backus, J. (1978) Can programming be liberated from the von Neumann style? A functional style and its algebra of programs. ACM Turing Lecture, *Communications of the Association for Computing Machinery,* **21**, no. 8, 613–641.

Backus, J., Williams, J.H. & Wimmers, E.L. (1990) An introduction to the programming language FL. In D.A. Turner (Ed.) *Research Topics in Functional Programming,* Chapter 9. Reading, Massachusetts: Addison-Wesley.

Barendregt, H.P. (1984) *The Lambda Calculus: Its Syntax and Semantics,* revised edition. Amsterdam: North-Holland.

Beeson, M.J. (1985) *Foundations of Constructive Mathematics: Metamathematical Studies.* Berlin: Springer-Verlag.

Benacerraf, P. & Putnam, H. (Eds) (1964) *Philosophy of Mathematics: Selected Readings.* Oxford: Blackwell.

Bernays, P. (1935) Sur le Platonisme dans les mathématiques. *L'enseignement Mathématique,* **34**, 52–69. English translation in Benacerraf & Putnam (1964), pp. 274–286.

Bishop, E. (1967) *Foundations of Constructive Analysis.* New York: McGraw-Hill.

Bishop, E. (1970) Mathematics as a numerical language. In Kino, Myhill & Vesley (1970), pp. 53–71.

Brookes, S. & Geva, S. (1992) Computational comonads and intensional semantics. In M.P. Fourman, P.T. Johnstone & A.M. Pitts (Eds) *Applications of Categories in Computer Science,* pp. 1–44. London Mathematical Society Lecture Note Series, **177**. Cambridge: Cambridge University Press.

Brouwer, L.E.J. (1905) *Leven, Kunst en Mystiek* (Life, Art and Mysticism). Waltman, Delft. English translation of selected parts in Heyting (1975), pp. 1–9.

Brouwer, L.E.J. (1907) *Over de grondslagen der wiskunde. Academisch proefschrift.* Amsterdam: Maas & van Suchtelen. English translation, *On the Foundations of Mathematics,* dissertation, in Heyting (1975), pp. 11–101.

Brouwer, L.E.J. (1908) The unreliability of the logical principles. In Heyting (1975), pp. 107–111.

Brouwer, L.E.J. (1912) *Intuitionism and Formalism.* Inaugural address at the University of Amsterdam. English translation in the *Bulletin of the American Mathematical Society,* **20**, 81–96. Reprinted in Benacerraf & Putnam (1964), pp. 66–77.

Brouwer, L.E.J. (1927) Intuitionistic reflections on formalism. In van Heijenoort (1967), pp. 490–492.

Brouwer, L.E.J. (1933a) Volition, knowlege, language, §3. In Heyting (1975), pp. 443–446. (This is a different translation of §3 of Brouwer (1933b.))

Brouwer, L.E.J. (1933b) Will, knowledge and speech. In van Stigt (1990), appendix 5.

Brouwer, L.E.J. (1947a) *Short retrospective notes on the course 'Intuitionist Mathematics'*. In van Stigt (1990), appendix 6.

Brouwer, L.E.J. (1947b) Guidelines of intuitionistic mathematics. In Heyting (1975), p. 477.

Brouwer, L.E.J. (1948) Consciousness, philosophy, and mathematics. In Heyting (1975), pp. 480–494.

Brouwer, L.E.J. (1952) Historical background, principles and methods of intuitionism. In Heyting (1975), pp. 508–515.

Brouwer, L.E.J. (1954) Points and spaces. In Heyting (1975), pp. 522–538.

Brouwer, L.E.J. (1955) The effect of intuitionism on classical algebra of logic. In Heyting (1975), pp. 551–554.

Brouwer, L.E.J. (1981) *Brouwer's Cambridge lectures on intuitionism*, edited by D. van Dalen. Cambridge: Cambridge University Press.

Burstall, R.M. & Darlington, J. (1977) A transformation system for developing recursive programs. *Journal of the Association for Computing Machinery*, **24**, no. 1, 44–67.

Burstall, R.M., MacQueen, D.B. & Sannella, D.T. (1980) HOPE: An Experimental Applicative Language. *Proceedings of the 1980 ACM LISP Conference*, Stanford, California, pp. 136–143.

Cantor, G. (1882) Ueber unendliche, lineare Punktmannigfaltigkeiten, Part 3. *Mathematische Annalen*, **20**, 113–121.

Cantor, G. (1883) *Grundlagen einer allgemeinen Mannigfaltigkeitslehre: Ein mathematisch-philosophischer Versuch in der Lehre des Unendlichen*. Leipzig: B.G. Teubner.

Cantor, G. (1887–8) Mitteilungen zur Lehre vom Transfiniten. *Zeitschrift für Philosophie und philosophiche Kritik*, **91**, 81–125, and **92**, 240–265.

Cantor, G. (1895) Beiträge zur Begrundung der transfiniten Mengenlehre, Part I. *Mathematische Annalen*, **46**, 481–512.

Cantor, G. (1899) Letter to Dedekind. In van Heijenoort (1967), pp. 113–117.

Cardelli, L. & Longo, G. (1991) A semantic basis for Quest. *Journal of Functional Programming*, **1**, no. 4, 417–458.

Carnap, R. (1937) *The Logical Syntax of Language*. London: Kegan Paul, Trench, Trubner & Co.

Carroll, L. (1895) What the tortoise said to Achilles. *Mind*, new series **4**, 278–280.

Chihara, C.S. (1982) A Gödelian thesis regarding mathematical objects: do they exist? and can we perceive them? *Philosophical Review*, **91**, no. 2, 211–227.

Chihara, C.S. (1989) Tharp's 'Myth and Mathematics'. *Synthese*, **81**, no. 2, 153–165.

Church, A. (1962) Mathematics and logic. In E. Nagel, P. Suppes & A. Tarski (Eds) *Logic, Methodology and Philosophy of Science*, pp. 181–186. Proceedings of the 1960 conference at Stanford University. Stanford: Stanford University Press.

Curry, H.B. (1954) Remarks on the definition and nature of mathematics. *Dialectica*, **48**, 228–233. Reprinted in Benacerraf & Putnam (1964), pp. 152–156.

Darlington, J. (1987) Software development in declarative languages. In S. Eisenbach (Ed.) *Functional Programming: Languages, Tools and Architectures*, Chapter 5. Chichester & New York: Ellis Horwood & Wiley.

Davie, A.J.T (1992) *An Introduction to Functional Programming Systems using Haskell.* Cambridge: Cambridge University Press.

Davis, M. (1965) (Ed.) *The Undecidable: Basic Papers on Undecidable Propositions, Unsolvable Problems and Computable Functions.* Hewlett, New York: Raven Press.

Davis, P.J. & Hersh, R. (1981) *The Mathematical Experience.* Boston: Birkhäuser.

Dauben, J.W. (1979) *Georg Cantor: His Mathematics and Philosophy of the Infinite.* Cambridge, Massachusetts: Harvard University Press.

Dedekind, R. (1890) Letter to Keferstein. In van Heijenoort (1967), pp. 98–103.

Detlefsen, M. (1979) On interpreting Gödel's second theorem. *Journal of Philosophical Logic*, **8**, no. 3, 297–313.

Detlefsen, M. (1986) *Hilbert's Program: An Essay on Mathematical Instrumentalism.* Dordrecht: Reidel.

Detlefsen, M. (1990a) Brouwerian intuitionism. *Mind*, **99**, no. 396, 501–534.

Detlefsen, M. (1990b) On an alleged refutation of Hilbert's program using Gödel's first incompleteness theorem. *Journal of Philosophical Logic*, **19**, no. 4, 343–377.

Diller, A. (1988) *Compiling Functional Languages.* Chichester: Wiley.

Diller, J. & Troelstra, A.S. (1984) Realizability and intuitionistic logic. *Synthese*, **60**, no. 2, 253–282.

Dummett, M. (1959) Wittgenstein's philosophy of mathematics. *Philosophical Review*, **68**, 324–348. Reprinted in Benacerraf & Putnam (1964), pp. 491–509.

Dummett, M. (1967) Frege's philosophy. In Dummett (1978), pp. 87–115.

Dummett, M. (1975) The philosophical basis of intuitionistic logic. In H.E. Rose & J.C. Shepherdson (Eds) *Logic Colloquium '73* (Bristol), pp. 5–40. Amsterdam: North-Holland. Reprinted in Dummett (1978), pp. 215–247.

Dummett, M. (1977) *Elements of Intuitionism.* Oxford: Oxford University Press.

Dummett, M, (1978) *Truth and Other Enigmas.* London: Duckworth.

Dummett, M. (1982) Realism. *Synthese*, **52**, no. 1, 55–112.

Dummett, M. (1991) *Frege: Philosophy of Mathematics.* Cambridge, Massachusetts: Harvard University Press.

Eilenberg, S. & Elgot, C.C. (1970) *Recursiveness.* New York: Academic Press.

Feferman, S. (1988) Hilbert's program relativized: proof-theoretical and foundational reductions. *Journal of Symbolic Logic*, **53**, no. 2, 364–384.

Feferman, S. (1991) Reflecting on incompleteness. *Journal of Symbolic Logic*, **56**, no. 1, 1–49.

Fletcher, P. (1988) *Truth, Proof and Infinity.* Ph.D. thesis, Mathematics Department, University of Bristol.

Fraenkel, A.A., Bar-Hillel, Y. & Levy, A. (1973) *Foundations of Set Theory*, second revised edition. Amsterdam: North-Holland.

Frege, G. (1884) *Die Grundlagen der Arithmetik, eine logisch mathematische Untersuchung über den Begriff der Zahl.* Breslau: Verlag von Wilhelm Koebner. English translation by J.L. Austin (1953) *The Foundations of Arithmetic*, 2nd revised edition. Oxford: Blackwell.

Frege, G. (1885) On formal theories of arithmetic. In McGuinness (1984), pp. 112–121.

Frege, G. (1892) Über sinn und bedeutung. *Zeitschrift für Philosophie und philosophische Kritik*, **100**, 25–50. English translation, On sense and reference, in Geach & Black (1970), pp. 56–78.

Frege, G. (1893) *Grundgesetze der Arithmetik, begriffsschriftlich abgeleitet*, vol. 1. Jena: Verlag Hermann Pohle. English translation of §§0–52 by M. Furth (1967) *The Basic Laws of Arithmetic: Exposition of the System.* Berkeley: University of California Press.

Frege, G. (1903) *Grundgesetze der Arithmetik, begriffsschriftlich abgeleitet*, vol. 2. Jena: Verlag Hermann Pohle. English translation of selected sections in Geach & Black (1970), pp. 159–244.

Frege, G. (1906) Reply to Mr Thomae's holiday *causerie*. In McGuinness (1984), pp. 341–345.

Geach, P. & Black, M. (Eds) (1970) *Translations from the Philosophical Writings of Gottlob Frege.* Oxford: Blackwell.

Gentzen, G. (1933) On the relation between intuitionist and classical arithmetic. In M.E. Szabo (1969) *The Collected Papers of Gerhard Gentzen*, pp. 53–67. Amsterdam: North-Holland.

George, A. (1988) The conveyability of intuitionism, an essay on mathematical cognition. *Journal of Philosophical Logic*, **17**, no. 2, 133–156.

George, A. (1993) How not to refute realism. *Journal of Philosophy*, **90**, no. 2, 53–72.

Giaquinto, M. (1983) Hilbert's philosophy of mathematics. *British Journal for the Philosophy of Science*, **34**, no. 2, 119–132.

Gielen, W., de Swart, H. & Veldman, W. (1981) The continuum hypothesis in intuitionism. *Journal of Symbolic Logic*, **46**, no. 1, 121–136.

Girard, J.-Y., Lafont, Y. & Taylor, P. (1989) *Proofs and Types.* Cambridge: Cambridge University Press.

Gödel, K. (1931) Über formal unentscheidbare Sätze der Principia Mathematica und verwandter Systeme I, *Monatschefte für Mathematik und Physik*, **38**, 173–198. English translation: On formally undecidable propositions of Principia Mathematica and related systems I, in Davis (1965) pp. 5–38.

Gödel, K. (1933a) Zur intuitionistischen Arithmetik und Zahlentheorie. *Ergebnisse eines mathematischen Kolloquiums, Heft* **4**, 34–38. English translation: On intuitionistic arithmetic and number theory, in Davis (1965), pp. 75–81.

Gödel, K. (1933b) The present situation in the foundations of mathematics. In S. Feferman (Ed.) *Kurt Gödel: Collected Works*, Vol III, pp. 45–53. New York: Oxford University Press.

Gödel, K. (1938) Lecture at Zilsel's. In S. Feferman (Ed.) *Kurt Gödel: Collected Works*, Vol III, pp. 86–113. New York: Oxford University Press.

Gödel, K. (1941) In what sense is intuitionistic logic constructive? In S. Feferman (Ed.) *Kurt Gödel: Collected Works*, Vol III, pp. 189–200. New York: Oxford University Press.

Gödel, K. (1944) Russell's mathematical logic. In P.A. Schlipp (Ed.) *The Philosophy of Bertrand Russell*, pp. 125–153. New York: The Tudor Publishing Company. Reprinted in Benacerraf & Putnam (1964), pp. 211–232.

Gödel, K. (1946) Remarks before the Princeton bicentennial conference on problems in mathematics. In Davis (1965), pp. 84–88.

Gödel, K. (1958) Über eine bisher noch nicht benützte Erweiterung des finiten Standpunktes. *Dialectica*, **12**, 280–287. English translation in S. Feferman (1990) (Ed.) *Kurt Gödel: Collected Works*, volume II, pp. 240–251. New York: Oxford University Press.

Gödel, K. (1964) What is Cantor's continuum problem? In Benacerraf & Putnam (1964), pp. 258–273.

Gödel, K. (1972) On an extension of finitary mathematics which has not yet been used. (A revised and expanded version of Gödel (1958).) In S. Feferman (1990) (Ed.) *Kurt Gödel: Collected Works*, volume II, pp. 271–280. New York: Oxford University Press.

Goguen, J.A. (1990) Higher-order functions considered unnecessary for higher-order programming. In D.A. Turner (Ed.) *Research Topics in Functional Programming*, Chapter 12. Reading, Massachusetts: Addison-Wesley.

Gonzalez, W.J. (1991) Intuitionistic mathematics and Wittgenstein. *History and Philosophy of Logic*, **12**, 167–183.

Goodman, N.D. (1970) A theory of constructions equivalent to arithmetic. In Kino, Myhill & Vesley (1970), pp. 101–120.

Goodman, N.D. (1972) A simplification of combinatory logic. *Journal of Symbolic Logic*, **37**, no. 2, 225–246.

Goodman, N.D. (1973a) The arithmetic theory of constructions. In A.R.D. Mathias & H. Rogers (Eds) *Cambridge Summer School in Mathematical Logic* (Cambridge, 1971), pp. 274–298. Lecture Notes in Mathematics, **337**. Berlin: Springer-Verlag.

Goodman, N.D. (1973b) The faithfulness of the interpretation of arithmetic in the theory of constructions. *Journal of Symbolic Logic*, **38**, no. 3, 453–459.

Goodman, N.D. (1979a) Review of Dummett (1977). *Journal of Symbolic Logic*, **44**, no. 2, 276–277.

Goodman, N.D. (1979b) Mathematics as an objective science. *American Mathematical Monthly*, **86**, 540–551.

Goodman, N.D. (1990) Mathematics as natural science. *Journal of Symbolic Logic*, **55**, no. 1, 182–193.

Goodman, N.D. (1991) Modernizing the philosophy of mathematics. *Synthese*, **88**, no. 2, 119–126.

Hallett, M. (1984) *Cantorian Set Theory and Limitation of Size*. Oxford: Clarendon Press.

Hannan, J. (1993) Extended natural semantics. *Journal of Functional Programming*, **3**, no. 2, 123–152.

Harrison, P. & Khoshnevisan, H. (1987) A functional algebra and its application to program transformation. In S. Eisenbach (Ed.) *Functional Programming: Languages, Tools and Architectures,* Chapter 6. Chichester & New York: Ellis Horwood & Wiley.

Hempel, C.G. (1945) On the nature of mathematical truth. *American Mathematical Monthly,* **52**, 543–556. Reprinted in Benacerraf & Putnam (1964), pp. 366–381.

Herbrand, J. (1931) Sur la non-contradiction de l'arithmétique. *Journal für die reine angewandte Mathematik*, **166**, 1–8. English translation, On the consistency of arithmetic, in van Heijenoort (1967), pp 618–628.

Hersh, R. (1991) Mathematics has a front and a back. *Synthese*, **88**, no. 2, 127–133.

Heyting, A. (1956) *Intuitionism: An Introduction*. Amsterdam: North-Holland.

Heyting, A. (1974) Intuitionistic views on the nature of mathematics. *Synthese,* **27**, nos 1/2, 79–91.

Heyting, A. (Ed.) (1975) *L.E.J. Brouwer: Collected Works,* vol. I. Amsterdam: North-Holland.

Hilbert, D. (1925) On the infinite. In van Heijenoort (1967), pp. 367–392.

Hilbert, D. (1927) The foundations of mathematics. In van Heijenoort (1967), pp. 464–479.

Hodes, H.T (1982) Review of Dummett (1977). *Philosophical Review,* **91**, no. 2, 253–262.

Holyer, I.J. (1991) *Functional Programming with Miranda.* London: Pitman.

Hughes, J. (1990) Why functional programming matters. In D.A. Turner (Ed.) *Research Topics in Functional Programming,* Chapter 2. Reading, Massachusetts: Addison-Wesley.

Hyland, J.M.E. (1991) Computing and foundations. In J.H. Johnson & M.J. Loomes (Eds) *The Mathematical Revolution Inspired by Computing,* pp. 269–284. Oxford: Clarendon Press.

Ignjatović, A. (1994) Hilbert's program and the omega-rule. *Journal of Symbolic Logic,* **59**, no. 1, 322–343.

Kino, A., Myhill, J. & Vesley, R.E. (Eds) (1970) *Intuitionism and Proof Theory.* Proceedings of the summer conference at Buffalo, New York, 1968. Amsterdam: North-Holland.

Kleene, S.C. (1945) On the interpretation of intuitionistic number theory. *Journal of Symbolic Logic*, **10**, no. 4, 109–124.

Kolmogorov, A.N. (1925) On the principle of excluded middle. In van Heijenoort (1967), pp. 414–437.

Kolmogorov, A.N. (1932) Zur deutung der intuitionistischen Logik, *Mathematische Zeitschrift,* **35**, 58–65.

Körner, S. (1960) *The Philosophy of Mathematics.* London: Hutchinson.

Kreisel, G. (1962) Foundations of intuitionistic logic. In E. Nagel, P. Suppes & A. Tarski (Eds) *Logic, Methodology and Philosophy of Science*, pp. 198–210. Stanford: Stanford University Press.

Kreisel, G. (1965) Mathematical logic. In T.L. Saaty (Ed.) *Lectures on Modern Mathematics,* volume III, pp. 95–195. New York: Wiley.

Kreisel, G. (1968) Lawless sequences of natural numbers. In D. van Dalen, J.G. Dijkman, S.C. Kleene & A.S. Troelstra (Eds) *Logic and Foundations of Mathematics: Dedicated to Prof. A. Heyting on his 70th Birthday,* pp. 222-248. Groningen: Wolters-Noordhoff.

Kreisel, G. (1970) Church's thesis: a kind of reducibility axiom for constructive mathematics. In Kino, Myhill & Vesley (1970), pp. 121–150.

Lalement, R. (1993) *Computation as logic.* Paris & Hemel Hempstead: Masson & Prentice-Hall.

Lavine, S. (1994) *Understanding the infinite.* Cambridge, Massachusetts: Harvard University Press.

Loeckx, J. & Sieber, K. (1984) *The Foundations of Program Verification.* Stuttgart: Wiley-Teubner.

Maddy, P. (1988) Believing the axioms, I, II. *Journal of Symbolic Logic,* **53**, no. 2, 481–511, and no. 3, 736–764.

Maddy, P. (1989) The roots of contemporary Platonism. *Journal of Symbolic Logic,* **54**, no. 4, 1121–1144.

Maddy, P. (1990) *Realism in mathematics.* Oxford: Clarendon Press.

Maddy, P. (1991) Philosophy of mathematics: prospects for the 1990s. *Synthese,* **88**, no. 2, 155–164.

Martini, S. (1992) Categorical models for non-extensional λ-calculi and combinatory logic. *Mathematical Structures in Computer Science,* **2**, no. 3, 327–357.

Martin-Löf, P. (1975) An intuitionistic theory of types: predicative part. In H.E. Rose & J.C. Shepherdson (Eds) *Logic Colloquium '73* (Bristol), pp. 73–118. Amsterdam: North-Holland.

Martin-Löf, P. (1982) Constructive mathematics and computer programming. In L.J. Cohen, J. Łoś, H. Pfeiffer & K.-P. Podewski (Eds) *Logic, Methodology and Philosophy of Science, VI,* Hanover, 1979, 153–175. Studies in Logic and the Foundations of Mathematics, **104**. Amsterdam: North-Holland.

Martin-Löf, P. (1984) *Intuitionistic Type Theory.* Napoli: Bibliopolis.

Martin-Löf, P. (1987) Truth of a proposition, evidence of a judgement, validity of a proof. *Synthese,* **73**, no. 3, 407–420.

Mayberry, J. (1977) On the consistency problem for set theory: an essay on the Cantorian foundations of classical mathematics, I, II. *British Journal for the Philosophy of Science,* **28**, 1–34, 137–170.

McCarthy, J. (1960) Recursive functions of symbolic expressions and their computation by machine. *Communications of the Association for Computing Machinery,* **3**, no. 4, 184–195.

McCarty, C. (1983) Intuitionism: an introduction to a seminar. *Journal of Philosophical Logic,* **12**, no. 2, 105–149.

McGuinness, B. (Ed.) (1984) *Gottlob Frege: Collected Papers on Mathematics, Logic, and Philosophy.* Oxford: Blackwell.

Meyer, A.R. (1982) What is a model of the lambda calculus? *Information and Control,* **52**, 87–122.

Milner, R. & Tofte, M. (1991) *Commentary on Standard ML.* Cambridge, Massachusetts: MIT Press.

Milner, R., Tofte, M. & Harper, R. (1990) *The Definition of Standard ML.* Cambridge, Massachusetts: MIT Press.

Moor, I. (1987) Realistic functional programming. In S. Eisenbach (ed) *Functional Programming: Languages, Tools and Architectures,* Chapter 7. Chichester & New York: Ellis Horwood & Wiley.

Musgrave, A. (1977) Logicism revisited. *British Journal for the Philosophy of Science,* **28**, 99–127.

Myers, C., Clack, C. & Poon, E. (1993) *Programming with Standard ML.* New York: Prentice-Hall.

Nagel, E. (1944) Logic without ontology. In Y.H. Krikorian (Ed.) *Naturalism and the Human Spirit.* New York: Columbia University Press. Reprinted in Benacerraf & Putnam (1964), pp. 302–321.

Nordström, B., Petersson, K. & Smith, J.M. (1990) *Programming in Martin-Löf's Type Theory: An Introduction.* Oxford: Clarendon Press.

Orwell, G. (1949) *Nineteen Eighty-Four: A Novel.* London: Secker & Warburg.

Parikh, R. (1971) Existence and feasibility in arithmetic. *Journal of Symbolic Logic,* **36**, no. 3, 494–508.

Parsons, C. (1979–80) Mathematical intuition. *Proceedings of the Aristotelian Society,* new series, **80**, 145–168.

Paulson, L.C. (1991) *ML for the Working Programmer.* Cambridge: Cambridge University Press.

Posy, C.J. (1974) Brouwer's constructivism. *Synthese,* **27**, nos 1/2, 125–159.

Posy, C.J. (1976) Varieties of indeterminacy in the theory of general choice sequences. *Journal of Philosophical Logic,* **5**, no. 1, 91–132.

Posy, C.J. (1977) The theory of empirical sequences. *Journal of Philosophical Logic,* **6**, no. 1, 47–81.

Putnam, H. (1980) Models and reality. *Journal of Symbolic Logic,* **45**, no. 3, 464–482.

Quine, W.V.O. (1937) New foundations for mathematical logic. *American Mathematical Monthly,* **44**, 70–80.

Quine, W.V.O. (1951) *Mathematical Logic,* revised edition. Cambridge: Harvard University Press.

Quine, W.V.O. (1953) Two dogmas of empiricism. In W.V.O. Quine (1953) *From a Logical Point of View*, pp. 20–46. Cambridge, Massachusetts: Harvard University Press. Reprinted in Benacerraf & Putnam (1964), pp. 346–365.

Quine, W.V.O. (1964) Truth by convention. In Benacerraf & Putnam (1964), pp. 322–345.

Quine, W.V.O. (1969) *Set Theory and its Logic,* revised edition. Cambridge, Massachusetts: Harvard University Press.

Revesz, G. (1988) *Lambda-Calculus, Combinators, and Functional Programming.* Cambridge: Cambridge University Press.

Russell, B. (1908) Mathematical logic as based on the theory of types. *American Journal of Mathematics*, **30**, 222–262. Reprinted in van Heijenoort (1967), pp. 150–182.

Russell, B. (1919) *Introduction to Mathematical Philosophy.* New York: Macmillan.

Russell, B. (1935–6) The limits of empiricism. *Proceedings of the Aristotelian Society*, new series, **36**, 131–150.

Scott, D.S. (1970) Constructive validity. In M. Laudet, D. Lacombe, L. Nolin & M. Schützenberger (Eds) *Symposium on Automatic Demonstration* (Versailles, December 1968), 237–275. Lecture Notes in Mathematics, **125**. Berlin: Springer-Verlag.

Shapiro, S. (1992) Foundationalism and foundations of mathematics. In M. Detlefsen (Ed.) *Proof and Knowledge in Mathematics*, pp. 171–207. London: Routledge.

Sieg, W. (1988) Hilbert's program sixty years later. *Journal of Symbolic Logic*, **53**, no. 2, 338–348.

Simpson, S.G. (1988) Partial realizations of Hilbert's program. *Journal of Symbolic Logic*, **53**, no. 2, 349–363.

Skolem, T. (1922) Some remarks on axiomatized set theory. In van Heijenoort (1967), pp. 290–301.

Skolem, T. (1923) The foundations of elementary arithmetic established by means of the recursive mode of thought, without the use of apparent variables ranging over infinite domains. In van Heijenoort (1967), pp. 302–333.

Sundholm, G. (1983) Constructions, proofs and the meaning of the logical constants. *Journal of Philosophical Logic*, **12**, no. 2, 151–172.

Tait, W.W. (1981) Finitism. *Journal of Philosophy*, **78**, no. 9, 524–546.

Tait, W.W. (1983) Against intuitionism: constructive mathematics is part of classical mathematics. *Journal of Philosophical Logic*, **12**, no. 2, 173–195.

Tait, W.W. (1986) Truth and proof: the Platonism of mathematics. *Synthese*, **69**, no. 3, 341–370.

Tharp, L.H. (1989) Myth and mathematics: a conceptualistic philosophy of mathematics I. *Synthese*, **81**, no. 2, 167–201.

Tharp, L.H. (1991) Myth and math, part II (preliminary draft). *Synthese*, **88**, no. 2, 179–199.

Thompson, S. (1991) *Type Theory and Functional Programming*. Wokingham: Addison-Wesley.

Tragessor, R.S. (1992) Three insufficiently attended to aspects of most mathematical proofs: phenomenological studies. In M. Detlefsen (Ed.) *Proof, Logic and Formalization*, pp. 162–198. London: Routledge.

Troelstra, A.S. (1977) *Choice Sequences: A Chapter of Intuitionistic Mathematics*. Oxford: Clarendon Press.

Troelstra, A.S. (1983) Analysing choice sequences. *Journal of Philosophical Logic*, **12**, no. 2, 197–260.

Troelstra, A.S. (1985) Choice sequences and informal rigour. *Synthese*, **62**, no. 2, 217–227.

Troelstra, A.S. & van Dalen, D. (1988) *Constructivism in Mathematics: An Introduction*, vol. 1. Studies in Logic and the Foundations of Mathematics, **121**. Amsterdam: North-Holland.

Turing, A.M. (1937) On computable numbers, with an application to the Entscheidungsproblem. *Proceedings of the London Mathematical Society*, series 2, **42**, 230–265.

Turing, A.M. (1939) Systems of logic based on ordinals. *Proceedings of the London Mathematical Society*, series 2, **45**, 161–228. Reprinted in Davis (1965), pp. 155–222.

Turing, A.M. (1950) Computing machinery and intelligence. *Mind*, **59**, no. 236, 433–460.

Tymoczko, T. (1979) The four-color problem and its philosophical significance. *Journal of Philosophy*, **76**, no. 2, 57–83.

Tymoczko, T. (1991) Mathematics, science and ontology. *Synthese*, **88**, no. 2, 201–228.

van Dalen, D. (1973) Lectures on intuitionism. In A.R.D. Mathias & H. Rogers (Eds) *Cambridge Summer School in Mathematical Logic* (Cambridge, 1971), pp. 1–94. Lecture Notes in Mathematics, **337**. Berlin: Springer-Verlag.

van Heijenoort, J. (1967) *From Frege to Gödel: A Source Book in Mathematical Logic, 1879–1931*. Cambridge, Massachusetts: Harvard University Press.

van Stigt, W.P. (1990) *Brouwer's Intuitionism*. Amsterdam: North-Holland.

Vesley, R.E. (1979) Review of Troesltra (1977). *Journal of Symbolic Logic*, **44**, no. 2, 275–276.

Wang, H. (1974) *From Mathematics to Philosophy*. London: Routledge & Kegan Paul.

Weinstein, S. (1983) The intended interpretation of intuitionistic logic. *Journal of Philosophical Logic*, **12**, no. 2, 261–270.

Weyl, H. (1921) Über die neue Grundlagenkrise der Mathematik (On the new crisis in the foundations of mathematics). *Mathematische Zeitschrift*, **10**, 39–79.

Wittgenstein, L. (1953) *Philosophical Investigations*. Translated by G.E.M. Anscombe. Oxford: Blackwell.

Wittgenstein, L. (1978) *Remarks on the Foundations of Mathematics* (third edition, revised and with new numbering scheme). Edited by G.H. von Wright, R. Rhees & G.E.M. Anscombe, translated by G.E.M. Anscombe. Oxford: Blackwell.

Wright, C. (1980) *Wittgenstein on the Foundations of Mathematics*. London: Duckworth.

Wright, C. (1982) Strict finitism. *Synthese*, **51**, no. 2, 203–282.

Yessenin-Volpin, A.S. (1970) The ultra-intuitionistic criticism and the antitraditional program for foundations of mathematics. In Kino, Myhill & Vesley (1970), pp. 3–45.

Zermelo, E. (1908) Investigations in the foundations of set theory I. In van Heijenoort (1967), pp. 199–215.

INDEX

INDEX OF SYMBOLS

INDEX OF AXIOMS, THEOREMS AND RULES OF INFERENCE

INDEX OF NAMES

INDEX OF TOPICS

GPSR Compliance

The European Union's (EU) General Product Safety Regulation (GPSR) is a set of rules that requires consumer products to be safe and our obligations to ensure this.

If you have any concerns about our products, you can contact us on ProductSafety@springernature.com

In case Publisher is established outside the EU, the EU authorized representative is:

Springer Nature Customer Service Center GmbH
Europaplatz 3
69115 Heidelberg, Germany

Batch number: 10755284

Printed by Printforce, the Netherlands